Regional Geology Reviews

The book-series presents the geology of each country, region and continent on Earth in a systematical way. Each book provides the reader with the state-of-the-art understanding of a regions geology. The books form the basis of understanding that students, researchers and professional geologists require when beginning investigations in a particular area and include as much information as possible such as: maps and cross-sections, past and current models, geophysical investigations, geochemical datasets, economic geology, geotourism (geoparks etc), geo-environmental/ecological concerns, etc.

Fadhil N. Sadooni · Hamad Al-Saad Al-Kuwari

Geological Evolution of Qatar and the Arabian Peninsula

Springer

Fadhil N. Sadooni
Environmental Science Center
Qatar University
Doha, Qatar

Hamad Al-Saad Al-Kuwari
Environmental Science Center
Qatar University
Doha, Qatar

ISSN 2364-6438 ISSN 2364-6446 (electronic)
Regional Geology Reviews
ISBN 978-3-031-94393-5 ISBN 978-3-031-94394-2 (eBook)
https://doi.org/10.1007/978-3-031-94394-2

Cover credit: Reproduced from the bp archive

This Springer imprint is published by the registered company Springer Nature Switzerland AG
The registered company address is: Gewerbestrasse 11, 6330 Cham, Switzerland

This book is dedicated to the memory of:

Sheikh Mansoor Bin Khalil Al-Shahwani Al-Hajiri
The blind who showed us the way

هذا الكتاب مكرس لذكرى:
الشيخ منصور بن خليل الشهواني الهاجري
الأعمى الذي دلنا على الطريق

Shaikh Mansoor Al-Shahwani (right) with the geologist George Lees on a field trip in 1934.

(Reproduced from the BP archive with permission)

Preface

Writing a book on the geologic evolution of Qatar within the context of the Arabian Peninsula has been a recurring subject of discussion for several years. The rapid urbanization of Qatar underscores the critical need for a deeper understanding of its geological landscape and natural resources. The task, however, appeared formidable due to the scarcity of shared-domain data and the variable quality of existing information. The closure of the Geology Department at Qatar University twenty-five years ago further exacerbated the situation by halting the production of new Qatari geology graduates and the retirement of the old generation Qatari geologists from the field or switching to managerial jobs.

Despite these challenges, the convergence of unique circumstances and our collective expertise, spanning over four decades in Qatar's geology, compelled us to undertake this endeavor. Our privileged access to rare rock materials and data, combined with our cumulative experience of approximately eighty years, positions us uniquely to document Qatar's geological narrative. This opportunity might not present itself to future generations.

This volume is not merely our work; it encapsulates the passion and dedication of countless individuals who have contributed to understanding Qatar's geology. While we have acknowledged many of these contributors, it's crucial to recognize that this book synthesizes the extensive body of work on Qatar's geological history that is accessible to us. The content distribution across chapters reflects the inherent focus areas within the region's geology, with hydrocarbon exploration significantly influencing the depth of research available on oil and gas reservoir rocks. Moreover, the accessible Tertiary rocks have been examined more thoroughly than subsurface strata, attracting amateur fossil hunters and professional geologists alike.

Our reliance on foundational works such as the "*Petrographic Atlas of Qatar*" by Focke et al. (1986), the research conducted by colleagues at *ExxonMobil Research Qatar*, and Jacque LeBlanc's invaluable contributions to revising the geological lexicon related to the Eocene rocks have been instrumental. Their collective work has significantly informed our understanding and interpretation of Qatar's geological framework.

The journey of compiling this book has been longer and more demanding than anticipated, balancing our daily professional responsibilities with the writing and meticulous preparation of maps and figures. We extend our heartfelt gratitude to Qatar University and the Qatar Foundation for their unwavering support, funding, and resources that have been crucial in bringing this project to fruition.

Embarking on such a mission is challenging, and we humbly acknowledge the possibility of errors within this work. We apologize for any inaccuracies and hope they do not detract from the value and insights this book aims to provide. We hope this book enriches the reader's understanding of Qatar's geology, fostering a deeper appreciation for this nation's geological complexities and beauty.

Doha, Qatar
March 2024

Fadhil N. Sadooni
Hamad Al-Saad Al-Kuwari

Reference

Focke J, Munn D, Al Kuwari S, Frikken H, Frei H (1986) Petrographic atlas of rock types, common in the subsurface of Qatar and some recent equivalents. In: General petroleum corporation (offshore operations), Qatar, pp 110–189

Acknowledgments

Writing a book of this magnitude requires drawing upon the wisdom, efforts, and support of numerous individuals. This support has manifested in various ways, including research collaboration, resource sharing, fieldwork, and engaging in deep, meaningful conversations. Considering this, we owe our gratitude to many of whom we hope to acknowledge here. We extend our heartfelt thanks to Ahmad Abdulaziz, Osman Abdulghani, Ghada Abrahim, Amer Jassim Al-Khafaji, A. S. Alsharhan, Ismail Al-Sheikh, Essam Attia, Tomaso R. R. Bontognali, Mike Bowman, James Duggan, Ali Galambour, Jeremy Goff, Indra Gunawan, Ahmed Hamza, Andrew Horbury, Ibnelwaleed Hussein, Jerry Jameson, Hamdi Kandil, Chris Kendall, Syed Oubee Khadri, Abdul Aziz bin Laboun, Jacques LeBlanc, Wael Mattar, Judith McKenzie, P. J. Moore, Issa Makhlouf, Quintin Morton, Sobhi Nasir, John Rivers, Abdulali Sadiq, Ahmad Sakhaee-Pour, Thomas Seers, Sabrina L. Skeat, Joyce Standring, Christian Strohmenger, Ali Tarabulsi, Christopher Tiratsoo, Elnaiem Ubaid, Frans Van Buchem, Sarah Vest, and Ruqaiya Yousif.

We are also immensely grateful to our colleagues at the Environmental Science Center, including Zulfa Al-Disi, Jassim Abdulla A. Al-Khayat, Ebrahim Al-Ansari, Alexandra Leitão-Ben Hamadou, Rajendran Sankaran, Sally Saliba, Veerasingam Subramanian, and Ponnumony Vethamony for their collaboration and support.

Thanks to Qatar University's leadership for their support throughout this project.

This book is a testament to the collective knowledge and dedication of all those mentioned, and we profoundly appreciate their contributions.

Competing Interests The authors have no competing interests to declare that are relevant to the content of this manuscript.

Contents

Historical Background and Early Geologic Exploration

1

Abstract

The early mapping of the Arabian Gulf, including Qatar, was part of the early Western interest in the region. One of the earliest geographic expeditions of the area was conducted by the Greek explorer Nearchus during the fourth century BC, which was part of the Alexander the Great campaign that occupied Babylon in present-day Iraq. The next phase of exploration was carried out by the Portuguese traveler Afonso de Albuquerque as part of an effort to establish trading posts in the region. During the nineteenth and twentieth centuries, the region attracted the attention of Europeans, particularly British explorers. In 1820, the British Indian Navy executed a hydrographic survey of the Gulf water and mapped its water and coastlines. These efforts were needed to facilitate the safe navigation of the British ships. Later, the British Royal Air Force (RAF) conducted aerial surveys of the area, generating the first detailed maps and charts of its topography and natural resources. The discovery of oil in Iran, Iraq, and the neighboring Bahrain, and the possible competition from the American oil companies working in Saudi Arabia, led the Anglo-Persian Oil Company (the Iraq Petroleum Company—IPC later) to negotiate and sign a deal with the Amir of Qatar. The first IPC geologic expedition arrived in Qatar on the first of March 1926. A 75-year concession was signed between the two sides on May 17th, 1935. On October 6th, 1938, the first exploration well, Dukhan-1 was drilled, and oil was encountered in the Jurassic Arab Formation. The discovery of the North Gas field was achieved by Shell in 1971, which signaled the start of the gas age of Qatar. Among the prominent geologists who contributed to a better understanding of Qatar's geology are George Martin Lees, Walter Sugden, Anthony John Standring, and Jacque LeBlanc.

1.1 Historical Background

The Arabian Gulf has long been a significant hub for trade, commerce, and cultural exchange, with a rich history of exploration and discovery. As early as the fifth century BC, Greek explorers were drawn to the region, seeking new lands, resources, and trading partners. Over the centuries, the Gulf has been traversed by many explorers and navigators from diverse cultural backgrounds, each contributing to our understanding of the region's geography, geology, and cultural diversity.

The early mapping efforts of the Arabian Gulf were critical in shaping our understanding of the region's geography and political and economic importance. The Gulf region has been the subject of cartographic exploration and mapping for centuries, with early efforts dating back to the ancient Greeks and Romans. One of the most significant early geographic expeditions in the Arabian Gulf was led by the Greek explorer Nearchus in the fourth century BC. Nearchus was part of the army of Alexander the Great, which had conquered much of the Persian Empire and was eager to explore and expand into new territories. Nearchus led an expedition in the Indus River Delta in modern-day Pakistan and traveled through the Arabian Gulf, exploring many coastal regions, islands, and ports. He documented his observations and insights in a detailed account of his travels, which provided valuable information about the region's geography, fauna, and cultures.

During the age of European exploration and colonization, the Arabian Gulf also drew the attention of several prominent explorers and adventurers. In the early sixteenth century, the Portuguese explorer Afonso de Albuquerque led a series of expeditions to the Gulf, hoping to establish Portuguese trading posts and expand their empire. Albuquerque successfully established several forts and trading posts along the Gulf

F. N. Sadooni and H. A. Al-Kuwari, *Geological Evolution of Qatar and the Arabian Peninsula*, Regional Geology Reviews,
https://doi.org/10.1007/978-3-031-94394-2_1

Fig. 1.1 Explorer William Gifford Palgrave (Wikipedia 2025a)

coast. Still, his ambitions were ultimately thwarted by the Ottomans, who drove the Portuguese out of the region in the late sixteenth century.

Later, in the nineteenth and twentieth centuries, European explorers played a significant role in exploring and mapping the Arabian Gulf. The British explorer William Gifford Palgrave (Fig. 1.1) spent several years traveling and studying the cultures and languages of the Gulf region, producing a valuable ethnographic account of his experiences. Similarly, the French explorer René Caillié journeyed to the Gulf in the mid-nineteenth century, visiting many major ports and cities and documenting his travels in a detailed account.

The British played a significant role in exploring and mapping the Arabian Gulf in the nineteenth and twentieth centuries. In the 1820s, the British Indian Navy began conducting hydrographic surveys of the Gulf to create accurate charts and maps of the region's waters and coastline. These surveys were critical for the safe navigation of British ships and protecting British interests in the Gulf.

In addition to hydrographic surveys, the British undertook several expeditions to explore the interior of the Gulf region. One of the most famous of these expeditions was led by Lieutenant-Colonel Lewis Pelly in the 1860s. Pelly was a political agent of the British colonial government in India, and he was tasked with establishing friendly relations with the tribes of the Gulf region and gathering intelligence on their political and military activities.

Pelly's expedition was challenging, as the interior of the Gulf was largely uncharted, and the tribes were often hostile to outsiders. Nevertheless, he successfully established diplomatic relations with many of the region's tribes, including the sheikhdoms of Abu Dhabi and Dubai. Pelly's explorations helped lay the groundwork for establishing British influence in the region and the eventual creation of the Trucial States, which became the United Arab Emirates in 1971 (Heard-Bey 2005).

Another important British expedition to the Gulf was led by Major Percy Cox in the early twentieth century. Cox was a political officer of the British Empire in Mesopotamia (modern-day Iraq) and was instrumental in negotiating the borders and treaties that would shape the region for decades. Cox's expeditions took him to many parts of the Gulf, including Bahrain, where he was key in negotiating the Bahrain Petroleum Company's oil concession (Fieldhouse 2006).

The Danish explorer Carsten Niebuhr led an early expedition to Qatar in the eighteenth century (around 1778). Niebuhr was part of a larger scientific expedition sent by the King of Denmark to explore and study the natural history, geology, and cultures of the Middle East. Niebuhr's explorations in Qatar took him to several of the region's islands, including Halul Island, which he named "The Blue Island" due to its distinctive blue-green vegetation. Niebuhr's detailed observations and drawings provided valuable insights into the flora and fauna of the Gulf islands (LeBlanc 2015) (Fig. 1.2).

Early expeditions in the Arabian Gulf also included exploring and mapping Qatar and its surrounding islands. One of the first known expeditions to Qatar was carried out by the British explorer James Raymond Wellsted in 1835. This effort was part of the British Indian Navy's hydrographic survey of the Gulf, and he was tasked with exploring and mapping the waters and coastline of Qatar. His explorations took him to several parts of the Qatari coast, including the city of Doha and the island of Al-Bidaa. The detailed charts and maps he produced were instrumental in establishing the safe navigation of British ships in Qatari waters.

In the early twentieth century, Qatar also drew the attention of several other explorers and scientists. One of the most prominent of these was the British archaeologist and explorer C. Leonard Woolley, who conducted excavations on the island of Al-Bidaa in 1914. Woolley's excavations uncovered several important artifacts and structures from the Islamic period, including a large fortification wall and a mosque. These discoveries helped to shed light on the history and culture of Qatar and the surrounding islands.

In the early twentieth century, the British Royal Air Force (RAF) also significantly mapped the Gulf region. The RAF conducted aerial surveys of the Gulf, producing detailed maps and charts that provided valuable information about the region's topography and natural resources. These surveys were critical for developing infrastructure, including the construction of airports and transportation networks, and they helped establish a more comprehensive understanding of the region's geography.

In addition to exploration and mapping, the Gulf region also drew the attention of several artists and writers in the nineteenth and twentieth centuries. One of the most famous was the British painter John Frederick Lewis (Fig. 1.3), who spent several years living and working in the Qatari city of Doha in the mid-nineteenth century. Lewis's paintings of Qatari landscapes, people, and architecture were some of the first to depict the region's rich cultural heritage and natural beauty, and they continue to be admired by art lovers and scholars today.

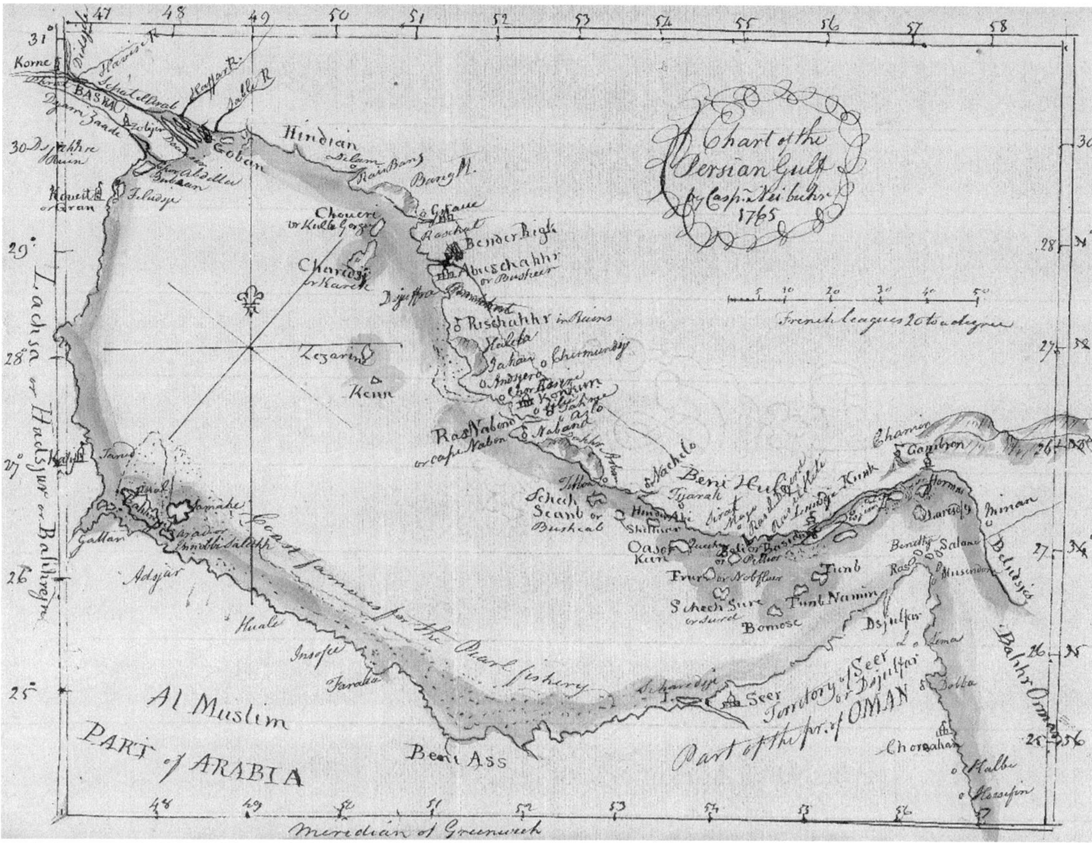

Fig. 1.2 Map of the Arabian Gulf and its islands by the Danish explorer (Carsten 1720)

Fig. 1.3 British artist John Frederick Lewis, who lived and worked in Doha (by Elliott & Fry 1860s) (Wikipedia 2025b)

In recent years, modern mapping technologies have continued to advance our understanding of the Gulf's geography. Satellite imagery and remote sensing technologies have enabled us to create highly detailed and accurate maps of the region's topography, geology, and land use patterns. These maps have provided valuable information for scientific research, urban planning, and environmental management, and they continue to be an essential tool for understanding and managing the Gulf region's natural resources and infrastructure.

The first mention of Qatar on maps was on the D'Anville-Niebuhr map, which was constructed in 1794, where there is a mention of an island called "Deh Rogn." The word *deh* is Persian, meaning a village, probably the present-day "Ras Rakan" in northern Qatar. However, the map of Niebur constructed in 1765 does not show this area but Huali and Faraha (Al-Huwela and Al-Fareha) (Johnstone and Wilkinson 1960; Zwemer 1900). Furthermore, Niebuhr, in his book (Voyage en Arabie, (Niebuhr 1780)), marked the land east of Salwa estuary as "Gattar." A new map by D'Anville from nearly 1794 was recently published. It shows Qatar as a point on the coast with the name Catura. The word "Khau" also

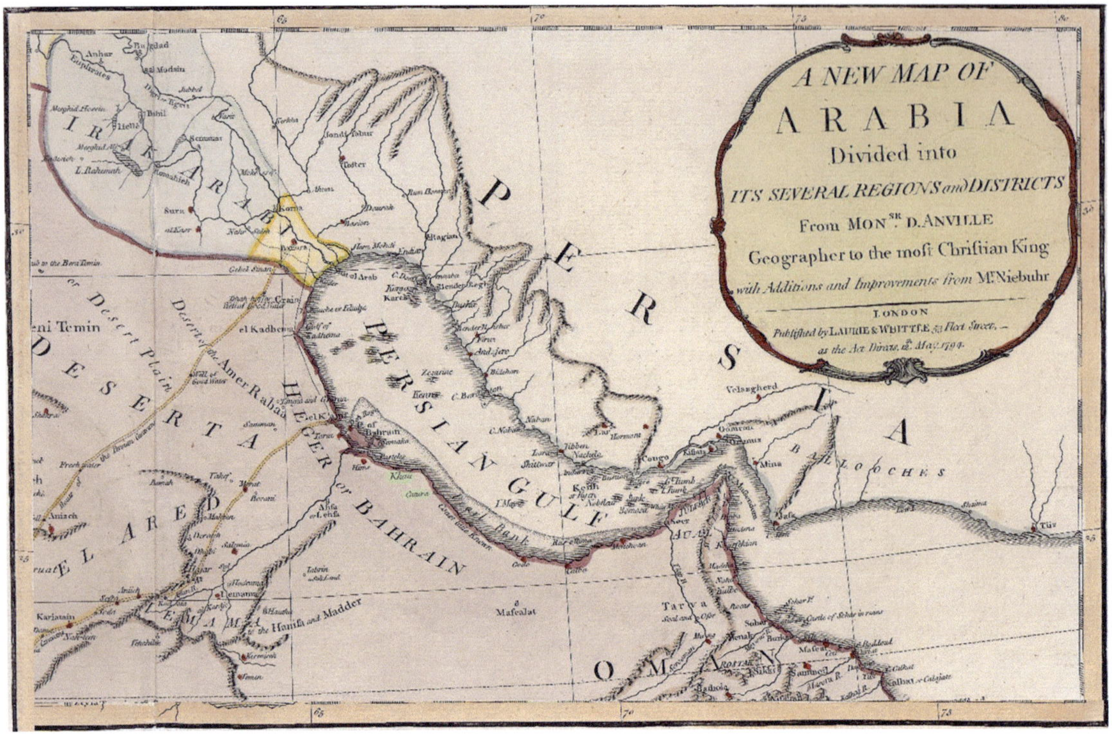

Fig. 1.4 Part of the Arabian Gulf by the French cartographer and geographer Jean-Baptiste Bourguignon D'Anville (1697–1782) showing the word Catura and Khau in green (Whittle 1794)

appears, which probably means "Khor" about the present-day city (Fig. 1.4).

More details about the country were published on a map with a scale of 1:1,641,836 that appeared in an atlas published by Vandermaelen in 1827. The map mentions four places, two on the northern tip of Qatar-Core Hessan (Khor Hassan, known today as Al-Khuwer) and Jumale (Al-Jumel). The first proper configuration of Qatar as a peninsula appeared in 1825 on a map constructed by Arrowsmith. The map shows a small island with the word "Ras Reccan" on its west coast.

1.2 Geological Expeditions

The Anglo-Persian Oil Company (changed name later to Iraq Petroleum Company, IPC), the oil company that drilled the first oil well in the Middle East, the Masjed Suleyman in Iran in 1908, and then the Kirkuk field in Iraq in 1927, considered Qatar as of "marginal interest" in terms of petroleum prospects (Morton 2020). However, two emerging factors changed the Company's attitude. First, on June 2nd, 1932, Standard Oil Company of California (Socal) found oil in Bahrain (Vassiliou, 2018). The second alarming factor for the IPC is that Socal considered looking for oil in Saudi Arabia and eventually won the bid there. The IPC's many attempts to have a share of Saudi oil failed. These new circumstances forced the IPC to target Qatar as the next potential concession.

The Anglo-Persian Company's request to survey the peninsula was approved by the British Colonial Office in 1925, making it possible for the first geologists to travel there (Fig. 1.5). Geologist George Lees led the survey party, while Haji Williamson was a guide. They were joined by Captain Smith from the British Indian Army. On March 1st, 1926, Lees rented a 35-ton motor launch and departed from Bahrain. However, before they could reach Doha, bad weather and turbulent seas compelled them to spend the night in a small cave close to Fuwairat on Qatar's northeastern coast. The following morning, Lees strolled along the rocky shore as the sun rose. He stopped when he spotted something dark and shiny in the intense sunlight and picked it up, turning it over in his hands. It was a bitumen fragment that had washed ashore in the ocean, most likely because of oil seepage under the sea (LeBlanc 2015; Morton 2010, 2013).

LeBlanc (2015) provided a summary of Lees's report from the 1926 survey:

> The Qatar peninsula is made up of a highly gentle fold in Eocene rocks, likely older than the rocks exposed in Bahrain. My collection of nummulites will determine the exact age, but I anticipate that they will show to be Middle Eocene. In the peninsula's middle part, the ground rises almost imperceptibly to only 91 m. Although there aren't many good exposures where dips can be measured, the average dip from the center cannot be more than 1 to 2 degrees in either direction. No signs of oil of any kind were discovered. Unfortunately, storms prevented

Fig. 1.5 First geologic survey team visited Qatar in October 1934. G. M. Lees (with the Panama hat in the middle), to his left, is Haji Williamson, then Peter Cox, and the local guide Sheikh Mansour. IPC Society Newsletter, No 105, January 2000, p. 32. Qatar 1934. Reproduced from the BP archive with permission

> me from traveling to the islands, where more proof might be found. A petroliferous development in the underlying Cretaceous rocks at Qatar is highly improbable due to the location's inferior structural quality. However, if drilling in Bahrain goes well and is successful, Qatar might experience similar circumstances. I recently received a copy of Dr. Heim's Geological Report on Kuwait, Hasa, and Bahrain from the Chief Geologist. His conclusions are unequivocal. A test in Bahrain would be a complete gamble and is not advised because Kuwait and Hasa have no prospects. Suppose the Eastern and General Syndicate chooses to drill in Bahrain. In that case, they must consider that since we competed for the concession, we must have strong supporting evidence from the Persian side that was unavailable to Dr. Heim.

The report was inconclusive, but the discovery of oil in Bahrain seven years later accelerated the Anglo-Persian Company's efforts to explore the geology of Qatar.

The first documented written communication between the Anglo-Persian Oil Company and the Amir of Qatar, Shaikh Abdulla Bin Jassim Al-Thani (Fig. 1.6), was a delegation consisting of C. C. Mylles and Haji Abdullah Williamson carrying a letter signed by N. A. Gass, the Deputy General Manager of the company, dated August 20th, 1932, and directed to the Amir. The letter asks the Amir for permission to conduct a geologic survey of the land of Qatar. The letter also briefly introduced Mr. Mylles (Fig. 1.7).

Fig. 1.6 Shaikh Abdulla Bin Jassim Al-Thani, Amir of Qatar (1880–1957) (Amiri-Diwan)

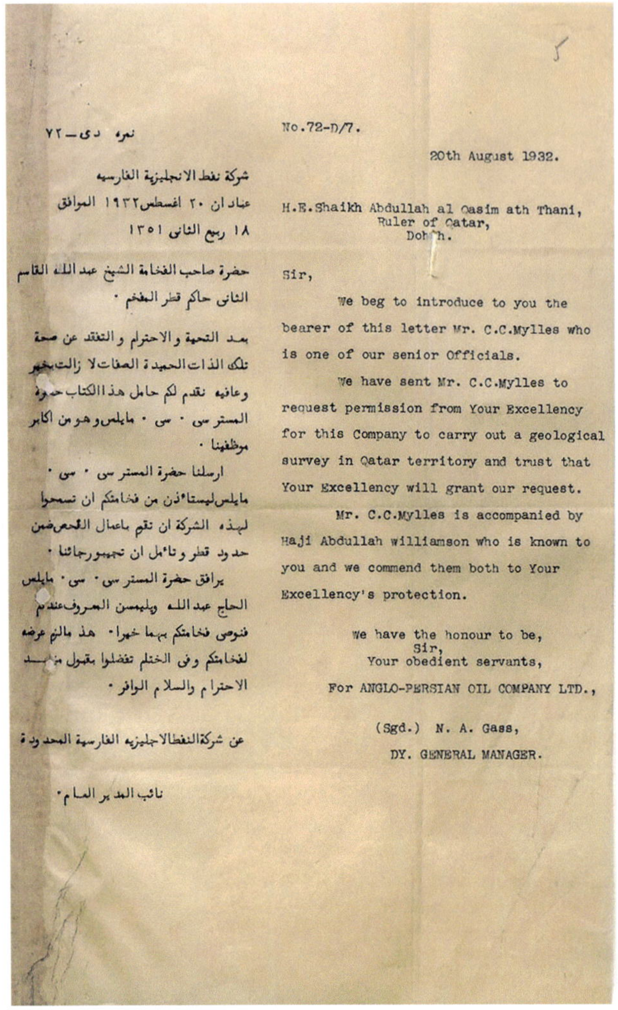
نمره دى — ٧٢

شركة نفط الانجليزية الفارسيه
عبادان ٢٠ اغسطس١٩٣٢ الموافق
١٨ ربيع الثانى ١٣٥١

حضرة صاحب الفخامة الشيخ عبد الله القاسم
الثانى حاكم قطر المفخم ·

بعد التحية و الاحترام و التفقد عن صحة
تلك الذات الحميدة الصفات لا زالت بخير
وعافيه نقدم لكم حامل هذا الكتاب حضرة
المستر سى · سى · مايلس وهو من اكابر
موظفينا ·

ارسلنا حضرة المستر سى · سى ·
مايلس ليستأذن من فخامتكم ان تسمحوا
لهذه الشركة ان تقوم باعمال الفحص ضمن
حدود قطر و نأمل ان تجيبو رجائنا ·

يرافق حضرة المستر سى · سى · مايلس
الحاج عبد الله ويليمسن المعروف عندكم
فنوصى فخامتكم بهما خيرا · هذ ما لزم عرضه
لفخامتكم وفى الختام تفضلوا بقبول مزيد
الاحترام والسلام الوافر ·

عن شركة النفط الانجليزيه الفارسية المحدودة

نائب المدير العام ·

No.72-D/7.

20th August 1932.

H.E.Shaikh Abdullah al Qasim ath Thani,
Ruler of Qatar,
Doh h.

Sir,

We beg to introduce to you the bearer of this letter Mr. C.C.Mylles who is one of our senior Officials.

We have sent Mr. C.C.Mylles to request permission from Your Excellency for this Company to carry out a geological survey in Qatar territory and trust that Your Excellency will grant our request.

Mr. C.C.Mylles is accompanied by Haji Abdullah Williamson who is known to you and we commend them both to Your Excellency's protection.

We have the honour to be,
Sir,
Your obedient servants,
For ANGLO-PERSIAN OIL COMPANY LTD.,

(Sgd.) N. A. Gass,
DY. GENERAL MANAGER.

Fig. 1.7 Introduction letter from the Deputy General Manager of the Anglo-Persian Company to the Amir of Qatar introducing Mylles, a senior officer of the Company ("File 10/3 I Qatar Oil Concession")

On August 25th, 1932, there was another letter, this time from Mylles, mentioning a previous agreement between the two sides in 1926 and asking for renewal and extension of that previous arrangement. That was a verbal agreement between the Amir of Qatar and a concession hunter named Frank Holmes. This letter detailed the requirements of the company from the Qatari side.

These include (1) giving the company the sole right to carry out a geologic survey of the country for two years, (2) accepting to receive and consider oil concession offers, and (3) not considering offers from any other third party (Fig. 1.8).

The Amir replied to the Deputy General Manager on August 26th, 1932, stating that he met with Mr. Mylles and reached a written agreement that was given to him (Fig. 1.9). On the same date, the Amir wrote a second letter detailing the reached agreement. He accepted granting the Company "a concession for the examination of the country of Qatar for the signs indicating the presence of oil" for two years as of August 1932. He also mentioned that the two parties will meet after the expiry of the concession, and it is up to him to accept or reject the submitted offer. He also confirmed that Qatar will not consider any third party for the same during the concession period (Fig. 1.10).

On December 3rd, the General Manager of the Anglo-Persian Company, E. H. O. Elkington, wrote to the Amir asking his permission to send two geologists with Haji Abdullah Williamson and some camp workers. They would arrive on the ship (Khuzestan) on the January 4th, 1933 (Fig. 1.11). The Amir replied on the 30th December, welcoming the team and promising to render the needed assistance and protection (Fig. 1.12).

In January 1933, the new geologist team consisting of Peter Cox and Eugene W. Shaw, with Haji Williamson as their guide, started a new survey using an admiralty map of the coastline. They landed, paid their respects to the Amir, and then began to map the peninsula using a prismatic compass and car kilometers. During this expedition, the team constructed the map that eventually served as the

نمره دي-72

شركة نفط الإنجليزية الفارسية

عبادان 20 أغسطس 1932 الموافق 18 ربيع الثاني 1351

حضرة صاحب الفخامة الشيخ عبد الله القاسم الثاني حاكم قطر المفخم.

بعد التحية والاحترام والتفقد عن صحة تلك الذات الحميدة الصفات لا زالت بخير وعافية نقدم لكم حامل هذا الكتاب حضرة المستر سي. سي. مايلس وهو من اكابر موظفينا.

أرسلنا حضرة المستر س. سي. مايلس ليستأذن من فخامتكم أن تسمحوا لهذه الشركة أن تقوم بأعمال الفحص ضمن حدود قطر وتأمل أن تجيبوا رجائنا.

يرافق حضرة المستر سي. سي. مايلس الحاج عبد الله ويليمسن المعروف عندكم فنوصي فخامتكم بهما خيرا. هذا ما تم عرضه لفخامتكم وفي الختام تفضلوا بقبول مزيد الاحترام والسلام الوافر.

عن شركة النفط الإنجليزية الفارسية المحدودة

نائب المدير العام

No. 72-D/7

20 August 1932

H. E. Shaikh Abdullah al Qasim ath Thani,

Ruler of Qatar, Doha

Sir,

We beg to introduce to you the bearer of this letter Mr. C. C. Mylles who is one of our senior officials.

We have sent Mr. C. C. Mylles to request permission from your Excellency for this Company to carry out a geological survey in Qatar territory and trust that your Excellency will grant our request.

Mr. C. C. Mylles is accompanied by Haji Abdullah Williamson who is known to you, and we commend them both to Your Excellency's protection.

We have the honour to be,

Sir,

Your obedient servants,

For ANGLO-PERSIN OIL COMPANY LTD,

(sgd.) N. A. Gass,

DY. GENERAL MANAGER.

Fig. 1.7 (continued)

foundation for the future concession. The admiralty map was a blown-up photostat of the coastal area. It was accurate to a certain extent, but it was later discovered to be 68 km south of Bahrain, where the coast had been indicated with a broken line. They were fortunate to have an excellent *Alveolina*-limestone bed to follow because it persisted across a large portion of the country. The map also showed the locations of all the water wells that were visited, in addition to the topographical and geological characteristics of the area. At that time, the locals manually dug all the water wells to a depth of up to 100 ft. (33 m). The eastern and northern ends of the peninsula were surveyed from Doha and a camp et al. Ga'ara between February 5th and February 14th. The Dukhan anticline was identified during this trip because it had steeper dips than the main peninsula. Once it was found, their efforts were focused on better defining it. They finally sketched the general surface stratigraphy of the peninsula after touring the entire country and carefully studying the Dukhan anticline, which is formed of the same rocks found in Bahrain. The team concluded that there is a "fair chance" of finding a commercial oilfield in the area (LeBlanc 2015; Morton 2013).

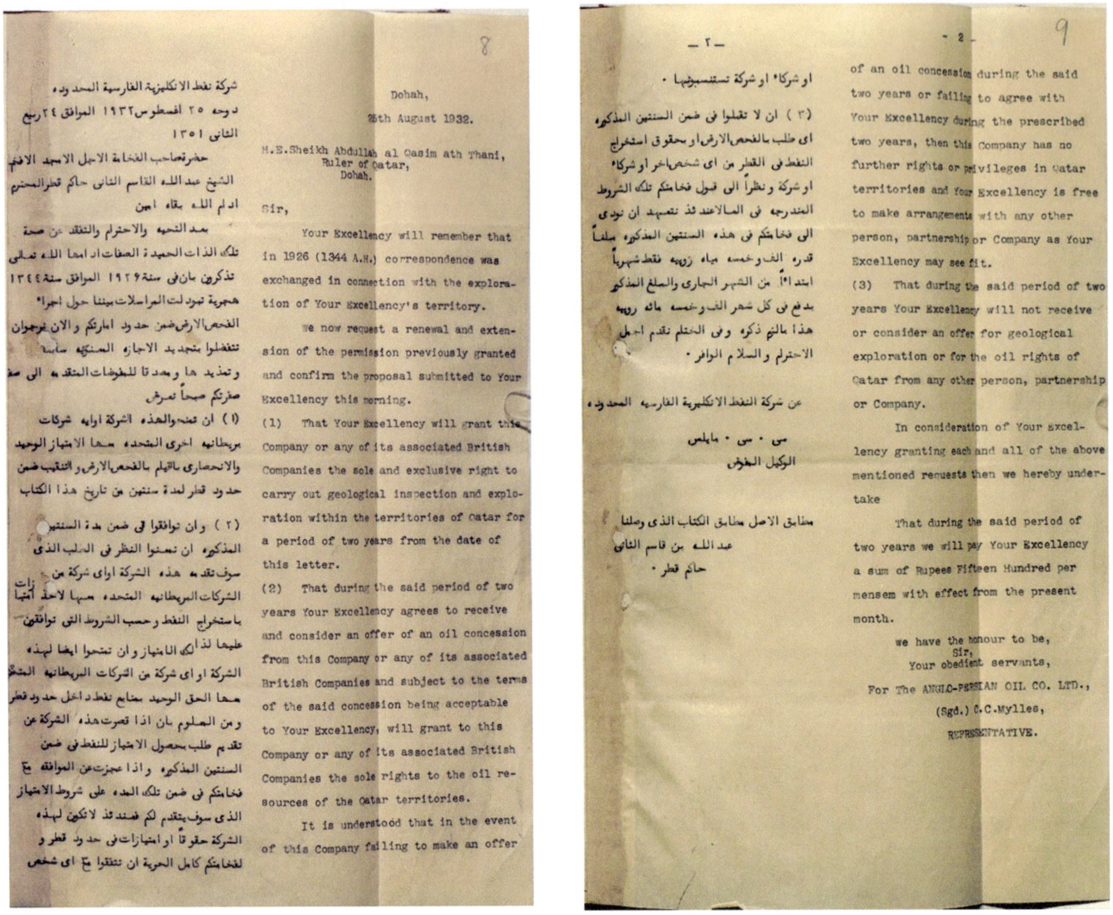

8

Dohah,

25th August 1932.

H.E.Sheikh Abdullah al Qasim ath Thani,
Ruler of Qatar,
Dohah.

Sir,

Your Excellency will remember that in 1926 (1344 A.H.) correspondence was exchanged in connection with the exploration of Your Excellency's territory.

We now request a renewal and extension of the permission previously granted and confirm the proposal submitted to Your Excellency this morning.

(1) That Your Excellency will grant this Company or any of its associated British Companies the sole and exclusive right to carry out geological inspection and exploration within the territories of Qatar for a period of two years from the date of this letter.

(2) That during the said period of two years Your Excellency agrees to receive and consider an offer of an oil concession from this Company or any of its associated British Companies and subject to the terms of the said concession being acceptable to Your Excellency, will grant to this Company or any of its associated British Companies the sole rights to the oil resources of the Qatar territories.

It is understood that in the event of this Company failing to make an offer

- 2 - 9

of an oil concession during the said two years or failing to agree with Your Excellency during the prescribed two years, then this Company has no further rights or privileges in Qatar territories and Your Excellency is free to make arrangements with any other person, partnership or Company as Your Excellency may see fit.

(3) That during the said period of two years Your Excellency will not receive or consider an offer for geological exploration or for the oil rights of Qatar from any other person, partnership or Company.

In consideration of Your Excellency granting each and all of the above mentioned requests then we hereby undertake

That during the said period of two years we will pay Your Excellency a sum of Rupees Fifteen Hundred per mensem with effect from the present month.

We have the honour to be,
Sir,
Your obedient servants,
For The ANGLO-PERSIAN OIL CO. LTD.,
(Sgd.) C.C.Mylles,
REPRESENTATIVE.

Fig. 1.8 Letter exchange between Mylles of the Anglo-Persian Oil Company and Sheikh Abdulla Bin Jasim Al-Thani about permission to allow a geologic expedition to work in Qatar ("File 10/3 I Qatar Oil Concession")

In the end, they concluded that while there is a decent chance that a profitable field may be developed in the Dukhan structure, the chances of drilling success in Qatar are much less favorable than in comparable circumstances within the normal folded zone of the Iraq-Persian basin. However, the prospects for the peninsula's main structure are more hypothetical, but if subsurface conditions are favorable, it would have a much larger accumulation of oil. To economically exploit an oil field in Qatar, a detailed geological mapping would be required. Still, additional surface evidence from the peninsula will not help clarify whether oil is present in commercial quantities. They suggested that a well should be located in the Da'aisa on the peak of the Dukhan structure. Before locating a test well on that anticline, detailed mapping of the main structure should be done if this well proves successful. It would take two geologists an entire six-month field season to determine this structure's precise shape. The party at the time was unaware of stratigraphy and deep-water structure. Still, since Bahrain's surface rocks were clearly of Eocene age, both anticlines appeared to be excellent prospects for Cretaceous oil. The amplitude of the main anticline of the peninsula, which Shaw and Cox had outlined only by aneroid readings in their surveillance, was to be ascertained in the winter of 1933–1934 by a precise survey of a critical traverse. The survey was conducted on a line chosen by Cox by chief topographical surveyor William E. Browne and Anglo-Persian geologist D. C. Ion.

On January 15th, 1935, Browne presented the report he had written. Mylles (along with "Haji" Williamson") visited Qatar several times in the years 1934 and 1935 to discuss a long-term concession with the Amir; Mylles signed a concession agreement for a 75-year period that was signed in Doha on May 17th, 1935.

T. F. Williamson and Rene Pomeyrol conducted a detailed geologic survey covering the entire country from November 1937 to May 20th, 1938. In addition to better understanding the already identified Dukhan structure and constructing a more detailed geologic map of the peninsula, finding fresh groundwater was a major issue in the Gulf region and was essential for the coming exploration and production efforts.

شركة نفط الإنكليزية الفارسية المحدودة

دوحة 25 اغسطوس 1932 الموافق 24 ربيع الثاني 1351

حضرة صاحب الفخامة الأجل الأمجد الأفخم الشيخ عبد الله القاسم الثاني حاكم قطر المحترم

أدام الله بقاه امين

بعد التحية والاحترام والتفقد عن صحة تلك الذات الحميدة الصفات ادامها الله تعالى تذكرون بأن في سنة 1926 الموافق سنة 1344 هجرية تبودلت المراسلات بيننا حول اجراء الفحص الأرض ضمن حدود امارتكم والآن نرجو أن تتفضلوا بتجديد الاجازة الممنوحة سابقا وتمديدها ومصدقا للمفوضات (كذا) المتقدمة الى صفرتكم (؟) صبحا لغرض

(1) ان تمنحوا لهذه الشركة أو أية شركات بريطانية أخرى المتحدة معها الامتياز الوحيد والانحصاري بالقيام بالفحص الأرض والتنقيب ضمن حدود قطر لمدة سنتين من تاريخ هذا الكتاب

(2) وان توافقوا في ضمن مدة السنتين المذكورة ان تمعنوا النظر في الطلب الذي سوف تقدمه هذه الشركة أو أي شركة من الشركات البريطانية المتحدة معها لأخذ امتيازات باستخراج النفط وحسب الشروط التي توافقون عليها لذلك الامتياز وأن تمنحوا أيضا لهذه الشركة أو أي شركة من الشركات البريطانية المتحدة معها الحق الوحيد بمنابع النفط داخل حدود قطر ومن المعلوم بان اذا قصرت هذه الشركة عن تقديم طلب بحصول الامتياز للنفط في ضمن السنتين المذكورة واذا عجزت عن الموافقة مع فخامتكم في ضمن تلك المدة على شروط الامتياز الذي سوف يتقدم لكم فعندئذ لا تكون لهذه الشركة حقوقا أو امتيازات في حدود قطر ولفخامتكم كامل الحرية أن تتفقوا مع أي شخص او شركاء او شركة تستنسبونها (كذا).

(3) أن لا تقبلوا في ضمن السنتين المذكورة أي طلب بالفحص الأرض أو بحقوق استخراج النفط في القطر من أي شخص آخر أو شركاء أو شركة ونظرا الى قبول فخامتكم تلك الشروط المندرجة في العالا (؟) عندئذ نتعهد أن نودي الى فخامتكم في هذه السنتين المذكورة سلفا مبلغا قدره الف وخمسه مياه (كذا) روبية فقط شهريا ابتداء من الشهر الجاري والمبلغ المذكور بدفع في كل شهر الف وخمسه مائه روبية هذا ما لزم ذكره وفي الختام نقدم أجل الاحترام والسلام الوافر.

عن شركة النفط الإنكليزية الفارسية المحدودة

سي. سي. مايلس

الوكيل المفوض

مطابق الأصل مطابق الكتاب الذي وصلنا

عبد الله بن قاسم الثاني

حاكم قطر

Doha, 25th August 1932.

H.E. Sheikh Abdulla al Qasim ath Thani (sic)

Ruler of Qatar,

Doha

Sir,

Your Excellency will remember that in 1926 (1344 A.H.), correspondence was exchanged in connection with the exploration of your Excellency's territory.

We now request a renewal and extension of the permission previously granted and confirm the proposal submitted to your Excellency this morning.

(1) That your Excellency will grant this company or any of its associated British companies the sole and exclusive right to carry out geological inspection and exploration within the territories of Qatar for a period of two years from the date of this letter.

(2) That during the said period of two years your Excellency agrees to receive and consider an offer of an oil concession from this company or any of its associated British companies and subject to the terms of the said concession on being acceptable to Your Excellency will grant this company or any of its associated British companies the sole rights to the oil resources of Qatar territories.
It is understood that in the event of this company failing to make an offer of an oil concession during the said two years or failing to agree with Your Excellency during the prescribed two years, then the company has no further rights or privileges in Qatar territories and Your Excellency is free to make arrangement with any other person, partnership or company as Your Excellency may see fit.

(3) That during the said period of two years Your Excellency will not receive or consider an offer for geological exploration or for the oil rights of Qatar from any other Person, partnership or company.

In consideration of Your Excellency granting each and all of the above mentioned requests then we hereby undertake

That during the said period of two years will pay Your Excellency a sum of Rupees Fifteen Hundred per month (?) with effect from the present month.

We have the honour to be,
Sir,
Your obedient servants,
For the ANGLO-PERSIAN OIL CO. LTD.,
(Sgd.) C.C. Mylles,
REPRESENTATIVE.

Fig. 1.8 (continued)

A detailed map of the 1:50,000 scale was constructed for the Dukhan structure using the conventional plan-table triangulation and stadia methods. These were the basic tools for mapping and oil well location in the Anglo-Persian Company then. All elevation was measured relative to the mean sea level near Zekrit Pier (LeBlanc 2015). A total area of 850 km^2 was mapped with this scale because it was considered of interest; the remainder of the country was mapped with a scale of 1:10,000.

Regarding water resources, the team examined the already drilled water wells and selected three wells, depending on water level and salinity, to be the future supply for the proposed Dukhan Camp. They concluded that there was a limited supply of fresh water in the area.

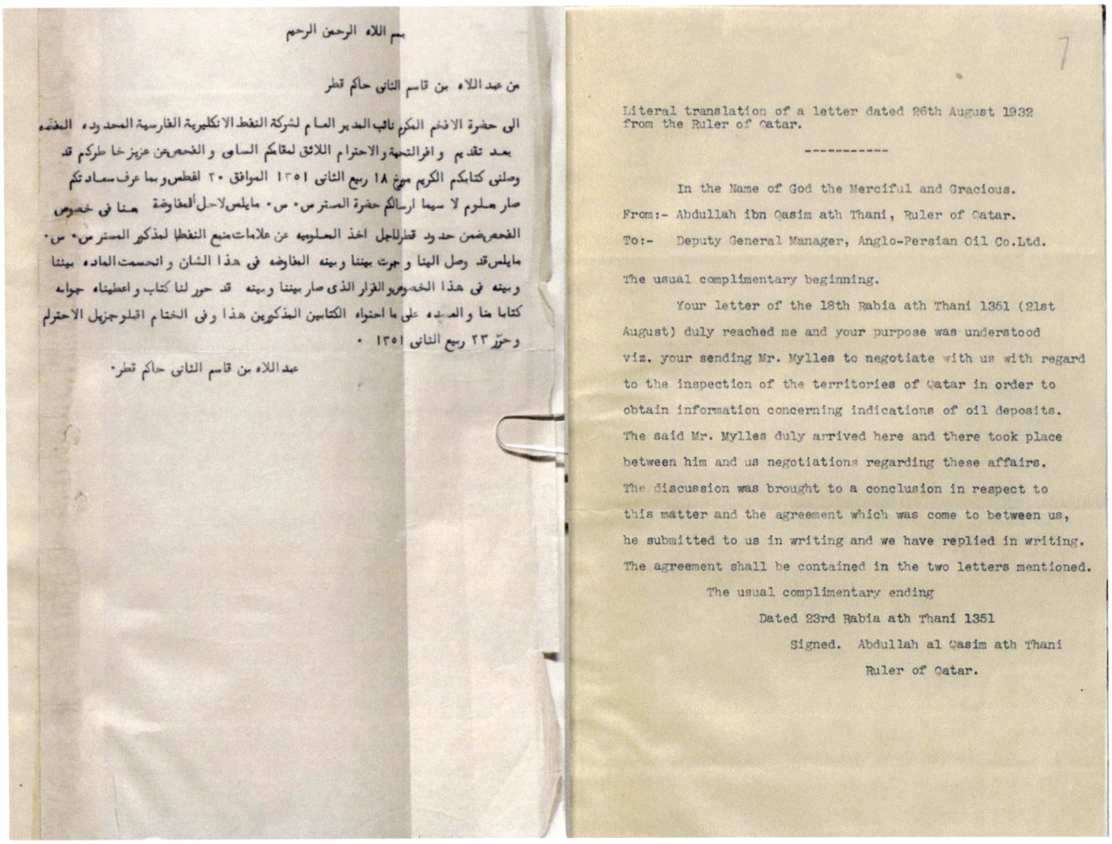

بسم اللاه الرحمن الرحيم

من عبد اللاه بن قاسم الثانى حاكم قطر

الى حضرة الافخم المكرم نائب المدير العام لشركة النفط الانكليزية الفارسية المحدوده المفخم
بعد تقديم وافر التحية والاحترام اللائق لمقامكم السامى والفحص عن عزيز خاطركم قد
وصلنى كتابكم الكريم مؤرخ ١٨ ربيع الثانى ١٣٥١ الموافق ٢٠ اغسطس وبما عرف سعادتكم
صار معلوم لا سيما ارسالكم حضرة المستر س٠ س٠ مايلس لاجل المفاوضة هنا فى خصوص
الفحص ضمن حدود قطر لاجل اخذ المعلوميه عن علامات منبع النفط المذكور المستر س٠ س٠
مايلس قد وصل الينا وجرت بيننا وبينه المفاوضه فى هذا الشان وانحسمت الماده بيننا
وبينه فى هذا الخصوص والقرار الذى صار بيننا وبينه قد حرر لنا كتاب واعطيناه جوابه
كتابا منا والعهده على ما احتواه الكتابين المذكورين هذا وفى الختام اقبلوا جزيل الاحترام
وحرر ٢٣ ربيع الثانى ١٣٥١ ٠

عبد اللاه بن قاسم الثانى حاكم قطر٠

7

Literal translation of a letter dated 26th August 1932 from the Ruler of Qatar.

In the Name of God the Merciful and Gracious.

From:- Abdullah ibn Qasim ath Thani, Ruler of Qatar.

To:- Deputy General Manager, Anglo-Persian Oil Co.Ltd.

The usual complimentary beginning.

Your letter of the 18th Rabia ath Thani 1351 (21st August) duly reached me and your purpose was understood viz. your sending Mr. Mylles to negotiate with us with regard to the inspection of the territories of Qatar in order to obtain information concerning indications of oil deposits. The said Mr. Mylles duly arrived here and there took place between him and us negotiations regarding these affairs. The discussion was brought to a conclusion in respect to this matter and the agreement which was come to between us, he submitted to us in writing and we have replied in writing. The agreement shall be contained in the two letters mentioned.

The usual complimentary ending

Dated 23rd Rabia ath Thani 1351

Signed. Abdullah al Qasim ath Thani

Ruler of Qatar.

Fig. 1.9 Reply of the Amir of Qatar to the Deputy General Manager of the Anglo-Persian Company confirming the arrival of Mr. Mylles ("File 10/3 I Qatar Oil Concession")

A final 75-year concession agreement was signed by Sheikh Abdullah bin Jassim Al-Thani, the Amir of Qatar and Iraq Petroleum Company, on May 17th, 1935. The concession was transferred to an associate company, Petroleum Development (Qatar) Ltd (PDQ), which later changed its name to Qatar Petroleum Company (QPC).

1.3 Drilling Dukhan-1

In 1938, Norval Baker, the chief geologist of the IPC at that time, selected the location of the first well, and the company started drilling. Drilling in the Dukhan-1 well started in October 1938. LeBlanc (2015) compiled a lively history of the drilling of Dukhan-1 based on the final well report written by the two IPC rig geologists who supervised the drilling process, E. J. Daniel and N. T. Langham. The Dukhan concession was drilled by a newly established *Iraq Petroleum Company* associated company, the Petroleum Development (Qatar) Limited (PDL), managed from the IPC offices in Kirkuk, northern Iraq. The PDL has the same shareholders as the original IPC (Morton 2013). Those were:

Royal Dutch Shell 23.7%
British Petroleum 23.75%
Compagnie Francais des Petroles 23.7%
Standard Oil of New Jersey 11.87%
Mobil Corporation 11.87%
Partex (Gulbenkian Foundation) 5.1%

On October 6th, 1938, the PDL started drilling the first well, Dukhan No. 1 [DK-0001], at Baker's chosen location. The rig was a fixed structure assembled on the spot using bolts and welding. Most supplies, including flat-top barges, had to be transported from Bahrain to Dukhan by boat, along with all drilling and construction materials. A jetty was constructed in the Bay of Zekrit to receive supplies. Bad weather, such as gales and intense sandstorms, frequently delayed transport. Dukhan No. 1, completed at a depth of 1733 m (5686 ft.), was eventually handed over to the Production Department as a producing oil well with an output of 4480 barrels per day after 459 days on the January 8th, 1940. Drilling and coring took 172 out of 459 days from spudding to handover. During the drilling process, the well was logged by E. J. Daniel from 0 to 2326 ft., by E. J. Daniel and N. T. Langham from 2326

بسم اللاه الرحمن الرحيم

من عبد اللاه بن قاسم الثاني حاكم قطر

الى حضرة الأفخم المكرم نائب المدير العام لشركة النفط الإنكليزية الفارسية المحدودة المفخمة

بعد تقديم وافر التحية والاحترام اللائق لمقامكم السامي والفحص عن عزيز خاطركم قد وصلني كتابكم الكريم مؤرخ 18 ربيع الثاني 1351 الموافق 20 أغسطس وبما عرف سعادتكم صار معلوم لا سيما ارسالكم حضرة المستر س. س. مايلس لأجل المفاوضة معنا في خصوص الفحص ضمن حدود قطر لأجل أخذ المعلومية عن علامات منبع النفط المذكور س. س. مايلس قد وصل الينا وجرت بيننا وبينه المفاوضة في هذا الشأن وانحسمت المادة بيننا وبينه في هذا الخصوص والقرار الذي صار بيننا وبينه قد حرر لنا كتاب واعطيناه جوابه كتابا منا والمقيدة على ما احتواه الكتابين المذكورين هذا وفي الختام اقبلو جزيل الاحترام

وحرر 23 ربيع الثاني 1351

عبد اللاه بن قاسم الثاني حاكم قطر

Literal translation of a letter dated 26th August 1932 from the Ruler of Qatar

----

In the Name of God the Merciful and Gracious.

From: Abdullah ibn Qasim ath Thani, Ruler of Qatar.

To: Deputy General Manager, Anglo-Persian Oil Co. Ltd.

The usual complimentary beginning.

Your letter of the 18th Rabia ath Thani 1351 (21st August) duly reached me and your purpose was understood viz. your sending Mr. Mylles to negotiate with us with regard to the inspection of the territories of Qatar in order to obtain information concerning indications of oil deposits. The said Mr. Mylles duly arrived here and there took place between him and us negotiations regarding these affairs. The discussion was brought to a conclusion in respect to this matter and the agreement which was come to between us, he submitted to us in writing and we have replied in writing. The agreement shall be contained in the two letters mentioned.

The usual complimentary ending.

Dated 23rd Rabia at Thani 1351

Signed. Abdullah al Qasim ath Thani

Ruler of Qatar

Fig. 1.9 (continued)

to 2912 ft., and by N. T. Langham from 2912 to 5686 ft. These two "resident" or well-site geologists oversaw almost all tasks. They served as mud engineers and gathered geological information. The Amir of Qatar was informed about the good news (Fig. 1.13).

The drilling faced serious problems and frequently despair due to the technical challenges they encountered and what they did to overcome them. For instance, circulation was lost after just five days of drilling. They continued drilling with seawater without getting any returns until they lowered the 16″ casing at 284.8 m (934 ft.) on November 2nd, 1938. On November 17th, two weeks later, drilling recommenced using a drilling mud made of seawater combined with salt and mud. The mud was prepared using the local Midra shale (an exposed member of the Dammam Formation). The well encountered several technical problems, including stuck pipes and bridging; at 1732.8 m (5685 ft.), they were forced to kill it.

The stratigraphic column down to around 304.8 m (1000 ft.) was known from the nearby Khatiyah water well, and the well program below that depth was based on predictions from the wells drilled earlier in Bahrain.

Oil was found not in the Cretaceous Mauddud Formation as in Bahrain. Still, in the Jurassic rocks (These were

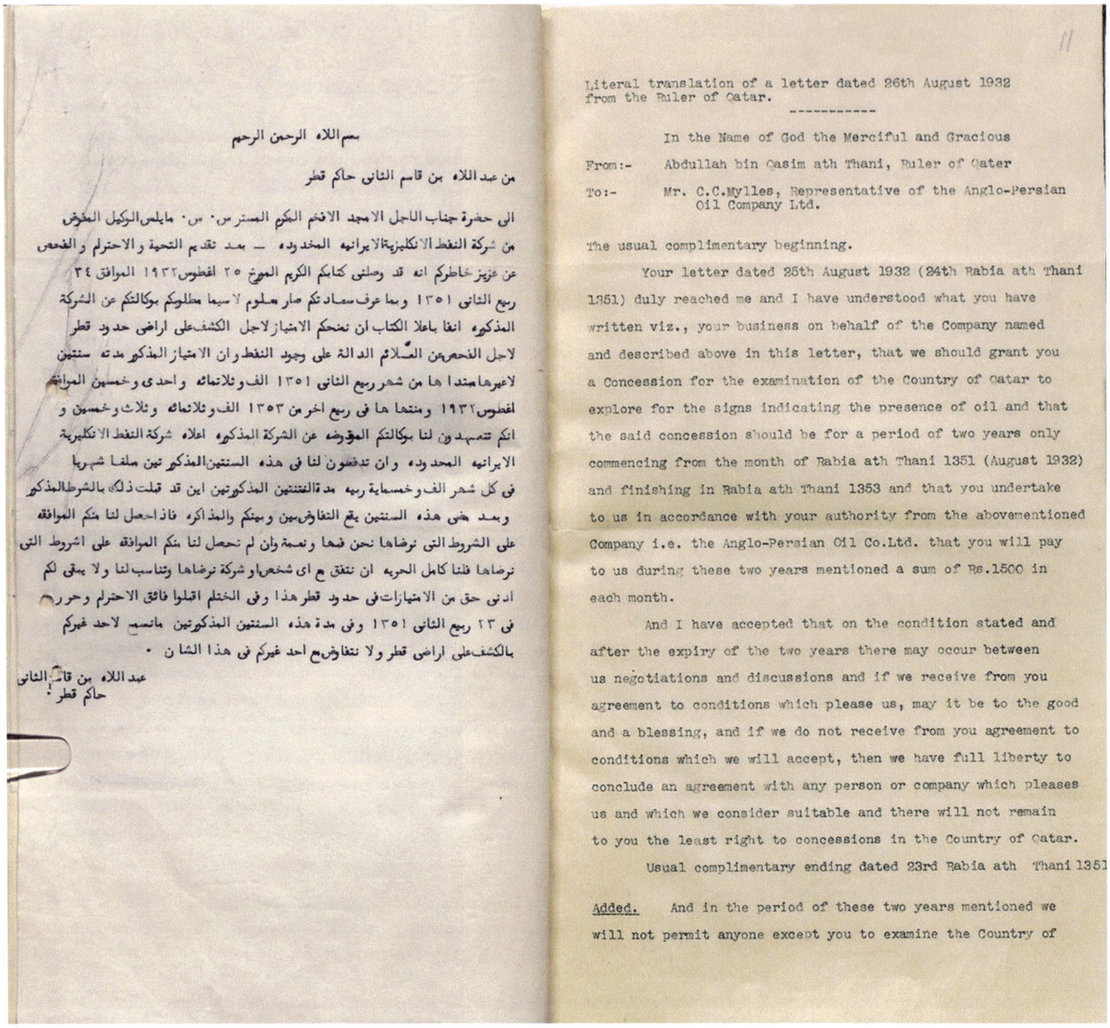

Literal translation of a letter dated 26th August 1932 from the Ruler of Qatar.

In the Name of God the Merciful and Gracious

From:- Abdullah bin Qasim ath Thani, Ruler of Qater

To:- Mr. C.C.Mylles, Representative of the Anglo-Persian Oil Company Ltd.

The usual complimentary beginning.

Your letter dated 25th August 1932 (24th Rabia ath Thani 1351) duly reached me and I have understood what you have written viz., your business on behalf of the Company named and described above in this letter, that we should grant you a Concession for the examination of the Country of Qatar to explore for the signs indicating the presence of oil and that the said concession should be for a period of two years only commencing from the month of Rabia ath Thani 1351 (August 1932) and finishing in Rabia ath Thani 1353 and that you undertake to us in accordance with your authority from the abovementioned Company i.e. the Anglo-Persian Oil Co.Ltd. that you will pay to us during these two years mentioned a sum of Rs.1500 in each month.

And I have accepted that on the condition stated and after the expiry of the two years there may occur between us negotiations and discussions and if we receive from you agreement to conditions which please us, may it be to the good and a blessing, and if we do not receive from you agreement to conditions which we will accept, then we have full liberty to conclude an agreement with any person or company which pleases us and which we consider suitable and there will not remain to you the least right to concessions in the Country of Qatar.

Usual complimentary ending dated 23rd Rabia ath Thani 1351

Added. And in the period of these two years mentioned we will not permit anyone except you to examine the Country of

Fig. 1.10 Approval of the Amir of Qatar to give the Anglo-Persian Company a two-year concession to examine the land of Qatar for oil potential ("File 10/3 I Qatar Oil Concession")

called the Zekrit Formation but changed name later to the Arab Formation to be compatible with the Saudi nomenclature.) in January 1940. This well became the first exploration well in the country after reaching a depth of 1777 m (5829 ft) in January 1940 and producing 4480 barrels of oil per day. Oil was found in the Upper Jurassic Number 3 Limestone Member of the Arab Formation. Afterward, oil was found also in the Number 4 Limestone Member and the Middle Jurassic Uwainat Member of the Araej Formation. Gas was found in the Permian Khuff Formation. The company drilled another two wells to delineate and confirm the producing structure, but the last one was drilled outside the structure area. Table 1.1 shows the identified stratigraphic succession in the well Dukhan-1 as identified by the rig geologists at the time compared with the prevailing stratigraphic nomenclature now (Figs. 1.14 and 1.15) (LeBlanc 2015).

Late in 1947, operations were restarted, and Qatar underwent its first-ever aerial photo survey. These aerial photographs were later used to precisely map the land features across the entire peninsula. Three drilling rigs were continuously used for the next three years to drill 25 wells to secure the first 1.5 million t/y of production. However, relatively quick pressure depth of about 1829 m (6000 ft.). After two years of work, eight wells using the no. 3 limestone drops after the start of production led to the development of the underlying No. 4 limestone, and by 1951, this member was supplying three-quarters of Qatar's oil production. The Dukhan structural area's surface mapping was resumed in 1948 with more thorough detail. During the months of February to April 1948, Max Chatton undertook the mapping of the northern part of the anticline from Ras Dukhan (tip of the peninsula) to an East–West

بسم اللاه الرحمن الرحيم

من عبد اللاه بن قاسم الثاني حاكم قطر

الى حضرة جناب الأجل الأمجد الأفخم المكرم المستر س. س. مايلس الوكيل المفوض من شركة النفط الإنكليزية الايرانية المحدودة- بعد تقديم التحية والاحترام والفحص عن عزيز خاطركم انه قد وصلني كتابكم المؤرخ 25 اغسطوس 1932 الموافق 24 ربيع الثاني 1351 وبما عرف سعادتكم صار معلوم لا سيما مطلوبكم بوكالتكم عن الشركة المذكورة انفا باعلا الكتاب ان نمنحكم الامتياز لأجل الكشف على أراضي حدود قطر لأجل الفحص عن العلائم الدالة على وجود النفط وان الامتياز المذكور مدته سنتين لا غيرها مبتداها من شهر ربيع الثاني 1351 الف وثلاثمائة واحدى وخمسين الموافقة اغسطوس 1932 ومنتهاها في ربيع آخر من 1353 الف وثلاثمائة وثلاث وخمسين انكم تتعهدون لنا بوكالتكم المفوضة عن الشركة المذكورة أعلاه شركة النفط الإنكليزية الإيرانية المحدودة وان تدفعون لنا في هذه السنتين المذكورتين مبلغا شهريا في كل شهر الف وخمسماية ربية مدة السنتين المذكورتين اين (أني؟) قد قبلت ذلك بالشرط المذكور وبعد مضي هذه السنتين يقع التفاوض بين (بيني؟) وبينكم والمذاكرة فاذا حصل لنا منكم الموافقة على الشروط التي نرضاها نحن فبها ونعمه وان لم تحصل لنا منكم الموافقة على الشروط التي نرضاها فلنا كامل الحرية ان نتفق مع أي شخص او شركة نرضاها وتناسب لنا ولا يبقى لكم ادنى حق من الامتيازات في حدود قطر هذا وفي الختام اقبلوا فائق الاحترام وحرر في 23 ربيع الثاني 1351 وفي مدة هذه السنتين المذكورتين ما نسمح لاحد غيركم بالكشف على أراضي قطر ولا نتفاوض مع احد غيركم في هذا الشأن.

عبد اللاه بن قاسم الثاني

حاكم قطر

Literal translation of a letter dated 26th August 1932 from the 'Mier of Qatar.

in the Name of God the Merciful and Gracious

From: Abdullah bin Qasim ath Thani, Ruler of Oater

To: Mr. C. C. Mylles, Representative of the Anglo-Persian Oil Company Ltd.

The usual complimentary beginning.

Your letter dated 25th August 1932 (24th Rabia ath Thani 1351) duly reached me and I have understood what you have written viz., your business on behalf of the Company named and described above in this letter, that we should grant you a Concession for the examination of the Country of Qatar to explore for the signs indicating the presence of oil and that the said concession should be for a period of two years only commencing from the month of Rabia ath Thani 1351 (August 1932) and finishing in Rabia ath Thani 1353 and that you undertake to us in accordance with your authority from the abovementioned Company i.e. the Anglo-Persian Oil Co. Ltd. that you will pay to us during these two years mentioned a sum of Rs.1500 in each month.

And I have accented that on the condition stated and after the expiry of the two years there may occur between us negotiations and discussions and if we receive from you agreement to conditions which please us, may it be to the good and a blessing, and if we do not receive from you agreement to conditions which we will accept, then we have full liberty to conclude an agreement with any person or company which pleases us and which we consider suitable and there -will not remain to you the least right to concessions in the Country of Qatar.

Usual complimentary ending dated 23rd Rabia ath Thani 1351

Added. And in the period of these two years mentioned we will not permit anyone except you to examine the Country of (Qatar).

Fig. 1.10 (continued)

line at roughly the level of well DK-0002 (LeBlanc 2015). On the other hand, from December 1948 to May 1949, D. M. Morton mapped the region south of Chatton's southernmost mapping limit (i.e., from DK-0002 to the Saudi Arabian border). A Robert Ray Gravity party finished the survey of Qatar (which had been started in the North in 1947) between January and June 1949, but not at any point across the Concession boundary. In the meantime, a 120-km pipeline from the Dukhan field to the port of Umm Said in southeast Qatar (also known as Messaieed) was constructed. There, terminal facilities were installed. The first shipment of Dukhan crude was made from Messaieed on December 31st, 1949. 80,000 tons of oil were transported to Europe by the tanker *S.S. President Manny*. Also, in 1949, the

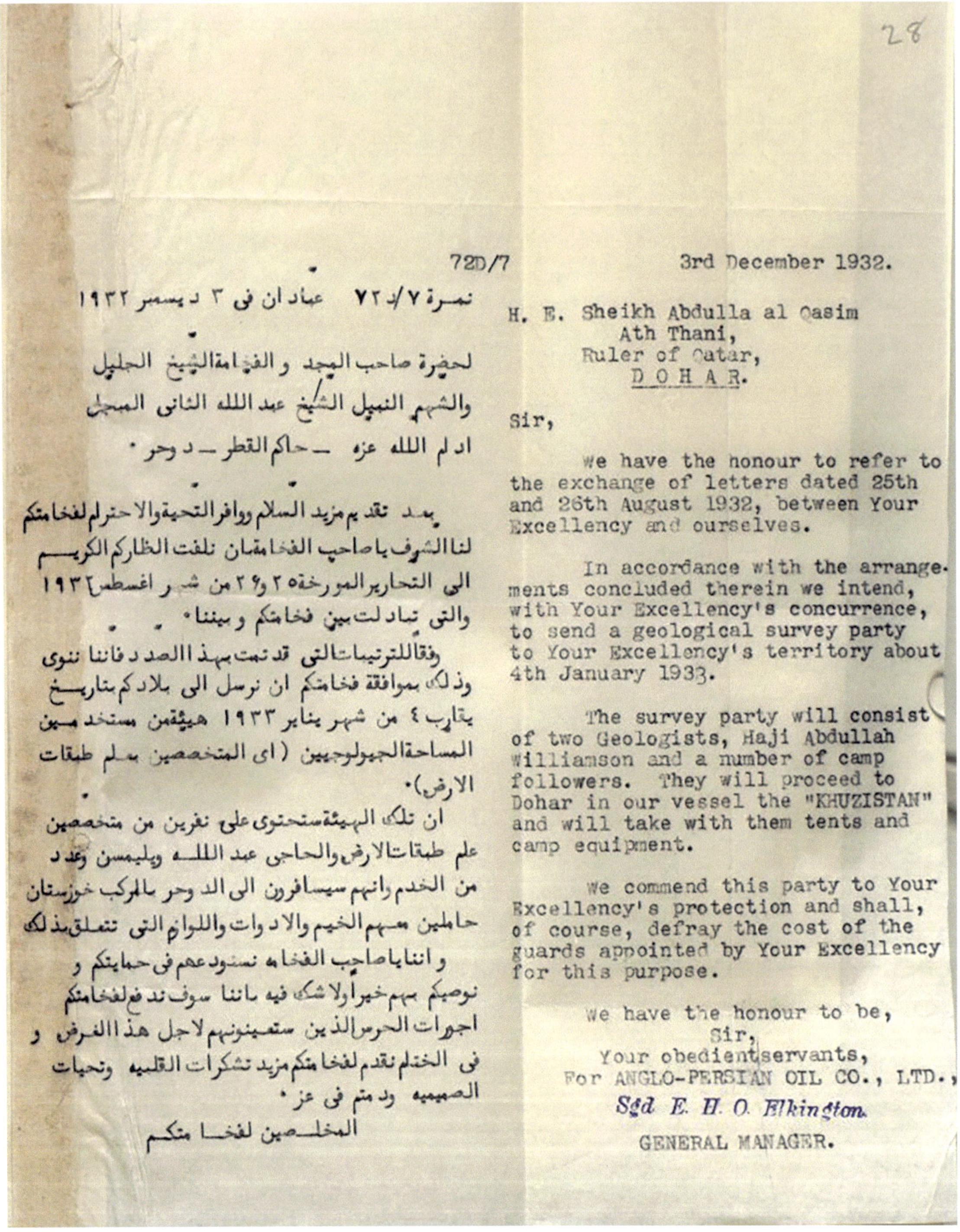

28

72D/7 3rd December 1932.

نمرة ٧/د/٧٢ عبادان في ٣ ديسمبر ١٩٣٢

لحضرة صاحب المجد والفخامة الشيخ الجليل
والشهم النبيل الشيخ عبد الله الثاني المبجل
ادام الله عزه — حاكم القطر — الدوحة.

بعد تقديم مزيد السلام ووافر التحية والاحترام لفخامتكم
لنا الشرف يا صاحب الفخامة ان نلفت انظاركم الكريم
الى التحارير المورخة ٢٥ و٢٦ من شهر اغسطس ١٩٣٢
والتي تبادلت بين فخامتكم وبيننا.
وفقا للترتيبات التي قد تمت بهذا الصدد فاننا ننوي
وذلك بموافقة فخامتكم ان نرسل الى بلادكم بتاريخ
يقارب ٤ من شهر يناير ١٩٣٣ هيئة من مستخدمين
المساحة الجيولوجيين (اي المتخصصين بعلم طبقات
الارض).
ان تلك الهيئة ستحتوي على نفرين من متخصصين
علم طبقات الارض والحاجي عبد الله ويليمسن وعدد
من الخدم وانهم سيسافرون الى الدوحة بالمركب خوزستان
حاملين معهم الخيم والادوات واللوازم التي تتعلق بذلك
واننا يا صاحب الفخامة نستودعهم في حمايتكم و
نوصيكم بهم خيرا ولا شك فيه باننا سوف ندفع لفخامتكم
اجورات الحرس الذين ستعينونهم لاجل هذا الغرض و
في الختام نقدم لفخامتكم مزيد تشكرات القلبية وتحيات
الصميمية ودمتم في عز.
المخلصين لفخامتكم

H. E. Sheikh Abdulla al Qasim
Ath Thani,
Ruler of Qatar,
D O H A R.

Sir,

We have the honour to refer to the exchange of letters dated 25th and 26th August 1932, between Your Excellency and ourselves.

In accordance with the arrangements concluded therein we intend, with Your Excellency's concurrence, to send a geological survey party to Your Excellency's territory about 4th January 1933.

The survey party will consist of two Geologists, Haji Abdullah Williamson and a number of camp followers. They will proceed to Dohar in our vessel the "KHUZISTAN" and will take with them tents and camp equipment.

We commend this party to Your Excellency's protection and shall, of course, defray the cost of the guards appointed by Your Excellency for this purpose.

We have the honour to be,
Sir,
Your obedient servants,
For ANGLO-PERSIAN OIL CO., LTD.,
Sgd E. H. O. Elkington.
GENERAL MANAGER.

Fig. 1.11 A letter from the General Manager of the Anglo-Persian Company to the Amir of Qatar asking for his approval to send a team of geologists ("File 10/3 I Qatar Oil Concession")

Amir gave the *Superior Oil Company* a fresh, independent, but undefined concession for his "Continental Shelf" lands (LeBlanc 2015; Uthman 1984). They conducted geophysical research on the seabed for a few months before leaving the field in 1952 because the prospects seemed dim, and the costs and difficulties seemed all too likely.

The well Dukhan 35 was Qatar's first blowout well. It was spudded on February 16th, 1953, and blew back during the coring of the Arab D reservoirs due to a leak in the preventer, which caught fire. Morton (2013) provided more details about issues associated with the safety of the Dukhan wells during the unrest that followed the Second World War.

نمرة 7/د 72 عبادان في 3 ديسمبر 1932

لحضرة صاحب المجد والفخامة الشيخ الجليل والشهم النبيل الشيخ عبد الله الثاني المبجل ادام الله عزه – حاكم القطر (كذا) دوحر (كذا)

بعد تقديم مزيد السلام ووافر التحية والاحترام لفخامتكم لنا الشرف يا صاحب الفخامة بان نلفت انظاركم الكريم الى التحارير المورخة 25 و 24 من شهر أغسطس 1932 والتي تبادلت بين فخامتكم وبيننا.

ووفقا للترتيبات التي قد تمت بهذا الصدد فاننا ننوي وذلك بموافقة فخامتكم ان نرسل الى بلادكم بتاريخ يقارب 4 من شهر يناير 1933 هيئة من مستخدمين المساحة الجيولوجيين (أي المتخصصين بعلم طبقات الأرض).

ان تلك الهيئة ستحتوي على نفرين من متخصصين علم طبقات الأرض والحاجي عبد الله ويليمسن وعدد من الخدم وانهم سيسافرون الى الدوحر (كذا) بالمركب خوزستان حاملين معهم الخيم والأدوات واللوازم التي تتعلق بذلك

وانا يا صاحب الفخامة نستودعهم في حمايتكم ونوصيكم بهم خيرا ولا شك فيه باننا سوف ندفع لفخامتكم اجورات الحرس الذين ستعينونهم لاجل هذا الغرض وفي الختام نقدم لفخامتكم مزيد تشكرات القلبية وتحيات الصميمية ودمتم في عز.

المخلص لفخامتكم

H. E. Sheikh Abdulla al Qasim

Ath Thani

Ruler of Qatar,

DOHAR (sic)

Sir,

We have the honour to refer to the exchange of letters dated 25th and 26th August 1932, between Your Excellency and ourselves

In accordance with the arrangements concluded therein we intend, with Your Excellency's concurrence, to send a geological survey party to Your Excellency's territory about 4th January 1933.

The survey party will consist of two Geologists, Haji Abdullah Williamson and a number of camp followers. They will proceed to Dohar (sic) in our vessel the "KHUZISTAN" and will take with them tents and camp equipment.

we commend this party to Your Excellency's protection and shall, of course, defray the cost of the guards appointed by Your Excellency for this purpose.

We have the honour to be,

Sir,

Your obedient servants

For ANGLO-PERSIAN OIL CO., LTD

Sgd. E.H.O. Elkington

GENERAL MANAGER

Fig. 1.11 (continued)

In November 1952, an offshore grant was awarded to *Shell Overseas Exploration Company*, renamed the Shell Company (Qatar) in August 1954. A gravity survey was started in June 1963, followed six months later by a seismic-reflection survey.

The company started exploration in 1953, leading to the discovery of the main offshore oilfields of Qatar. In February 1955, Shell spudded its first well, Matbakh-1 (M-1), with a mobile drilling platform MU-1. The well was abandoned as a dry hole in August of the same year after reaching a depth of 2044 m (6706 ft.). Then, the exploration effort of the company led to the discovery of Idd El Shargi (1960), Maydan Mahzam (1963), and Bul Hanine (1965) and the El Bunduq field that straddles the borders between Qatar and the United Arab Emirates. In 1971, Shell (Qatar) discovered the North Field, the world's largest gas field.

The structural interpretation of the seismic record showed that salt-tectonic structures dominated the eastern area, whereas the northern area was flat and featureless. In January

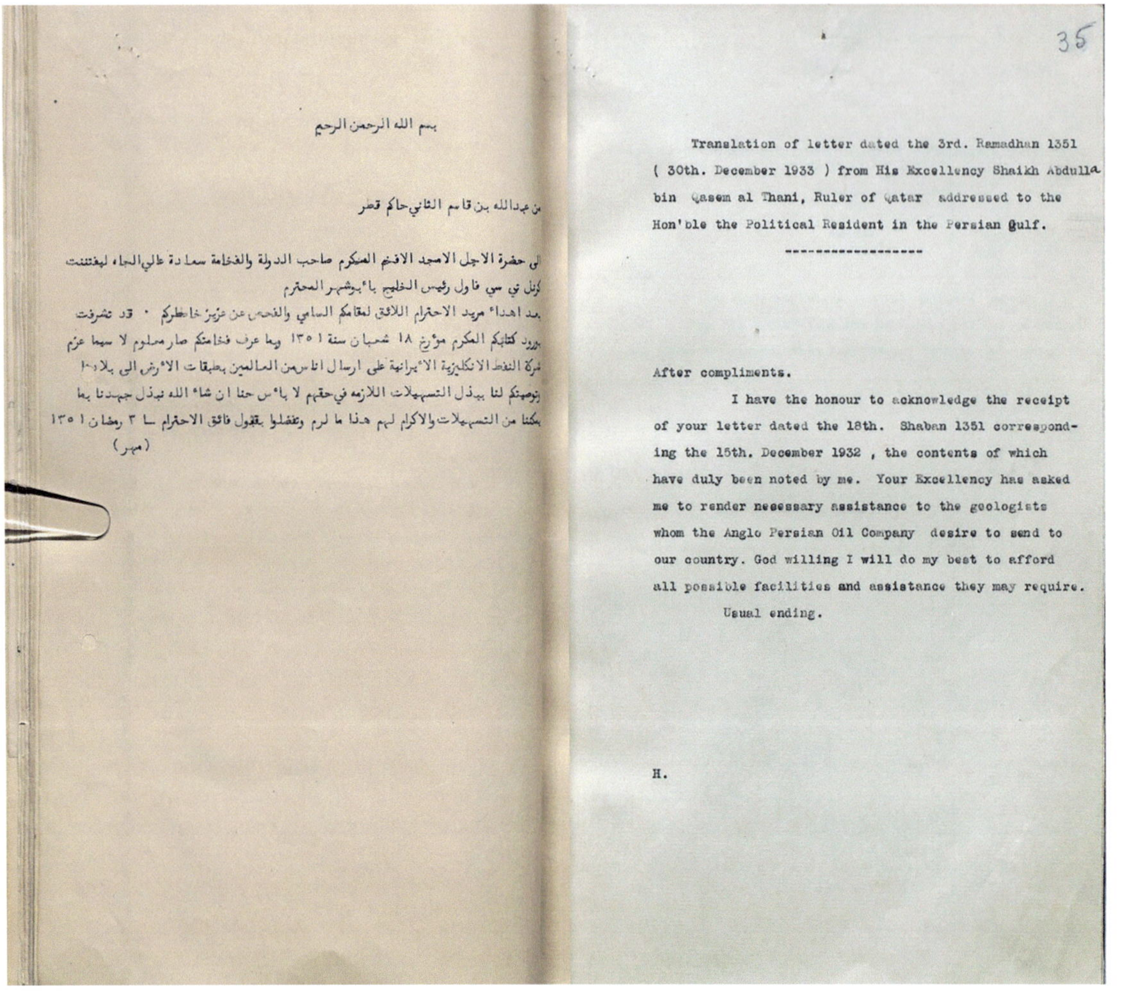

بسم الله الرحمن الرحيم

من عبدالله بن قاسم الثاني حاكم قطر

الى حضرة الاجل الامجد الافخم المكرم صاحب الدولة والفخامة سعادة عالي الجاه ليفتننت
كرنل تي سي فاول رئيس الخليج بابوشهر المحترم
بعد اهداء مزيد الاحترام اللائق لمقامكم السامي والفحص عن عزيز خاطركم . قد تشرفت
بورود كتابكم المكرم مؤرخ ١٨ شعبان سنة ١٣٥١ وبما عرف فخامتكم صار معلوم لا سيما عزم
شركة النفط الانكليزية الايرانية على ارسال اناس من العالمين بطبقات الارض الى بلادنا
وتوصيتكم لنا ببذل التسهيلات اللازمه في حقهم لا بأس حنا ان شاء الله نبذل جهدنا بما
يمكنا من التسهيلات والاكرام لهم هذا ما لزم وتفضلوا بقبول فائق الاحترام ٣ رمضان ١٣٥١
(مهر)

35

Translation of letter dated the 3rd. Ramadhan 1351 (30th. December 1933) from His Excellency Shaikh Abdulla bin Qasem al Thani, Ruler of Qatar addressed to the Hon'ble the Political Resident in the Persian Gulf.

After compliments.

I have the honour to acknowledge the receipt of your letter dated the 18th. Shaban 1351 corresponding the 15th. December 1932 , the contents of which have duly been noted by me. Your Excellency has asked me to render nessessary assistance to the geologists whom the Anglo Persian Oil Company desire to send to our country. God willing I will do my best to afford all possible facilities and assistance they may require.

Usual ending.

H.

Fig. 1.12 A letter from the Amir of Qatar to the General Manager of the Anglo-Persian Company welcoming the geologist team and promising assistance and protection ("File 10/3 I Qatar Oil Concession")

1956, the first offshore well was drilled on the crestal part of the elongated, hourglass-shaped Idd El Shargi structure. It reached a depth of 3714 m (12,182 ft.) but was dry. A second well drilled in December 1959 (Idd El Shargi-2) to a depth of 2600 m (8528 ft.) into the northern dome of the structure proved oil in the Middle Jurassic Araej Formation and the Upper Jurassic Arab II and Arab IV reservoirs of the equivalent Qatar Formation in onshore. In March 1961, well No. 4 on the southern dome discovered oil in the Lower Cretaceous Shuaiba Formation. In 1964, a second and more prolific offshore discovery, Maydan Mahzam, was developed some 30 km northeast of Idd El Shargi. A discovery well, drilled in July 1963 to a total depth of 2811 m (9220 ft), encountered oil in the Upper Jurassic Qatar (Arab) Formation, and the El Bunduq Field, straddling the Qatar-U.A.E. boundary, was discovered in the following year by Abu Dhabi Marine Areas Company (ADMA).

The Qatari geologic lexicon was compiled by Walter Sugden, A. J. Standring, and Claude Cavelier (1975) and published by the International Stratigraphic Commission. With few exceptions, most of the nomenclature used is derived from neighboring countries such as Saudi Arabia, Kuwait, and Iraq.

1.4 Deep Exploration

In 1954, Qatar Petroleum developed plans to drill a deep well on the Dukhan structure to explore formations predating the known Jurassic Arab limestone-producing formations to find more reserves. A decision was made in January 1955 to drill a deep well to formations older than any previously explored in Qatar. The Khuff Formation of the Permian age was the best prospect for the well after geological predictions based on the information from Qatar, Saudi Arabia, and Bahrain were made. In these countries, the Khuff Formation was known to have reservoir qualities. The well was intended to investigate this horizon

بسم الله الرحمن الرحيم

من عبد الله بن قاسم الثاني حاكم قطر

الى حضرة الأجل الأمجد الأفخم المكرم صاحب الدولة والفخامة سعادة عالي الجاه ليفتنت كرنل تي سي فاول رئيس الخليج بأبو شهر المحترم

بعد اهداء مزيد الاحترام اللائق لمقامكم السامي والفحص عن عزيز خاطركم. قد تشرفت بورود كتابكم المكرم مؤرخ 18 شعبان سنة 1351 وبما عرف فخامتكم صار معلوم لا سيما عزم شركة النفط الإنكليزية الإيرانية على ارسال أناس من العالمين بطبقات الأرض الى بلادنا وتوصيتكم لنا ببذل التسهيلات اللازمة في حقهم لا بأس حنا ان شاء الله نبذل جهدنا بما يمكننا من التسهيلات والاكرام لهم هذا ما لزم وتفضلوا بقبول فائق الاحترام – 31 رمضان 1351

(مهر)

Translation of letter dated the 3rd Ramadhan 1351 (30th December 1933) from His Excellency Shaikh Abdulla bin Qasem al Thani, Ruler of Qatar addressed to the Hon'able the political Resident in the Persian Gulf

----

After compliments.

I have the honour to acknowledge the receipt of your letter dated the 18th Shaban 1351 corresponding the 15th December 1932, the contents of which have duly been noted by me. Your Excellency has asked me to render necessary assistance to the geologists whom the Anglo Persian Oil Company desire to send to our country. God Willing I will do my best to afford all possible facilities and assistance they may require.

Usual ending.

Fig. 1.12 (continued)

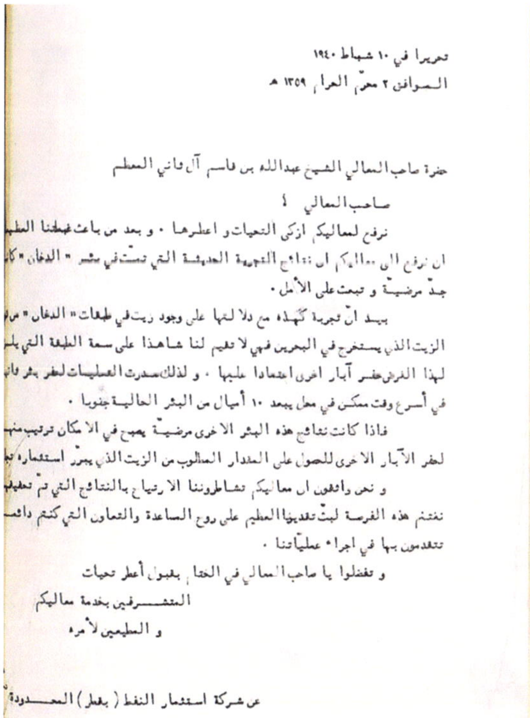

تحريرا في ١٠ شباط ١٩٤٠

الموافق ٢ محرم الحرام ١٣٥٩ هـ

حضرة صاحب المعالي الشيخ عبدالله بن قاسم آل ثاني المعظم

صاحب المعالي !

نرفع لمعاليكم ازكى التحيات و اعطرها . و بعد من باعث غبطتنا العظيمة ان نرفع الى معاليكم ان نتائج التجربة الحديثة التي تمت في بئر " الدخان " كانت جد مرضية و تبعث على الأمل .

بيد ان تجربة كهذه مع دلالتها على وجود زيت في طبقات " الدخان " من الزيت الذي يستخرج في البحرين فهي لا تقيم لنا شاهدا على سعة الطبقة التي يلزم لهذا الغرض حفر آبار اخرى اعتمادا عليها . و لذلك صدرت التعليمات لحفر بئر ثانية في أسرع وقت ممكن في محل يبعد ١٠ أميال من البئر الحالية جنوبا .

فاذا كانت نتائج هذه البئر الاخرى مرضية يصبح في الامكان ترتيب منهج لحفر الآبار الاخرى للحصول على المقدار المطلوب من الزيت الذي يبرر استثماره تجاريا

و نحن واثقون ان معاليكم تشاطروننا الارتياح بالنتائج التي تم تحقيقها و نغتنم هذه الفرصة لبث تقديرنا العظيم على روح المساعدة والتعاون التي كنتم دائما تتقدمون بها في اجراء عملياتنا .

و تفضلوا يا صاحب المعالي في الختام بقبول أعطر تحيات

المتشرفين بخدمة معاليكم

و المطيعين لأمره

عن شركة استثمار النفط (بقطر) المحدودة

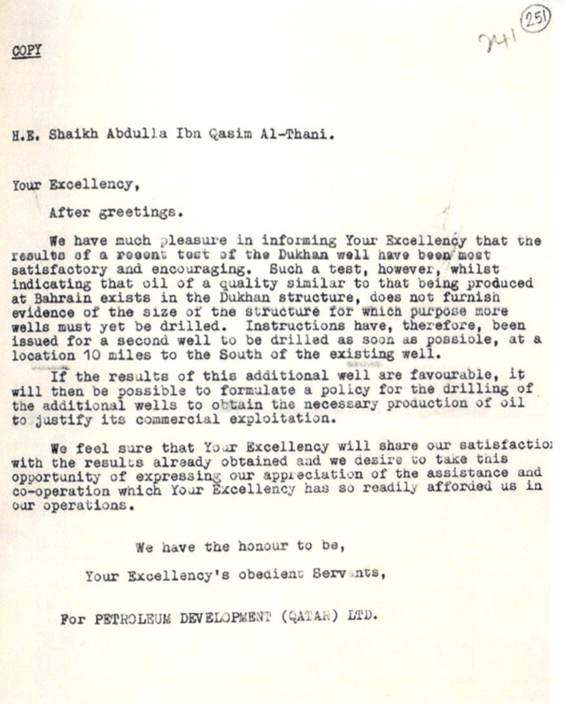

COPY

941 (251)

H.E. Shaikh Abdulla Ibn Qasim Al-Thani.

Your Excellency,

After greetings.

We have much pleasure in informing Your Excellency that the results of a recent test of the Dukhan well have been most satisfactory and encouraging. Such a test, however, whilst indicating that oil of a quality similar to that being produced at Bahrain exists in the Dukhan structure, does not furnish evidence of the size of the structure for which purpose more wells must yet be drilled. Instructions have, therefore, been issued for a second well to be drilled as soon as possible, at a location 10 miles to the South of the existing well.

If the results of this additional well are favourable, it will then be possible to formulate a policy for the drilling of the additional wells to obtain the necessary production of oil to justify its commercial exploitation.

We feel sure that Your Excellency will share our satisfaction with the results already obtained and we desire to take this opportunity of expressing our appreciation of the assistance and co-operation which Your Excellency has so readily afforded us in our operations.

We have the honour to be,

Your Excellency's obedient Servants,

For PETROLEUM DEVELOPMENT (QATAR) LTD.

Fig. 1.13 A letter from the Petroleum Development (Qatar) Ltd. to the Amir of Qatar reporting the presence of oil in Dukhan structure ("File 10/3 I Qatar Oil Concession")

تحريرا في 10 شباط 1940

الموافق 2 محرم الحرام 1359 هجرية

حضرة صاحب المعالي الشيخ عبد الله بن قاسم آل ثاني المعظم

صاحب المعالي،

نرفع لمعاليكم ازكى التحيات واعطرها، وبعد من باعث غبطتنا العظيم ان نرفع الى معاليكم ان نتائج التجربة الحديثة التي تمت في بئر "الدخان" كانت جدّ مرضيّة وتبعث على الأمل.

بيد أن تجربة كهذه مع دلالتها على وجود زيت في طبقات "الدخان" من نوع الزيت الذي يستخرج في البحرين فهي لا تقيم لنا شاهدا على سعة الطبقة التي يلزم لهذا الغرض حفر آبار أخرى اعتمادا عليها. ولذلك صدرت التعليمات لحفر بئر ثانية في أسرع وقت ممكن في محل يبعد 10 اميال م البئر الحالية جنوبا.

فاذا كانت نتائج هذه البئر الأخرى مرضية يصبح في الإمكان ترتيب منها لحفر الآبار الأخرى للحصول على المقدار المطلوب من الزيت الي يبرر استثماره تجاريا.

ونحن واثقون ان معاليكم تشاطروننا الارتياح بالنتائج التي تم تحقيقها نغتنم هذه الفرصة لبث تقديرنا العظيم على روح المساعدة والتعاون التي كنتم دائما تتقدمون بها في اجراء عملياتنا.

وتفضلوا يا صاحب المعالي في الختام بقبول أعطر التحيات

المتشرفين بخدمة معاليكم

والمطيعين لأمره

H.E. Shaikh Abdulla Ibn Qasim Al-Thani

Your Excellency,

After greetings.

We have much pleasure in informing Your Excellency that the results of a recent test of the Dukhan well have been most satisfactory and encouraging. Such a test, however, whilst indicating that oil of a quality similar to that being produced at Bahrain exists in the Dukhan structure, does not furnish evidence of the size of the structure for which purpose more wells must yet be drilled. Instructions have, therefore, been issued for a second well to be drilled as soon as possible, at a location 10 miles to the South of the existing well.

If the results of this additional well are favourable, it will then be possible to formulate a policy for the drilling of the additional wells to obtain the necessary production of oil to justify its commercial exploitation.

We feel sure that Your Excellency will share our satisfaction with the results already obtained and we desire to take this opportunity of expressing our appreciation of the assistance and co-operation which Your Excellency has so readily afford us in our operations.

We have the honour to be,

Your Excellency's obedient Servants,

For PETROLEUM DEVELOPMENT (QATAR) LTD.

Fig. 1.13 (continued)

and the underlying unexplored section to roughly 5029 m (6500 ft.). Qatar Petroleum spudded the well DK-0065 on July 20th, 1958. The well penetrated a total depth of 4042 m (13,261 ft.). Several cores were cut at different intervals (LeBlanc 2015).

1.5 The Discovery of the North Field

The discovery of the North Dome Gas Field (commonly referred to as "North Field") by Shell in 1971 was the peak of the exploration activities in Qatar during the 1970s. The field occupies an area of about 6000 km^2, or roughly half the size of the State of Qatar and is located off the northeastern coast of the Qatar peninsula. The structure also crosses the Iranian border, which is called the South Pars Field, and was discovered by the National Iranian Oil Company in 1990. Production in the field started in 1989, producing gas from the Permo-Triassic Khuff Formation. The Khuff Formation is formed of dolomitic limestone with an approximate thickness of 823 m (2700 ft.), of which some 274.3 m (900 ft.) can be considered as a reservoir. With total proven reserves of 900 trillion standard cubic feet, it is the world's largest single non-associated gas reservoir.

Table 1.1 Stratigraphy of the well Dukhan-1 as recorded by the well-site geologists compared to the present-day stratigraphic nomenclature (after LeBlanc 2015)

Depth (ft)	Formation names (1938)	Names (today equivalent)
0000–1010	Bahrein	Hasa Group (includes Dammam, Rus, and Umm er Radhuma Formations)
1010–1656	Tayarat Limestone	Simsima Ruilat Halul
1656–2408	Khatiyah	Laffan Shale Mishrif Ahmadi
1656–2072	Upper Khatiyah beds	
2072–2408	Lower Khatiyah beds	
2408–2589	Mauddud Limestone	Mauddud
2589–3038	Rutbah Sandstone	Nahr Umr
3038–5047	Musandam	Thamama Group
3038–4151	Upper Musandam beds	
4151–5047	Lower Musandam beds	
5047–5686	Zekrit	Hith Qatar Fahahil

Fig. 1.14 Dukhan-1 well drilling rig (LeBlanc 2015)

Fig. 1.15 Christmas tree valves cover Dukhan well-1, the first well drilled in Qatar in the Dukhan area (LeBlanc 2015)

1.6 Prominent Geologists

George Martin Lees (1898–1955)

(*Source* The Royal Publishing Society)

https://royalsocietypublishing.org/doi/pdf/https://doi.org/10.1098/rsbm.1955.0012

Lees was born in Dundalk, County Louth, Ireland, on April 16, 1898. He attended St Andrews College, Dublin, from 1906 to 1915, and then the Royal Military Academy to prepare for service in the then-ongoing First World War. He joined the Royal Artillery, where he served in France before becoming a flying instructor in Egypt, where he gained his first taste of the Middle East. He moved from Egypt to Mesopotamia (now Iraq) and was involved in capturing Kirkuk from the Turks.

Lees stayed in Iraq after the war to work in civil administration for a while (1919–1921). In January 1921, he returned to England on leave, resigning from politics and deciding to work in geology. In the same year (1921), he began working as an Assistant Geologist for the Anglo-Persian (later Anglo-Iranian) Oil Company. He stayed with this company until his health forced him to retire in 1953. He remained an adviser to the company (later renamed the British Petroleum Company) until he died in 1955.

Lees returned to the Middle East in 1922 and 1923, and in the winter of 1924–1925, he was chosen to lead an oil-finding tour of southwest Persia. As a result of this season's work, several structures were recommended for exploration, four of which (Haft Kei, Gach Saran, Agha Jari, and Pazanun) were later drilled and proved to be important producing fields. The following season, 1925–1926, Lees conducted a pioneer geological survey of Oman with K. Washington Gray. The papers from that work were presented to the Geological Society (1928) and the Geographical Society (1928).

From 1926 to 1928, he was on study leave at the University of Vienna, where he attended lectures by F. E. Suess and L. Kober. He received a Ph.D. from the University of Vienna for his thesis on the geology of Oman.

Lees investigated oil prospects and oil company organization and administration on the geological side in the USA, Canada, Egypt, and Germany between 1928 and 1930.

Lees was appointed Chief Geologist of the Anglo-Iranian Oil Company on November 1, 1930. He held this important position for nearly twenty-three years, when the company's activities in many parts of the world expanded dramatically.

Most major oil fields in Persia were discovered between 1930 and 1940, as were the massive fields at Kirkuk in Iraq and Kuwait and the smaller one in the Qatar peninsula in Arabia, where Lees had obtained a two-year option from the Sheikh in 1926.

The Geological Society awarded Lees the Bigsby Medal in 1943 to recognize his leading role in the discovery of oil in Britain, as well as the importance of his geological work in this area and Persia and Oman. Candidates for this medal must be under the age of 45, and given the magnitude of his accomplishments, it is remarkable that he still qualified, albeit by a few weeks. He was elected a Fellow of the Royal Society five years later, in 1948. Finally, he served as President of the Geological Society for two years, 1951–1952. He was the first geologist to achieve this distinction or to be elected as a fellow of the Royal Society while working in industry. The American Association of Petroleum Geologists awarded Lees the Sidney Powers Memorial Medal in April 1954, his final public recognition. This medal is regarded as the highest honor American oil geologists can receive, and it had never previously been awarded to a non-American.

He died on January 25, 1955.

Walter Sugden
(1923–1977)

(Courtesy of Jacques LeBlanc)

Walter "Bill" Sugden was born in the Yorkshire district, UK. In January 1944, he received first-class honors and a degree in oil technology from the *Royal School of Mines* in London. After working as a paleontologist with North Venezuelan Petroleum Co. Ltd. For three years, he started a job as a subsurface geologist with the Iraq Petroleum Company in 1947 and remained with the company until 1959. His first assignment was to northern Iraq, where he was tasked with overseeing the rapid post-war development of the Kirkuk field as well as the discovery and appraisal of the Bai Hassan oilfield. He was then assigned to Dukhan in Qatar as a resident geologist, where he arrived on March 19, 1951. Sugden was promoted to area geologist on April 1, 1953. He was involved in field mapping and stratigraphical appreciation of early Trucial Coast wells, as well as much of the location, control, and structural and reservoir interpretation of early Dukhan field wells. He spent much of his spare time studying modern sedimentation and organic geochemistry in the sabkhas and lagoons around Qatar and researching Arabian Gulf hydrology. He also prepared the Qatar fascicule of the "*International Lexique Stratigraphique*." On January 9, 1957, he left Qatar and returned to Iraq as a Divisional paleontologist for the Basrah Petroleum Company.

He resigned from his job in 1959 and enrolled at Oxford University. Unfortunately, he could not finish his studies. He then accepted a position as a lecturer in the Geology Department at Townsville University, an extension college of the University of Queensland. His publications covered a wide range of topics, including structural analysis, recent sedimentation, Arabian Gulf hydrology, and the development of pyrite, with a detailed report on the Lower Fars Formation in Iraq. He died in 1977 (LeBlanc 2015).

Anthony John Standring
(1927–2021)

(Courtesy of Mrs. Joyce Standring)

A. J. Standring was born on the June 12th, 1927. He received his B.Sc. in 1948 from the University of Bristol (UK) and his M.Sc. in 1976 from the same university for his thesis on the Thamama Group. After graduating in 1948, he stayed in Bristol to work on Ordovician biostratigraphy. He took part in the Lauge Koch East Greenland Expedition in 1952. John was appointed geologist with the Falkland Islands Dependencies Survey (later the British Antarctic Survey) in Hope Bay, north of the Antarctic Peninsula, at the end of that year. He led survey parties in 1953 and 1954, examining and collecting from available outcrops. In 1955, he was awarded the Polar Medal, which was presented to him by HM the Queen. Also, *Standring Inlet* was named after him (see https://data.aad.gov.au/aadc/gaz/display_name.cfm?gaz_id=111361). He returned to the United Kingdom to work on his collection at Bristol University. He was hired by the Iraq Petroleum Company in

1956 as an Assistant Paleontologist based in Dukhan, Qatar (1956–1960) at the headquarters of the Exploration Division, which oversaw field parties and exploration drilling in the Trucial States (now UAE), Oman, the Aden Protectorate (now southern Yemen), and Qatar. On October 23, 1956, he arrived in Dukhan. The Exploration Department in Qatar focused primarily on deep drilling, such as the Dukhan-65 well. Exploratory wells on the main Qatar Arch were drilled due to periodic reviews of potential prospects. There was an extensive collaboration among departmental staff on regional geological and stratigraphic interpretation of geophysics, as well as relating this to known surface and subsurface geology. As part of this, he collaborated with the Divisional Paleontologist at the time (Max Chatton) on naming and dating the region's fauna. He became more involved with regional synthesis after becoming a Divisional Paleontologist and working on the fauna of DK-65's deep section. All of the fossils studied came from wells and field samples. Important newly identified specimens, such as the Devonian fish sample from the well Dukhan-65, were sent to London for verification and were completed at the British Museum (*Natural History*). While in Dukhan, he was promoted to Divisional Paleontologist and was also appointed Divisional Stratigrapher at the same time. He held this position while stationed in Bahrain (1961–1966) and Abu Dhabi (1966–1971), when the concession was limited to onshore Abu Dhabi and Qatar. He took a leave of absence from the Gulf in 1971 and was made redundant, eventually leaving I.P.C. in 1972. He then moved to the UK to pursue his M.Sc. in Lower Cretaceous biostratigraphy at Bristol University (1972–1975). From 1976 to 1982, he was the manager of *Palaeoservices Ltd*., a company that provided geological and biostratigraphic services to industry worldwide. From 1982 to 1986, he worked as an independent oil company consultant, primarily in the Arabian Gulf, North Africa, and Bangladesh. He left the company in 1986. The Qatar Lexicon was his sole published work, but he also presented a paper to an Aramco conference in the early 1960s when he was based in Bahrain. Otherwise, his work was documented in several internal Company reports on well and field material, as well as stratigraphic syntheses [Standring (1959) (1961) (1962), White (1969), LeBlanc (2012)], including reports presented to "The Stratigraphic Liaison Meetings." (LeBlanc 2015; Mrs. Joyce Standring, Personal Communications 1923).

Jacques LeBlanc

(Photo by Fadhil Sadooni)

Jacques on the Dam Formation (Middle Miocene) outcrops, Qatar

Jacques (originally from Alberta, Canada) has been a geologist since 1986, specializing in mining, oil and gas, data management, surface mapping, stratigraphy, and assisting public museums with their geological and paleontological resources. He worked in various countries, including Canada, Chad, Colombia, Libya, Niger, and Qatar. He is fluent in French, English, and Spanish. For over 13 years, he was in charge of a team of six individuals responsible for managing data for Qatar's massive Dukhan oil field. Jacques authored numerous publications on the surface geology and paleontology of Qatar and other countries. He conducted a Master Class on Stratigraphy within the organization he worked for, Qatar Petroleum (currently known as Qatar Energy). Between 2007 and 2020, the Qatar Museums (QM) utilized his expertise to curate its geological and paleontological archives to showcase Qatar-specific specimens in its National Museum. While conducting fieldwork in Qatar, he discovered a burial site containing the remains of Miocene dugongs. He then joined forces with the

National Museum and paleontologists from the Smithsonian Museum to meticulously analyze and study the findings. He compiled a comprehensive review (more than 1000 pages) on the stratigraphy and paleontology of Qatar. Jacques also reviewed the part of the Qatar Geologic Lexicon on the Tertiary strata, introducing a new member to the Dammam Formation.

References

Amiri-Diwan. Retrieved 08-02 from https://www.diwan.gov.qa/about-qatar/qatars-rulers/sheikh-abdullah-bin-jassim-al-thani?sc_lang=ar-QA

Carsten N (1720) Chart of the Persian Gulf. https://ediscovery.qnl.qa/islandora/object/QNL%3A00039745

Fieldhouse DK (2006) Western Imperialism in the Middle East 1914–1958. OUP Oxford. https://books.google.com.qa/books?id=K28TDAAAQBAJ

File 10/3 I Qatar Oil Concession. In. Qatar Digital Library: British Library

Heard-Bey F (2005) From Trucial States to United Arab Emirates: A Society in Transition. Motivate. https://books.google.com.qa/books?id=O6CIPwAACAAJ

Johnstone T, Wilkinson C (1960) Some geographical aspects of Qatar. An original article from the Geographical Journal. Geographical Journal

LeBlanc J (2015) A historical account of the stratigraphy of Qatar, Middle-East

Morton MQ (2010) Black Gold and Frankincense

Morton MQ (2013) The British India Line in the Arabian Gulf, 1862–1982. Liwa J 5(10)

Morton MQ (2020) Masters of the Pearl: a history of Qatar

Niebuhr C (1780) Voyage en Arabie & en d'autres Pays circonvoisins

Uthman NM (1984) With their bare hands: the story of the oil industry in Qatar. Longman

Vassiliou MS (2018) Historical dictionary of the petroleum industry (Second Edition). Rowman & Littlefield Publishers. https://doi.org/10.5771/9781538111604

Whittle Ls (1794) A new map of Arabia divided into its several regions and districts: from Monsr. D. Anville Geographer to the most Christian King; with additions and improvements from Mr. Niebuhr. https://ediscovery.qnl.qa/islandora/object/QNL%3A00004387#page/1/mode/1up

Wikipedia (2025a) Gifford Palgrave. Retrieved 08-02 from https://en.wikipedia.org/wiki/Gifford_Palgrave

Wikipedia (2025b) John Frederick Lewis. https://en.wikipedia.org/wiki/John_Frederick_Lewis

Zwemer SM (1900) Arabia; the cradle of Islam. F. H. Revell company

Surface Geology 2

Abstract

Qatar is a 11,610 km^2 peninsula. It represents a low deflation surface shaped by the persistent Shamal wind with low relief ranging from 103 m above sea level in the SW to around 6 m below sea level in the Dukhan area to the west. The topography is characterized by distinctive features formed during the last few thousand years by surface geologic processes. Generally, the country is divided into three topographic subdivisions. These are the limestone plateau, the sand south, and the coastal areas. The limestone plateau is formed of Tertiary carbonates and evaporites of the lower Eocene Dammam Formation that covers around 80% of the surface area with some outcrops of the Rus Formation, the middle Miocene Dam Formation, and small outcrops of the Quaternary clastics of the Hofuf Formation. The plateau is characterized by extensive karst features called locally (Duhool). Sand sheets and dunes, mostly of crescent and barchan types, cover a considerable part of the country's southern parts, oriented from NNW to NW, representing the prevalent wind direction. The sand consists mainly of quartz with carbonate fragments. Heavy minerals include apatite, collophane, anhydrite, and kyanite. The sand grains are mainly of eolian origin, although some were of coastal or marine origin. This sand originates from Saudi Arabia, east Iraq, or the Zagros Mountains, east Iran. Some inland and coastal areas are punctuated by salt-encrusted flat areas, the sabkhas and salinas. They are mostly formed due to the upward movement of groundwater to compensate for the loss of surface moisture in the surface sediments due to extensive evaporation. They are characterized by the presence of halite, anhydrite, and gypsum. The most important sabkhas in Qatar are found in Dukhan, Sawda Nathil, Umm Saeed, Fishakh and Hussain, and Khor Al-Adaid.

2.1 Introduction

Qatar is a peninsula with an approximate area of 11,610 km^2. The country is a deflation surface shaped by the persistent Shamal wind, except for some low hills in the northwest. The surface topography is generally characterized by low relief ranging from around 103 m above sea level in the southwest to nearly 6 m below Dukhan Sabkha. However, most of the country is less than 40 m above sea level (Fig. 2.1). A remarkably uniform limestone bed underlies the larger part of the country. In some places, the limestone is overlain by younger strata that form several mesa-type hills. The oldest exposed rocks are the Lower Eocene Rus Formation, which consists mainly of dolomite and limestone with scattered outcrops of Miocene (covering around 8%) of the country's total area.

The topographic surface of Qatar is covered by lower Eocene carbonates and evaporites of the Dammam Formation, which are overlain unconformably in the south and southwestern parts by Miocene marls and limestone of the Dam Formation. The country is nearly flat, punctuated by two main folds affecting its topography. These are the Dukhan anticline in the west, with a height of around 60 m, 56 km long, and 4.8–6.5 km cross. Nearly in the middle of the country, there is a gradual swell known as the mid-peninsular rise, separating a carbonate platform to the north from carbonate-evaporite terrain in the south. The area to the south, southern end, and eastern flank of the Dukhan anticline is characterized by mesas (devil's tables) and wind-eroded hills. The southern and southern areas are also characterized by dunes that resulted from the impact of the most prevalent Shamal wind. The general surface geologic map of Qatar appears in Fig. 2.2.

F. N. Sadooni and H. A. Al-Kuwari, *Geological Evolution of Qatar and the Arabian Peninsula*, Regional Geology Reviews,
https://doi.org/10.1007/978-3-031-94394-2_2

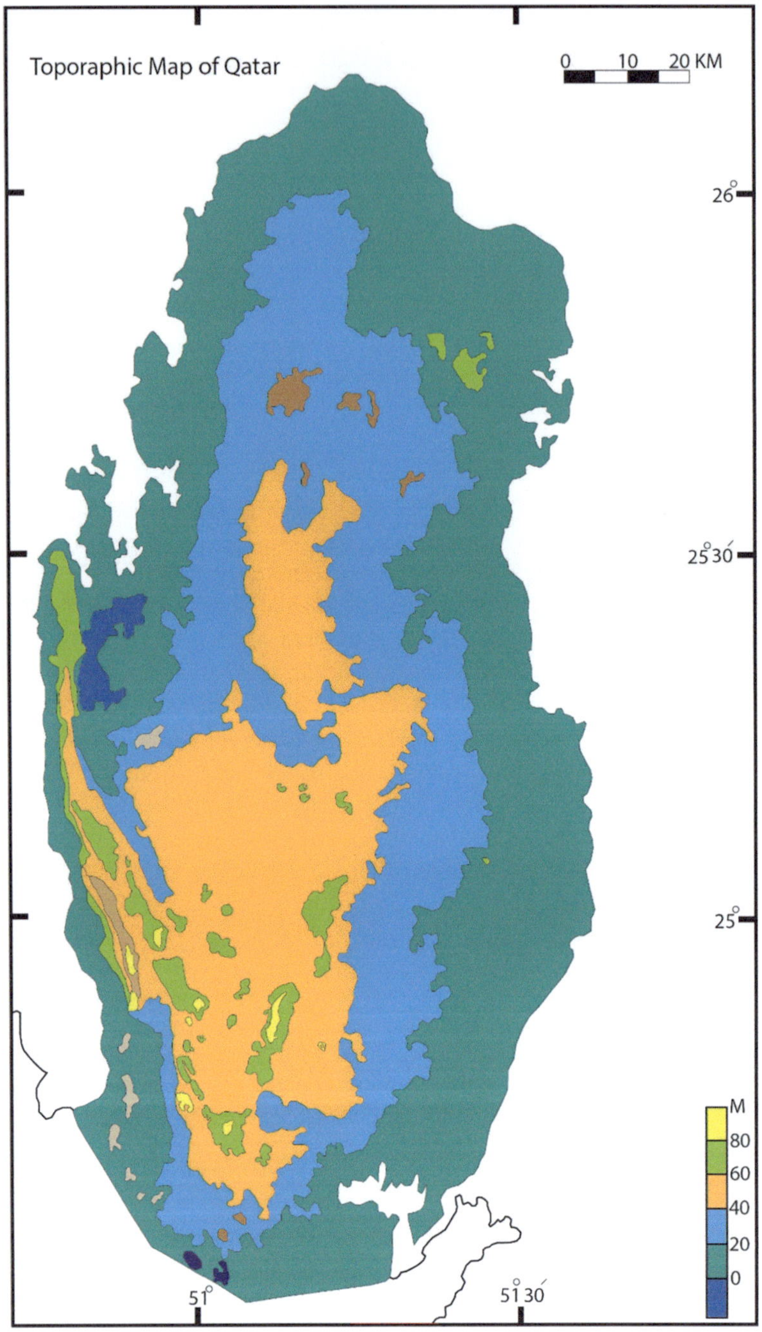

Fig. 2.1 Topographic map of Qatar (Environmental Science Center archives)

2.2 Physiographic Subdivisions

Qatar's surface geography is characterized by a set of distinctive physiographic units that developed during the last few thousand years by surface geologic processes such as weathering, erosion, and transportation. Several different classifications exist for such features, but (Johnstone and Wilkinson 1960) classification is more detailed and will be adapted in this section.

Johnstone and Wilkinson (1960) Qatar's topographic surface is divided into three distinctive areas: the main limestone plateau, the sandy south, and the coasts. These features are described using many local names. To preserve the country's old heritage, these terms will be used here for future generations of geographers and geologists.

2.2.1 The Limestone Plateau

This area is described locally as *Barr Gatar*. It is characterized by karst features and consists of swells (*hazim, pl. hezum*) and depression (Fig. 2.3). Its surface is covered with silicified, dolomitized carbonate debris known locally as *hasa* or *hajar* (Fig. 2.4a). Some karst lands have dissolved

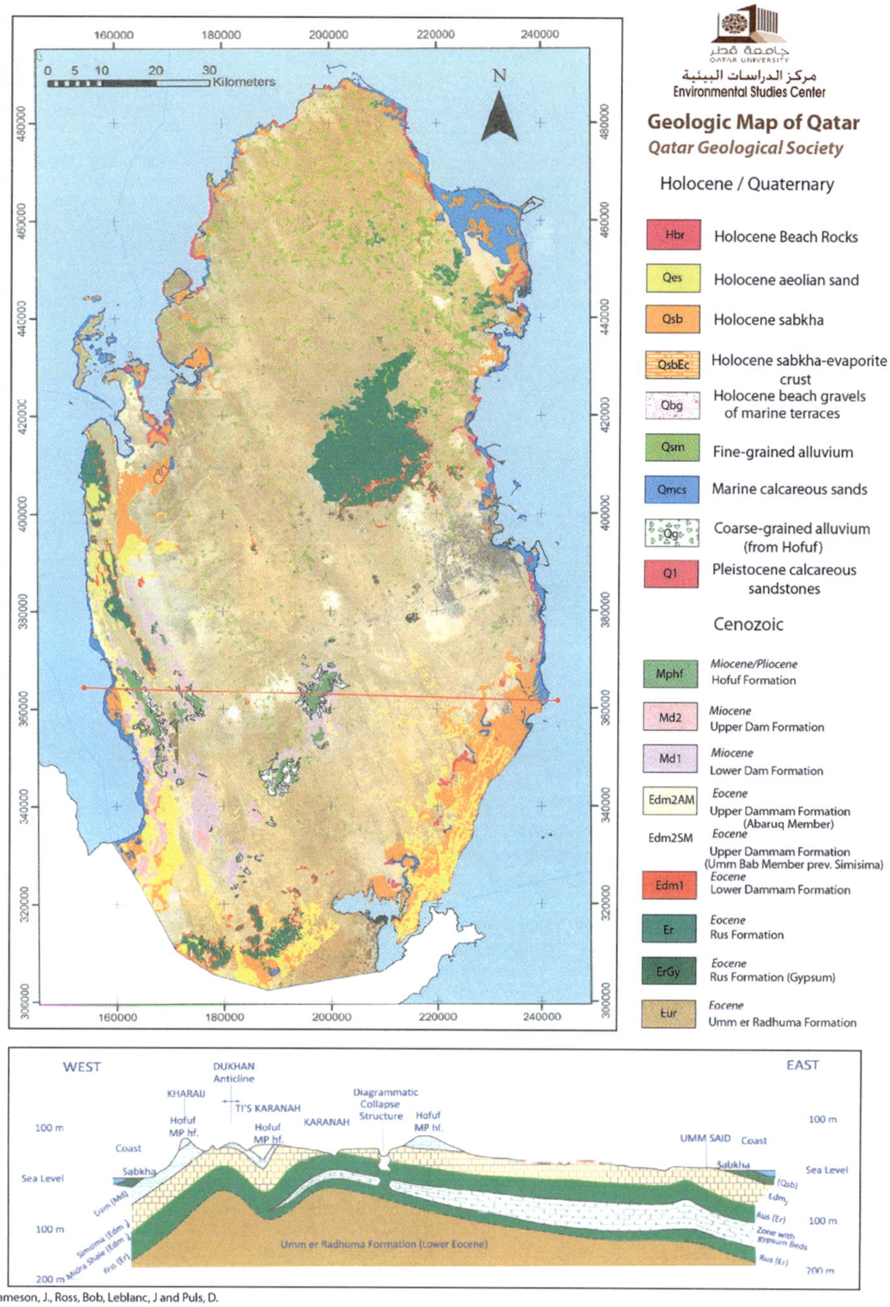

Fig. 2.2 Surface geologic map of Qatar (Environmental Science Center archives)

and collapsed, forming caves and caverns called locally (*dahl, pl dehul*) (Fig. 2.4b). If the area is not fully collapsed, there would be a depression where pools of water can accumulate during wet seasons. These are called (*kharsa, pl kherasi*) (Fig. 2.4c). A high rocky hill is known locally as *iihzila* (Fig. 2.5a), while a small low mound is called *hdiba* (hunchback) (Fig. 2.5b). The whole of Dukhan anticline is called *mashabiyya*, while the hills around it are called (*jabal, pl jibal, or jabalat*). The wind-eroded hills found in the Miocene rocks of the eastern flank of the Dukhan structure are called *gehab* (Fig. 2.5c). The hills characterized by scarp with thin beds of soft, white, chalky limestone are called (*birga, pl. birag*) because of the bright color of the chalky limestone (Fig. 2.5d). *Nakhash* means nose, which indicates the area characterized by a conspicuous spur, such as the southern plunging end of the Dukhan anticline.

In areas of this plateau where a considerable thickness of soil is developed that sustains perennial vegetation cover, the area is called (roda, pl rodat, *or riyad*) (Fig. 2.6). Some of these represent good grazing areas and are covered by low grass

Fig. 2.3 Barr Gattar: the limestone plateau, which is the main topographic feature of Qatar, consists of swells (Hazim) and depressions

Fig. 2.4 Topographic features of Qatar **a** the rocky surface of the plateau formed of pebbles (Hasa or Hajar) **b** karst (Dahl) in carbonate rocks **c** a karst depression filled with water (Kharsa)

Fig. 2.5 Topographic features of Qatar **a** rocky hills (iihzila) **b** low mound hill (Hdiba) **c** Eroded hill (Gehab) **d** a hill with white, soft, chalky rocks (Birga)

Fig. 2.6 A typical Roda or Rawdha: a relatively thick-soil-covered shallow karst depression characterized by relatively abundant vegetation

Fig. 2.7 Short, ephemeral stream with some vegetation (shi'b)

called Naim. The other grass is called (*Jiri, pl jiryan*). The central area of the roda where water collects is called *manga'* (soaked). Around this wet depression is a hard pan called (*Mistah*), which means flat. The part without vegetation is called *suna'*. Feeding these depressions during rainy seasons are small, shallow valleys (*wadi pl. widyan*). Short streams characterized by some vegetation are called (*shi'b pl. She'ub*) (Fig. 2.7).

Smooth parts of the plateau east of the limestone scarp are called (*rigga*). If it is covered with finer stones, then it is called (*hashu*); if covered with reddish dust, then it is *nega*.

2.2.1.1 Karst Features

Karst (from the 19th-century German word *der Karst*, the name of a limestone area in Slovenia) includes many dissolution features, including expansive caves and depressions. They are locally called "Dahl pl. Duhool." Karst topography is a characteristic of Eocene exposure surfaces in Qatar and large portions of the Arabian Peninsula (Abul-Haggag 1965; Duggan et al. 2014; Edgell 1990; Emil et al. 2021; Howari et al. 2016; Mukhopadhyay et al. 1996; Orndorff et al. 2018; Sadiq and Nasir 2002). Despite having an arid climate with an average annual precipitation of 7.5 cm, the country's 11,586 km^2 of karst topography is well-developed, with more than 9700 depressions identified (Orndorff et al. 2018). Several significant caves exist in Qatar, but many have likely been filled with blown sand or collapsed to form some of the thousands of depressions or dolines. A large number (9736) of shallow depressions are responsible for Qatar's most prominent topographical features (Embabi and Ali 1990; Sadiq and Nasir 2002).

Chalky dolostone and limestone of the Eocene Dammam Formation are exposed on about 80% of the land's surface (Fig. 2.2). The Miocene Dam Formation's limestone, dolostone, and evaporite rocks are disconformably overlain by 40–80 m of the Dam Formation in southern Qatar. Most frequently, the erosional caps of the higher elevations are covered by the Dam Formation. These shallow-water carbonate rocks frequently alternate with anhydrite, gypsum, and clay deposits and are covered in some places by Quaternary deposits (Cavelier 1970; Orndorff et al. 2018; Sadooni 2014). A stratigraphic setting of rocks with different coherence stacked vertically represents the ideal setting for karst development. Humid climate regimes are linked to active karst regions, where the organic matter in the soil and precipitation combine to produce carbonic acid, which, over time, dissolves carbonate and evaporite rocks both above and below the surface.

Karstification in Qatar and other regions of the Arabian Gulf is linked to the calcareous, dolomitic, and gypsiferous Dammam Formation, which dates to the Eocene (Abul-Haggag 1965; Al-Sayari and Zötl 1978; Jado and Zötl 1984; Mukhopadhyay et al. 1996). Local stratigraphic and petrographic studies (Sadiq and Nasir 2002) indicate that a selective dissolution of rock, fueled by rainwater and groundwater, is the main karst-forming process affecting the Dammam Formation. Such karstification occurs at the subsurface interface between the upper dolomite layers and the lower limestone of the Dammam Formation, which contains gypsum. According to Sadiq and Nasir (2002). Most sinkholes resulted from preferential dissolution caused by the differences in composition between dolomitic, calcitic, and gypsum rocks.

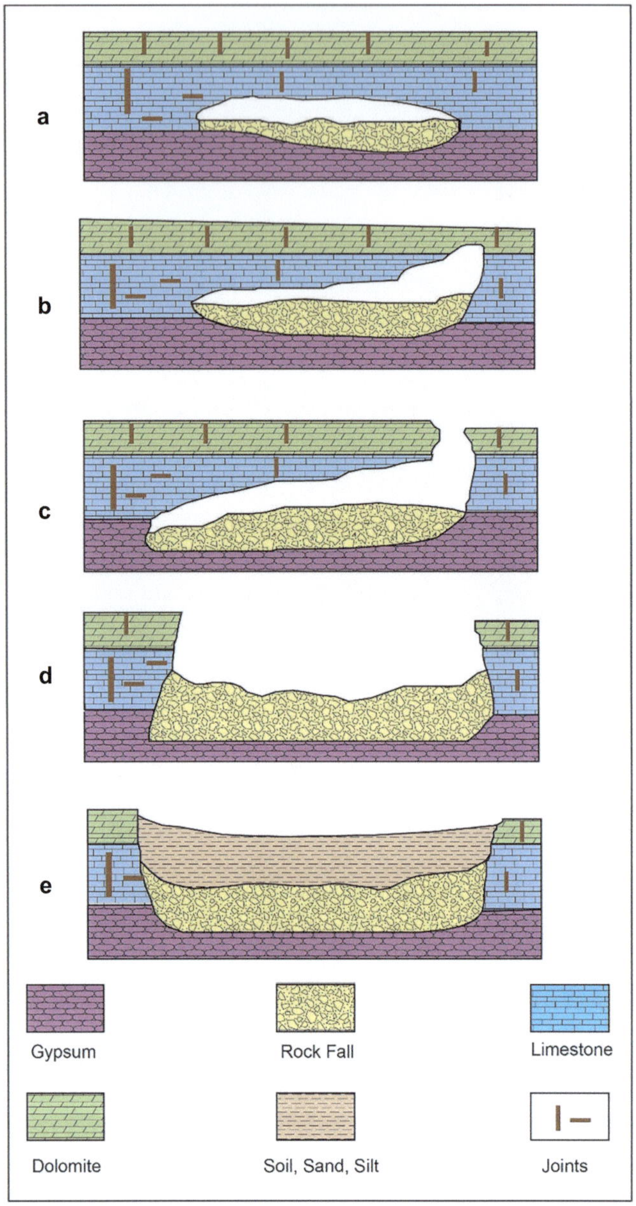

Fig. 2.8 A proposed model for the formation of sinkholes in the central part of Qatar (Sadiq and Nasir 2002)

Mechanism

Karstification developed during the Middle Pleistocene, between 325,000 and 560,000 BP (Edgell 1990; Rauert et al. 1988). The Arabian Gulf region had a wet climate during this time (Edgell 1990). Additionally, the massive amounts of Neogene sediment deposited show that the Miocene and Pliocene experienced a pluvial and humid climate (Whybrow and McClure 1980). Similar evidence can be seen in the large Early Quaternary alluvial fans (Al-Saad et al. 2002). The amount of gypsum and limestone that dissolves within the surface-level meteorological environment depends on several variables, such as the amount and distribution of soil cover, temperature, biological activity, and the structural fragility and lithology of the carbonate substrate (Hill 2000). The main way that temperature affects limestone dissolution is through its impact on biological activity. Karstification is significantly impacted by the lithology of the carbonate substrate, with calcite

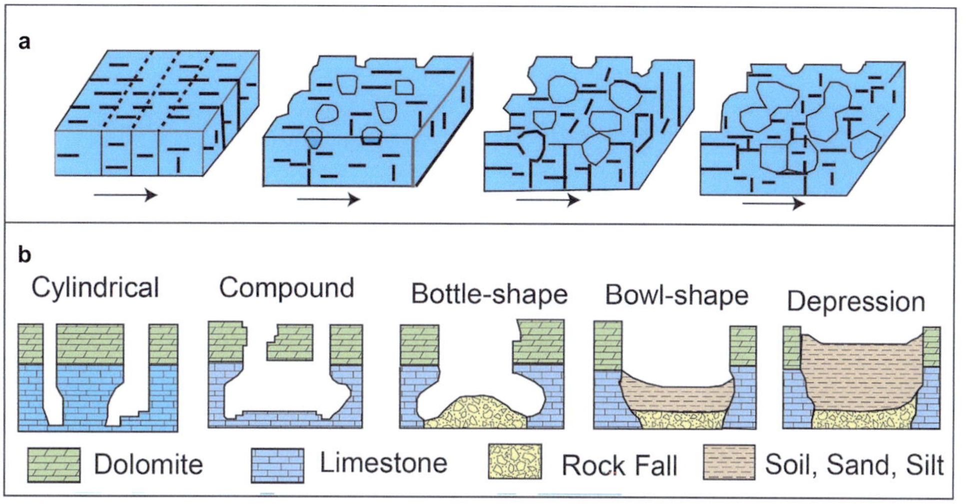

Fig. 2.9 Proposed model for the formation of sinkholes in the northern parts of Qatar (Sadiq and Nasir 2002)

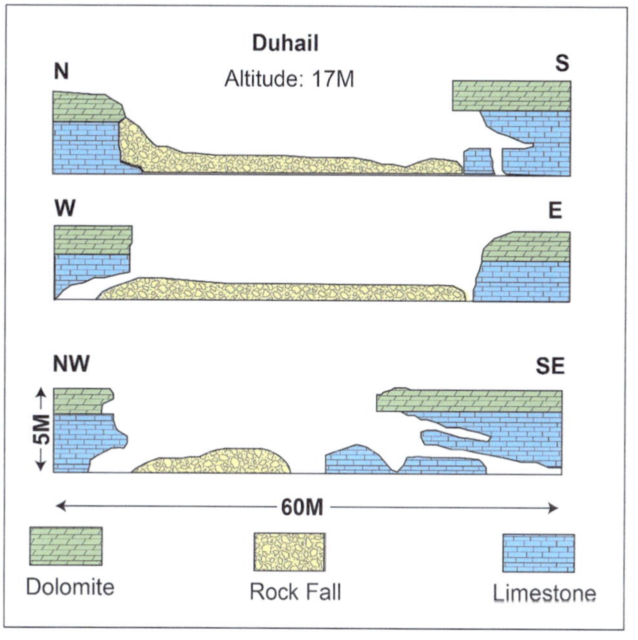

Fig. 2.10 Ideal lithologic profiles of the Duhail sinkhole, Doha, Qatar (Sadiq and Nasir 2002)

dissolving more easily than dolomite due to its higher solubility (Martinez and White 1999). Where dolomite and gypsum are in contact with the same aquifer, karstification can occur (Bischoff et al. 1994). Calcite precipitates because of gypsum dissolution, consuming carbonate ions released by dolomite.

The karst pits that make up Qatar's sinkholes and depressions are structurally controlled and appear to have started as joint-flow drainage fractures (Sadiq and Nasir 2002). Strong vertical control is indicated by the profiles of all sinkholes in northern Qatar, which is attributed to the bedrock's numerous fractures and joints. Based on sinkhole long-axis data, such lines of structural weakness in the bedrock affect sinkhole orientation. Sinkholes frequently develop along high-permeability routes through the vadose zone, locations of fracture concentrations, and intersections (White 1988). Long major weak points, sinkholes may also lengthen due to a combination of dissolution and collapse (Reeder et al. 1996). If fracture zones help it, collapse may start at a depth in the bedrock (Fig. 2.8). Joints and fractures divert rainwater to specific locations on the emergent surface, where it concentrates and causes dissolution along joint intersections, resulting in cylindrical pits that eventually propagate vertically downward (Fig. 2.9a). The initial perpendicular orientation of the pits is also explained by gravity-driven drainage. In addition, the similar width of shallow and deep cylindrical pits is explained by joint-flow drainage. Differences in depth may be due to how long drainage was concentrated at a specific location. Some pits could only spread downward a few meters, and lateral amalgamation of nearby

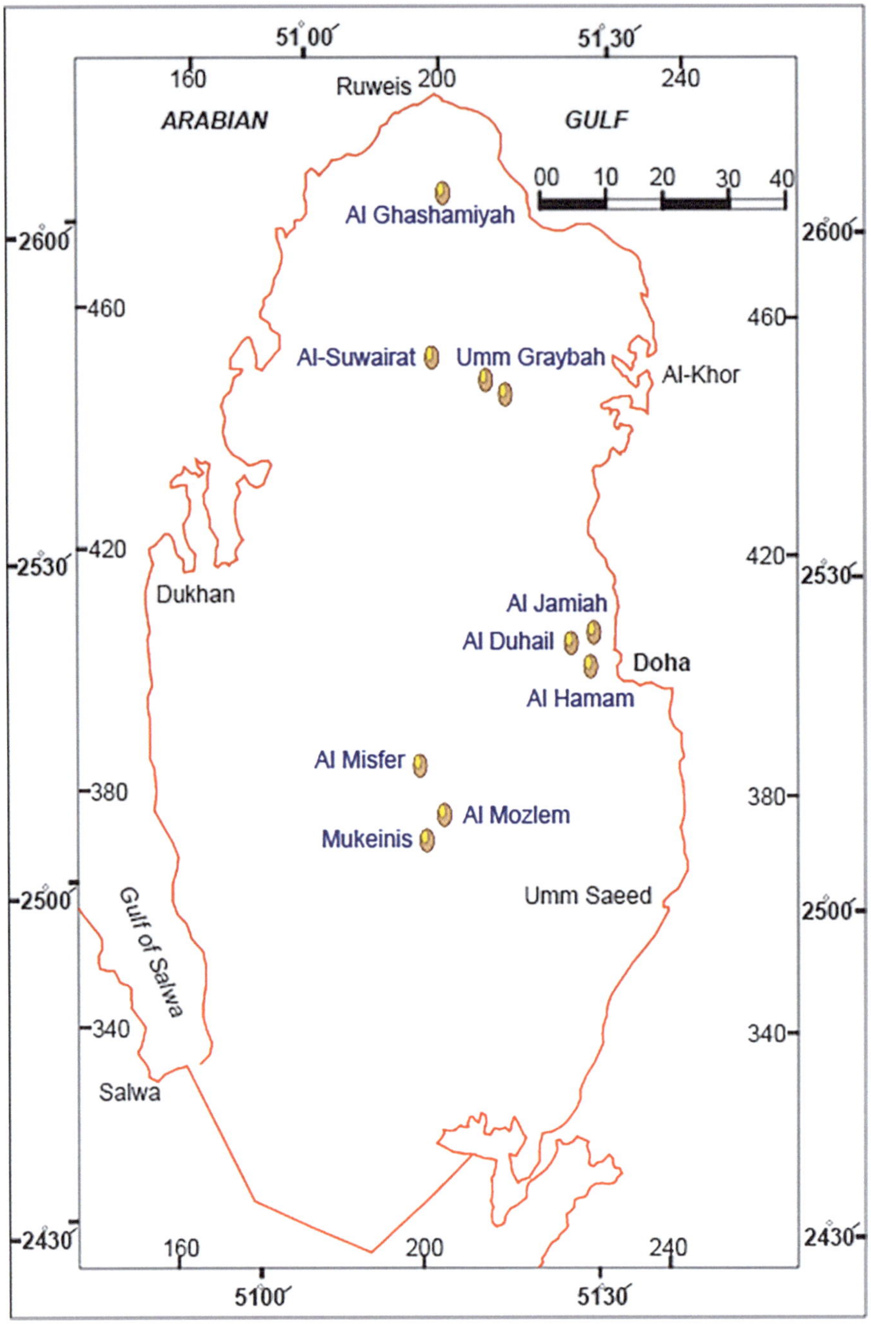

Fig. 2.11 Distribution of the main "duhool" in Qatar

pits was the main growth method. The sinkholes' formation would have continued once it was started. The investigated karst pits are similarly smooth and rounded compared to contemporary karst terrains (James and Choquette 1984; Sweeting 1973). Qatar is home to a wide variety of karst-pit forms. Round pits change into depressions and bowl-shaped ones through several bottles and compound intermediates (Fig. 2.9b). This suggests that each form began around a cylindrical pit and represented a genetic sequence. Sinkholes may serve as a metaphor for the initial stages of depression development, as shown in Fig. 2.8b. The lower part of the pit wall's limestone and gypsum were removed preferentially due to dissolution and collapse. Most likely, a collapse caused the upper portion of the pit to enlarge and turn into a bowl-shaped depression (Fig. 2.10). The initial density of the cylindrical pits determined the direction along which additional pit enlargement took place. Pits that were closer together gradually merged to form bowl-shaped pits by gradually lowering the height of the interpit area where the pits were more closely packed (Fig. 2.8b). Further dissolution at this location seems to have been minimal once the interpit area had completely been denuded, with pit enlargement occurring primarily through lateral dissolution and amalgamation.

Numerous karst features support (Walkden 1974) hypothesis that a shallow-water table existed during development. Particularly, the presence of flat-based pits and the enlargement of karst pits by lateral amalgamation rather than vertical deepening (Fig. 2.8b) indicate the existence of a water-saturated or carbonate-cemented zone a few meters beneath the sediment surface. The above-discussed progressive formation of sinkholes in northern and central Qatar is summarized in Figs. 2.8 and 2.9, respectively. The sinkholes/depressions under investigation only formed due to collapse. Other explanations, though, could exist.

Distribution

Many of these depressions express subsurface collapse structures on the surface. The majority are circular, with diameters ranging from a few hundred meters to 3 km. Some are as deep as 25 m, while others are as shallow as a few centimeters. Embabi and Ali (1990) associated many sinkholes with these depressions. Analysis of the morphometric and spatial distribution parameters of karst depressions reveals that the Qatari karst comprises broad, shallow depressions with an average density of one depression per kilometer squared (Embabi and Ali 1990). The distribution of the main "duhool" is shown in Fig. 2.11.

Duggan et al. (2014) showed that the mapping method impacts the quantity and size of depressions or closed drainages. His study developed GIS workflows to find depressions with significant drainage basins that were most likely to be linked to collapse. In the Rus Formation, the size and depth of depressions were directly related to the presence or absence of gypsum layers (Fig. 2.12). Larger depressions were discovered where gypsum from the Rus Formation was probably dissolved.

Controls over the various morphologies and density of features can be determined by examining the relationship between geologic and hydrologic parameters and the spatial distribution of karst features (Orndorff et al. 2000; Weary and Orndorff 2002). These include controls based on stratigraphy, lithology, structure, topography, and hydrogeology. Before using Landsat satellite imagery, air photo analysis was used to locate depressions and sinkholes in Qatar. By creating a more thorough spatial distribution and database of karst, high-resolution digital elevation models made from LiDAR have recently greatly improved the identification of controls on karst processes. The three geographical areas of depression that Duggan et al. (2014) identified have different characteristics and are each connected to the Eccleston et al. (1981) model of deposition and gypsum dissolution in the Rus Formation (DC, RS, and DS in Fig. 2.12).

The central area (RS) is associated with residual sulfate (residual gypsum) with elongated depressions and a southern depositional sulfate area (DS) (bedded gypsum) with deep circular depressions. The northern area (DC) is associated with depositional carbonate (no gypsum) with shallow, low-lying, irregular-shaped depressions (Fig. 2.1). The southern region is connected to bedded gypsum and appears to be the intrastratal karst that causes collapse sinkholes. Numerous known caves are found along or close to the crest of the Qatar Arch, suggesting that karst development is controlled structurally (Sadiq and Nasir 2002).

Most depressions and sinkholes are located south of latitude 25° 20′, the northernmost extent of the deep gypsum and anhydrite horizons from the Eocene period. Qatar has a dry, arid climate today, but during the Miocene and Pleistocene epochs, the climate alternated between being moist and dry (Al-Saad et al. 2002; Butzer 1961). The Upper Dammam Unit limestone underwent karstification, creating simpler groundwater pathways. The primary aquifers containing potable groundwater in Qatar today are Tertiary sedimentary rocks, which were replenished by sporadic rainstorms on outcrops of the same rocks in Saudi Arabia. Figure 2.12 shows the general distribution of karst features in Qatar in association with facies distribution, drainage, and structural trends.

Classification

According to aerial photographs, there are three types of karst in Qatar: sinkholes, simple depressions, and compound depressions. One-center depressions are referred to as simple depressions. Compound depressions are large, irregularly shaped depressions with multiple centers. These depressions appear to have formed by combining several smaller depressions. The areas between the depressions are heavily sylolitized, according to field research, but they were likely characterized initially by small-scale, fretted, and sculpted karren forms typical of subaerial karstification. The northern and central regions of Qatar are where sinkholes are most prevalent (Fig. 2.13). Structural control on the karst development can be seen in the analysis of sinkholes and the major axis orientations of depressions. The joint and fracture orientations in the study area correspond with the prominent NW–SE and NE–SW long-axis orientations of sinkholes. Karst areas in central Qatar are characterized by calcrete surfaces that are 0.1–1 m thick. Sinkhole wall rocks have numerous stylolites, voids, and cavities from selective dissolution, and they are highly jointed and fractured. The upper 1–2 m of the walls are composed of dolomitic rock, according to thin-section studies of the wall rocks from various sinkholes in northern Qatar.

In contrast, the lower part of the walls is composed of biomicritic to biosparitic limestone with a sizeable percentage of dissolution pores that are primarily filled with

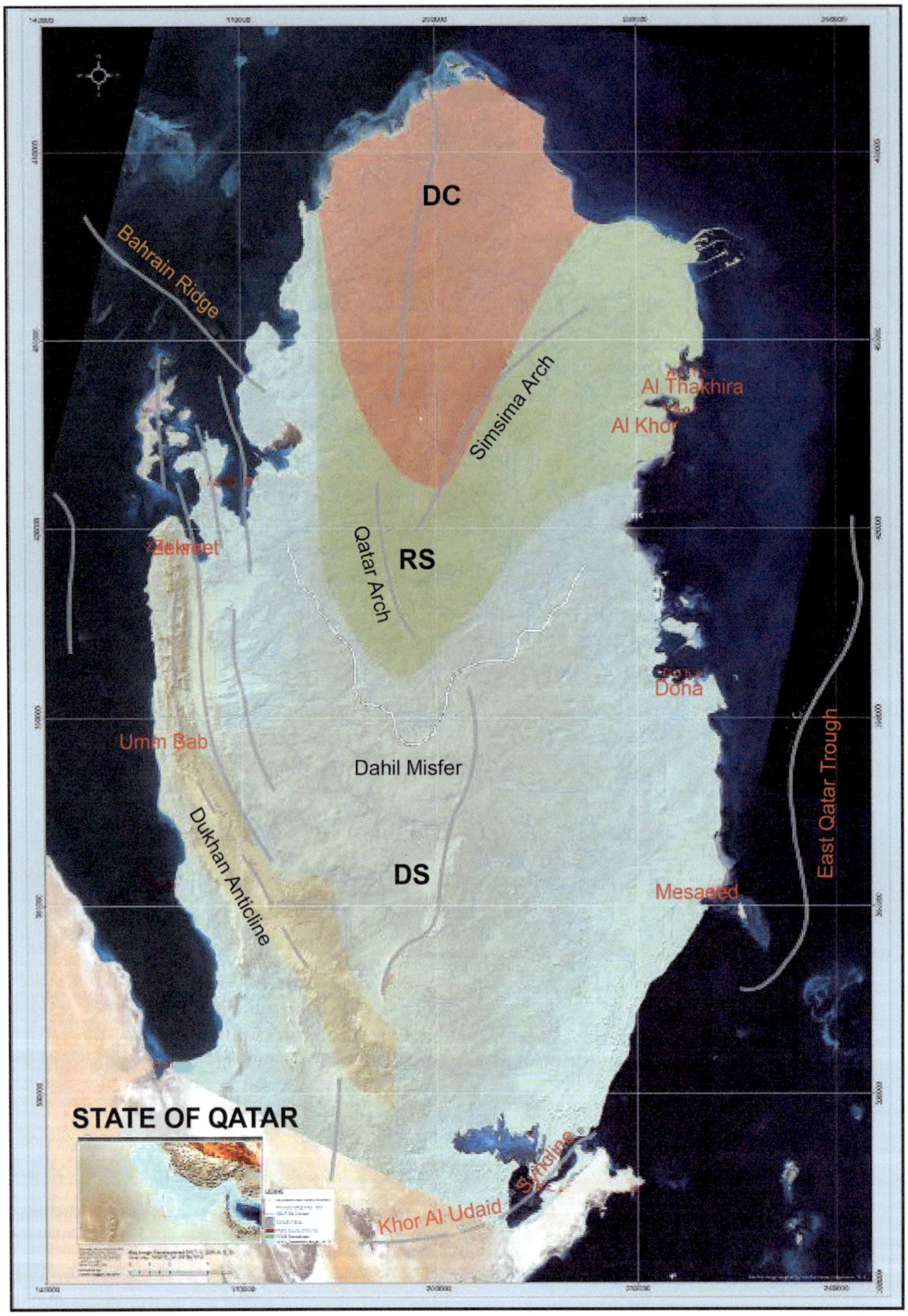

Fig. 2.12 Regional association of facies, drainage, and structure (DC: Depositional carbonates; RS: Residual sulfates; DS: Depositional sulfates) (Duggan et al. 2014)

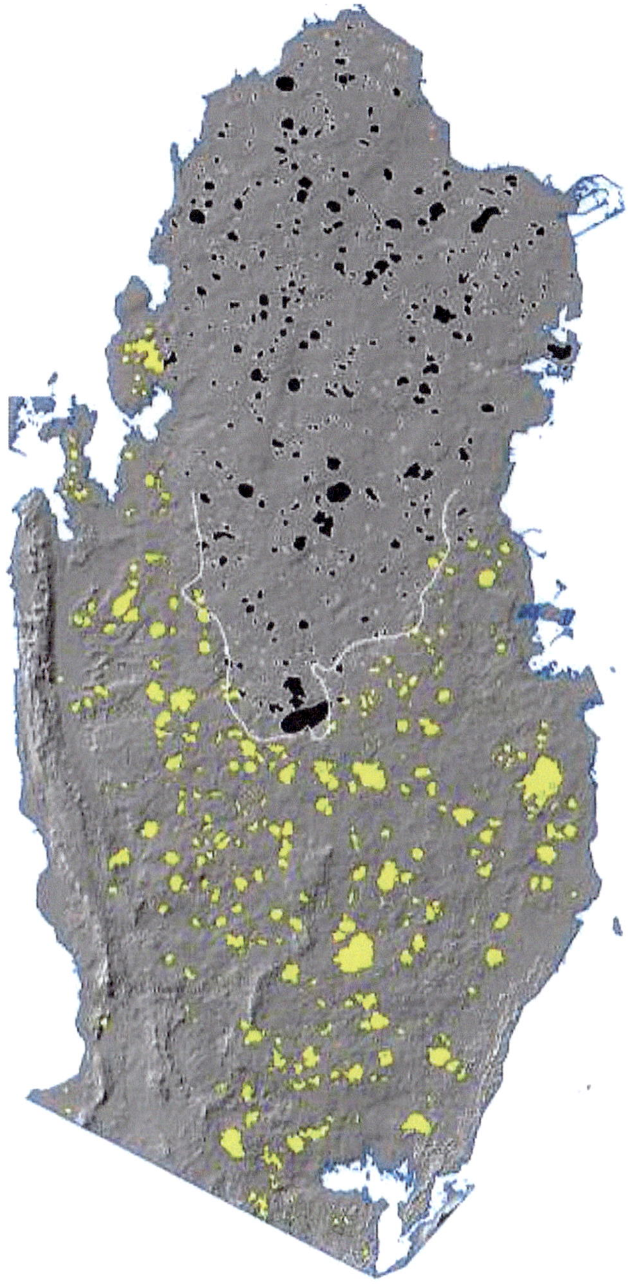

Fig. 2.13 Regional distribution of possible karst features in Qatar (Duggan et al. 2014)

gypsum. The rocks are from the Upper Dammam Unit. In central Qatar, thick gypsum beds underlie the limestone, while marl underlies the upper calcrete. The rocks that make up the sinkholes in central Qatar are found in the Upper and Lower Dammam units.

Morphology

Several types of karst-pit morphology exist. Different morphotypes illustrate variations in karst-pit shape, as shown in Fig. 2.9:

Cylindrical karst pits have a narrow tubular form with vertical or steeply inward inclined pit walls.

Bottle-shaped karst pits comprise cylindrical pits in which the lower portion of the pit wall has been removed to produce a beveled wall that slopes outwards.

Compound karst pits comprise two or more adjacent cylindrical or bottle pits with a common interpret area.

Bowl-shaped karst pits have steep sides with a gently concave, irregular, or flat base.

Depressions can be former sinkholes partly or completely filled with autochthonous material.

2.2.2 The Sandy South

Sand accumulations and dunes that cover this area are largely the product of the northwest Shamal wind, the most prevalent wind in Qatar. Sand usually accumulates when it is baffled by a barrier such as a tuft, shrub, or grass (Fig. 2.14). These deposits will be relatively thin sheets and are locally called *Ramla* (Fig. 2.15). Where the sand accumulates in a shallow depression, then it is called *Bratha.* With a continuous supply of sand by the blowing wind, the R*amla* develops into longitudinal and narrow dunes called (Irg pl. urug or (*Sahib pl. Sihhib*). The long, low dune is called *Urgub.* Sand deposits that grow larger than the *Bratha* in the valleys, forming low mounds, are called (*Zubar pl. Zebayir*) or (*Hdiba, pl. Hdab*). The crescent-shaped dunes or barkhan are called *(Ti's pl. Te'us*) or (*Khet pl. Kheyut).*

The apex of the crescent dunes points at different angles ranging from NNW to NW, which is the prevalent direction of the wind. The dune's windward side, characterized by a low slope and packed sand deposits, is known as Zahr al Khet, and the steep, leeward side is *Sayydl.* Toward the sea, these dunes become interlinked to form a sand sea. These areas are called *Nigyan.*

The singular *niga* means a dune of unspecific type. Some of these sand dunes areas are characterized by relatively large, relatively high mesas with parallel rock strata called *Tewar.* Isolated hills that are horn-shaped or small mesas are called (*Gurn pl. Gurun*).

Most of the sand dune deposits are concentrated mostly in the southern part of the country due to the wind dynamic and directions (Figs. 2.16 and 2.26). They are found in five areas. These are (a) the northern and NW sand fields, (b) the central fields to the NW and W of Umm Said, (c) the coastal field SE of Umm Said, (d) Rawdat al Faras belt, and (e) Sawdat Nathil field. The second and the third areas host the major fields. They cover an area of about 1000 km^2 (Ashour 1985).

Fig. 2.14 Start of sand dune formation. Baffling of sand by a barrier such as this plant

2.2.2.1 Formation Mechanism

Investigation of the sand accumulation mechanism (Ashour 1985; Embabi and Ashour 1993) has indicated that sand starts accumulating as single barchans in the central parts of the country, about 20 km southwest of Doha. These isolated barchans may form due to baffling objects such as shrubs, vegetation, or topographic tuft. They range in height between 0.8 and 40 m. Their width varies between 15 and 1000 m. The azimuth of their longitudinal axis of the windward slope varies between 138° and 165°, with a mean of 152°. As the barchans move in a southwest direction under the action of the prevailing shamal wind, they merge, forming more complex barchanoid forms (such as compound, complex, longitudinal, and transverse dunes). These types are found south of Umm Said and the southern sabkhas along the Qatari-Saudi border. Ashour (1985) suggested that the more complex dune forms result from the accumulation of barchans in the low moist topographic area of the sabkhas. These relatively wet areas will provide the needed moisture to delay or even fix the movement of these sand bodies. It is also possible that halophytes fringing the sabkhas, the topographic mounds, and depression may slow or stop the dune movement at the end. Much of the sand dunes end at the coastline of the Arabian Gulf, sink and lose their horns to the marine water, and gradually become digested in the marine environment under wave and tide action (Fig. 2.17). This is reflected in the relatively high quartz content of the marine coastal sands south of Umm Said.

2.2.2.2 Sand Composition

The dune sands are composed of quartz grains, carbonate fragments, which vary between 9 and 23%, argillaceous dust, and trace amounts of heavy minerals. A higher amount of carbonate materials is found closer to the northwest coast. Dolomite ($CaMg(CO_3)_2$ forms a significant carbonate grain fraction. These may be derived from the weathering of the carbonate surface cover of the country, which formed mainly of the Eocene Dammam Formation. Samples with low dolomite contents, as reflected by low Mg contents, may indicate that the carbonate fraction of these sands is derived from an outside source and was transported with the quartz (Ashour 1985; Curtis 1983).

There are two heavy mineral assemblages in the dune sands. These are apatite, collophane, anhydrite, and kyanite, which may be of local origin, while garnets, diopside, rutile, zircon, sillimanite, and epidote from an exogenous origin (Ashour 1985; Briggs 1983). A study of sand surface texture using the electron microscope (Embabi and Ashour 1993) suggested that the surface textures have a fresh appearance due to recent eolian processes acting in the area from which the samples were collected. However, some sand grains are from former coastal or marine environments (Ashour 1985).

In a review paper on the Arabian dune sands, (Garzanti et al. 2013) studied samples from two locations in Qatar, one from Niqyan, near Doha, and the second from Khor Al-Adaid.

Fig. 2.15 Thin sand sheet (Ramla) with ripple marks generated by wind action

Fig. 2.16 A typical sand dune, Khor Al-Adaid, southern Qatar

Samples from the Niqyan area are formed mainly of carbonates with volcanic low-rank metavolcanic rocks with plagioclase and feldspar. Heavy mineral assemblages are formed of epidote, garnet, amphibole, pyroxene, and others. The Khor Al-Adaid samples are formed of carbonates, low-rank metamorphic rocks, volcanic and low-rank metavolcanics,

Fig. 2.17 Sinking of the sand sediments into the waters of the Arabian Gulf, Sealine

plagioclase, and feldspar. Heavy mineral assemblages include quartz, pyroxene, amphibole, olivine, spinel, and epidote (Fig. 2.18).

2.2.2.3 Textural Properties

The grain size properties and composition are the results of wind action, source materials, and the topography of the hosting areas.

A scatter plot diagram of the mean size versus standard deviation of the dune sands is shown in Fig. 2.19. It shows that the best sorting values are found in the medium to fine sands at a mean size of 1.26 Φ and 2.24 Φ. Sorting declines as the sand gets finer or coarser. This size range may represent the wind-lifting capacity for the higher size. Due to abrasion and progressive coring processes, the grain size and sorting values would decrease in the downwind direction. However, local fluctuations may be attributed to one or more of the following environmental processes: (a) an increase in the supply of materials derived from the local surface, (b) topography of the initial surface, and (c) a change of wind direction and intensity. It was noticed that wind with strong opposing SE wind may modify the upper part of the slip faces of the dunes and even transport sand grains northwards; (d) secondary coating of gypsum or other minerals found in dune sands located in the sabkha areas in the south and southeast regions of Qatar (Ashour 1985).

Roundness and Sphericity

Sand grains of the dune sands are generally rounded and subrounded. Table 2.1 shows the distribution of individual grains according to their size and roundness. Around 73–84% of the grains are well-rounded, rounded, and subrounded, and about 55% are highly spherical (Ashour 1985; Embabi and Ashour 1993).

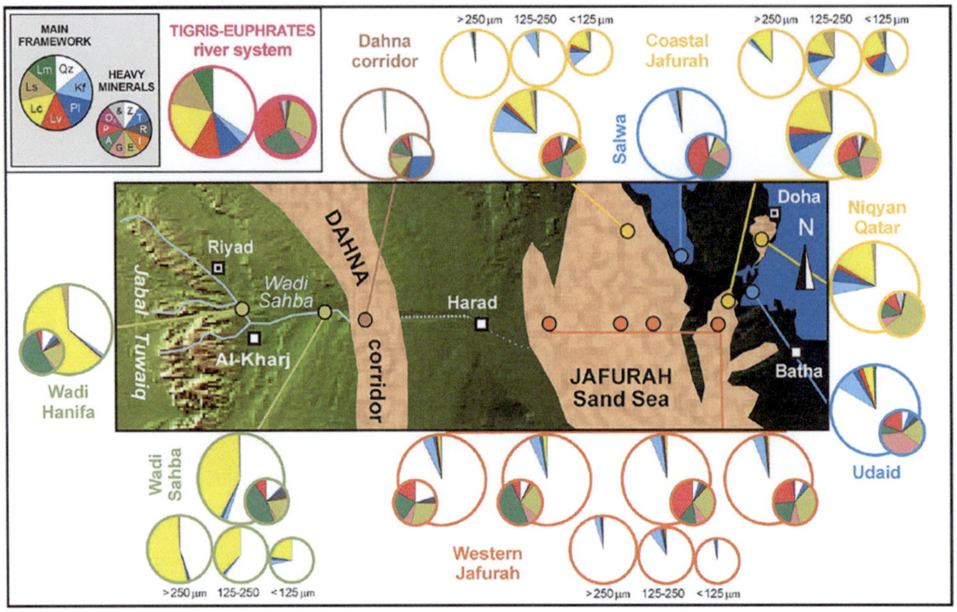

Fig. 2.18 Main composition and heavy mineral assemblages of dune sands from the Arabian Peninsula (including two locations in Qatar) [After (Garzanti et al. 2013)]

A comparison was made between samples taken from the tail and the crest of the dunes to check whether the roundness was related to the sample's position in the dune. The average mean roundness of different grain sizes suggests that the differences in roundness between crest and tail samples are negligible (Fig. 2.20). Individual profiles show that in 48% of the studied profiles, the roundness at the crest was higher than the roundness at the tail; in 43%, the roundness at the tail was higher than at the crest. In 9%, the roundness was equal. This also indicates that the difference in roundness between crest and tail samples, as shown in individual profiles, is negligible (Fig. 2.21). This may suggest that the relatively short length of the dune is not enough to cause a significant change in the grain's roundness because the roundness is not a function of size. Data suggest that roundness is irregular across the study area (Ashour 1985).

2.2.2.4 Origin

Several hypotheses were introduced to explain their origin because the sand dunes were not fully studied. One of the earlier hypotheses was laid down by Cavelier (1970), who suggested that eolian sands are derived from coastal marine sands, which are periodically reworked by the shamal wind. The prevalence of Quartz as the mineral-forming mineral of these deposits works against this assumption. In a report by Seltrust Engineering Limited (1980). It was assumed that these sands were derived from the Saudi Arabian mainland to the northwest during the Quaternary when the Gulf of Bahrain and Salwa Bay were dry lands. Then, sand deposits started migrating gradually southeastward. Ashour (1985) noticed that an extension of the azimuth 152° (average dune direction) from the main dune fields does not reach Saudi Arabia's dune fields but instead falls somewhat to the east in Iran or east Iraq. He emphasized that there is no evidence to suggest that there has been a significant shift in wind direction since the beginning of the Quaternary.

Ashour (1985) indicated that the heavy mineral suites, the interpretation of the depositional environment using the scanning electron microscope, and the geochemical and textural analyses suggest that sand deposits are derived mainly from the Arabian mainland, the sands currently being blown off the Arabian land mass into the Gulf of Bahrain, some of this sand is being deposited on the NW coasts of Qatar by tidal action. This is confirmed by the presence of exogenous heavy minerals similar to those of Saudi Arabia. It is also believed that the sands blown off into the Gulf are recirculated by anticlockwise currents and tidal action back to the northwest coast. This is supported by heavy minerals of local origin in NW samples. Furthermore, there is evidence on some sand grains of former coastal or marine environments as shown by the subdued "V"-shaped pits (Ashour 1985; Griffin 1983).

Garzanti et al. (2013) suggested two origins for the dune sands in Arabia. These are endogenic and result from the recycling of sands in Arabia by weathering and transporting sands from the local barren areas in Arabia and exogenic sources from outside Arabia, such as the Zagros mountains and the southern Iraq deltas and fluvial systems. They suggested that the Arabian Gulf was exposed during the Pleistocene lowland. As a result, carbonate-rich sands rich with quartz and feldspar were derived from the weathering and erosion of the Bitlis-Zagros orogen. These deposits were transported via the

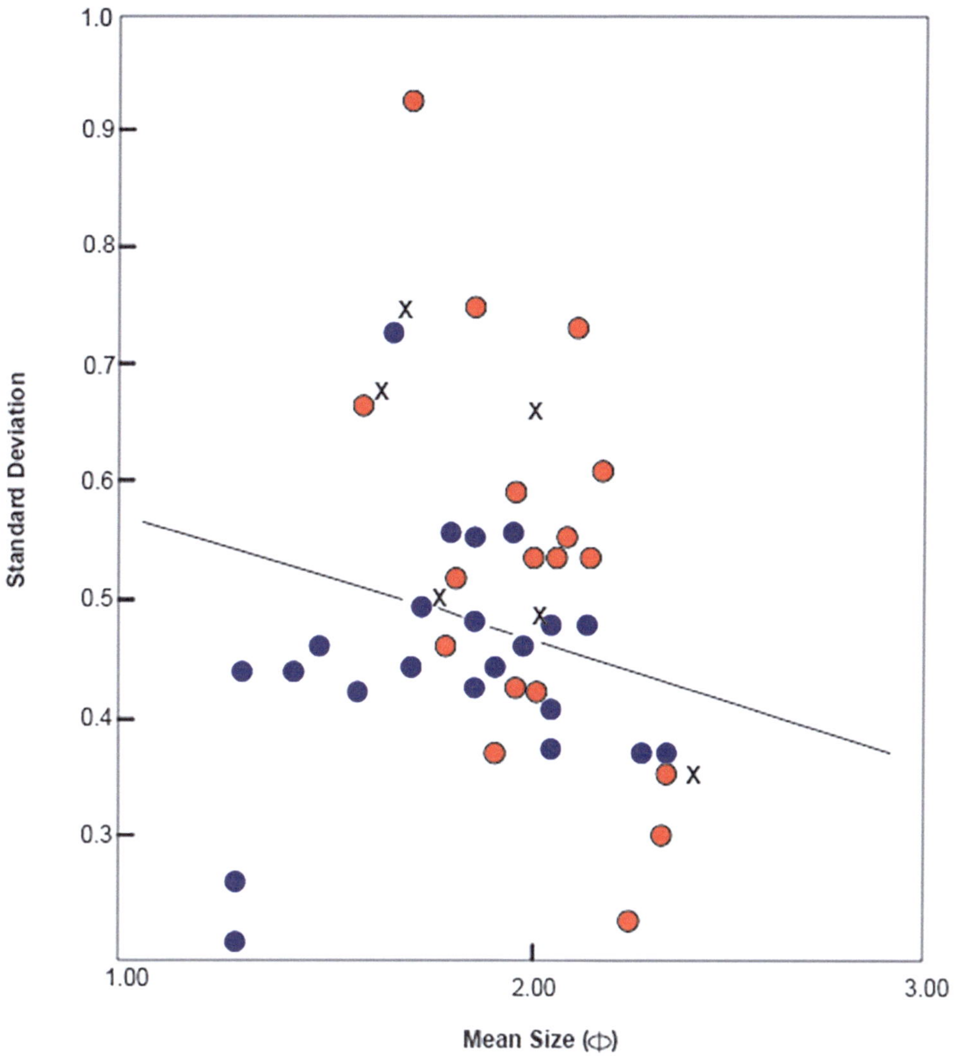

Fig. 2.19 A scatter plot diagram of the mean size versus standard deviation of the dune sands (Ashour 1985)

Table 2.1 Percentage distribution of roundness (Ashour 1985)

Roundness	Size (μm)				
	500	355	250	180	Total
Well-rounded	20.35	23.79	23.85	30.09	24.52
Rounded	30.64	26.02	22.17	27.15	26.50
Subrounded	29.50	33.98	27.49	16.52	26.82
Subangular	11.18	10.87	17.49	11.76	12.85
Angular	5.36	4.17	6.18	7.01	5.68
Very angular	3.18	1.07	2.82	7.46	3.63

Tigris–Euphrates axial river system and were blown inland by northwesterly Shamal winds and mixed with polycyclic quartzose sand from Interior Arabia. Coastal dunes accumulated during each arid glacial stage (aeolianites) and were recycled repeatedly through the Quaternary (Fig. 2.22).

2.2.2.5 Nebkhas

They are structureless phreatophyte mounds formed by trapping sands and silts by shrubs and plants, sometimes covered with a thin salt veneer (Fig. 2.23). They have wedge-like shapes that narrow on the windward side. Their height ranges from a few centimeters to 150 cm, while their length ranges from a few centimeters to 5 m. Field observation revealed that these nebkhas develop in three stages:

(1) In its early stages, a tiny plant grows on the surface of the sabkhas (plant type, size, and shape are a function of soil texture, salt content, and groundwater level).

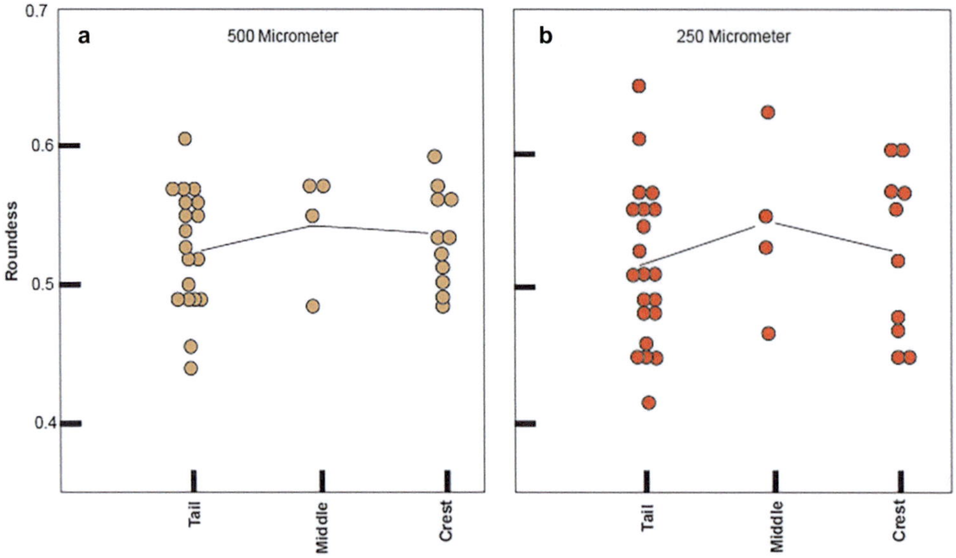

Fig. 2.20 Roundness values for sand grains collected from different parts of the dunes. The dots represent the roundness value for each sample, and the line represents the mean value by averaging the sample means (Ashour 1985)

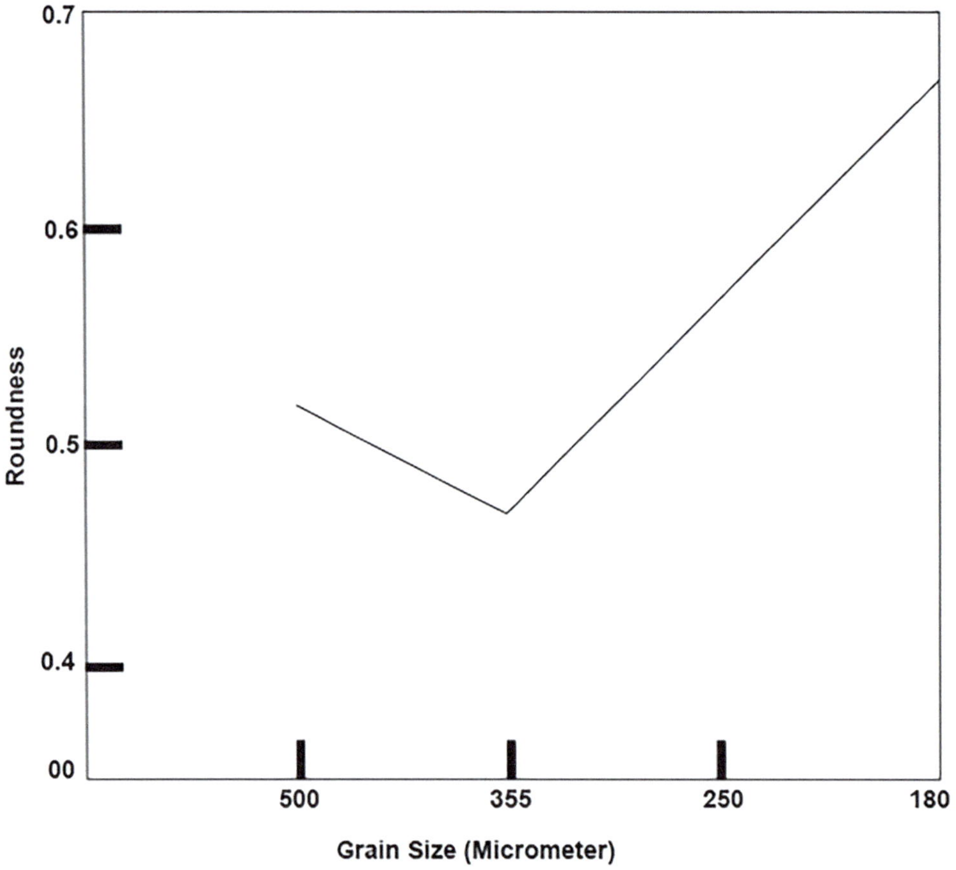

Fig. 2.21 Variation of roundness with grain size in individual profiles (Ashour 1985)

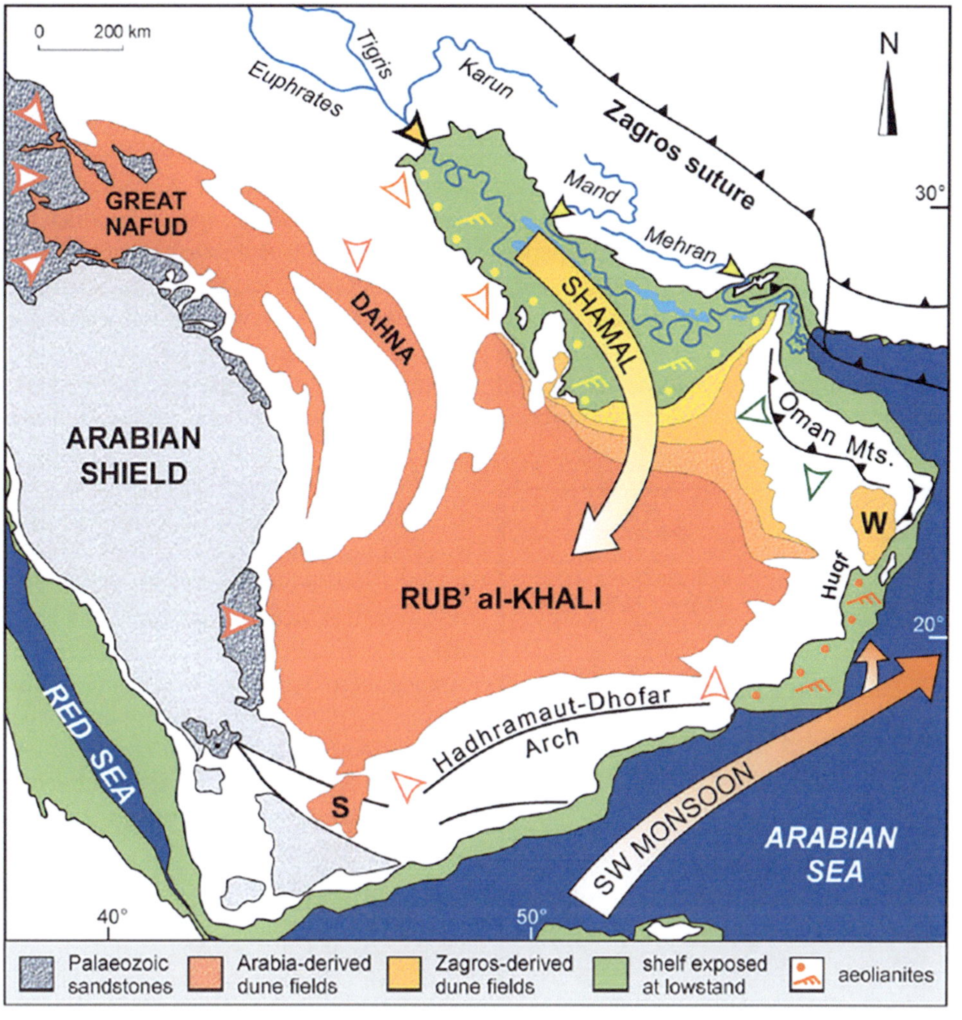

Fig. 2.22 Modes of transportation and major provenances of dune sands in the Arabian Peninsula (Garzanti et al. 2013)

Fig. 2.23 Entrapment of sand by plants (Nebkha), Sauda Nathil Sabkha

The growing plant traps moving sands and silts, forming a mound; (2) As the mound grows, the plant continues to grow, extending its root and branch system in all directions to catch water and sunlight, trapping more sand and silts. This is the mature stage; (3) In the old stage, the mound grows to the point where water can no longer reach the branches, the plant gradually deteriorates, and the salt crust is dissected, leaving the friable material of the mound to be driven by the prevailing wind. Finally, the mound collapses gradually, leaving behind dried and decayed roots and small sand heaps (Ashour 1991; Ashour et al. 2013).

2.2.3 Sabkhas and Salinas

Some inland and coastal areas of Qatar are punctuated with salt-encrusted flat areas. These are sabkhas and salinas. The term sabkha is an Arabic word that entered the English dictionary after the pioneering work of Shearman (1963) and other workers from the Imperial College team on the Abu Dhabi sabkhas during the sixties of the last century. Classic Arabic dictionaries, such as Lisan Al-Arab by Ibn Mandhoor (1232–1311), provide a comprehensive definition of the word and give the basic components of the sabkhas as it is known now. It states that the sabkha is salty land with scarce vegetation and algae covering its surface (Fig. 2.24).

Yechieli and Wood (2002) suggested that all evaporite-forming bodies, such as pans, playas, sabkhas, salinas, saline lakes, and salt flats, have the same hydrological system. Contrary to the old research (e.g., Evans et al. 1964), which assumed that marine water seeps or spills to the coastal sabkha, causing the formation of evaporites, they suggested that sabkha is a continental hydrological regime that results from the upward movement of rainwater and groundwater to compensate for moisture loss in the upper sediments or soils under extremely arid climatic conditions. The contribution of seawater, according to them, is limited to a narrow zone near the shoreline of the coastal sabkha. The so-called sabkha cycle or sabkha-sequence, which became so popular during the sixties and seventies of the last century to interpret old carbonate-evaporite succession in the geologic record (e.g., Wood and Wolfe 1969) amalgamates the continental sabkha (or the supratidal zone) with the marine intertidal and subtidal sediments. Although some workers divided sabkhas into continental and coastal, the mechanism remains the same (Fig. 2.25).

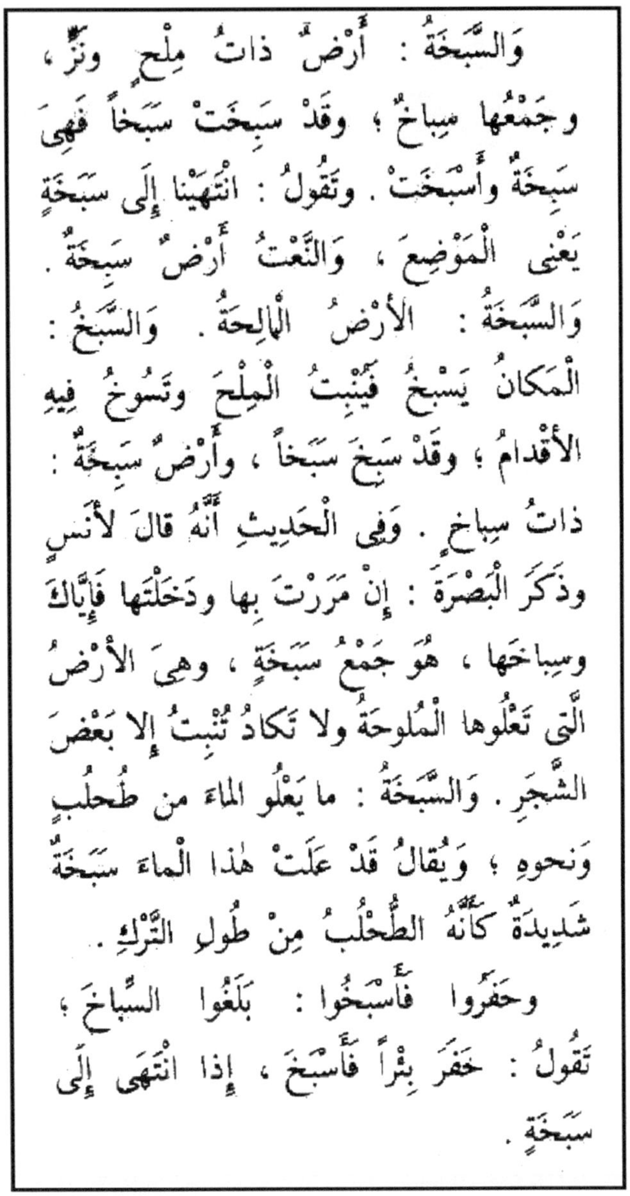
وَالسَّبَخَةُ : أَرْضٌ ذاتُ مِلْحٍ ونَزٍّ ،
وجَمْعُها سِباخٌ ؛ وقَدْ سَبِخَتْ سَبَخاً فَهِي
سَبِخَةٌ وأَسْبَخَتْ . وتَقُولُ : انْتَهَيْنا إِلَى سَبَخَةٍ
يَعْنِي الْمَوْضِعَ ، وَالنَّعْتُ أَرْضٌ سَبِخَةٌ .
وَالسَّبَخَةُ : الأَرْضُ الْمالِحَةُ . وَالسَّبَخُ :
الْمَكانُ يَسْبَخُ فَيُنْبِتُ الْمِلْحَ وتَسُوخُ فِيهِ
الأَقْدامُ ؛ وقَدْ سَبِخَ سَبَخاً ، وأَرْضٌ سَبِخَةٌ :
ذاتُ سِباخٍ . وَفِي الْحَدِيثِ أَنَّهُ قالَ لأَنَسٍ
وذَكَرَ الْبَصْرَةَ : إِنْ مَرَرْتَ بِها ودَخَلْتَها فَإِيَّاكَ
وسِباخَها ، هُوَ جَمْعُ سَبَخَةٍ ، وهِيَ الأَرْضُ
الَّتِي تَعْلُوها الْمُلُوحَةُ ولا تَكادُ تُنْبِتُ إِلا بَعْضَ
الشَّجَرِ . وَالسَّبَخَةُ : ما يَعْلُو الماءَ مِنْ طُحْلُبٍ
ونَحْوِهِ ؛ وَيُقالُ قَدْ عَلَتْ هٰذا الْماءَ سَبَخَةٌ
شَدِيدَةٌ كَأَنَّهُ الطُّحْلُبُ مِنْ طُولِ التَّرْكِ .
وحَفَرُوا فَأَسْبَخُوا : بَلَغُوا السِّباخَ ؛
تَقُولُ : حَفَرَ بِئْراً فَأَسْبَخَ ، إِذا انْتَهَى إِلَى
سَبَخَةٍ .

Fig. 2.24 Definition of the Arabic word "sabkha" as appeared in the dictionary from Lisan Al-Arab Dictionary

Works by Warren (1996) on the coastal sabkha of Abu Dhabi and Al-Yousef et al. (2006) on Qatar indicate that most of the water budget in the sabkhas is derived from shallow groundwater and surface runoff into the inland basin. Evaporite minerals such as gypsum and salt are precipitated from the concentration of such water rather than from seawater. Among these, the milky white, soft anhydrite described by Perthuisot (1977) from the upper portion of the gypsum layers in the vadose zone 5–10 cm below the surface (Puls et al. 2009). He suggested that this anhydrite resulted from the early hydration of gypsum. This process is enhanced by elevated surface temperatures.

The most prominent sabkhas are those found along the coastal areas of the Arabian Gulf. The Matti sabkha in Abu Dhabi was the first sabkha studied by Shearman (1963), Shearman (1966) and Evans et al. (1969). Other sabkhas were documented from south Iraq (Aqrawi and Sadooni 1987; Ghazanfar 2006; Aqrawi and Sadooni 1987; Ghazanfar 2006), Kuwait (Gunatilaka and Mwango 1987), Qatar (Al-Youssef 2015), and Oman (Mettraux et al. 2010).

Ashour et al. (2013) subdivided the sabkhas of Qatar into inland and coastal sabkhas. He reported that both types cover around 7% of the country's surface area. Figure 2.26 shows the aerial distribution of Sabkhas in the Qatar Peninsula. Inland sabkhas are those that are not connected to the sea. He reported that around 78 such inland sabkhas cover an area of around 205 km^2. They covered relatively low areas ranging in elevation between 3 and 4 m. He suggested that these sabkhas occupy two belts; one is around 10 km wide parallel to the shoreline, and the second is a curvilinear belt that extends from Khor Al-Adaid in the east to the southern tip of Salwa Bay in the west at the southern part of the country.

2.2.3.1 Continental Sabkha

This is a common topographic feature in many parts of the Middle East, from which the term sabkha was originally derived. Farmers of the region define sabkha as a relatively

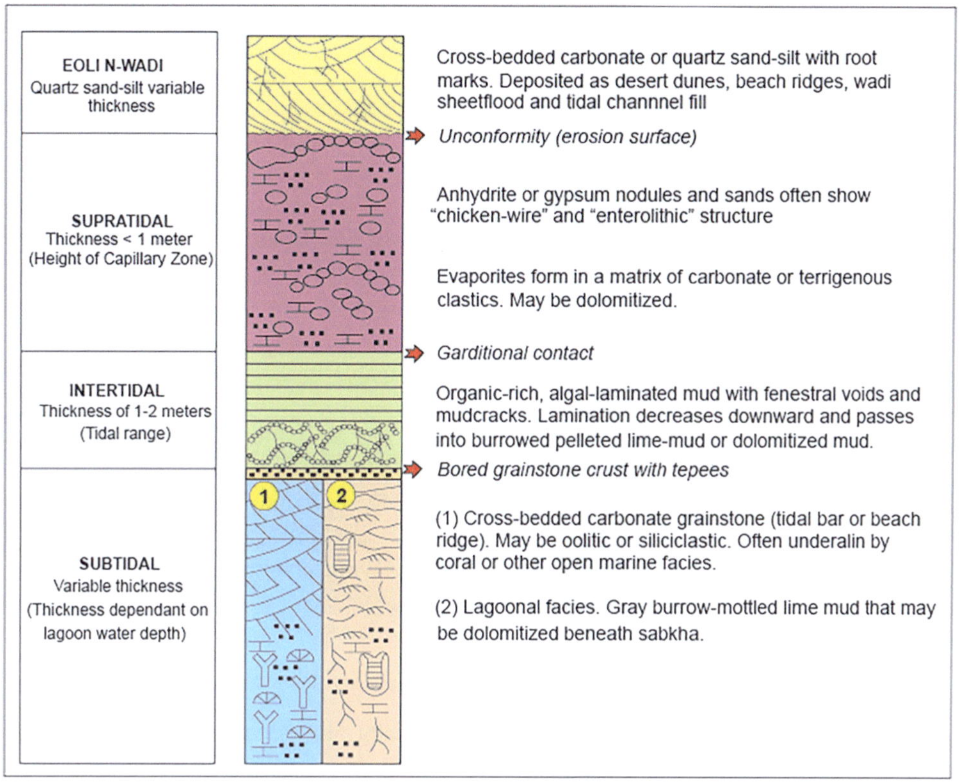

Fig. 2.25 Holocene sequence of coastal sabkha in the Arabian Gulf region, or the so-called sabkha cycle (Warren 1991)

low area with dark and salty soil unsuitable for agriculture. Only scarce halophyte plants can survive in such harsh conditions. Al-Yousef (2003) described an example of such sabkha in Qatar, the Upper Dukhan Sabkha. This sabkha is a continental evaporitic basin (Salina) around 5 m below sea level. The surface sediments are thick crusts of gypsum with salt brine (Fig. 2.27). The eastern part of this sabkha is a salt basin that represents the lowest point of the area. Puls et al. (2009) documented the presence of around 30 cm of gypsum, anhydrite, and halite. The surface sediments are characterized by a concentric pattern with a salt pan in its middle part. This is followed by gypsum or anhydrite crusts and microbial mats. The southern end is covered by thin sheets of sand and small nebkhas.

According to Al-Yousef (2003), the Sabkha depression is filled with eolian sand deposited during the late Pleistocene lowstand, followed by transgressive marine carbonates of the Holocene. Based on chemical analysis of the sabkha waters, she postulated that groundwater and surface water are the primary sources of these evaporites caused by precipitation in the capillary zone.

2.2.3.2 Coastal Sabkha

The coastal sabkha usually fringes complex coastal depositional regimes such as lagoons, barriers, tidal platforms, and shelves. They are influenced by marine estuaries, as in the case of Khor Al-Adaid, where the sabkha sediments are affected by water circulation and sediment dispersal patterns, which are influenced, in turn, by wind regime and topography. Such a complex interplay of these factors generates more diverse sediments. Grain-supported sediments occur on windward beaches, while mud-supported sediments are found in protected subtidal lagoons. Although marine processes are important in sediment production, the arid climate and the extremely high temperature quickly enhance sabkha processes. As stated earlier, "sabkha sediments" include an amalgamation of all these types. Puls et al. (2009) described the following sediment types from the Dukhan coastal sabkha based on texture, grain type and size, and depositional processes. All or some of these facies can be found in other coastal sabkhas. This summary is based on their work.

Topographic Features of the Coastal Sabkha

Exposed Bedrocks

The outcrops neighboring the Dukhan sabkha and other sabkhas in Qatar consist of limestone and dolomite of the Eocene Dammam Formation (see Chap. 9). They form hard, resistant ridges and mesas. In areas where the bedrocks are removed by erosion and weathering, the surface is covered by a thin soil veneer formed mainly of calcareous materials with some silica and iron oxide nodules.

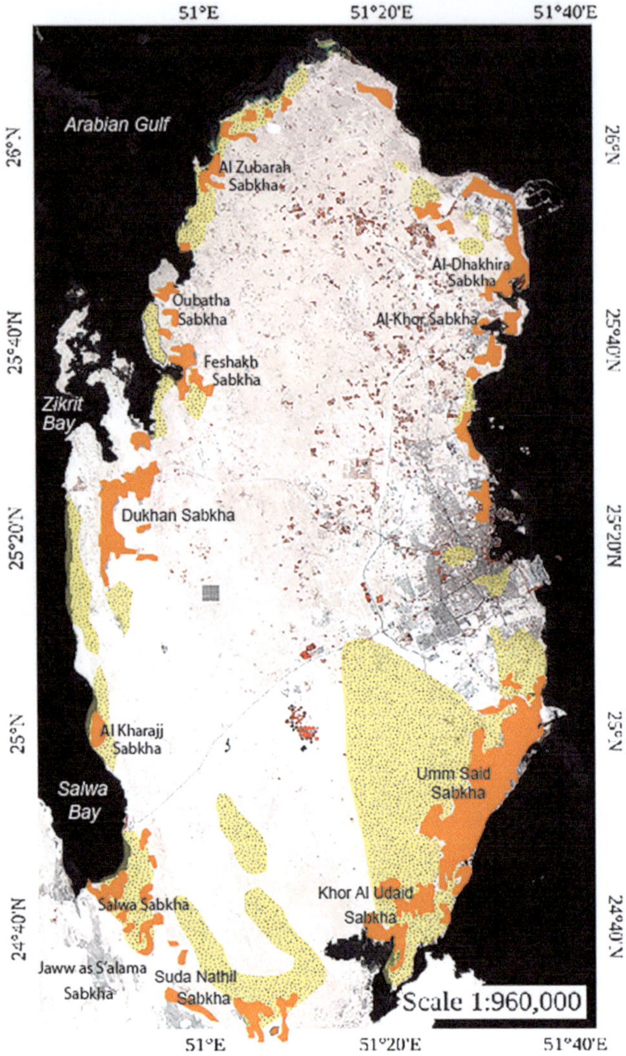

Fig. 2.26 Aerial distribution of the major sand fields and sabkhas in Qatar

Fig. 2.27 Surface evaporitic sediments of Dukhan continental sabkha (Rajendran Sankaran, ESC)

Sand Dunes

Sand dunes are encroaching on the Sabkha edges, covering a considerable part of the southern Upper Sabkha. They consist of fine to medium quartz sand with fewer calcareous sand, forming low and mobile dunes ranging in height between 0.5 and 1 m. The persistent northeasterly wind transported these materials, which are usually moderate to well-sorted. They are intercalated with sparse vegetation, particularly during the wet season (see the sand dune section above in this chapter for more details).

Salina

Salina is a temporary feature that develops mainly during the short wet season. It is formed of sheets of gypsum crust and halite. According to Puls et al. (2009), the most common minerals are anhydrite, halite, and gypsum. These evaporitic sediments are underlain by microbial mats, suggesting that this area was part of the upper intertidal zone when it was under more marine influence.

Gypsum Flat

This is the main sediment type in continental and coastal sabkhas. It consists of a deformed gypsum crust due to different growth stages and diagenesis. This crust is formed by the action of capillary rise over a relatively long and stable period. The thickness of the crust depends on the length of this period. This sediment gypsification prevents surface deflation by wind and makes such topographic features distinctive from the surrounding areas.

Microbial Mat

Microbial mats are the most spectacular topographic feature of the sabkha areas. They are black or dark brown layers, several centimeters thick, of bacteria (cyano and aerobic) with algae and lichen and may be other microorganisms. They are characteristic features of the upper intertidal zone, where they are subjected to a wet period during the high tide and an exposure period during the low tide. Their extension to the lower intertidal zone is limited by the feeding of grazing gastropods. Under long exposure periods, microbial mats desiccate into polygonal structures. This process will include the precipitation of gypsum and halite due to extensive evaporation under arid climate conditions. They may have different textures and structures, but this subject is beyond the scope of this work. For more details, see Rivers et al. (2020).

Tidal Channels and Bars

Qatar's coastal areas are characterized by shallow tidal channels where seawater flows through the intertidal zone. Some of these are drowned old fluvial channels or estuaries. They are usually sinuous in pattern and contain light-colored sediments. The tidal range is microtidal, and since the coastal sabkha is broad, the water movement is slow, and the channels are shallow. This leads to the deposition of medium to fine calcarenite sand such as gastropods and other bivalves.

Intertidal Mud and Sand Flats

This wide and flat area comprises the middle to lower tidal range and is punctuated by the tidal channels. Its sediment is mainly silt and mud, but sand-sized grains, mostly calcareous, are present, too. These include skeletal grains of foraminifera, shells, and peloids.

Beach Sediments

The type and thickness of beach sediments in the coastal sabkhas of Qatar may vary depending on the wind direction and the nature of marine organisms that are transported after their demise by waves and currents to the beach area. In the Dukhan coastal sabkha, these deposits are thin and discontinuous. They are formed of fine- to medium-grained sediments formed mainly of foraminifera skeleton. In the northern and eastern coasts of the country, the sediments are composed of shell debris and red algae. Beach rocks extend along the seaward sides of the beaches, and they help in sediment accumulation by baffling the transported sediments.

Marine Hardgrounds

Marine hardgrounds were described for the first time by Shinn (1969) from Zikrit Bay. They are formed of calcarenite cemented at the water–sediment interface where there is little or no disturbance. The discovery of submarine cementation was an important milestone at that time. The process occurs at depths of 1–3 m and leads to the stabilization of carbonate sediments and the formation of 5–20 cm thick lithified surfaces. These surfaces may be continuous or broken due to mechanical stress.

Subtidal Lagoon

These are sheltered depressions below low-tide limits and formed by different mechanisms, including evaporite dissolution, karstification, and tectonism. They are composed of foraminifera-bearing sand and mud. Water depth is less than 3 m in Dukhan Sabkha, where the lagoon is protected from the wind by Hawar Islands. Some subtidal hardgrounds are in the area where the sedimentation rate is low.

Water Chemistry

Al-Youssef (2015) conducted detailed chemical analyses for 22 water samples collected from the Dukhan Sabkha and one sample of normal seawater in her attempt to understand the formation of gypsum in the studied sabkha (Table 2.2, Fig. 2.28). The mean pH of the brine samples is 6.8, with a range of 6.4–7.3. In station D27, an artificial pit located in northern Sabkha yielded the highest value. At 30 and 50 cm depths, stations D6 and D16 in the northeastern extension of the sabkha recorded the lowest values. The mean for Na^+ is 38.6 ppt (range: 11.3, 91.1 ppt), and the mean for Cl^- is 62.1 ppt (range 20.2–146.8 ppt). The highest Na^+ and Cl^- concentrations were detected in a sample from station D27. The lowest concentrations of the two elements were found in a sample from station D21 near the eastern margin of the middle sabkha (at a depth of 50 cm). She concluded that the proportion of Na^+ and Cl^- decreases gradually from north to south and is primarily concentrated in the northeastern portion of the sabkha. The mean for Ca^{2+} is 1.5 ppt (range: 0.2–1.5 ppt) and 6.6 ppt for SO_4^{2-} (range 2.8–12.5 ppt). In samples of shallow subsurface brine (at depths of 0.5 and 1 m) from the southeastern and western parts of the Sabkha, Ca^{2+} concentrations were highest. In the southeast portion of the sabkha, at a depth of approximately 90 cm, the lowest Ca^{2+} concentration was measured. The highest concentration of SO_4^{2-} was found in brine from station D27, while the lowest concentration was found in a sample from station D48 in the southern portion of the sabkha at a depth of approximately 1 cm.

The mean Mg^{2+} level is 3.8 ppt (1.3–9.7 ppt). The highest proportion was found in the northern portion of the sabkha, while the lowest was found in a sample from station D14 in the southern portion at a depth of approximately 60 cm. In general, Mg^{2+} concentration is greatest in the northeastern portion of the sabkha and decreases as one travels south. The average level of Sr^{2+} is 0.03 ppt (range from 0.02 to 0.07 ppt). There is no difference in the mean Sr^{2+} concentration between the selected samples from the sabkha. The northeastern portion of the sabkha contains the highest concentration of sabkha brines. The mean HCO_3^{2-} concentration in brines is 0.1 ppt (range 0.01–0.17 ppt). Within the sabkha, the concentration of HCO_3^{2-} increases to the north east and south. The average TDS value is 112.8 ppt (41–261.1 ppt). Samples from station D27 in the northern portion of the sabkha contained the highest proportion (at a depth of about 50 cm). At a depth of approximately 50 cm, station D21 in the southern portion of the sabkha recorded the lowest proportion. The concentration of TDS is greatest in the northeastern portion of the sabkha and decreases toward the south. The statistical analysis revealed a strong relationship between Na^+ and Cl^- and Cl^- and Mg^{2+} in Dukhan Sabkha brines, with R^2 values of 0.97 and 0.96, respectively. Ca^{2+} and SO_4^{2-}, Ca^{2+} and Cl^-, Cl^- and Ca^{2+}, and Mg^{2+} and Ca^{2+} have no apparent relationship, but SO_4^{2-} has a relatively strong relationship with Mg^{2+} ($R^2 = 0.63$). She concluded that the sabkha water is derived mainly from groundwater and rainwater seepage.

Table 2.2 Chemical analyses of 22 samples collected from Dukhan Sabkha (after Al-Yousef 2015)

Location													
Station	Longitude	Latitude	Dep (Cm)	pH	Cond.	Na^+ (ppt.)	Cl^- (ppt)	Ca^{2+} (ppt)	SO_4^{2-} (ppt)	Mg^{2+} (ppt)	Sr^{2+} (ppt)	HCO_3^{2-} (ppt)	TDS (ppt)
D1	N = 25° 25′ 48″	E = 50° 49′ 97″	110	7.1	121.5	33.4	55.8	1.6	9.4	3.6	0.03	0.08	103.9
D5	N = 25° 23′ 76″	E = 50° 49′ 88″	100	6.5	131.3	36.5	58.6	2	4.7	3.1	0.04	0.07	104.9
D6	N = 25° 23′ 56‴	E = 50° 50′ 70″	30	6.4	155.8	47	82	1.6	8.1	4.4	0.04	0.07	143.2
D7	N = 25° 23′ 57″	E = 50° 51′ 48″	30	6.6	127.2	34.9	58.6	1.6	7.4	3.8	0.03	0.07	106.5
D8	N = 25° 23′ 99″	E = 50° 51′ 51″	25	6.9	167.9	53.3	84.1	1.6	8.3	5.4	0.03	0.08	152.7
D10	N = 25° 24′ 78″	E = 50° 51′ 16″	80	6.8	138.1	39.7	60.7	1.7	5.2	3.9	0.03	0.08	111.3
D11	N = 25° 25′ 20″	E = 50° 51′ 10″	120	6.9	131.4	36.9	58	1.8	6.5	3.6	0.03	0.04	106.9
D12	N = 25° 25′ 33″	E = 50° 50′ 58″	60	6.8	151.4	46.5	67.3	1.7	6.9	4.6	0.04	0.06	127.1
D14	N = 25° 15′ 85″	E = 50° 49′ 58″	60	7.2	58.8	12.3	22.6	1.6	3.3	1.3	0.04	0.14	41.1
D16	N = 25° 24′ 65″	E = 50° 53′ 40″	50	6.4	192.8	67.5	99	1.4	6.9	6.3	0.03	0.06	181.3
D17	N = 25° 26′ 69″	E = 50° 53′ 69″	50	7.3	141.8	46.3	69.4	1.7	7.7	4.6	0.03	0.13	129.7
D20	N = 25° 22′ 73″	E = 50° 49′ 86″	120	6.9	139.9	39.9	57.8	1.6	5.3	4	0.03	0.1	108.7
D21	N = 25° 21′ 26″	E = 50° 50′ 48″	50	7.1	55.7	11.3	20.2	1.4	6.5	1.5	0.03	0.11	41
D22	N = 25°18′ 67″	E = 50° 51′ 90″	90	7.2	81.3	19	31.8	0.8	3.9	2.3	0.04	0.08	57.9
D23	N = 5° 18′ 95″	E = 50° 51′ 38″	50	6.7	148.8	44.5	67.1	1.5	7.7	3.8	0.03	0.11	124.7
D24	N = 25° 19′ 07″	E = 50° 50′ 78″	60	6.8	129.1	36.1	54.7	1.6	5.6	3.1	0.04	0.08	101.8

(continued)

Table 2.2 (continued)

Location													
Station	Longitude	Latitude	Dep (Cm)	pH	Cond.	Na^+ (ppt.)	Cl^- (ppt)	Ca^{2+} (ppt)	SO_4^{2-} (ppt)	Mg^{2+} (ppt)	Sr^{2+} (ppt)	HCO_3^{2-} (ppt)	TDS (ppt)
D33	N = 25° 23′ 62″	E = 50° 50′ 93″	30	6.6	185.1	61.5	112.9	1.4	10.8	6.2	0.03	0.06	192.9
D34	N = 25° 21′ 28″	E = 50° 50′ 58″	80	6.7	145.7	42.7	74.3	1.6	8.7	4.3	0.04	0.07	131.7
D37	N = 25° 23′ 72″	E = 50° 49′ 69″	50	6.8	128.1	36	57.8	1.1	6.6	3.7	0.02	0.1	105.4
D38	N = 25° 24′ 58″	E = 50° 52′ 64″	50	6.8	143.1	41	70.9	1.2	5.8	4.2	0.02	0.1	123.2
D48	N = 25° 17′ 11″	E = 50° 49′ 44″	50	7.2	98.9	25.2	40.7	2	2.8	2.1	0.05	0.01	72.8
D27	N = 25° 25′ 40″	E = 50° 51′ 06″	Art. Pit	7.2	220	91.1	146.8	0.9	12.5	9.7	0.07	0.17	261.1
Mean				6.8	132.1	38.6	62.1	1.5	6.6	3.8	0.03	0.1	112.8
Standard seawater in grams/liter				–	–	10.56	18.98	0.40	2.65	1.27	0.008	–	34.3

Fig. 2.28 Sample locations in Dukhan Sabkha, Al-Youssef (2015)

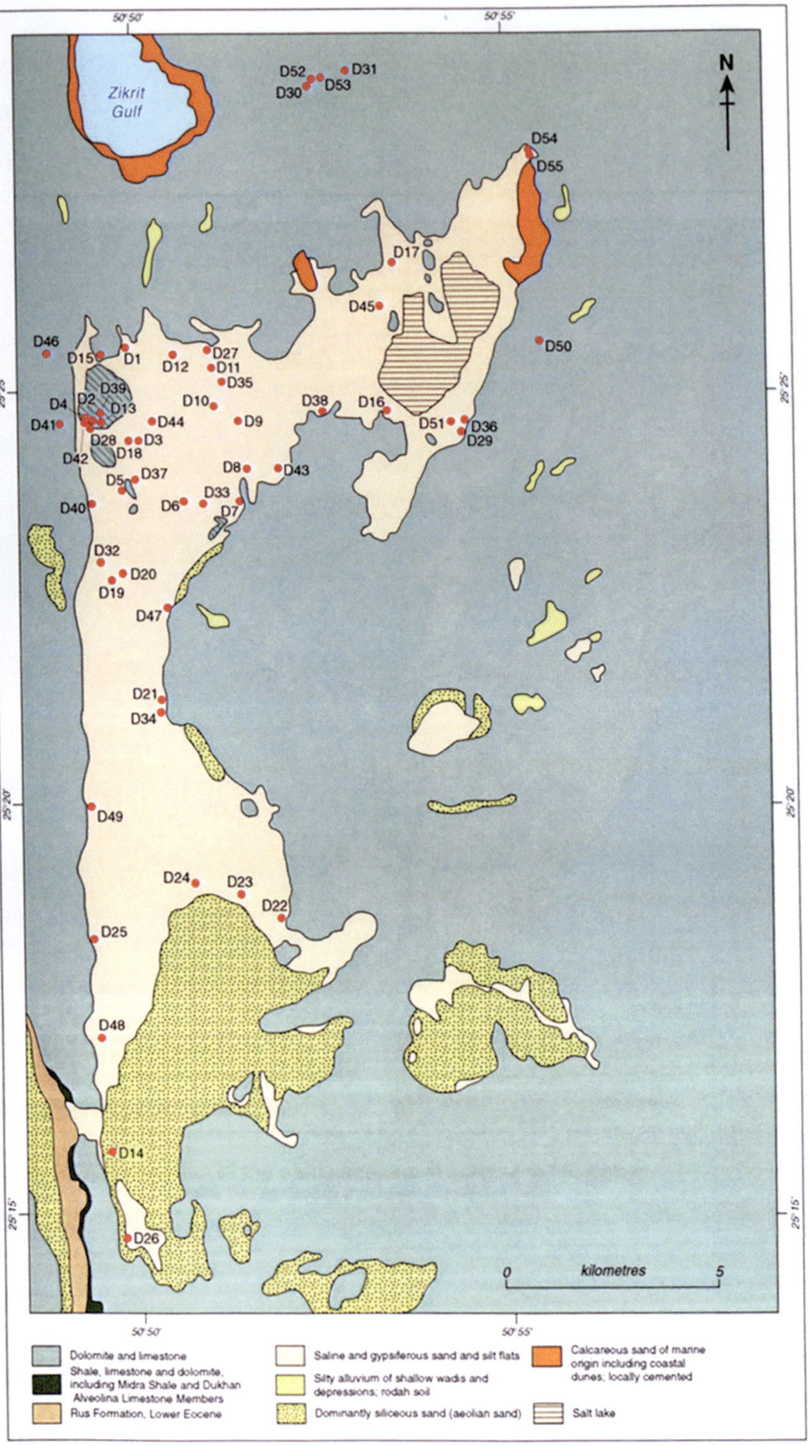

Sediment Analyses

Ashour et al. (2013) conducted X-ray diffraction analyses on sediments smaller than 0.063 mm in 69 samples representing 14 profiles in the four primary sabkhas: Dukhan, Suda Nethil, Umm Said, and Khor Al-Adaid. He concluded the following (Table 2.3):

- *Quartz* is the predominant mineral in all sections and horizons, except for the uppermost horizon. This is because the mineral is abundant in both solid rocks and surface deposits in the region. Gypsum, anhydrite, and halite are common, especially in the uppermost horizons. The limestone rocks that comprise a large portion of Qatar's geological section or the sulfur- and chloride-rich underground water could be the source of these minerals.

- *Feldspars* are abundant in all horizons because of the weathering of sandy and calcareous deposits containing the mineral.
- *Magnetite* is occasionally discovered in specific locations due to chemical weathering processes.
- The presence of *goethite* in some horizons of a few profiles indicates that magnetite was hydrated under humid conditions.

Sediment Geochemistry

Ashour et al. (2013) suggested that the major chemical elements forming the sabkha sediments in Qatar include the following (Table 2.4):

- *Silica* SiO_2 varies between 18 and 79%, is highly concentrated in the lower horizons of all profiles, and is less concentrated in the upper horizons. There is a wide variation in the silica concentration; this could be attributed to the irregular wind regime during deposition.
- *Aluminum* in the form of Al_2O_3 ranges between 0.56 and 6.55%. Ashour et al. (2013) suggested a cumulative correlation between the percentages of silica and aluminum, irrespective of the sabkha type or sample depth. Sabkhas in the shadow of sand dunes or sand sheets, such as the Umm Said in the southwest, contain a higher proportion of aluminum.
- *Calcium Carbonates*, $CaCO_3$, have a concentration ranging from 6 to 18.2%. The source of calcium carbonates is mainly the Tertiary carbonate cover that is exposed all over the country. Ashour et al. (2013) noticed that the $CaCO_3$ percentage is higher in coastal sabkhas and increases downwards in most studied profiles.
- *Sulfates* SO_4: Sulfate is formed by evaporation of sulfate-rich water; it is considered the initial stage, whereas sulfide is a late stage due to reduction factors. The most significant of such factors are the decomposition of organic organisms and the presence of non-gaseous bacteria. These factors, among others, result in the formation of hydrogen sulfide. There are two primary sources of sulfates in the studied sabkhas: the gypsum found in sediments of the Rus Formation and seawater that usually contains between 3 and 6 g of sulfates per liter (Ashour et al. 2013). Gypsum and anhydrite are the most abundant sulfate minerals in the sabkha sediments. The average annual ratio for Dukhan Sabkha (averaged from 39 samples in five profiles) is approximately 14%. It is also observed that the percentage of sulfate increases with calcium to form calcium sulfates or gypsum and decreases with silica. Sulfates in sabkhas are less prevalent beneath the underground table. The high percentage of sulfates in inner sabkhas compared to coastal sabkhas may indicate a rich source of sulfates that impacts inner sabkhas. This fact emphasizes the significance of groundwater in the formation of evaporites in the inner sabkhas (Ashour et al. 2013).

Chlorides: According to Ashour et al. (2013), the percentage of chlorides varies greatly from one sector to the next and from one sabkha to the next, indicating that these salts are significantly influenced by climate factors since they require long-term dryness.

The average of chloride in inner sabkhas and coastal sabkhas is as follows:

- Dukhan (39 sample) 4.4%.
- Sauda-Nithil (8 samples) 6.2%.

Table 2.3 Major minerals in the different sabkhas of Qatar (after Ashour, 2013)

Sabkha		Quartz	Halite	Gypsum	Feldspar	Calcite	Dolomite	Anhydrite	Other
Dukhan	s	24	42	27	06	01	00	01	00
	m	8	04	60	27	01	00	00	00
	b	00	16	42	23	05	tr	00	00
Sauda Nethil	s	48	08	00	01	07	03	33	00
	m	19	05	42	29	01	02	00	tr
	b	60	09	06	02	14	02	01	00
Um Said	s	82	02	00	07	04	01	00	00
	m	54	01	00	35	07	00	00	00
	b	67	tr	00	24	06	01	00	00
Khor Al-Adaid	s	34	8	tr	17	04	00	00	00
	m	47	04	00	34	11	01	00	00
	b	70	03	00	14	10	00	00	00

s = Surface, m = Middle, b = Bottom

Table 2.4 Major chemical elements of the sabkha sediments in Qatar (Ashour et al. 2013)

Elements/Sabkhas		SiO_2	Al_2O_3	Fe_2O_3	MnO	CaO	MgO	Na_2O	K_2O	CO_2	SO_3	Cl	H_2O	Depth
Dukhan	s	24.88	2.03	0.51	0.01	19.27	4.95	13.08	0.12	1.51	10.49	21.67	1.87	2 cm
	m	41.49	2.39	0.51	0.01	15.85	4.13	1.28	0.12	1.39	21.48	2.24	7.34	0.26
	b	61.85	3.50	0.90	0.01	7.81	2.42	1.49	0.26	5.18	5.39	3.09	6.48	1.04
Sauda Nethil	s	79.39	1.16	0.43	0.01	3.97	0.50	0.44	0.35	5.83	0.80	0.82	4.51	0.05
	m	68.26	4.00	1.09	0.02	7.37	0.75	0.31	0.41	9.86	0.80	0.70	5.04	16.50
	b	61.06	2.08	1.28	0.02	10.97	1.31	2.44	0.33	12.29	1.60	4.84	2.06	40.00
Khor Al-Adaid	s	37.99	3.81	1.31	0.02	16.56	1.93	2.36	0.35	10.19	14.29	4.77	6.01	0.15
	m	64.91	4.88	1.22	0.02	7.94	1.29	0.98	0.46	9.08	1.30	2.18	2.11	14.00
	b	68.88	4.57	1.05	0.01	7.56	1.00	0.87	0.45	8.27	1.30	1.92	287	53.00

s = Surface, m = Middle, b = Bottom

- Umm Said (23 samples) 7.8%.
- Khor Al-Adaid(4 samples) 2.7%.

Magnesium: The percentage of magnesium in inland sabkhas is greater than in coastal sabkhas. The lower percentage of magnesium in Umm Said Sabkha is due to the slow formation of dolomite, according to De Groot (1973).

The average of magnesium oxide is as follows:

- Dukhan 3.3%.
- Sauda-Nithil 2.4%.
- Umm Said 1.2%.
- Khor Al-Adaid 1.2%.

Fig. 2.29 Halite crust on Dukhan sabkha during summertime

2.2.3.3 Main Sabkhas in Qatar

The following is a short description of the major sabkhas of Qatar. This is based on the works of Ashour (1991), Ashour (1987), Ashour et al. (2013), Al-Yousef et al. (2006), Illing et al. (1965), Jameson et al. (2009), Perthuisot (1977), Puls et al. (2009), Shinn (1973a, b), Perthuisot and Jauzein (1978), Al-Youssef (2015), Engel et al. (2022), Nasir et al. (1999), Rajendran et al. (2021), Rivers et al. (2020).

Dukhan Sabkha

Dukhan sabkha is the largest inland sabkha and salina close to the coastal area. The area is a depression with an upturned L shape neighboring the Dukhan anticline. The length of its north–south arm is around 20 km, and its width is 3 km in the north, 2 km in the middle, and 4 km in the south, being reduced to around 1 km at the southern end. The length of the second arm is 8.5 km with a width of 4 km. The total area is around 73 km^2 with a width that ranges between 2 and 4 km. The sabkha surface is located around 1 m below sea level. The area is surrounded by hills and mesa structures of dolomitic limestone of the Upper Dammam Formation. The eastern side of the sabkha is linear and overlooks the sabkha on a steep slope. The northern side ranges in elevation between 3 and 12 m punctuated with some Tertiary rocks hills up to 27 m; Ashour (1987) suggested that the present-day Dukhan sabkha was part of the Zikrit Bay and became exposed after the drop in sea level 3000 years ago. The groundwater table is between 2 and 120 cm below the surface. Sabkha is considered a deflation surface where its surface features are impacted by the presence of evaporites in the surface sediments, the water table, and the wind action. During the dry season, the sabkha surface may develop shrinkage cracks, nebkhas (sand accumulations baffled by shrubs or plants), channels, and fissures. The Salina part of the sabkha is in the south, around 6 m below sea level (Al-Youssef 2015; Ashour 1987).

During summertime, the sabkha surface is covered with a crust of halite (3 mm–2 cm) that overlain a sand deposit usually intercalated with gypsum (Fig. 2.29). A gypsum bed up to 2 m thick is found in the central part of the sabkha, while the marginal part is covered with sand and silt of eolian and fluvial origin. According to (Al-Youssef 2015), anhydrite and halite are present in the northeastern part of the Sabkha area. Nodular anhydrite is around 20 cm deep and mixed with fine gypsum crystals, halite, and microbial mats. The Sabkha area is encroached by eolian sand accumulation in the western fringe.

Puls et al. (2009) divided the Dukhan sabkha into the Upper (inland) and the Lower (coastal) sabkha. They suggested that the sabkha stretches for around 50 km along the eastern flank of the Dukhan anticline and is transected by a series of east–west structural lineaments. These are probably associated with deeper faults generated during the development of the Dukhan structure. These faults brought the Eocene strata to the surface, thus generating a sill that prevented seawater circulation between the lower coastal and the upper (supratidal) sabkha. In some locations, the sill is around 1.45 m above the mean sea level.

Sawda Nathil Sabkha

This is inland sabkha situated on the southwestern borders of Qatar with Saudi Arabia. It is formed of a semi-closed depression of around 22.5 km^2 that is aligned in the NW–SE direction and is bordered by Tertiary hills formed of limestone and dolomite in the north and sand dunes in the other directions. The surface elevation ranges between -1 and $+$ 1 m. The sediment of this sabkha has more sand than evaporites compared to other sabkhas in Qatar. Ashour et al. (2013) suggested that this sabkha may be connected with Salwa Bay before the last sea level drop.

Strohmenger and Jameson (2018) suggested that Sawda Nathil sabkha is one of a series of depressions extending from Umm Bab along the Saudi Arabia–Qatar border to Khor Al-Adaid. Before six to eight thousand years, these depressions were embayments filled with marine water and separated the Qatari peninsula from Saudi Arabia. These embayments were quickly filled with dune sands carried southward from Qatar. Since then, they have gradually become more evaporitic. In Qatar, inland depressions such as Sawda Nathil represent spectacular environments (Fig. 2.30). Many depressions are near or below sea level. This brought the groundwater table nearer to the surface. Extensive evaporation led to salt saturation, producing thick gypsum and salt crusts (sabkhas) and shallow hypersaline ponds (salinas) with unique microbial gypsum stromatolites. The current surface of the ground is a combination of relict marine facies, deflated dune sands, inland sabkhas, and salinas.

Radiocarbon dating for marine shells and the sediments enclosing them was conducted by Strohmenger and Jameson (2018) revealed an approximate age of 6000 years before the present (yr BP). This coincides with a well-documented Highstand phase in Qatar when the sea level was 2–4 m higher than the present. These beach deposits are probably remnants of the post-glacial Holocene transgression and predecessor of the Sabkha regime. The sabkhas (gypsum and salt flats) and salinas (saline ponds) are younger and resulted from the infilling of the embayments from the time of the sea level drop to the present day.

Umm Saeed Sabkha (Mesaieed)

Umm Saeed sabkha is located on the southeastern coast of Qatar, covering an area of 350 km^2 or around 60% of the total sabkha areas of the country. The northern portion of Qatar's largest coastal sabkha spans about 60 km south toward the Saudi Arabian border. According to Ashour et al. (2013), Umm Saeed sabkha is an example of a coastal sabkha that forms in the shadow of sand dunes, implying that the major budget of the sediment in such sabkha is derived from the sand dunes. Jameson et al. (2009) and Slowakiewicz et al. (2014) suggested that this sabkha is located in the transition between leeward and oblique coastline in relation to the prevalent Shamal wind. Hence, these

Fig. 2.30 A general view of the inland sabkha of Suda Nathil

Fig. 2.31 Microbial mats bordering clastic-dominated supratidal zone in Umm Saeed Sabkha

winds are the primary source of the siliciclastic sediments that comprise most of the sabkha. Ashour et al. (2013) reported that the total thickness of the sediment there is around 30 m and is formed mostly of quartz and feldspar, with debris from dolomitic limestone and some evaporite minerals. Slowakiewicz et al. (2014) suggested that siliciclastic sediments comprise most of the sabkha, but detrital carbonate, shell material, and pedogenic grain coatings are also present. The sabkha surface is flat, ranging in elevation from zero at the coast up to 3 m on the west (Fig. 2.31).

Fishakh and Hussain Sabkhas

This sabkha is located on the western coast of the country near Fishakh Bay. Its total area is around 18.5 km^2. The surface of the sabkha is flat, ranging in elevation between zero and 2 m on the landward part. It is surrounded by a 0.5 m cliff, which may represent the old coastline (Ashour et al. 2013; Embabi and Ali 1990).

Illing and Taylor (1993) studied both sabkhas. They noticed that each sabkha is a flat supratidal expanse that extends laterally via algal mats near high tide levels into the intertidal zone bordering the shallow lagoons that connect to the shallow sea region west of Qatar (Fig. 2.32). The average tidal range in the lagoons is less than half a meter, and water movement is frequently impacted equally by wind and tide. The floor of the lagoons is composed of lime mud with some soft pellets and contains small lamellibranchs, gastropods, and *Peneroplid* foraminifera. The intertidal zones are dominated by the pellets of borrowing gastropods. Fine, translucent gypsum crystals precipitated from concentrating lagoon water are prevalent in the upper intertidal layer. They are covered by and interwoven with dark green and black microbial mat layers. These are up to 300 m wide and 30 cm thick in Faishakh. The microbial mats exhibit a range of growth morphologies, including irregularly bulbous shapes, ridges, and thin, flaky sheets, which are typical of various regions of the flats. The microbial flats span the normal range of high tide levels so that their upper portion is relatively dry and blends imperceptibly into the 3 km wide supratidal sabkha in Faishakh.

The Faishakh sabkha has progressed into the lagoon, with microbial flats expanding across the advancing intertidal belt and being covered by supratidal sabkha sediment. This ranges from muddy sand to sandy mud, with no discernible stratification beyond thin, weakly cemented layers and the remnants of buried microbial mats. Its brownish hue extends to the water table, which is approximately 30 cm below the surface in most of the sabkha. A blue-gray hue and strong hydrogen sulfide odor indicate reducing conditions below this level. Fine gypsum crystals, like those at the top of the intertidal zone, precipitate in elongated swaths where local shallow depressions in the sabkha surface allow the occasional influx of wind-driven water from the lagoon at high tide. These tiny clear euhedral gypsum crystals are elongated and about a centimeter in length. Underneath the surface of the majority of sabkha and increasing landward. Typically, these crystals are greater than 15 cm in diameter and contain carbonate inclusions. They are typically found in brown, structureless carbonate mud, which X-rays and SEM have revealed to be cryptocrystalline to microcrystalline dolomite (Illing and Taylor 1993; Illing et al. 1965).

Fig. 2.32 A general view of the intertidal zone of the Feshakh Sabkha (Zulfa Al-Disi)

Khor Al-Adaid Sabkha

The sabkha forms only one component of south Qatar's diverse and complicated depositional regime. Khor Al-Adaidis a tide-dominated estuary whose subtidal part has been divided into three parts: (1) Inner Lagoon, which has an irregular shape with an area of 73.5 km^2 and called "The Inner Sea." (2) Outer Lagoon, a smaller, elliptical shape in shape with an area of 53.5 km^2. This lagoon is separated from the Inner Lagoon by a sandy ridge probably of eolian origin (3) Entrance Channel, a narrow passage, 15 km long and 3 km wide, connecting the Outer Lagoon with the Arabian Gulf waters. Rivers et al. (2020) provided a comprehensive review of these environments.

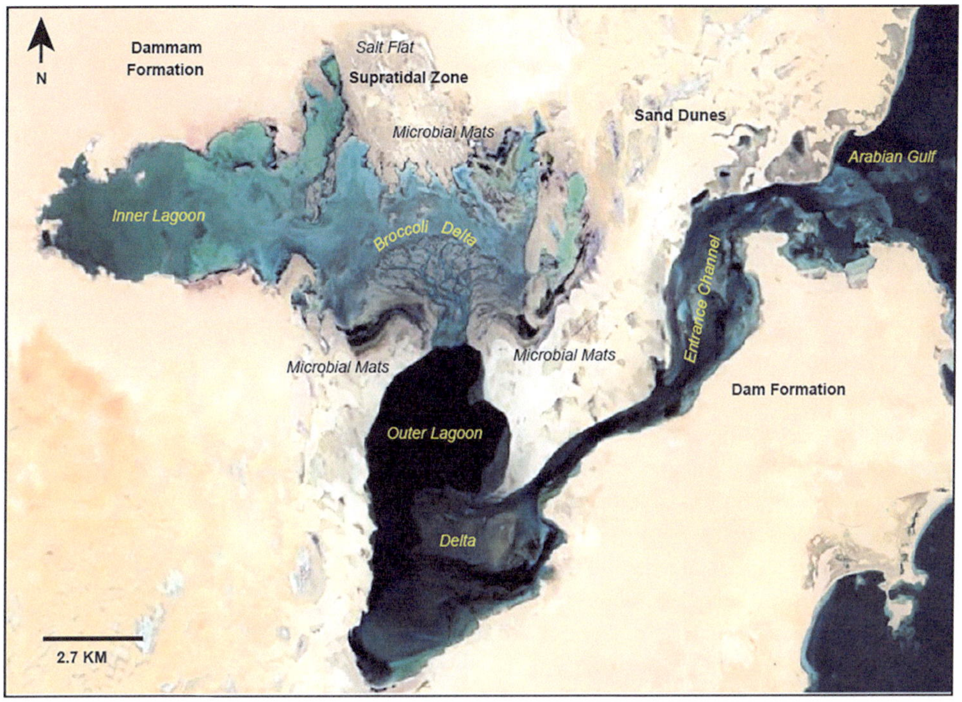

Fig. 2.33 Main depositional regimes of Khor Al-Adaid (Markings modified from Rivers et al. 2020)

Fig. 2.34 Different types of microbial mats at Khor Al-Adaid. **a** The usual blistered mats. **b** Polygonal mats. **c** Destroyed microbial mats by grazing gastropods. **d** Cat-ear microbial mat

The southern shoreline of the connecting channel is bordered by the Miocene Dam Formation outcrops. A flood-tidal delta, the Broccoli Delta, can be easily seen in the area's satellite images and also at the mouth of the Entrance Channel (Fig. 2.33). The Inner Lagoon is bordered by the Eocene Dammam Formation outcrops. Well-developed dune fields encroach from north, east, and west. The Khor represents a rare combination of marine and eolian systems working together.

Where there are no delimiting bedrocks or sand dunes, an intertidal zone develops around these water bodies. This area has well-developed microbial mats (black areas in Fig. 2.34). Microbial mats have developed different shapes and structures depending on the nature of the present sediments. In areas closer to the Supratidal Zone, the mat cover is only a few centimeters thick and has a blistered shape. In the transitional areas between the intertidal and the subtidal lagoons, the mats become decimeter-thick, and closer to the subtidal zone, they are polygonal. There are also rare "cat-ear" mats closer to the salt flat (Fig. 2.34). These are probably stressed types and may result from the extreme temperature and salinity or the influx of clastic materials from the neighboring areas. In most Khor areas, the microbial mats are surrounded by sand sheets or dunes, except in the northern part, where the intertidal zone gives way to a relatively wide and flat salt plane of the Supratidal Zone (Fig. 2.33).

References

Abul-Haggag A (1965) A recent karstic phenomenon in the limestone country of north central Najd, Saudi Arabia. Bull Soc Géog Egypte 38:73–80

Al-Saad H, Nasir S, Sadooni F, Alsharhan AS (2002) Stratigraphy and sedimentology of the Hofuf Formation in the State of Qatar in relation to the tectonic evolution of the East Arabian Block. Monatshefte 7. https://doi.org/10.1127/njgpm/2002/2002/426

Al-Sayari SS, Zötl JG (1978) Quaternary period in Saudi Arabia, 1, sedimentological, hydrogeological, hydrochemical, geomorphological and climatological investigations. In: Central and Eastern Saudi Arabia. Springer, New York

Al-Yousef MM (2003) Mineralogy, geochemistry and origin of Quaternary sabkhas in the Qatar Peninsula, Arabian Gulf. University of Southampton, UK

Al-Yousef M, Stow DAV, West IM (2006) Salt lake area, northeastern part of Dukhan Sabkha, Qatar. In: Khan MA, Böer B, Kust GS, Barth H-J (eds) Sabkha ecosystems: volume II: West and Central Asia. Springer Netherlands, pp 163–181. https://doi.org/10.1007/978-1-4020-5072-5_13

Al-Youssef M (2015) Gypsum crystals formation and habits, Dukhan Sabkha, Qatar. J Earth Sci Climatic Change 6(10):12. https://doi.org/10.4172/2157-7617.1000321

Aqrawi A, Sadooni F (1987) Recent tidal flat sediments of Khor Al-zubair, NW Arabian gulf, Iraq. J Water Resour 6(2):18–37

Ashour M (1985) Textural properties of Qatar dune-sands. J Arid Environ 8(1):1–14

Ashour MM (1987) Surficial deposits of Qatar Peninsula. Geol Soc London Spec Publ 35:361–367

Ashour M (1991) The sabkhas in the Peninsula of Qatar: biogeological and geomorphology study

Ashour MM, Mogheeth S, Metwali A, Al-Ghazali G, Ghafour S, Shaksipi R, Ali A (2013) Sabkhas in Qatar peninsula. Landscape Geodiversity 1:10–35

Bischoff JL, Juliá R, Shanks WC III, Rosenbauer RJ (1994) Karstification without carbonic acid: bedrock dissolution by gypsum-driven dedolomitization. Geology 22(11):995–998. https://doi.org/10.1130/0091-7613(1994)022%3c0995:Kwcabd%3e2.3.Co;2

Briggs J (1983) Heavy mineralogy of the Qatar desert sands. Unpublished report. Qatar Ministry of Public Works

Butzer KW (1961) Climatic changes in arid regions since the Pliocene, a history of land use in arid regions. In: Proceedings arid zone research, Paris

Cavelier C (1970) Geological description of the Qatar Peninsula (Arabian Gulf). Bureau de recherches géologiques et minières. https://books.google.com.qa/books?id=s2xpMgAACAAJ

Curtis CD (1983) Link between aluminum mobility and destruction of secondary porosity1: geologic notes. AAPG Bull 67(3):380–384. https://doi.org/10.1306/03b5ad22-16d1-11d7-8645000102c1865d

De Groot K (1973) Geochemistry of tidal flat brines at Umm Said, SE Qatar, Persian Gulf

Duggan J, Jameson J, Ross R, Strohmenger C (2014) Karst prediction—testing predictions against data, State of Qatar. University of Leeds, Leeds, United Kingdom, October 2014. ExxonMobil & Qatar Center for Coastal Research

Eccleston BL, Pike JG, Harhash I (1981) The water resources of qatar and their development. vol. I. Food and Agricultural Organization of the United Nations

Edgell HS (1990) Evolution of the Rub'al Khali desert. J King Abdel Aziz Univ: Earth Sci Special Issue. First Saudi Symposium on Earth Science, eddah, 1989

Embabi NS, Ali AA (1990) Geomorphology of depressions in the Qatar Peninsula (in Arabic). Al-Ahleia Press, p 357

Embabi N, Ashour M (1993) Barchan dunes in Qatar. J Arid Environ 25(1):49–69

Emil MK, Sultan M, Alakhras K, Sataer G, Gozi S, Al-Marri M, Gebremichael E (2021) Countrywide monitoring of ground deformation using InSAR time series: a case study from Qatar. Remote Sens 13(4):702. https://www.mdpi.com/2072-4292/13/4/702

Engel M, Strohmenger CJ, Peis KT, Pint A, Brill D, Brückner H (2022) High-resolution facies analysis of a coastal sabkha in the eastern Gulf of Salwa (Qatar): a spatio-temporal reconstruction. Sedimentology 69(3):1119–1150. https://doi.org/10.1111/sed.12938

Evans G, Kendall CGSC, Skipwith P (1964) Origin of the coastal flats, the sabkha, of the Trucial Coast, Persian Gulf. Nature 202(4934):759–761. https://doi.org/10.1038/202759a0

Evans G, Schmidt V, Bush P, Nelson H (1969) Stratigraphy and geologic history of The Sabkha, Abu Dhabi, Persian Gulf. Sedimentology 12(1–2):145–159. https://doi.org/10.1111/j.1365-3091.1969.tb00167.x

Garzanti E, Vermeesch P, Andò S, Vezzoli G, Valagussa M, Allen K, Kadi KA, Al-Juboury AIA (2013) Provenance and recycling of Arabian desert sand. Earth-Sci Revi 120:1–19. https://doi.org/10.1016/j.earscirev.2013.01.005

Ghazanfar SA (2006) Skhat regions of Iraq. In: Khan MA, Böer B, Kust GS, Barth H-J (eds) Sabkha ecosystems: volume II: West and Central Asia. Springer Netherlands, pp 211–217. https://doi.org/10.1007/978-1-4020-5072-5_17

Griffin C (1983) A palaeoenvironmental interpretation of Qatar dune sands, using the scanning electron microscope (unpublished report)

Gunatilaka A, Mwango S (1987) Continental sabkha pans and associated nebkhas in southern Kuwait, Arabian Gulf. Geol Soc London Spec Publ 35(1):187–203

https://hdl.handle.net/2027/uc1.31822013353982

Hill CA (2000) Overview of the geologic history of cave development in the Guadalupe Mountains, New Mexico. Journal of Cave and Karst Studies, 62(2):60–71. https://digitalcommons.usf.edu/kip_articles/4175

Howari FM, Aldouri R, Sadiq A (2016) Gravity investigations of recent sinkholes and karst pits of Dahal Al-Hamam, State of Qatar. Environ Earth Sci 75(5):440. https://doi.org/10.1007/s12665-016-5298-x

Illing LV, Taylor JCM (1993) Penecontemporaneous dolomitization in Sabkha Faishakh, Qatar; evidence from changes in the chemistry of the interstitial brines. J Sediment Res 63(6):1042–1048. https://doi.org/10.1306/d4267c8f-2b26-11d7-8648000102c1865d

Illing LV, Wells AJ, Taylor JCM, Pray LC, Murray RC (1965) Penecontemporary dolomite in the Persian Gulf1. In: Dolomitization and limestone diagenesis, vol 13. SEPM Society for Sedimentary Geology. https://doi.org/10.2110/pec.65.07.0089

Jado A, Zötl J (1984) Quaternary period in Saudi Arabia, vol 2. Springer, New York, pp 1–361

James NP, Choquette PW (1984) Diagenesis 9. Limestones—the meteoric diagenetic environment. Geosci Canada 11(4). https://journals.lib.unb.ca/index.php/GC/article/view/3395

Jameson J, Puls DD, Kozar MG (2009) Holocene sabkha and coastal systems of Qatar: process models for the interpretation of ancient Arabian Plate carbonate evaporite reservoirs. https://doi.org/10.3997/2214-4609-pdb.151.iptc13679

Johnstone T, Wilkinson C (1960) Some geographical aspects of Qatar. Geogr J

Martinez MI, White WB (1999) A laboratory investigation of the relative dissolution rates of the Lirio Limestone and the Isla de Mona Dolomite and implications for cave and karst development on Isla de Mona. J Cave Karst Stud 61(1):7–12

Mettraux M, Homewood PW, Kwarteng AY, Mattner J (2010) Coastal and continental sabkhas of Barr Al Hikman, Sultanate of Oman. In: Quaternary carbonate and evaporite sedimentary facies and their ancient analogues. pp 183–204. https://doi.org/10.1002/9781444392326.ch8

Mukhopadhyay A, Al-Sulaimi J, Al-Awadi E, Al-Ruwaih F (1996) An overview of the Tertiary geology and hydrogeology of the northern part of the Arabian Gulf region with special reference to Kuwait. Earth-Sci Rev 40(3):259–295. https://doi.org/10.1016/0012-8252(95)00068-2

Nasir SJ, El-Kassas IA, Sadiq AAM (1999) Mineralogy and genesis of heavy minerals in coastal dune sands, south eastern Qatar

Orndorff RC, Weary DJ, Lagueux KM (2000) Geographic information systems analysis of geologic controls on the distribution on dolines in the ozarks of South-Central Missouri, USA. Acta Carsologica 29(2). https://doi.org/10.3986/ac.v29i2.456

Orndorff RC, Knight MA, Krupansky JT, Al-Akhras KM, Stamm RG, Samad USA, Ahmed E (2018) Linking geology and geotechnical engineering in Karst: The Qatar geologic mapping project. In: 15th sinkhole conference

Perthuisot J-P (1977) La sebkha de Doukhane (Qatar) et la transformation: gypse—anhydrite plus water. [The sabkha of Dukhan (Qatar) and the transformation: gypsum to anhydrite plus water]. Bull Soc Geolo Fr 19(5):1145–1149

Perthuisot JP, Jauzein A (1978) Le Khour el Aadid, lagune sursalée de l'Emirat de Qatar. Revue de géographie physique et de géologie dynamique Paris 20(4):347–358

Puls D, Jameson J, Kozar M, Al-Ansi H, LeBlanc J (2009) The Dukhan sabkha: a modern analogue for the Arab C carbonate reservoir, Dukhan Field, Qatar. In: International petroleum technology conference

Rajendran S, Al-Kuwari HA-S, Sadooni FN, Nasir S, Govil H (2021) Remote sensing of inland Sabkha and a study of the salinity and temporal stability for sustainable development: a case study from the West coast of Qatar. Sci Total Environ 782:146932. https://doi.org/10.1016/j.scitotenv.2021.146932

Rauert W, Geyh MA, Henning GJ (1988) Results of 14C and U/Th dating of sinter samples from caves of the As Summan Plateau

Reeder P, Brinkmann R, Alt E (1996) Karstification on The Northern Vaca Plateau, Belize

Rivers J, Engel M, Dalrymple R, Yousif R, Strohmenger CJ, Al-Shaikh I (2020) Are carbonate barrier islands mobile? Insights from a mid to late-Holocene system, Al Ruwais, northern Qatar. Sedimentology 67(1):534–558

Sadiq AM, Nasir SJ (2002) Middle Pleistocene karst evolution in the State of Qatar, Arabian Gulf. J Cave Karst Stud 64(2):132–139

Sadooni F (2014) Geology of Qatar: short introduction, vol 1st edn. Environmental Studies Center, Qatar University

Seltrust Engineering Limited (1980) Qatar geological map, scale I: 100,000 and explanatory booklet

Shearman D (1963) Recent anhydrite, gypsum, dolomite and halite from the coastal flats of the Arabian shore of the Persian Gulf. Proc Geol Soc London 1607(63):5

Shearman D (1966) Origin of marine evaporites by diagenesis. Trans Inst Min Metall 75:208–215

Shinn EA (1969) Submarine lithification of Holocene carbonate sediments in the Persian Gulf. Sedimentology 12(1–2):109–144. https://doi.org/10.1111/j.1365-3091.1969.tb00166.x

Shinn EA (1973a, 1973//) Carbonate coastal accretion in an area of longshore transport, NE Qatar, Persian Gulf. The Persian Gulf, Berlin, Heidelberg

Shinn EA (1973b, 1973//) Sedimentary accretion along the leeward, SE coast of Qatar Peninsula, Persian Gulf. The Persian Gulf, Berlin, Heidelberg

Slowakiewicz M, Thomas L, Tucker M, Zheng Y, Whitaker F, Pancost R (2014) Biogeochemistry of Holocene intertidal microbial mats of Qatar: implications for petroleum source rock formation in carbonate-evaporite systems

Strohmenger CJ, Jameson J (2018) Gypsum stromatolites from Sawda Nathil: relicts from a southern coastline of Qatar. Carbonates Evaporites 33(2):169–186. https://doi.org/10.1007/s13146-017-0365-2

Sweeting MM (1973) Karst landforms. Columbia University Press. https://catalog.hathitrust.org/Record/001273090

Walkden GM (1974) Palaeokarstic surfaces in upper Visean (Carboniferous) limestones of the Derbyshire Block, England. J Sediment Res 44(4):1232–1247. https://doi.org/10.1306/212f6c80-2b24-11d7-8648000102c1865d

Warren JK (1991) Sulfate dominated sea-marginal and platform evaporative settings:: sabkhas and salinas, mudflats and salterns. In: Developments in sedimentology, vol 50. Elsevier, pp 69–187

Warren J (1996) Evaporites, brines and base metals: what is an evaporite? Defining the rock matrix. Aust J Earth Sci 43(2):115–132

Weary D, Orndorff R (2002) Physical controls on karst features in the Ozark Plateaus of Missouri, USA, as determined by multivariate analyses in a geographic information system (GIS). Acta Carsologica 3:181–194

White WB (1988) Geomorphology and hydrology of karst terrains

Whybrow PJ, McClure HA (1980) Fossil mangrove roots and palaeoenvironments of the miocene of the eastern Arabian Peninsula. Palaeogeogr Palaeoclimatol Palaeoecol 32:213–225. https://doi.org/10.1016/0031-0182(80)90041-3

Wood GV, Wolfe MJ (1969) Sabkha cycles in The Arab/Darb formation off thetrucial coast of Arabia. Sedimentology 12(3–4):165–191. https://doi.org/10.1111/j.1365-3091.1969.tb00873.x

Yechieli Y, Wood WW (2002) Hydrogeologic processes in saline systems: playas, sabkhas, and saline lakes. Earth-Sci Rev 58(3):343–365. https://doi.org/10.1016/S0012-8252(02)00067-3

Structural Configuration and Tectonic History

3

Abstract

The history of the Arabian Peninsula's tectonic development is intricately connected to worldwide geological occurrences and alterations in the configuration and location of the Arabian Plate. From the Proterozoic to the Paleozoic and into the Cenozoic, the region saw substantial structural alterations due to tectonic collisions, rifting, and compressional and extensional phenomena. Prominent tectonic formations, including the Arabian Shield, Najd Fault System, and Qatar Arch, developed due to these forces. The borders of the Arabian Plate, especially in the north and east, were delineated by interactions with the Eurasian Plate. This led to orogenic zones and rift systems like the Red Sea and Dead Sea Rift. The Mesozoic and Cenozoic eras were characterized by significant tectonic activity, particularly the Zagros collision and the Cretaceous-Tertiary transition, during which extensive ophiolite obduction and structural reactivation occurred. Every tectonic phase affected sedimentary processes, resulting in substantial carbonate and clastic deposits throughout the Arabian Plate. The Late Cretaceous epoch had the final event of the "Big Five" mass extinctions, driven by substantial climatic and tectonic alterations. The interaction of the Arabian Plate with other plates at the Eocene–Oligocene boundary resulted in significant climatic and sea level changes, signifying the onset of Antarctic ice and widespread volcanic activity. The tectonic history of the Arabian Peninsula demonstrates a complex interaction of geological processes that have influenced its present landscape, resource allocation, and structural frameworks, especially affecting hydrocarbon buildup in the area. This chapter thoroughly examines the tectonic components and geological development of the Arabian Plate, detailing its transition from the supercontinent Gondwana to its current arrangement.

3.1 Introduction

During most of the Phanerozoic era, Eastern Arabia, including Qatar, was located on the African Plate. The breakout initiated in the early Mesozoic era with the opening of the Neotethys, leading to the separation of the Iranian landmass. During the late Tertiary period, a spreading ridge emerged from the Indian Ocean. It extended further, forming the Gulf of Aden and the Red Sea Rift. Collectively, these boundaries constitute the southeastern and southwestern perimeters of the Arabian Plate. Both regions have recently formed and have limited coastal plain development. The similarity of the geological characteristics on both sides of the Gulf of Aden and the Red Sea, which display consistent geological structures, suggests that the separation occurred relatively recently and was accompanied by a minor yet noteworthy transcurrent movement. The shear fractures that resulted from the Red Sea opening extend onshore, continuing through the Gulf of Aqaba and forming the Dead Sea Rift. These fractures come into contact with the Taurus Mountains in the north, completing the region's periphery. On the eastern shores of the Red Sea, there are uplifted Precambrian rocks along with their sedimentary and volcanic cover. This area has more significant differences in elevation compared to the western side. The elevation gradually decreases in the northern and eastern directions. The west side of the Jordan Rift is bordered by the Levant uplands, which form the elevated rim of the area (Alsharhan and Nairn 1997; Beydoun 1988, 1991).

The tectonic history of the Arabian Peninsula was divided into four stages by Beydoun (1991). These are:

1. The continental basement formed during the late Proterozoic as the eastern part of the Nubo-Arabian craton extended to present-day Africa. Stern and Johnson (2010) comprehensively reviewed the geologic, petrologic, and

F. N. Sadooni and H. A. Al-Kuwari, *Geological Evolution of Qatar and the Arabian Peninsula*, Regional Geology Reviews,
https://doi.org/10.1007/978-3-031-94394-2_3

geophysical aspects of the Arabian Plate continental lithosphere.
2. Eastern Arabia, including Qatar, was part of a marginal shelf of the Gondwana supercontinent till the Permian.
3. At the end of the Paleozoic-early Mesozoic, a continental divergence created a "promontory" or a bridge, and the Arabian Plate started separating as a spur or a branch from the African Plate.
4. During the middle-late Tertiary, a collision occurred between the Arabian Plate and the Eurasian Plate, causing a wide-scale compression that modified the northern and northeastern margins of the plate. This collision was associated with a rotational movement of the plate that caused further rifting and completely separated the Arabian Plate from Africa, modifying the south, southwest, and newly formed northwest margins.

3.2 Structural Elements

Throughout its geologic history, the Arabian Plate and before that, the African Plate had developed a group of structural elements that reflect the tectonic history and basin development imposed on it during different geologic periods. Some elements are positive, such as shields, faults, and arches; others are negative, such as basins. Some of them are still present, withstanding the impacts of geologic processes, while others were destroyed, and there are only traces or indirect evidence of their existence. Both of these elements are necessary to reconstruct its history.

3.2.1 The Arabian Shield

The Arabian Shield has been a tectonically stable region since the Precambrian. Its rocks form the larger parts of the surface of the present-day Arabian Peninsula (Fig. 3.1). They are also the source of the sediment that accumulated in the Arabian Basin during geologic time. The Arabian Shield is around 770,000 km^2 (Gettings et al. 1986). It covers the areas of Najd, Hijaz, and Asir to the west and connects with the Yemeni mountains, extending on the Arabian Peninsula's southern shorelines. Before the opening of the Red Sea during the Tertiary, the Arabian Shield was part of the African Shield, which comprised the Nubian Shield in Egypt, Sudan, Eritrea, and Ethiopia. These complementary tracts comprise the Arabian-Nubian Shield (ANS) (Stern and Johnson 2010).

The original crust of the western parts of the Arabian Shield comprises a combination of at least three to five constructional blocks, three representing intra-oceanic island arcs formed of metavolcanic rocks and diorite. These are long and curved chains of oceanic islands associated with intense seismic and volcanic activities. They are of Tonian to Ediacaran age that originated in the Mozambique Ocean. The arc assemblages comprise juvenile suites of tholeiitic and calc-alkaline volcanic rocks and large amounts of volcaniclastic sedimentary and plutonic rocks. The shield was formed over about 450 million years, between ~ 1000 and 530 Ma, and its growth indicates a complex crustal evolution as part of the Rodinia-Gondwana supercontinent cycle (Johnson et al. 2013). The island arcs are Asir, Hijaz, and Midyan terranes. These islands started to merge together 950 million years ago and led to the formation of the neocraton by the accretion of these island arcs. The process of coalescence was associated with closure and arc collision. These activities caused extensive deformation and ophiolite obduction. Ophiolites are deformed parts of oceanic plates pushed over continental plates. They can be seen in the present day as sutures that desiccate the shield rocks into north–south units, indicating a general east–west convergence. According to Beydoun (1991), these intense deformation processes culminated with what he called microplate and (micro) continental collision, whose remains are preserved as small terranes in the eastern parts of the shield.

This period was followed by collision-related growth of elongated gneiss domes and other intracratonic magmatic intrusions. Some north–south granite intrusions followed the old structural terrains. Subsequently, the northwest-southeast strike-slip Najd Fault System offset these structures during the latest Proterozoic-earliest Paleozoic.

3.2.2 The Najd Fault System

The Najd Fault System is a prominent tectonic structure that traverses the central region of Saudi Arabia, extending from the Red Sea coastline to the Gulf coastline. It is visible on the surface of the Shield over an area of 300 km wide and 1200 km long. It is distinguished by a sequence of normal faults that trend northwest-southeast and have remained active for millions of years. The fault system is a component of the extensive Red Sea Rift System, which is believed to be caused by the divergent boundary between the African and Arabian Plate.

The Najd Fault System is comprised of three primary segments: the Red Sea segment, the Central segment, and the Gulf segment. The Red Sea segment is distinguished by a sequence of normal faults aligned parallel to the Red Sea coastline, and it is regarded as the most dynamic section of the fault system. The Central segment traverses the middle parts of Saudi Arabia. It is distinguished by a sequence of en-echelon normal faults, which have resulted in the uplift of the Arabian Shield. The Gulf segment is distinguished by a sequence of parallel normal faults along the Gulf coast, and it is regarded as the least dynamic portion of the fault system.

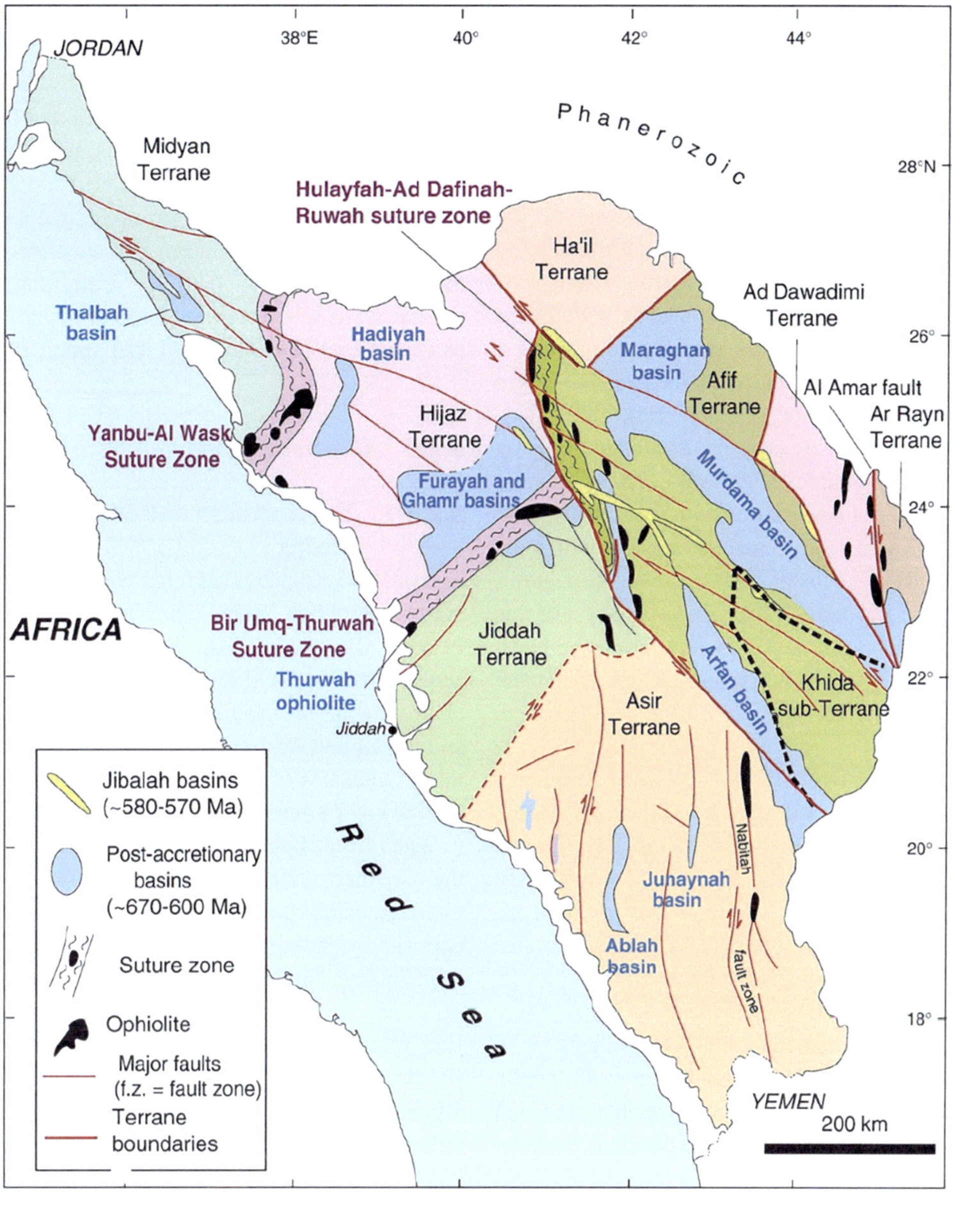

Fig. 3.1 Detailed map of the Arabian-Nubian Shield (Stern and Johnson 2010)

The Najd Fault System's activity is believed to be a consequence of the continuous collision between the African Plate and the Arabian Plate, which displaced the Arabian Plate northward. Consequently, the crust in the Gulf region is stretching and thinning, creating normal faults similar to those observed in the Najd Fault System.

The fault system has exerted a substantial influence on the geological and tectonic characteristics of the region, as well as significantly affecting the potential for hydrocarbon resources in the area. The faults have facilitated hydrocarbons' migration and resulted in the creation of different structural traps.

Many features resulting from the Najd stresses can be traced throughout the Shield, including some of the northeast-trending structures in the western parts, which neighbor the present-day Red Sea.

The cratonization of the Arabian Shield and the ending of the Najd Fault System occurred around 550 million years ago. This was followed by epicontinental subsidence, creating a differential surface upon which broad Paleozoic basins were initiated. Some workers (e.g., Greenwood et al. 1980) suggested that the impact of the Najd faulting was not restricted to the Shield only. Still, it may be responsible for forming the Infracambrian/Early Cambrian salt basins of southeast Oman. These salt basins contributed significantly to the structural development of the region.

3.2.3 The Arabian Platform

After the complete cratonization and consolidation and the decline of the Najd faulting activity, this vast region stabilized and witnessed only rare periods of epeirogenesis and some local taphrogenesis. During most of the latest Proterozoic to Late Paleozoic, Arabia, with Turkey, central and northwest Iran, and other areas, formed a long and wide passive margin of the Gondwana Supercontinent neighboring the Paleo-Tethys ocean. These areas were covered most of the time by shallow epeiric seas. Their areal extent was controlled by the periods of transgression and regression with the advancement of Paleozoic time. Their depositional settings also changed between tropical and temperate latitudes of the southern hemisphere. According to Beydoun (1991), the orientation of this margin changed from northwesterly facing during the latest Precambrian to westerly facing during the Cambrian and back to northwesterly and northerly facing during the Devonian (Fig. 3.2).

Climate-wise, this passive margin was under temperate conditions with warm periods during the Middle Cambrian, the Devonian, and the Late Permian, as reflected by the presence of warm water carbonates. Deposited clastic sediments reflect at least two glaciation periods during the Late Ordovician and Carboniferous-Permian periods. There is also evidence of late Proterozoic glaciation in Oman.

The Proterozoic-Phanerozoic sediments of this region were deposited in rift-generated basins and grabens or in broad intrashelf and epicratonic sags (Beydoun 1988, 1991). These were the depositional regimes all over the area, but these successions were rarely drilled due to the thick, younger sediments and the deep burial. The only areas where these sediments can be studied are those that escaped burial or were brought closer to the surface by tectonism. Among these exceptional areas are the Ghaba and Fahud salt basins that were formed during the Early Cambrian due to the left-lateral, strike-slip movement of the Najd Fault System. In these basins, the Huqf Supergroup sediments were deposited. These sediments represent syn-rift sequences and are assigned to the Neoproterozoic to "Early" Cambrian. They consist of marine clastics, evaporites, and carbonates that overly the basement rocks (AL-Johi and Al-Laboun 2015).

3.2.4 The Salt Basins

3.2.4.1 The Northern and Southern Salt Basins

The salt basins are divided into two main groups: the Northern and Southern Salt Basins. The Northern Salt Basins are found in Qatar, Kuwait, and Iraq, and they are characterized by thick sequences of salt and gypsum deposited in a shallow marine environment. The Southern Salt Basins are found in Oman and the United Arab Emirates, and they are characterized by thick sequences of salt and anhydrite deposited in a shallow marine environment (Fig. 3.3).

The Infracambrian Hormuz and Ara evaporites of the Southern Salt Basin were accumulated in a wide interconnected system of basins, with corresponding deposits extending from the Arabian Peninsula across Iran, Afghanistan, Pakistan, and India. There seems to be a connection between the Infracambrian salt basins of the Middle East (Oman, Hormuz) and southern Asia (Pakistan, India) (Gorin et al. 1982). The Ghaba and Fahud salt basins in Oman are subsidizing rift basins resulting from the left-lateral, strike-slip movement of the Najd transform fault system. The

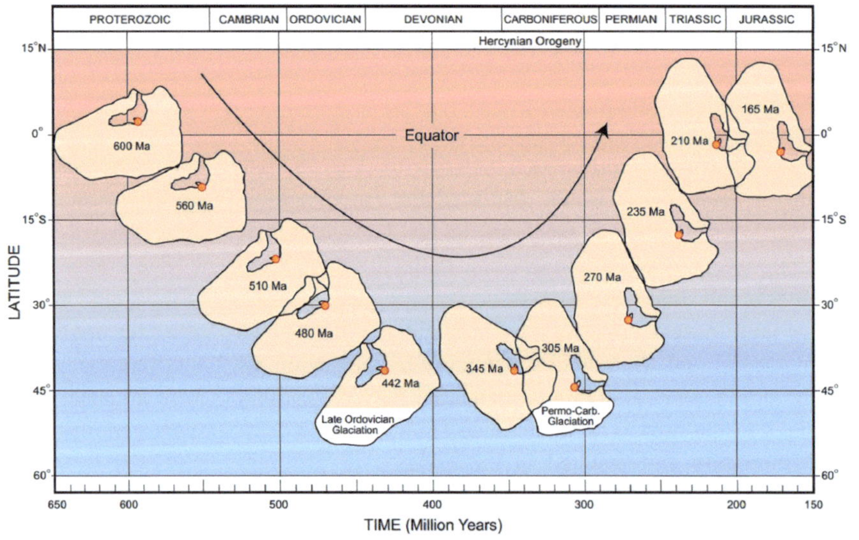

Fig. 3.2 Position of the Arabian Plate, including Qatar (shown as an orange circle) relative to the paleolatitudes during the Paleozoic (Konert et al. 2001)

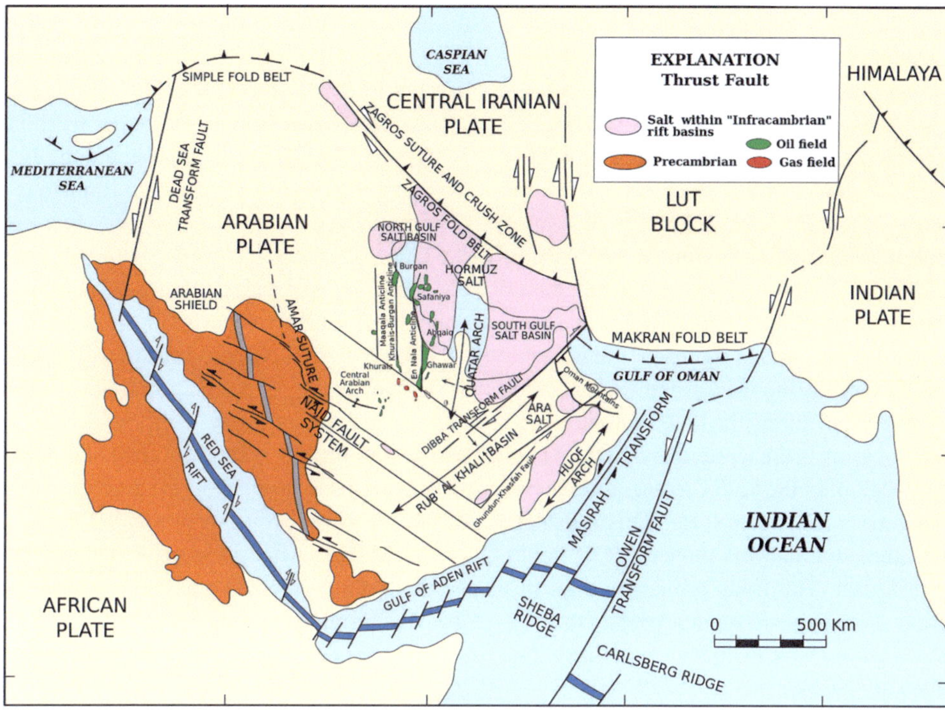

Fig. 3.3 Tectonic map of the Arabian Plate showing its main structural elements (Pollastro 2003)

Ara Group is the highest division of the Huqf Supergroup. The Huqf Supergroups are considered syn-rift deposits by Ahlbrandt et al. (2000). They comprise cycles of continental to marine clastics, evaporite, and carbonate deposits, suggesting a period of basin restriction in rapidly sinking basins.

It overlies the metamorphic basement and is unconformably overlain by the Haima Supergroup (AL-Johi and Al-Laboun 2015). The Ara Salt basins were created behind the volcanic margin of northeastern Gondwana. The accumulation of salt in these basins and the change to a drier climate in the region may have been caused by the uplift of the Andes mountains and their drying effect on the surrounding area behind the Cadomian active margin (Husseini 1988; Stern and Johnson 2010).

The evaporites of the Ara formation (of the Huqf Group) are believed to be equivalent to those of the Hormuz Salt Series in Iran and some parts of the Arabian Gulf, including Qatar (see Chap. 4). The Hormuz Salt Series consists of several cycles and belongs to the late Proterozoic to Early Cambrian (Stocklin 1968).

The Hormuz Series is composed of more than 3000 m of sedimentary rocks such as halite, gypsum, limestone, cherty dolomite, red sandstone, and shale, along with volcanic rocks that are likely to have formed at the same time as the sediments (Al-Husseini 2000; Kent 1970). Similar deposits were described from the Alborz Mountains, Iran, and assigned to the Soltanieh Formation. According to Husseini (1988), the Hormuz formation is considered a syn-rift sequence, while the Soltanieh sediments are platform sequences deposited on the Paleo-Tethys side of the Zagros Suture. This means that the Hormuz evaporite basins were intracontinental (Husseini 1988).

The Zagros and Dibba faults delineate the eastern boundary of the "Lower Gulf" Hormuz Salt Basin, while the northern and western boundaries are not clearly defined. Al-Husseini (2000) suggests no salt is present along the Qatar Arch and its northeastern extension toward the Zagros Suture.

The structures north of the Qatar Arch are identified as Hormuz Salt domes due to their oval shapes and the corresponding negative gravity anomalies observed. The Dammam and Bahrain domes serve as typical illustrations, while the Dukhan field of Qatar is understood to be a manifestation of a Hormuz Salt wall.

The complete distribution of the Hormuz Salt in the Arabian Plate has not been fully delineated. The southern region of Arabia stretches westward until it reaches the Najd Fault System (Al-Husseini 2000; Faqira and Al-Hauwaj 1998). The extent of the Hormuz formation along the western side of the Zagros Suture is also not well-defined.

3.2.4.2 The Rub' Al-Khali Basin

The Rub' Al-Khali basin spans an expanse of more than 300,000 km^2 in the southeastern region of Saudi Arabia. The area is situated beneath a Quaternary sand sea, and the changes in its structure from the Late Precambrian to Neogene periods are only documented through the analysis of reflection seismic, gravity, magnetic data, and well data. The gravity and magnetic data exhibit north–south and northwest-southeast patterns that align with the mapped Precambrian faults. The reflection seismic data reveals the presence of undrilled Precambrian rifts, which are the deepest structures observed. These rifts are filled with layered sediments and can be found at depths of up to 13 km. Ediacaran-Cambrian Ara/Hormuz mobile salt distribution is limited to a specific area within the eastern Rub' Al-Khali, known as an embayment. The Precambrian rifts exhibit regional uplift and were eroded to a nearly flat surface during the early Phanerozoic era. The Qatar Arch formed during the Carboniferous period and became more pronounced during the Late Triassic period. This fold separated subbasins in the western and eastern parts of the Rub' Al-Khali region. In the Late Cretaceous, a period of folding took place on a kilometer scale. This happened simultaneously with thrusting and the movement of ophiolite rocks in eastern Oman. The folds primarily align in a north–south direction, at an angle to the northwest compression direction, and sometimes have steep fault zones near their central surfaces (Stewart 2016).

3.2.5 The Arches

3.2.5.1 The Central Arabian Arch

This structural unit significantly impacted the depositional regimes of the area. It is the area where the northern and southern Tuwayq segments of the Interior Homocline change in strike. These two segments are part of the arch, which impacted the whole rock succession from the basement upward. Its effects can also be seen in eastern Qatar. The importance of this arch is due to its re-arking of the maximum curvature of the Interior Homocline in central Arabia and supporting the reversal between the north-dipping and south-dipping parts of the shelf, which underlie the Rub' Al Kahli (Fig. 3.4).

The earliest indication of the impact of this tectonic unit on the sedimentation regime can be found in the Middle Cretaceous Wasia Formation. In the southern outcrops of this formation, its sediments are continental, and they disconformably overlie the Upper Cretaceous continental sediments. On the arch and the areas north of it, the sediments are marine and transgress the older strata. The zone separating these two sedimentary units is aligned with the Central Arabian Arch. The same regime was consistent during the Upper Cretaceous Aruma Formation deposition.

3.2.5.2 The Hail Arch

This unit is similar to the Central Arabian Arch. It is located around Hail, northern Saudi Arabia, and marked by a shift in the strike of the Interior Homocline. In this area, the Lower Paleozoic rocks have a regional strike of around N 45° W in the northern Tuwayq segment but shift gradually westward in alignment with the underlying basement. The strike remains parallel to that of the basement margin, swinging west and then west-southwest to delineate the west flank of the arch.

The first indication of this arch activity comes from the Middle Cretaceous rocks, where sandstone sediments were continuously deposited across the arch crest. In contrast, the Upper Cretaceous Aruma sediments show a sharp facies change from marine carbonates on the eastern flank to continental sandstone across the arch crest.

The Hail Arch extends northward into Iraq and connects with the Rutbah High. Its formation was probably during the Cretaceous or early Tertiary period. Its development had a significant effect on the evolution of the Tethys. Before its formation, the strike of the Paleozoic and most of the Mesozoic strata were parallel to the northern margin of the Shield. After its appearance, the continuity across northern and western Arabia between the Arabian Gulf and the Mediterranean seas was broken.

3.2.5.3 The Qatar Arch

The Qatar Arch is a part of the Qatar–South Fars Arch. This large structure was more than 300 km long and 100 km wide and lasted several hundreds of Ma (Perotti et al. 2011). Qatar is considered an onshore, low-relief, topographic expression of this structure, and it protrudes some 150 km into the Arabian Gulf. The Tertiary strata that cover the surface of Qatar are gently deformed in a north–south regional anticline, which continues offshore in a north-northeast to northeast direction. Talbot and Alavi (1996) suggested that the Qatar–South Fars Arch continued further north into the Zagros belt. This zone remained a high area reflected by the thin sediments over the Fars Platform. It is also free of salt diapirs, which are extensive to the west and east (Fig. 3.5).

The Qatar arch is characterized by a thin Phanerozoic sedimentary cover draping Precambrian horst block that was rejuvenated frequently through geologic time. Such indications include:

1. The presence of Hormuz Salt Series sediments in the southeast and northeast of the Arch but not on it, which eventually delineates north-northeast to the northeast basement structural features aligned with other basement structures (Al-Husseini 2000; Edgell 1991; Perotti et al. 2011).
2. The relatively low thickness of the Mesozoic sedimentary cover over the Arch.
3. The impact of the emergence and submergence of the Arch on the nature of the sediments. The Arch was emergent, and from the end of the Triassic to the Early Jurassic, the Turonian and there were also minor uplifts during the Early-Middle Eocene and the Middle Eocene. A major uplift of the Arch occurred during the Late Eocene–Oligocene and is represented by a major unconformity in the region. A

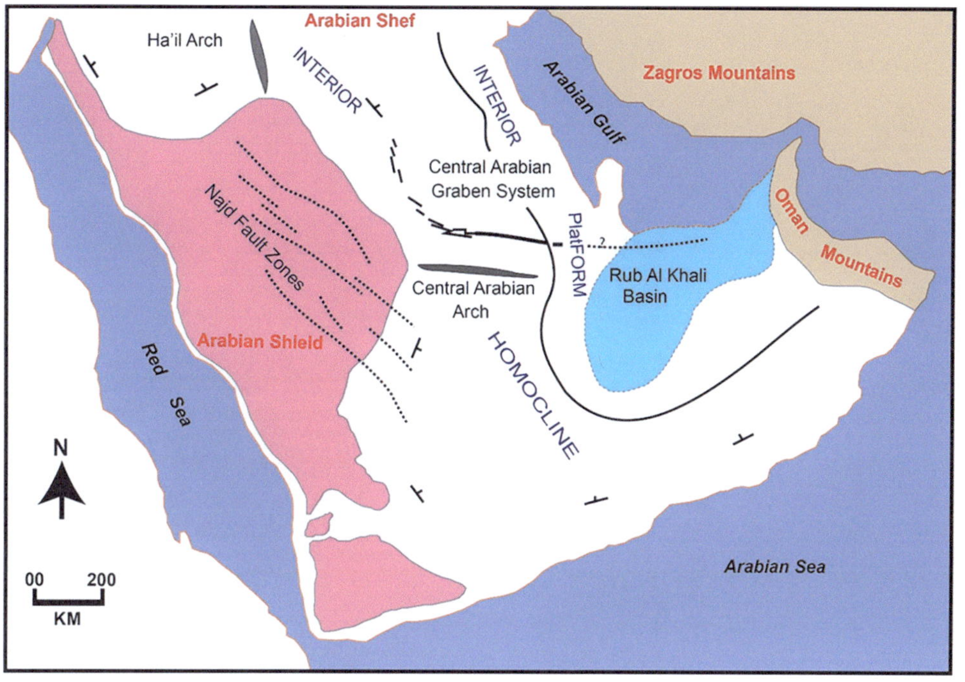

Fig. 3.4 Some of the main structural elements of the of the Arabian Peninsula (Hancock and Kadhi 1978)

more intense uplift happened during the Late Miocene-Pliocene. These emergence times were represented by missing sediments or erosion in Qatar and nearby areas (Alsharhan and Nairn 1997; Sadooni and Alsharhan 2019).

3.3 Tectonic Boundaries of the Arabian Plate

The Arabian Plate measures 2600 km north–south and 3000 km east–west and is bounded by six tectonic boundaries (Fig. 3.6):

1. *The Gulf of Aden* is a transform fault zone with a net movement toward the northeast.
2. *The Red Sea*: This extension center has seven transform faults with net movement toward the north.
3. *The Dead Sea*: A group of lateral transform faults with net movement toward the north.
4. *The northern boundary* is where the Arabian Plate contacts the Eastern Anatolian Fault. This is a collision zone forming thrust faults and folding.
5. *The eastern boundary* is a collision zone between the Arabian and Eurasian Plates. It is also called the Zagros Suture Zone. This boundary resulted from the rotation of the Arabian Plate anticlockwise between 4 and 20°. This region represents a subduction zone where the Arabian Plate subducts below the Eurasian Plate.
6. *The Owen Fracture Zone* is a transform fault extending from the southeast to the southern parts of Arabia. It forms the Arab Sea coastal faults that merge with the faults of the Aden Gulf.

The distribution of seismic activity provides evidence of the geological foundation that determines the boundaries of the region (Fig. 3.6). The marking is primarily concentrated along the fronts of the mountains and, to a lesser degree, along the ridge-rift system that constitutes the boundaries of the southwestern and northwestern regions. These lines are essentially plate boundaries. The western region of the Arabian Plate is rich in young volcanic activity, predominantly found in Yemen and extends into Saudi Arabia. These volcanic formations are commonly called "harrats." They were also reported from Jordan, Syria, and Turkey. The Red Sea and the Gulf of Aden are relatively new ocean basins that can be compared to the Neotethys, formed during the Late Triassic period by separating central Iran and Arabia.

The main geological features of the area were identified by Beydoun (1988) as being caused by three factors: the *rifting of the earth's crust in southern and eastern Arabia* due to the spreading of the seafloor in the Gulf of Aden and the Red Sea, resulting in the formation of new ocean basins; the *folding of rocks in the north and northeast* in the Taurus-Zagros-Oman Orogenic Zone as a result of the collision of continents; and the *strike-slip faulting* along the Dead Sea Rift or Levant Fracture Zone. Recognition of these essential features is essential for understanding the geological development of the region.

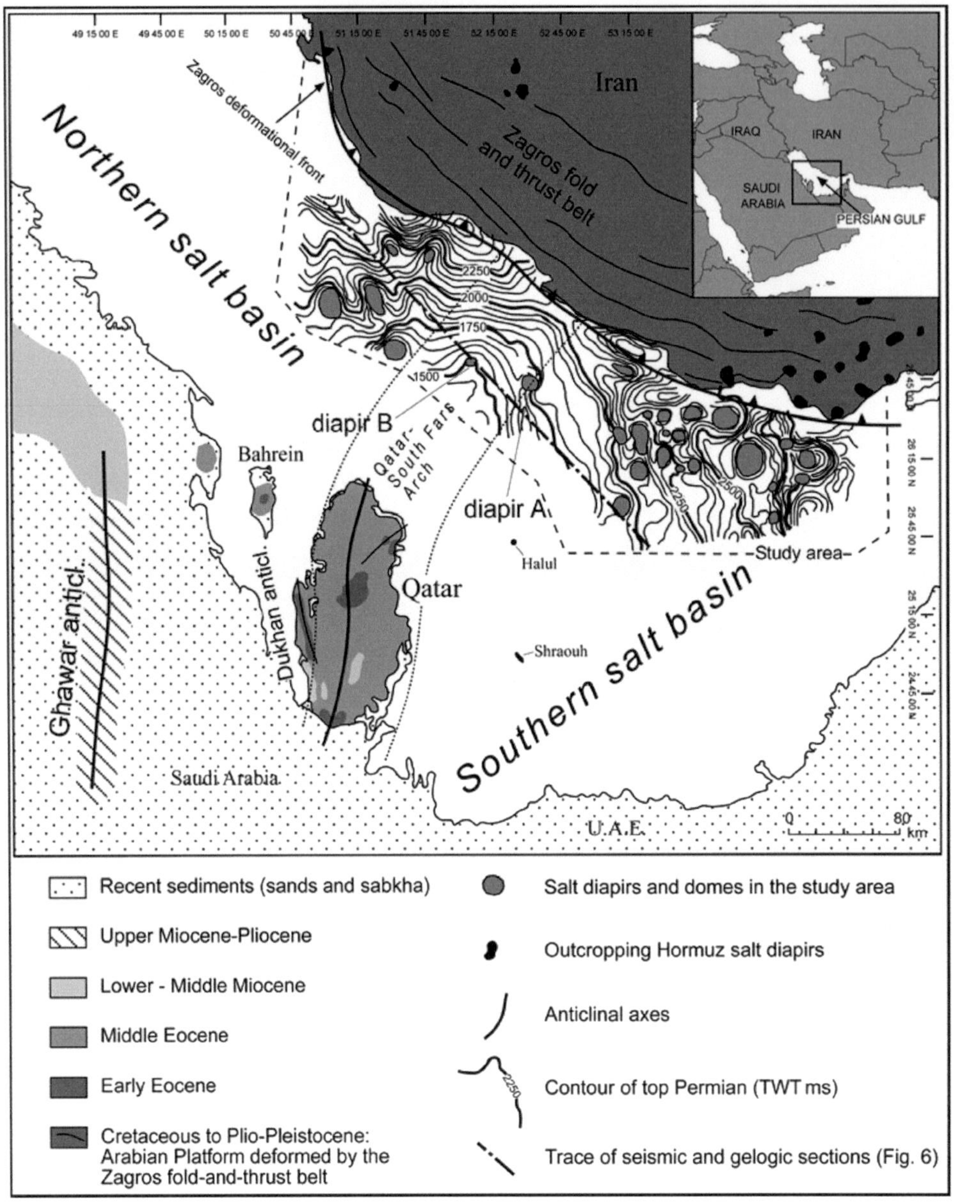

Fig. 3.5 Tectonic and seismic contour map of the Qatar-Fars Arch. The contour interval is in two ways travel time. The seismic map is near the top of the Permian (Perotti et al. 2011)

This is sometimes challenging due to frequent such events in or near areas of current tectonic activity.

3.4 Tectonic Evolution of the Arabian Plate

During most of the Phanerozoic, Arabia, including Qatar, was located on the northern margin of the Gondwana Supercontinent. This location strongly impacted the nature of sediments and their stratigraphic hierarchy in Arabia. Johnson (2008) provided a vivid history of Arabia during that time. According to him, the Arabic part of the Gondwana margin was surrounded by three successive ocean basins. These oceans were eventually destructed, and their products were eliminated during the successive subduction and collision events. Few such remains were documented from the allochthonous remnants found along the suture zone in the Alpine-Himalayan mountains. Recent advances in the work in these areas (e.g., Ruban et al. 2007; Stampfli et al. 2001, 1991; Stampfli 2000; Stampfli and Borel 2002; von Raumer et al. 2002, 2003) emphasized the important role of tectonic activity along the northern margins of Gondwana in shaping the stratigraphic configuration, depositional settings, and even hydrocarbon accumulation in Arabia.

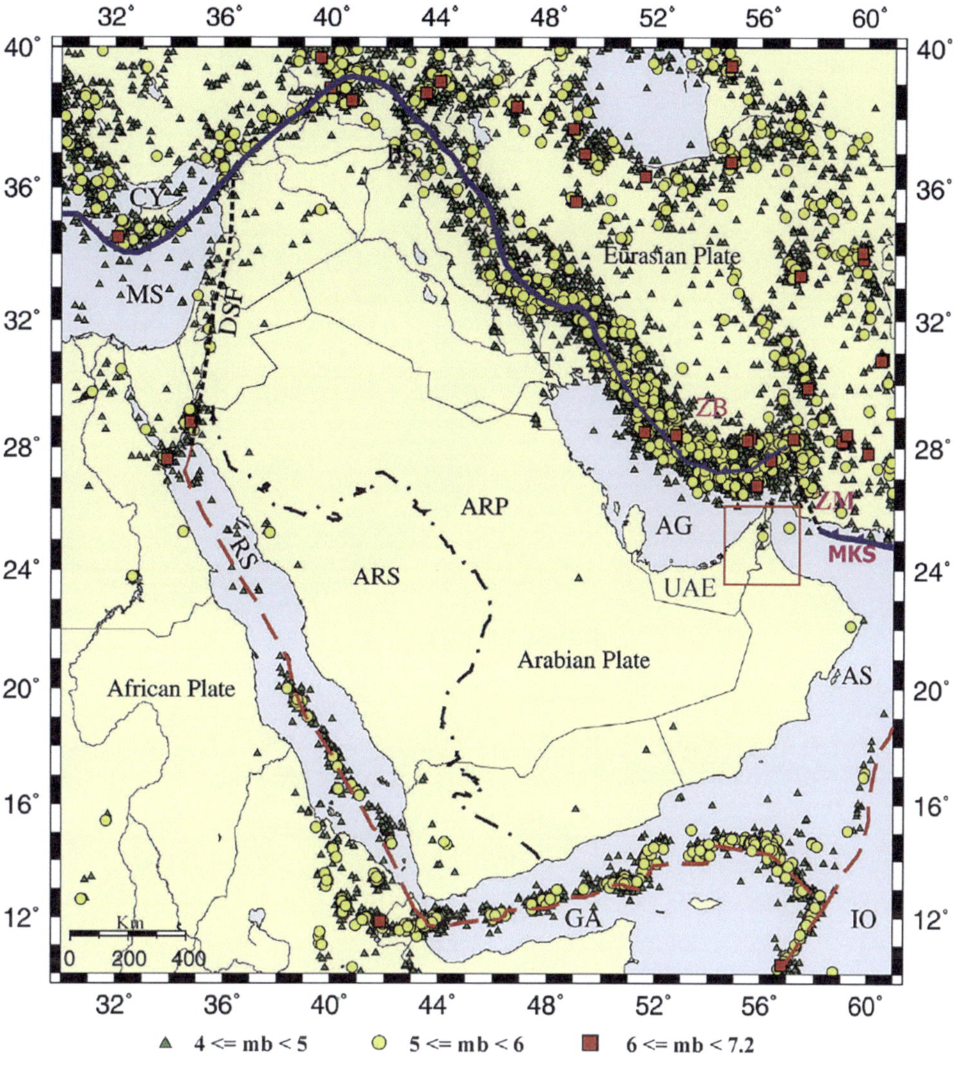

Fig. 3.6 Tectonic boundaries of eastern Arabia as delineated by recorded seismic activity (Al Khatibi et al. 2014)

3.4.1 The Cambrian (Late Cambrian, About 500 Ma)

During the Cambrian, most continents were located in the southern hemisphere. The Gondwana Supercontinent extended from the equator (Australia) to the South Pole (North Africa) (Fig. 3.7). Gondwana's tectonic movement was due to the relative rotation of its different cratons. The Cambrian extension that created the Hormuz Salts Basins of Arabia at the Precambrian-Cambrian boundary ended, and there was no major tectonic activity during the late Cambrian (500 Ma) (Berra and Angiolini 2014; Johnson 2008). During that time, clastic continental deposits fringed a shallow-water platform comprising shales and mixed siliciclastic and carbonate facies (Konert et al. 2001). In Saudi Arabia, these sediments belong to the Siq Formation, consisting of sandstone and pebbles derived from the Arabian Shield (see Chap. 4). In the northern parts of the Arabian Plate, these sediments are assigned to the Khabour Formation. It consists of alternations of sandstone and shale deposited in a shallow epeiric sea on the passive margin of northern Gondwana (Al-Juboury et al. 2020).

3.4.2 The Late Ordovician

Paleogeographic reconstructions from the Late Ordovician period reveal that most continental blocks were positioned in the southern hemisphere. This geographic distribution enabled biotic exchanges over comparatively limited oceanic distances, leading to the development of a cosmopolitan biota (Berra and Angiolini 2014; Johnson 2008; Ruban et al. 2007).

During the Ordovician era, the eastern parts of Arabia were subject to uplifting, rifting, and magmatism due to the Ordovician major extension associated with the collision between eastern Avalonian terranes and the Gondwana

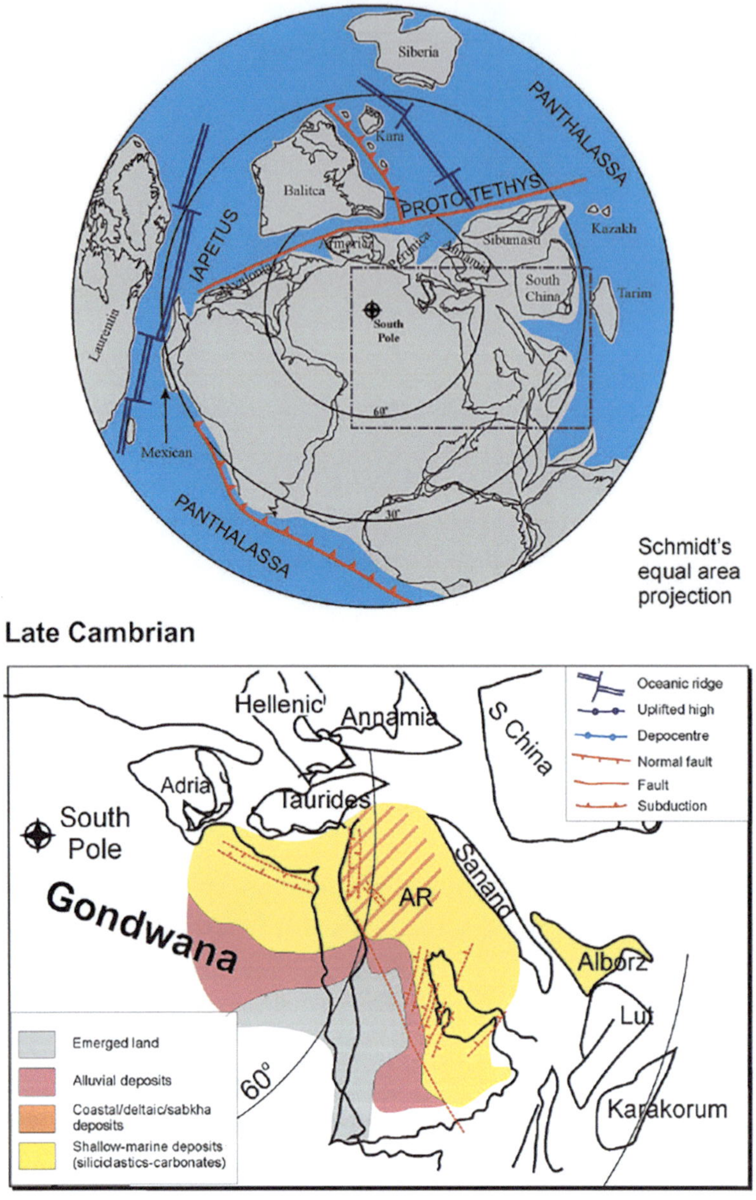

Fig. 3.7 Global paleogeography and depositional environments in Arabia during the Late Cambrian (Berra and Angiolini 2014)

margin north of Arabia. The reason for this extension is not fully understood, but it had a major impact on the Ordovician depositional patterns. Evidence for multiple tectonic cycles resulting from this extension can be traced in the Ordovician stratigraphy of Oman, starting during the Arenig period and continuing intermittently during the rest of the Ordovician period. This extension, however, did not generate new ocean basins; instead, it was a series of failed rifts. Examples of such basins are the extensional basins of Oman and the Tabas Graben in Iran (Berra and Angiolini 2014; Johnson 2008) (Fig. 3.8).

The global climate significantly deteriorated at the end of the Ordovician period, culminating in the Hirnantian glacial episode (Ashgillian cooling event), which persisted for approximately 0.5–1 million years. Glacial deposits attributable to this period have been identified across various regions of Peri-Gondwanan and Gondwana, including North Africa, South America, Arabia, and South Africa. This glaciation followed a phase of climatic improvement along the Northern Gondwana margin, characterized by the accumulation of temperate bioclastic limestones and bryozoan mud mounds atop an extensive Early-Middle Ordovician terrigenous succession. The transition from the pre-Hirnantian "greenhouse" conditions to the "ice-house" state of the Hirnantian period was abrupt, with no preceding climatic oscillations to facilitate biotic adaptation to the changing conditions (Brenchley et al. 1994).

During the Hirnantian glacial episode, Arabia was situated at relatively high latitudes in the southern region. This cooling event manifested in large ice sheets that generated continental

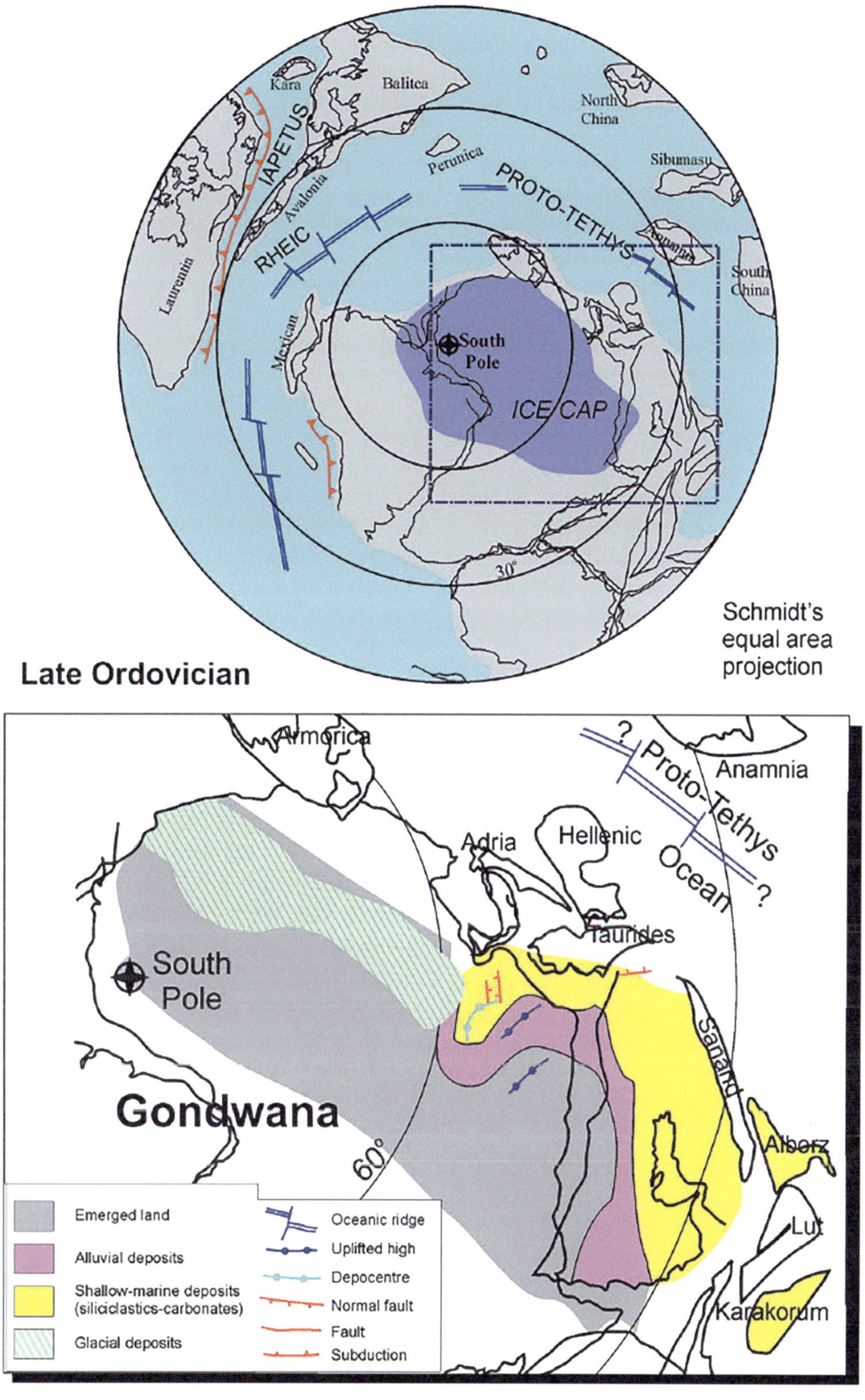

Fig. 3.8 Global paleogeography and depositional environments of Arabia during the Late Ordovician (Berra and Angiolini 2014)

glacial sediments in North Africa (El-ghali 2005), Jordan (Amireh et al. 2001), and western Saudi Arabia (Ziegler 2001). These glacial sediments cut deep valleys and generated grooved surfaces in the Ordovician Qasim and Saq strata of northern Saudi Arabia (see Chap. 4).

There were two significant periods of continental ice sheet cycles. Each cycle is marked by a distinct erosional unconformity beneath the deposits (Johnson 2008; Vaslet 1990). The ice advances have resulted in two distinct glacial deposits: the Zarqa Formation below and the Sarah Formation above. The Zarqa-Sarah section rests unconformably above the Quwarah Member of the Qasim Formation and below the Lower Silurian Qusaiba Formation. Occasionally, the glacial incision completely removed the pre-existing Paleozoic section, resulting in Upper Ordovician strata directly overlaying the Proterozoic basement.

The Ordovician glaciation on the Arabian Plate generated topographic terrains representing an important component of the region's hydrocarbon systems. The relief impacted the arrangement of different rock strata and consequently affected the distribution of source rocks during the subsequent Qusaiba transgression (Johnson 2008).

The melting of ice caps during the Hirnantian discharged immense volumes of water into the worldwide ocean. It led to a swift eustatic rise in sea level and the development of low-oxygen conditions on continental shelves. The following secondary transgression rapidly inundated the shallow clastic shelves of northern Gondwana. Clastic shorelines moved further inland during this period, creating a broad marine shelf where fine-grained clastics were deposited. This was culminated by the deposition of organic-rich shales (The Hot Shale), which are significant source rocks in North Africa and the Arabian Peninsula (Berra and Angiolini 2014; Johnson 2008; Lüning et al. 2000).

This period also witnessed the first of the "Big Five" Mass Extinctions of the Phanerozoic era, eradicating 85% of species, 61% of genera, and 12–24% of families, with the Ordovician extinction closely linked to the glacial events, suggesting climatic shifts as a primary cause. Yet, the extinction likely resulted from a multifaceted series of events, including sea-level fluctuations, oceanic restructuring, nutrient dynamics, and anoxia development, all interconnected with climatic transformations (Johnson 2008; Sepkoski 1982).

3.4.3 The Early Devonian Rifting

During early Devonian times, Gondwana (including Arabia) was located on the Southern Pole, while northern Africa and Antarctica were nearer the equator (Fig. 3.9) (Berra and Angiolini 2014). Deposition during the Devonian was strongly impacted by plate tectonics, as illustrated by the deposition of the Jauf Formation in Arabia. At that time, the margins of the Gondwana Supercontinent changed from collisional to extensional, eventually forming a passive margin. During the Early Devonian time, a global sea-level fall resulted in a depositional hiatus in most of the Middle East, probably enhanced by the tectonic uplift (Ruban et al. 2007). The middle latitude position of North Africa and part of Arabia favored the development of alluvial deposits. In Saudi Arabia, clastic sediments were deposited in the newly developed basin. These sediments were sourced from the Pan-African hinterlands to the southwest, the Late Ordovician rift shoulders of Oman to the southeast, and the active rift shoulders in Iran to the northeast.

A post-rift transgression covered large parts of the platform, depositing marine carbonates and distal clastic facies in the offshore marine environments near the passive margin of Jordan and Syria (Johnson 2008; Sharland et al. 2001). This was followed by a second marine transgression associated with the deposition of the Hammamiyat Limestone Member of the Jauf Formation (D2) (Sharland et al. 2001). This transgressive phase penetrated deeper into marine and marginal facies to the south and east of Arabia, reaching Oman. The southeast and southwest regions remained hinterlands that remained a source for the clastic influx to the marine shelf.

3.4.4 The Carboniferous Northern Gondwana Margin

Two different theories exist regarding the nature of the northern Gondwana margin, including Arabian during the Carboniferous. Stampfli (2000) suggested that it was a passive margin facing the Paleo-Tethys Ocean. Sengor et al. (1990) proposed it was an active volcanic margin when the southern part of the Paleotethys was subducting under Arabia. Johnson (2008) tried to merge the two theories, noticing enough evidence to support Sengor et al. (1990) conception as manifested by the deformation, metamorphism, and orogenesis along the northern margin of Gondwana. At the same time, the uplift and erosion may represent a back-arc indication of the start of subduction.

3.4.5 The Hercynian (Variscan) Orogeny

The Hercynian Orogeny and the Variscan Orogeny refer to the same geological events but are termed differently based on the regions in which they were studied and identified. The term "Hercynian" is derived from the Latin name for the Harz Mountains in Germany and is traditionally used in European geological literature. At the same time, "Variscan" comes from the Variscan Mountains in France and Germany or after a legendary Germanic tribe in northeastern Bavaria,

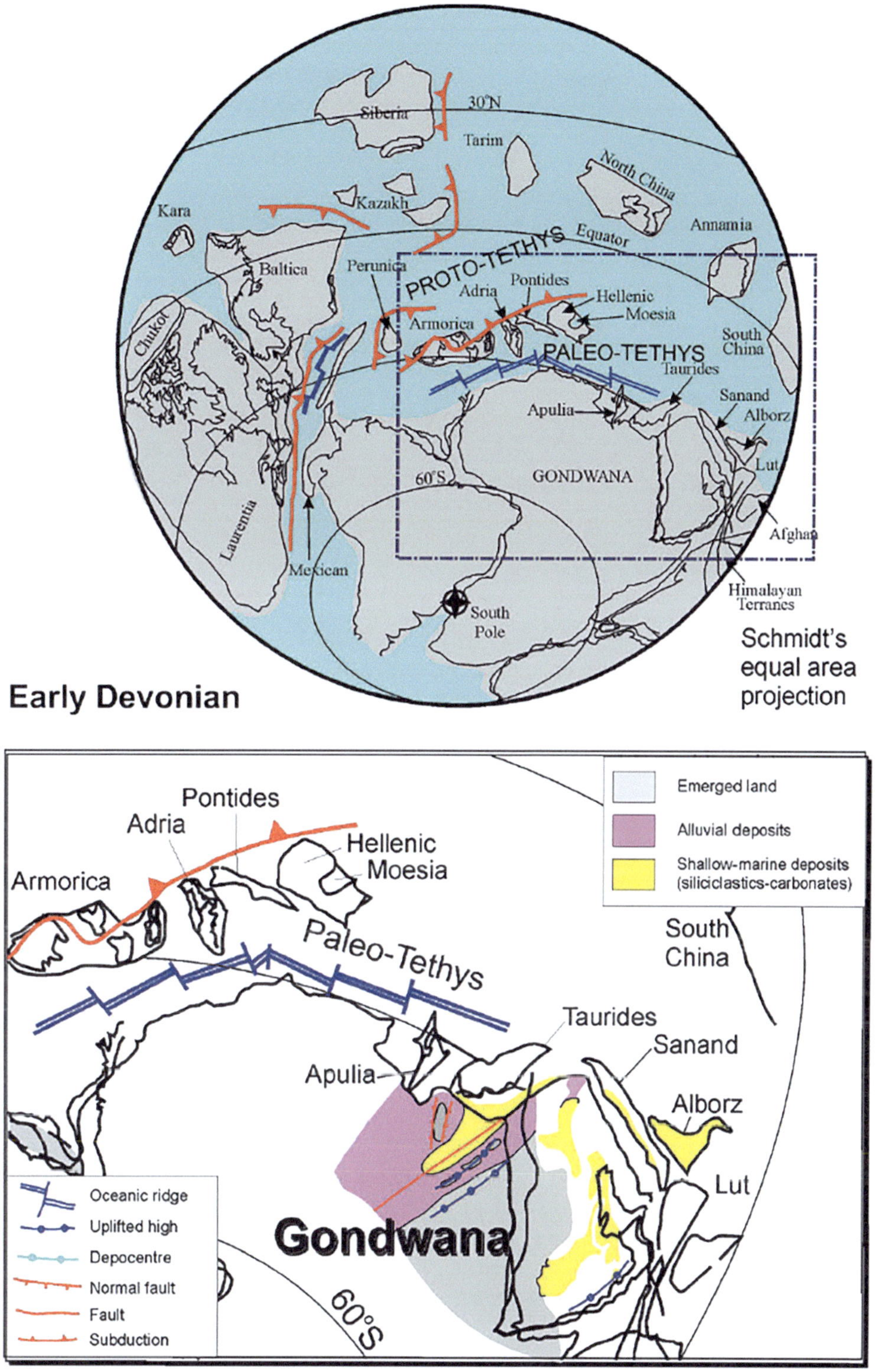

Fig. 3.9 Global paleogeography and depositional environments of Arabia during the Early Devonian (Berra and Angiolini 2014)

as mentioned by the Roman author Tacitus. "Hercynian" is commonly used in North American and Asian geological literature. Variscan" is predominantly used in European geological contexts.

Both terms describe the mountain-building events during the late Paleozoic era, approximately 370–290 million years ago, when the proto-European and North African (Gondwanan) tectonic plates collided. The orogeny affected large

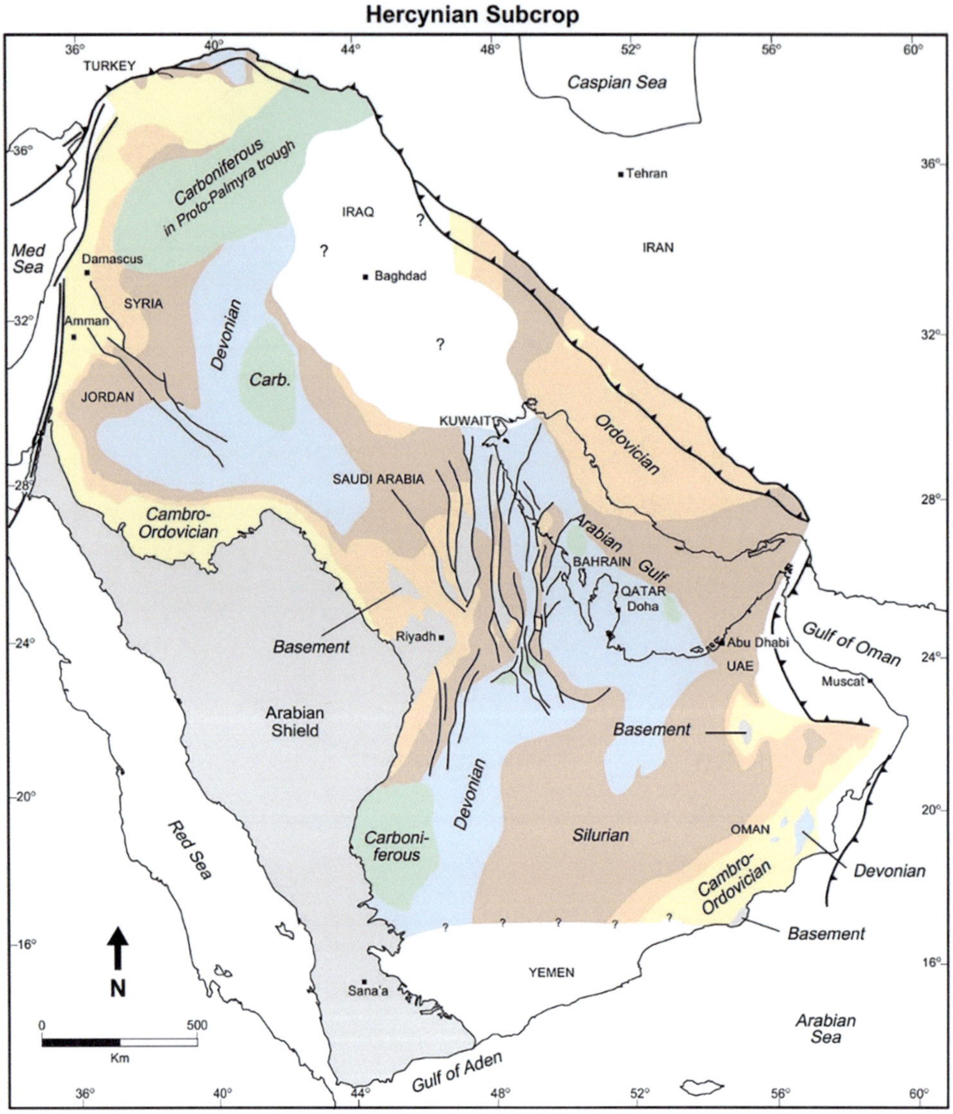

Fig. 3.10 Hercynian subcrop map shows only local Lower Carboniferous strata in Syria and some isolated wells in Saudi Arabia. Otherwise, the Carboniferous sediments were missing in most of Arabia due to uplift and erosion during the Orogeny (Konert et al. 2001)

parts of Europe, North Africa, and the eastern USA, forming mountain ranges and altering geological structures.

The Hercynian orogeny is a collisional orogeny of microcontinents that separated from the northern margin of Gondwana from the Cambrian to the Devonian and later involved in the collision of Gondwana and Laurussia during the Devonian and Carboniferous.

In Arabia, the Hercynian Orogeny was active from the Late Devonian to the Early Carboniferous and eroded several kilometers of sediments in the impacted areas. Other effects of the orogeny include changes in basin geometry, regional uplift, folding, and inversion tectonics. These features suggest that the Arabian Plate was impacted by a series of NW-directed compression phases during that time. Changes in the basin geometry in central Arabia were reported by Konert et al. (2001) (Fig. 3.10). His map shows an NE-trending basement high (aligned with the direction of the orogeny compression direction) protruding into the Central Arabian Arch. The Devonian-Silurian and older sediment distribution, facies patterns, and thickness variations suggest that the Arch was either formed or at least rejuvenated during the Hercynian Orogeny and persisted till the Mesozoic. This structural feature is overprinted by basement-cored uplifts such as the Ghawar anticline.

The north–south (NS) oriented Hercynian-generated uplifts, such as the Ghawar structure, are demarcated by reverse faults, which implies that these uplifts resulted from a regional compressive stress regime. The erosion that followed the Hercynian Orogeny and before the Permian period partially moderated the topographical relief, a phenomenon evidenced by the thickness and facies diversity observed in the Unayzah Formation. Several Hercynian faults, defining the major NS-oriented uplifts, underwent reactivation phases during the Triassic and notably in the Late Cretaceous period, as evinced by the pronounced thickening of the Aruma Group along the margins of these uplifts. However, not all the geological structures in the region resulted from the Hercynian event. For instance, the Harmaliyah anticline, situated east of the Ghawar field, retains the most intact Devonian stratigraphic sequence in Saudi Arabia, indicating its formation occurred after the Hercynian events (Konert et al. 2001).

The deformation and uplift during the Carboniferous period led to the disintegration of depositional environments, segmenting the Arabian shelf into smaller subbasins, demarcated by structurally elevated highs undergoing erosion. Prominent relief was formed on fault-bounded compressional structures, many interpreted as inverted Proterozoic grabens. Additionally, broader, unfaulted swells developed, generating significant relief to induce extensive erosion across vast areas. These extensive features are presumably the result of a confluence of factors, encompassing lithospheric folding and the reactivation of major basement terrane boundaries. Stratigraphically, the relief established during the Hercynian epoch played a pivotal role in determining the distribution and characteristics of facies within the clastics of the Upper Carboniferous and Lower Permian Unayzah Formation. The era witnessed recurrent fluvial incisions and reworking processes, leading to highly discontinuous stratigraphic architecture (Johnson 2008).

3.4.6 The Early Permian: Neotethys Breakup and Khuff Transgression

The late Paleozoic era experienced significant plate tectonic reconfiguration, as depicted in Fig. 3.11 Shortly following the culmination of the Hercynian orogeny during the Late Carboniferous, extension started along the northern Gondwana margin, from northwestern Australia through northern India and Arabia. This resulted in the amalgamation of Gondwana and Laurasia, forming the supercontinent known as Pangea. Adria and Apulia were previously separated and combined into a microplate known as Adria during the Early Permian period. The Cimmerian terranes (Iran, Central Afghanistan, Karakorum, Qiangtang) formed due to the Neo-Tethys Ocean opening along the eastern edge of Gondwana, spanning from Arabia to Australia. The terranes moved northward from southern Gondwanan paleolatitudes during the Early Permian period, eventually reaching subequatorial paleolatitudes by the Middle Permian to Early Triassic (Berra and Angiolini 2014; Muttoni et al. 2009; şengör 1979).

The Paleotethys reached its maximum development during the Late Devonian time and started to shrink during the rest of the Paleozoic. It got narrower due to initiating two subduction systems on the northern (Laurasian) and possibly the southern (Gondwanan) sides. This subduction continued to consume Paleotethys oceanic lithosphere until a Paleotethys oceanic spreading center encountered the subduction zone along southern Laurasia, on the northern side of Paleotethys.

Neotethys rifting along the eastern Gondwana margin, spanning from India to Oman, commenced during the Carboniferous period. Subsequently, the continental landmass was separated, and the oceanic crust was created in the Early Permian era. During the Permian–Triassic period, there was a significant area along the Eurasian margin where the Paleotethys oceanic crust was subducted northward. This activity continued for most of the period (Berra and Angiolini 2014; Metcalfe 2006).

During the Pennsylvanian-Early Permian period, ice covered much of Gondwana, forming glacial deposits in the southern latitudes. Due to a warm equatorial current that flowed westward, the tropical belt was limited to low latitudes. When this current reached the continental shelves of the western Tethys Gulf, it changed direction and flowed southeastward, carrying warm surface waters toward the northern corner of Arabia (Angiolini et al. 2007).

The development of the Neotethys passive margin triggered a significant tectonically induced marine transgression, resulting in the rapid inundation of the interior of the Arabian platform. By the mid-Permian period, approximately 265 million years ago, the entire region had transformed into a broad carbonate shelf, facilitating the deposition of the Khuff Formation throughout the remainder of the Permian era. This geological transition marked a shift in depositional environments on the Arabian platform, transitioning from arid conditions to a more humid climate due to the dissipation of the orographic effect caused by the rift-flank highlands.

Furthermore, the Khuff carbonates and evaporite deposition signified a pivotal change in sedimentation patterns as the Arabian plate gradually drifted northward into lower southern latitudes. Subsequently, Arabian stratigraphy became predominantly characterized by carbonate deposition, reflecting the evolving geological processes and paleoenvironmental conditions during this period (Johnson 2008).

3.4.7 The Permian–Triassic Boundary

Globally, the transition between the Permian and Triassic periods was characterized by low sea levels, as evidenced by the prevalence of continental or shallow marine facies (Fig. 3.12). A notable rise in sea level marked the onset of the Triassic period. The conclusion of the Permian period

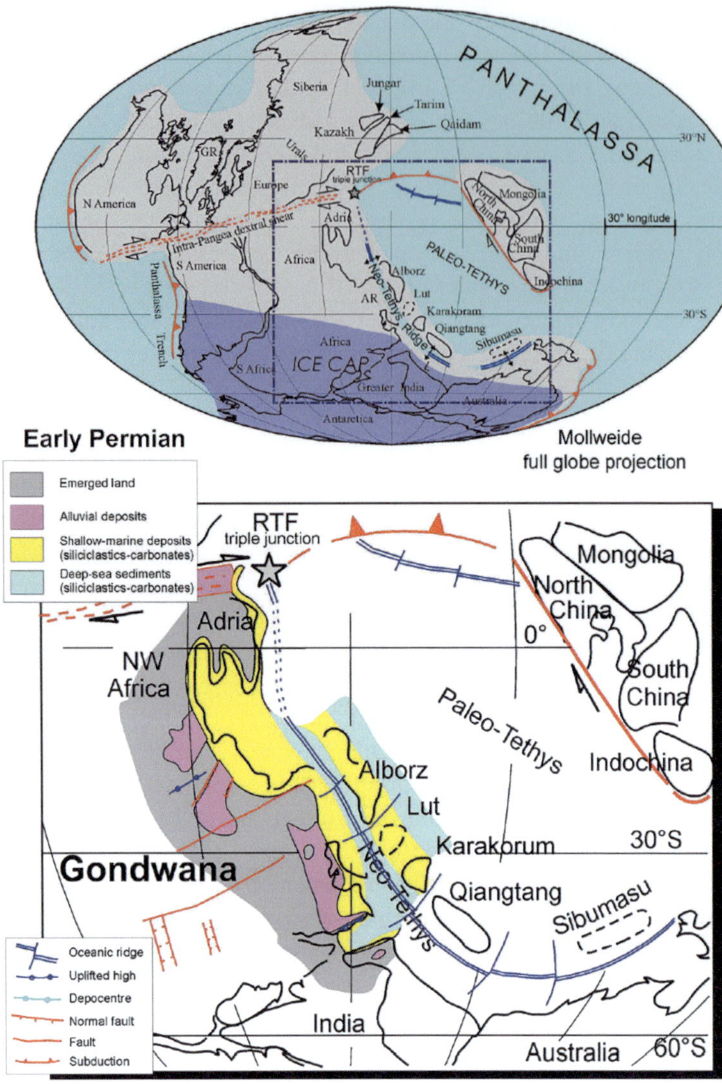

Fig. 3.11 Global paleogeography and depositional environments of Arabia during the Early Permian (Berra and Angiolini 2014)

is widely recognized as one of the most significant extinction events in the past 600 million years, leading to the 75–96% of species died (Erwin 2006). Various hypotheses have been proposed to explain the causes of this extinction event, including rapid sea-level fluctuations and/or anoxia (Hays et al. 2007), extraterrestrial impact (Becker et al. 2001), volcanic eruptions and/or extreme global warming (Reichow et al. 2009), or ozone layer collapse (Beerling et al. 2007).

Evaluating the pattern of the latest Permian extinction, Berra and Angiolini (2014) suggested that it occurred abruptly during rising sea levels. The sudden nature of the extinction event across different paleogeographic settings at the end of the Permian aligns with scenarios proposing climatic and environmental deterioration triggered by the Siberian Traps volcanism. This volcanic activity is believed to have significantly increased greenhouse gas emissions, contributing to global warming and ozone depletion. The occurrence of a pronounced negative $\delta^{13}C$ excursion worldwide near the end-Permian extinction event further supports this hypothesis (Horacek et al. 2007).

These findings underscore the complex interplay of geological, environmental, and climatic factors that contributed to the profound ecological changes during the Permian–Triassic boundary. They also highlight the significance of the Siberian Trap volcanism as a potential catalyst for these events.

3.4.8 The Mesozoic: Triassic

The Norian period is significant in the evolution of the Paleo-Tethys Ocean, as it marks a pivotal interval characterized by notable geological events. During this stage, the commencement of the Cimmerian orogeny along the southern margin of

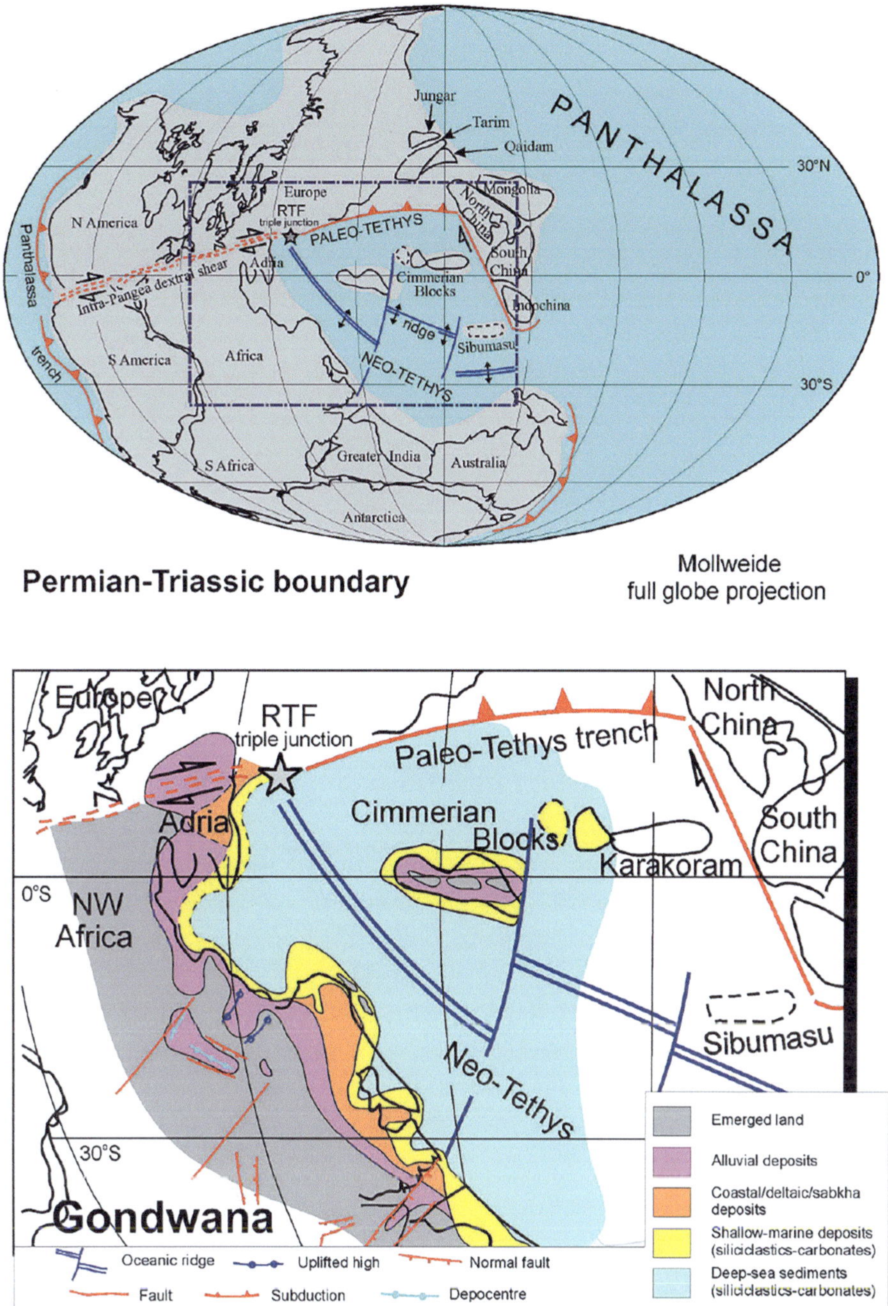

Fig. 3.12 Global paleogeography and depositional environments of Arabia during the Permian–Triassic boundary (Berra and Angiolini 2014)

Asia signifies the initial evidence of the closure of the Paleo-Tethys Ocean. This orogeny was initiated by the collision of the Iran blocks with the active margin of Turan (Fig. 3.13).

Meanwhile, the Neo-Tethys Ocean remained wide open, with clear paleogeographic indications observed in the central part of the Tethyan Gulf. Northern Africa and Arabia acted as the southern passive margin of the Neo-Tethys, featuring extensional basins in Libya and Egypt (Schandelmeier et al. 1997). Frequent unconformities associated with extension along the Neotethys margin, spanning from Syria to Libya through Arabia, can be interpreted as far-field effects resulting from the Cimmerian orogeny.

During the Norian period, the prevailing climate conditions in the Tethys latitudes were predominantly arid. This aridity is reflected in the deposition of carbonate facies along the Tethys margin, juxtaposed with evaporitic facies and coastal and continental fluvial to playa environments. The arid climatic conditions likely facilitated the early and widespread dolomitization observed on most Norian carbonate platforms (Iannace and Frisia 1994). The arid belt is believed to have

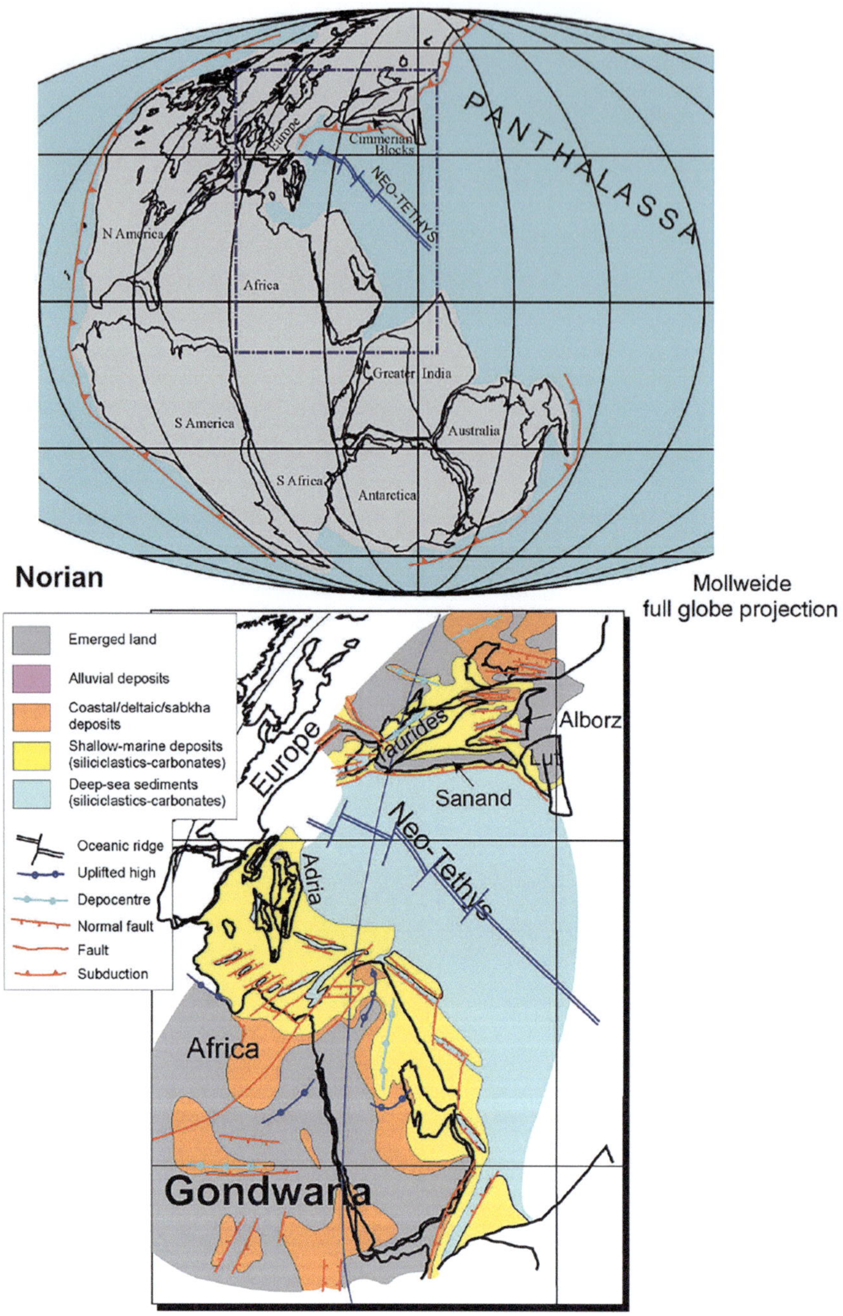

Fig. 3.13 Global paleogeography and depositional environments of Arabia during the Norian (Berra and Angiolini 2014)

extended beyond 40° latitude north and south, as evidenced by the distribution of climatically sensitive facies.

Climate reconstructions conducted by Ziegler et al. (2003) and Sellwood and Valdes (2006) indicate that the arid belt persisted up to the equator for much of the Triassic period, with a notable absence of a humid equatorial belt. Due to a shift toward humid conditions, the pervasive dolomitization of carbonate platforms ceased close to the Norian-Rhaetian boundary in the Western Tethys (Berra 2012). This transition is further substantiated by the increase in siliciclastic deposits along part of the northern margin of the Tethys.

During Callovian time, the formation of oceanic crust occurred between India and Arabia following an Early Jurassic rifting stage (Fig. 3.14). The opening of the Mesogea Ocean separated the margin of Arabia and North Africa from a major continental block comprising a large carbonate platform and related peri-platform basins. Coastal deposits were evident along the southern coast of the Mesogea Ocean in present-day northern Africa (Guiraud and Bosworth 1999), while shelf carbonates were deposited in most of eastern to southern Arabia. This geodynamic setting between two extensional basins likely facilitated the development of deep

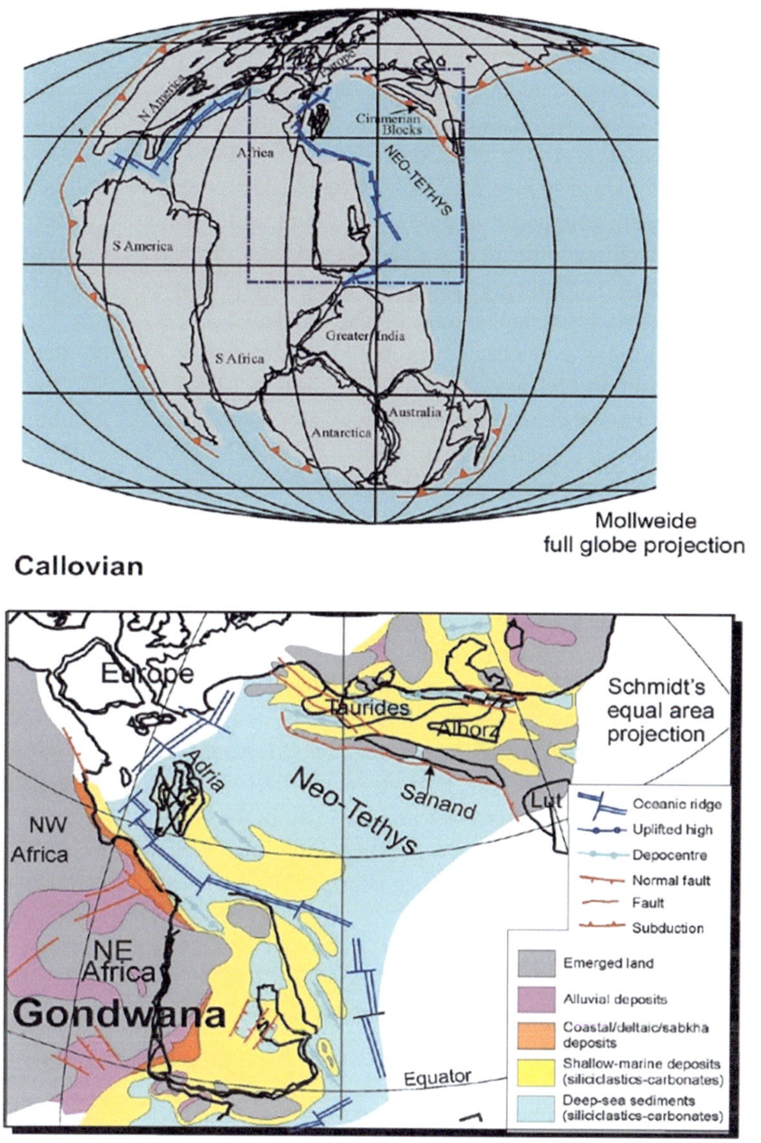

Fig. 3.14 Global paleogeography and depositional environments of Arabia during the Callovian (Berra and Angiolini 2014)

seaways on the Arabian shelf, partially controlled by extensional tectonics such as the Arabian Basin (Berra and Angiolini 2014).

3.4.9 The Late Cretaceous

During the Late Cretaceous period, Arabia underwent a significant phase of compression and structural growth, marking one of the most crucial periods of deformation on the Arabian Plate. This deformation event was pivotal in the formation or significant growth of major structural hydrocarbon traps. The onset of deformation can be traced back to mid-Cretaceous uplift, regression, and associated erosion. As the Late Cretaceous deformation peaked, ophiolites were obducted along the margins of Arabia. The end of deformation on most parts of the Arabian Plate is demarcated by an unconformity near the base of Cenozoic strata, typically at or beneath the base of the early Tertiary Umm Er Radhuma Formation (see Chap. 9).

This deformation event manifested across the entire Arabian Plate to varying degrees, indicating the selective reactivation of Proterozoic basement fabric, reminiscent of the Hercynian Orogeny. This deformation and structural reactivation is best understood within a regional tectonic framework. Temporally, ophiolite obduction, deformation in Abu Dhabi, and a phase of robust transpression along the transform plate boundary between India and Arabia coincide (Johnson et al. 2005; Johnson 2008). The transpression along the paleotransform was resolved through partitioning into N30E strike-slip and N60W compression. In response to this compression, the Arabian Plate experienced deformation at multiple scales, from basin-scale lithospheric buckling to field-scale inversion of Infracambrian grabens and seismic-scale fault networks that developed within the more brittle strata (Filbrandt et al. 2006; Johnson et al. 2005).

Ophiolite emplacement in Oman occurred in two distinct phases. Around 105 million years ago, relative motion between India and Arabia initiated rapid changes as India and Madagascar began to separate. This led to a major reorganization of plate boundaries in the region and placed a major oceanic transform northeast of Arabia under compression. By approximately 98 million years ago, this region experienced significant overthrusting that transitioned into active subduction, resembling a mechanism proposed by Hacker et al. (1996). Concurrently, the extension behind the trench resulted in rapid rollback and the formation of a new, supra-subduction oceanic crust (the future Samail ophiolite). By 94 million years ago, this ophiolite had rotated clockwise in response to plate boundary forces near the Arabian margin, leading to rapid obduction along the northern Oman margin. A second phase of ophiolite obduction in eastern Oman occurred later, during the earliest Tertiary period, which was completed 60 million years ago. Structural investigations in eastern Oman have correlated a systematic set of faults and joints with regional Late Aptian to Maastrichtian NW–SE compression and lithospheric buckling (Bertotti et al. 2005; Montenat et al. 2003).

All these geological activities occurred within the context of strong oblique convergence between India and Arabia, resulting in deformation of the plate interior at various scales (Johnson et al. 2005; Johnson 2008).

3.4.10 The Cretaceous-Tertiary Boundary

By the end of the Cretaceous period, the configuration of present-day continents had taken its final shape (Fig. 3.15), with only the northernmost Atlantic Ocean remaining incompletely opened due to ongoing rifting between Greenland and Canada. Notable tectonic events included Africa's separation from Antarctica and India's initiation of a northern trajectory, ultimately leading to the Himalayan orogeny. The Alpine Tethys and associated basins connected the Central Atlantic Ocean with Neo-Tethys were in a convergent plate regime. Concurrently, the intricate puzzle of Adria, Greece, and Turkey blocks was nearing the southern margin of Eurasia following the nearly complete subduction of the Neo-Tethys Ocean.

The progressive closure of Neo-Tethys also influenced the evolution of the passive margin of Arabia, where the Peri-Arabian Massif high contributed sediments in both northward (toward Neo-Tethys) and southward directions. The origin of this high was linked to the convergence of the lower plate (Arabia) with the southern margin of Laurasia, represented by the Sirjan blocks of central Iran, or to an intra-oceanic subduction zone (Stampfli and Borel 2002). The southern margin of Arabia was likely delineated by a transform fault, separating the Arabian plate from India. South of the Peri-Arabian Massif, sedimentation on the Arabian plate transitioned from prevailing deep-sea clastics and shallow-water carbonates to a vast coastal plain characterized by the deposition of alluvial sediments.

The end-Cretaceous period marked the occurrence of the last of the "Big Five" mass extinctions (Bambach et al. 2004), which had a profound impact on both marine and terrestrial ecosystems and left a lasting biogeographic imprint on modern biota (Krug et al. 2009). Extinction events affected various taxa, including marine reptiles, cephalopods, foraminifers, brachiopods, sharks, dinosaurs, pterosaurs, certain bird groups, and marsupial mammals. However, the pattern of this extinction event remains a subject of debate, with some interpretations suggesting a gradual decline leading up to the K-Pg boundary. In contrast, others propose catastrophic die-offs (Macleod et al. 1997).

One of the earliest proposed causal mechanisms for the end-Cretaceous mass extinction was a major asteroid impact (Alvarez et al. 1980; Ocampo et al. 2006), with potential impact craters such as the Chicxulub Crater in the Yucatan (Hildebrand et al. 1991) or Umm Mechemain Crater, Western Iraq (Fig. 3.16) being identified. Other proposed triggering mechanisms include global warming and flood basalts (Deccan Traps) (Courtillot 2005), highlighting the complexity of factors that may have contributed to this significant extinction event (Berra and Angiolini 2014).

3.4.11 The Zagros Collision

The Zagros fold and thrust belt is a geologically deformed zone of crustal rocks. It was formed in the foreland region where the Arabian Plate and the Eurasian Plate collided. It contains one of the largest petroleum provinces in the world (Alsharhan and Nairn 1997; Cooper 2007). The Zagros Mountain range spans approximately 1800 km, stretching from the Bitlis Suture Zone in the northwest to the boundary with the Makran Trench, located east of the Strait of Hormuz in the southeast. The belt exhibits variations in width, characterized by two prominent *salients* in the Lorestan and Fars regions, where the thrust belt protrudes toward the foreland. Additionally, two main embayments are located at Kirkuk and Dezful, between these bulges. The differences in shape along the direction of the fault are caused by the arrangement of the late Proterozoic to Early Cambrian Hormuz salt strata. The salt bed is thick and uninterrupted under the raised parts of the fault, while it is absent, thin, or irregularly distributed in the lower parts. The distribution of the Hormuz salt is determined by the size of the late Proterozoic basins. The Zagros fold belt is an integral part of the Alpine-Himalayan chain. The Zagros plate boundary is part of a wider plate boundary

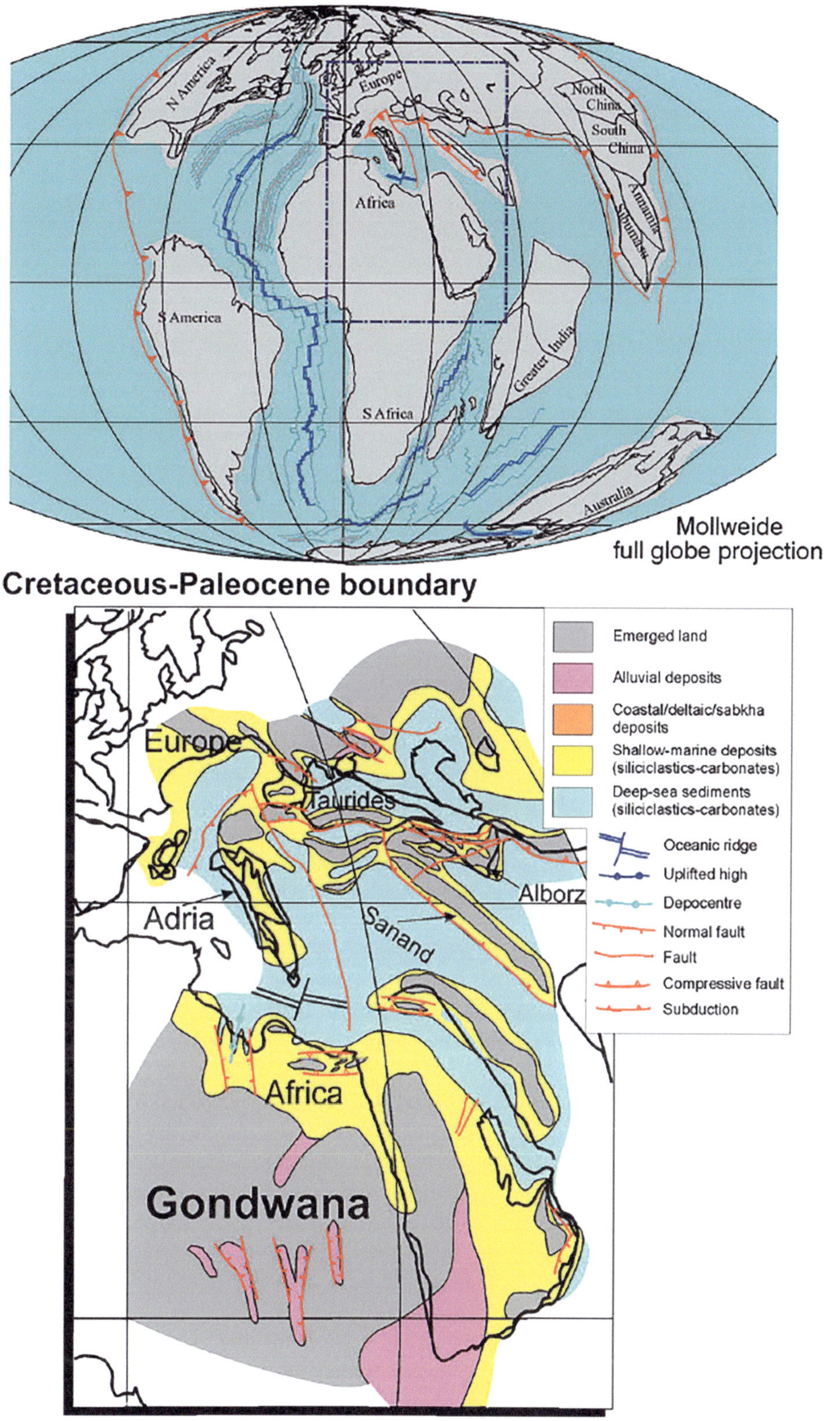

Fig. 3.15 Global paleogeography and depositional environments of Arabia during the Cretaceous-Tertiary Boundary (Berra and Angiolini 2014)

Fig. 3.16 Umm Mechemain Crater, western Iraq. One of the proposed mechanisms for its formation is being an impact crater

that covers most of Iran. The thick fold-thrust wedge in the Zagros fold belt represents a load contributing to the Arabian plate's northeastward tilt. Such Tertiary tilting was reported from the Ghawar oilfield (Abu-Ali et al. 1999). Vannucci et al. (2004) suggested that present stress in this area has an NE-SW direction and is caused by the partitioning along the plate boundary (Blanc et al. 2003) (Fig. 3.17).

The Zagros mountain range is created in a region where two tectonic plates, the Arabian Plate and the Eurasian Plate, are moving toward each other at an angle. The Arabian Plate is moving northwards relative to the Eurasian Plate at a rate of approximately 3 cm per year. The angle of obliqueness decreases as one moves southwards along the Zagros, and the collision becomes almost perpendicular within the Fars region.

The Tethyan oceanic crust was subducted toward the northeast along this section of the Eurasian Plate until the continental crust of the Arabian Plate started to be part of this convergent boundary. The precise timing of the subsequent collision's initiation is uncertain. Nevertheless, there is substantial evidence indicating that the Asmari Formation experienced deformation either during its deposition in the Oligocene period (Ahmadhadi et al. 2008; Sadooni and Alsharhan 2019) or potentially even earlier in the Late Eocene. Deformation in the Zagros Fold-Thrust Belt has persisted since then, with evidence of regional unconformities indicating multiple distinct phases of deformation. Unconformities have been identified at the lower and upper boundaries of the Asmari Formation (Late Eocene and Early Miocene), at the lower boundary of the Agha Jari Formation (late mid-Miocene), at the lower boundary of the Bakhtiari Formation (latest Pliocene), and during the mid-Pleistocene. These unconformities are interpreted as markers for dating these distinct phases. The deformation pulses were linked to the movement of the active deformation front toward the southwest (Hessami et al. 2001).

A manifestation of this collision is reflected in the gradual change from a thin thrust belt detached on Hormuz salt on the foreland side to more basement-associated deformation on the hinterland side (Sherkati et al. 2005). Evidence of the pre-Oligocene Zagros convergence and collision as deducted from thicknesses and facies changes was reported by Sherkati and Letouzey (2004). They noticed that the Paleocene-Eocene thickness and facies variations reflect indications of uplift and landward migration of depocenters along the High Zagros fault. These may represent an early indication of the Zagros collision. Also, the Eocene Pabdeh marls (the basinal equivalent of the Dammam Formation) increase in thickness toward the NE. This may be the tectonic load associated with the Sanandaj-Sirjan block. During the Early Oligocene, the thrust belt propagated southwestward along several Phanerozoic detachment horizons.

The presence of unconformities in the fold belt indicates that fold-thrust uplift has propagated from the northeast to the southwest. The rocks in the area have become younger toward the southwest, closer to the current thrust front (Hessami et al. 2001). The presence of extensive folding of the allochthonous and para-allochthonous sheets, as well as deep high-angle earthquake focal mechanisms, provides evidence of deeper faults and the involvement of the basement in the deformation (Vannucci et al. 2004).

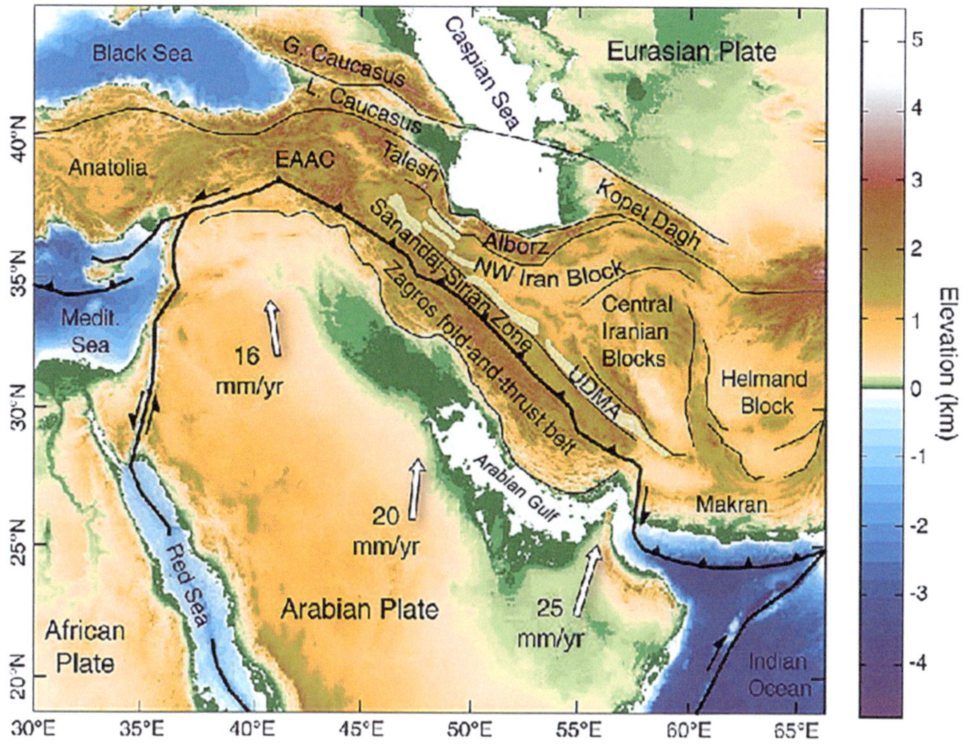

Fig. 3.17 Main tectonic units associated with the Zagros belt (Austermann and Iaffaldano 2013)

The deposition of the Gachsaran evaporites (a partial equivalent of the Dam Formation in Qatar) and their equivalents was due to the progressive restriction of depositional environments in the Zagros foredeep basin. This setting originated during the Oligocene in the southern region and became prevalent by the mid-Miocene era (Bahroudi and Koyi 2004). The fold-thrust deformation extended toward the southwest, causing the Gachsaran evaporites to become a significant upper detachment. This detachment was secondary to the main basal detachment in the Infracambrian Hormuz salt (Blanc et al. 2003).

3.4.12 The Eocene–Oligocene Boundary

The Eocene–Oligocene boundary, which occurred approximately 33.5 million years ago, marks a significant interface of major events that took place globally. This event represents the initial substantial decrease in global temperatures during the Cenozoic era and the commencement of Antarctic glaciation (Kargaranbafghi and Neubauer 2018). It is regarded as the pivotal time period in geological history following the extinction of dinosaurs. This event was marked by a substantial shift in climate, transitioning from the warm Eocene climate to a colder one. This temperature change led to the formation of ice sheets in Antarctica. The climatic changes caused significant geological processes such as the extinction of plants and animals, erosion, and changes in sea levels across various regions of the world (Cotton and Pearson 2011; Maravelis and Zelilidis 2012; Prothero 1994; Sarkar et al. 2003; Wade and Pearson 2008). Sheldon et al. (2002) observed significant climatic changes occurring at the Eocene–Oligocene boundary, characterized by a shift toward cooler and drier conditions compared to the previously dominant climate. During the Eocene epoch, the climate was characterized by abundant precipitation, with rainfall exceeding 1000 mm and a relatively high average temperature of over 12 °C. During the Oligocene period, the precipitation levels varied from 600 to 800 mm, while the average annual temperature ranged from 5 to 12 °C. Ivany et al. (2000) proposed an alternative perspective, arguing that fluctuations in temperature, rather than a shift in the average annual temperature, were the primary driver of the faunal changes observed during that time. In their research on the Eocene–Oligocene boundary in Kutch, western India, Sarkar et al. (2003) discovered several significant and sudden changes in climate and environment. These changes included a decrease in sea surface temperature, a decline in sea levels due to a lack of oxygen, and a reduction in phytoplankton production, which disrupted the existing ecological system. Cotton and Pearson (2011) conducted a study on three drill cores in Tanzania to investigate the extinction of the planktonic foraminifera Hantkeninidae. They discovered that this extinction event occurred in close proximity to the Eocene–Oligocene boundary. The extinction

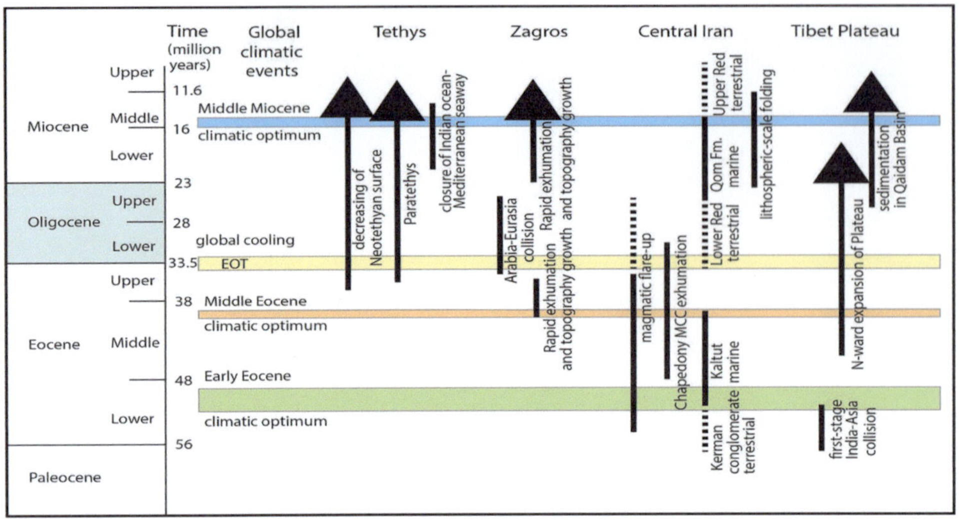

Fig. 3.18 Main climatic and tectonic events during the Eocene–Oligocene boundary (Sadooni and Alsharhan 2019)

was ascribed to potential climatic changes that initiated the processes, while a sudden influx of nutrients may have delivered the decisive blow. In their study, Karoui-Yaakoub et al. (2017) documented a significant decline in the population of planktic foraminifera at the Eocene–Oligocene boundary in northern Tunisia. This decline included the extinction of the turborotalids, followed by a substantial decrease in the size of *Pseudohastigerina*, and ultimately the extinction of the hantkeninids. Houben et al. (2012) discovered that the $\delta^{18}O$ values of benthic foraminifera during the Eocene–Oligocene boundary indicate the beginning of a significant glaciation period in Antarctica, albeit only partially. Maravelis and Zelilidis (2012) found that the Eocene–Oligocene boundary in the Thrace Basin, Northeast Aegean Sea in Greece, is marked by reduced water temperature, salinity, nutrient supply, and oxygen levels. There is a current dispute about the factors that led to the Eocene–Oligocene event and the magnitude of its impact. To obtain a thorough examination of the topic, refer to Prothero (1994) comprehensive analysis. Kargaran-bafghi and Neubauer (2018) proposed that this boundary was linked to significant occurrences in the area, such as the gradual cessation of the Neo-Tethys, the increasing construction and volcanic activity, and subsequently, the drying out of the Iranian Plateau. A significant correlation was discovered between the tectonic deformation and uplift of the Iranian Plateau and these climatic changes. In the northeastern region of the Arabian Plate, this boundary is linked to volcanic activity, the erosion of a significant portion of the Oligocene rock layers, the disappearance of certain bottom-dwelling organisms like the Hantkeninidae, the initiation of the Gulf of Aden's tectonic rift, the end of the Neo-Tethys oceanic basin, and the accumulation of freshwater and sedimentary deposits (Alsharhan and Nairn 1997) (Fig. 3.18).

An important concern regarding this boundary is the mechanism responsible for the significant shifts in animal species turnover. Certain researchers assert that these changes are catastrophic, whereas others contend that alterations in the plate configuration caused them during that time period.

References

Abu-Ali MA, Rudkiewicz J-LL, McGillivray JG, Behar F (1999) Paleozoic petroleum system of Central Saudi Arabia. GeoArabia 4(3):321–336. https://doi.org/10.2113/geoarabia0403321

Ahlbrandt TS, Pollastro RM, Klett TR, Schenk CJ, Lindquist S, Fox JE (2000) Region 2 assessment summary—Middle East and North Africa,—Description and results, U.S. Geological Survey Digital Data Series 60., Issue

Ahmadhadi F, Daniel J-M, Azzizadeh M, Lacombe, O (2008) Evidence for pre-folding vein development in the Oligo-Miocene Asmari formation in the Central Zagros Fold Belt, Iran. Tectonics 27(1). https://doi.org/10.1029/2006TC001978

Al Khatibi E, Abou Elenean KM, Megahed AS, El-Hussain I (2014) Improved characterization of local seismicity using the Dubai Seismic Network, United Arab Emirates. J Asian Earth Sci 90:34–44. https://doi.org/10.1016/j.jseaes.2014.04.009

Al-Husseini MI (2000) Origin of the Arabian Plate structures: Amar Collision and Najd Rift. GeoArabia 5(4):527–542. https://doi.org/10.2113/geoarabia0504527

AL-Johi A, Al-Laboun A (2015) Neoproterozoic (infracambrian)—Paleozoic total petroleum systems of Arabian Peninsula; an overview. Arab J Earth Sci 2(1):1–21

Al-Juboury AI, Howard J, Thusu B, Nichols G, Vincent SJ, Kaye MND, Vautravers B (2020) Mineralogy, geochemistry and source potential of shale in the Cambro-Ordovician Khabour formation, Iraq. J Afr Earth Sci 170:103901. https://doi.org/10.1016/j.jafrearsci.2020.103901

Alsharhan AS, Nairn AEM (1997) Sedimentary basins and petroleum geology of the Middle East. Elsevier

Alvarez LW, Alvarez W, Asaro F, Michel HV (1980) Extraterrestrial cause for the cretaceous-tertiary extinction. Science

208(4448):1095–1108. https://doi.org/10.1126/science.208.4448.1095

Amireh B, Schneider W, Abed A (2001) Fluvial-shallow marine-glaciofluvial depositional environments of the Ordovician System in Jordan. J Asian Earth Sci 19(1–2):45–60

Angiolini L, Gaetani M, Muttoni G, Stephenson MH, Zanchi A (2007) Tethyan oceanic currents and climate gradients 300 m.y. ago. Geology 35(12):1071–1074. https://doi.org/10.1130/g24031a.1

Austermann J, Iaffaldano G (2013) The role of the Zagros orogeny in slowing down Arabia-Eurasia convergence since ~5 Ma. Tectonics 32(3):351–363. https://doi.org/10.1002/tect.20027

Bahroudi A, Koyi HA (2004) Tectono-sedimentary framework of the Gachsaran formation in the Zagros foreland basin. Mar Pet Geol 21(10):1295–1310

Bambach RK, Knoll AH, Wang SC (2004) Origination, extinction, and mass depletions of marine diversity. Paleobiology 30(4):522–542. https://doi.org/10.1666/0094-8373(2004)030%3c0522:OEAMDO%3e2.0.CO;2

Becker L, Poreda RJ, Hunt AG, Bunch TE, Rampino M (2001) Impact event at the Permian-triassic boundary: evidence from extraterrestrial noble gases in fullerenes. Science 291(5508):1530–1533. https://doi.org/10.1126/science.1057243

Beerling DJ, Harfoot M, Lomax B, Pyle JA (2007) The stability of the stratospheric ozone layer during the end-Permian eruption of the Siberian Traps. Philos Trans R Soc A Math Phys Eng Sci 365(1856):1843–1866. https://doi.org/10.1098/rsta.2007.2046

Berra F (2012) Sea-level fall, carbonate production, rainy days: how do they relate? Insight from Triassic carbonate platforms (Western Tethys, Southern Alps, Italy). Geology 40(3):271–274

Berra F, Angiolini L (2014) The evolution of the Tethys region throughout the Phanerozoic: a brief tectonic reconstruction

Bertotti G, Immenhauser A, Koppen JKJT-V (2005) Stratigraphic and regional distribution of fractures in Barremian-Aptian carbonate rocks of Eastern Oman: outcrop data and their extrapolation to Interior Oman hydrocarbon reservoirs. Int J Earth Sci 94(3):447–461. https://doi.org/10.1007/s00531-005-0472-5

Beydoun ZR (1988) The Middle East: regional geology and petroleum resources. Scientific Press. https://books.google.com.qa/books?id=oU8ZAQAAIAAJ

Beydoun ZR (1991) Arabian plate hydrocarbon geology and potential—a plate tectonic approach. Am Assoc Pet Geol. https://doi.org/10.1306/St33533

Blanc E-P, Allen MB, Inger S, Hassani H (2003) Structural styles in the Zagros simple folded zone, Iran. J Geol Soc 160(3):401–412

Brenchley PJ, Marshall JD, Carden GAF, Robertson DBR, Long DGF, Meidla T, Hints L, Anderson TF (1994) Bathymetric and isotopic evidence for a short-lived Late Ordovician glaciation in a greenhouse period. Geology 22(4):295–298. https://doi.org/10.1130/0091-7613(1994)022%3c0295:Baiefa%3e2.3.Co;2

Cooper M (2007) Structural style and hydrocarbon prospectivity in fold and thrust belts: a global review

Cotton LJ, Pearson PN (2011) Extinction of larger benthic foraminifera at the eocene/oligocene boundary. Palaeogeogr Palaeoclimatol Palaeoecol 311(3):281–296. https://doi.org/10.1016/j.palaeo.2011.09.008

Courtillot V (2005) New evidence for massive pollution and mortality in Europe in 1783–1784 may have bearing on global change and mass extinctions. Comptes Rendus Géosci 337(7):635–637. https://doi.org/10.1016/j.crte.2005.03.001

Edgell HS (1991) Proterozoic salt basins of the Persian Gulf area and their role in hydrocarbon generation. Precambr Res 54(1):1–14. https://doi.org/10.1016/0301-9268(91)90065-I

El-ghali MAK (2005) Depositional environments and sequence stratigraphy of paralic glacial, paraglacial and postglacial Upper Ordovician siliciclastic deposits in the Murzuq Basin, SW Libya. Sed Geol 177(3):145–173. https://doi.org/10.1016/j.sedgeo.2005.02.006

Erwin DH (2006) Extinction: how life on Earth nearly ended 250 million years ago. Princeton University Press

Faqira M, Al-Hauwaj A (1998) Hydrocarbon potential of an Infracambrian salt basin in the western Rub'Al Khali, Saudi Arabia. In: Abstracts, 5th meeting of the Saudi Society for Earth Sciences, Dhahran

Filbrandt JB, Al-Dhahab S, Al-Habsy A, Harris K, Keating J, Al-Mahruqi S, Ozkaya SI, Richard PD, Robertson T (2006) Kinematic interpretation and structural evolution of North Oman, Block 6, since the Late Cretaceous and implications for timing of hydrocarbon migration into Cretaceous reservoirs. GeoArabia 11(1):97–140

Gettings ME, Blank Jr HR, Mooney WD, Healey JH (1986) Crustal structure of southwestern Saudi Arabia. J Geophys Res Solid Earth 91(B6):6491–6512. https://doi.org/10.1029/JB091iB06p06491

Gorin GE, Racz LG, Walter MR (1982) Late Precambrian-Cambrian Sediments of Huqf Group, Sultanate of Oman1. AAPG Bull 66(12):2609–2627. https://doi.org/10.1306/03b5ac82-16d1-11d7-8645000102c1865d

Greenwood WR, RE A, RJ R (1980) Precambrian geologic history and plate tectonic evolution of the Arabian Shield

Guiraud R, Bosworth W (1999) Phanerozoic geodynamic evolution of Northeastern Africa and the Northwestern Arabian platform. Tectonophysics 315(1–4):73–104

Hacker BR, Mosenfelder JL, Gnos E (1996) Rapid emplacement of the Oman ophiolite: thermal and geochronologic constraints. Tectonics 15(6):1230–1247. https://doi.org/10.1029/96TC01973

Hancock PL, Kadhi A (1978) Analysis of mesoscopic fractures in the Dhruma-Nisah segment of the central Arabian graben system. J Geol Soc 135(3):339–347. https://doi.org/10.1144/gsjgs.135.3.0339

Hays LE, Beatty T, Henderson CM, Love GD, Summons RE (2007) Evidence for photic zone euxinia through the end-Permian mass extinction in the Panthalassic Ocean (Peace River Basin, Western Canada). Palaeoworld 16(1–3):39–50

Hessami K, Koyi HA, Talbot CJ, Tabasi H, Shabanian EL (2001) Progressive unconformities within an evolving foreland fold–thrust belt, Zagros Mountains. J Geol Soc 158(6):969–981. https://doi.org/10.1144/0016-764901-007

Hildebrand AR, Penfield GT, Kring DA, Pilkington M, Camargo ZA, Jacobsen SB, Boynton WV (1991) Chicxulub crater: a possible cretaceous/tertiary boundary impact crater on the Yucatán Peninsula, Mexico. Geology 19(9):867–871.https://doi.org/10.1130/0091-7613(1991)019<0867:Ccapct>2.3.Co;2

Horacek M, Richoz S, Brandner R, Krystyn L, Spötl C (2007) Evidence for recurrent changes in Lower Triassic oceanic circulation of the Tethys: the $\delta 13C$ record from marine sections in Iran. Palaeogeogr Palaeoclimatol Palaeoecol 252(1):355–369. https://doi.org/10.1016/j.palaco.2006.11.052

Houben AJP, van Mourik CA, Montanari A, Coccioni R, Brinkhuis H (2012) The Eocene–Oligocene transition: changes in sea level, temperature or both? Palaeogeogr Palaeoclimatol Palaeoecol 335–336:75–83. https://doi.org/10.1016/j.palaeo.2011.04.008

Husseini M (1988) The Arabian infracambrian extensional system. Tectonophysics 148(1–2):93–103

Iannace A, Frisia S (1994) Changing dolomitization styles from Norian to Rhaetian in the Southern Tethys Realm. In: Dolomites, pp 75–89. https://doi.org/10.1002/9781444304077.ch6

Ivany LC, Patterson WP, Lohmann KC (2000) Cooler winters as a possible cause of mass extinctions at the Eocene/Oligocene boundary. Nature 407(6806):887–890. https://doi.org/10.1038/35038044

Johnson C, Hauge T, Al-Menhali S, Bin Sumaidaa S, Sabin B, West B (2005) Structural styles and tectonic evolution of onshore and offshore Abu Dhabi, UAE. International Petroleum Technology Conference

Johnson CA (2008) Phanerozoic plate reconstructions of the Middle East: Insights into the context of Arabian tectonics and sedimentation. Abu Dhabi International Petroleum Exhibition and Conference

Johnson PR, Halverson GP, Kusky TM, Stern RJ, Pease V (2013) Volcanosedimentary Basins in the Arabian-Nubian shield: markers of repeated exhumation and denudation in a neoproterozoic accretionary orogen. Geosciences 3(3):389–445. https://www.mdpi.com/2076-3263/3/3/389

Kargaranbafghi F, Neubauer F (2018) Tectonic forcing to global cooling and aridification at the Eocene-Oligocene transition in the Iranian plateau. Glob Planet Change 171:248–254

Karoui-Yaakoub N, Grira C, Mtimet MS, Negra MH, Molina E (2017) Planktic foraminiferal biostratigraphy, paleoecology and chronostratigraphy across the Eocene/Oligocene boundary in northern Tunisia. J Afr Earth Sci 125:126–136. https://doi.org/10.1016/j.jafrearsci.2016.11.009

Kent PE (1970) The salt of the Persian Gulf Region. Trans Leicester Literary Philos Soc 64:56–88

Konert G, Afifi AM, Al-Hajri S, i. A., & Droste HJ (2001) Paleozoic Stratigraphy and Hydrocarbon Habitat of the Arabian Plate. GeoArabia 6(3):407–442. https://doi.org/10.2113/geoarabia0603407

Krug AZ, Jablonski D, Valentine JW (2009) Signature of the end-cretaceous mass extinction in the Modern Biota. Science 323(5915):767–771. https://doi.org/10.1126/science.1164905

Lüning S, Craig J, Loydell DK, Štorch P, Fitches B (2000) Lower Silurian 'hot shales' in North Africa and Arabia: regional distribution and depositional model. Earth-Sci Rev 49(1):121–200. https://doi.org/10.1016/S0012-8252(99)00060-4

Macleod N, Rawson PF, Forey PL, Banner FT, Boudagher-Fadel MK, Bown PR, Burnett JA, Chambers P, Culver S, Evans SE, Jeffery C, Kaminski MA, Lord AR, Milner AC, Milner AR, Morris N, Owen E, Rosen BR, Smith AB et al (1997) The cretaceous-tertiary biotic transition. J Geol Soc 154(2):265–292. https://doi.org/10.1144/gsjgs.154.2.0265

Maravelis A, Zelilidis A (2012) Paleoclimatology and paleoecology across the Eocene/Oligocene boundary, thrace basin, Northeast Aegean Sea, Greece. Palaeogeogr Palaeoclimatol Palaeoecol 365–366:81–98. https://doi.org/10.1016/j.palaeo.2012.09.015

Metcalfe I (2006) Palaeozoic and Mesozoic tectonic evolution and palaeogeography of East Asian crustal fragments: the Korean Peninsula in context. Gondwana Res 9(1):24–46. https://doi.org/10.1016/j.gr.2005.04.002

Montenat C, Barrier P, Soudet HJ (2003) Aptian faulting in the Haushi-Huqf (Oman) and the tectonic evolution of the southeast Arabian platform-margin. GeoArabia 8(4):643–662. https://doi.org/10.2113/geoarabia0804643

Muttoni G, Gaetani M, Kent DV, Sciunnach D, Angiolini L, Berra F, Garzanti E, Mattei M, Zanchi A (2009) Opening of the Neo-Tethys Ocean and the Pangea B to Pangea A transformation during the Permian. GeoArabia 14(4):17–48. https://doi.org/10.2113/geoarabia140417

Ocampo A, Vajda V, Buffetaut E (2006) Unravelling the cretaceous-paleogene (KT) catastrophe: evidence from flora fauna and geology [Article]. Biological processes associated with impact events, pp 203–227. https://www.scopus.com/inward/record.uri?eid=2-s2.0-41549087569&partnerID=40&md5=fe4432bfc0020554e677d5bc08caf9f6

Perotti CR, Carruba S, Rinaldi M, Bertozzi G, Feltre L, Rahimi M (2011) The Qatar–South fars arch development (Arabian Platform, Persian Gulf): insights from seismic interpretation and analogue modelling. In: Uri S (ed) New frontiers in tectonic research. IntechOpen, pp Ch. 13. https://doi.org/10.5772/20299

Pollastro RM (2003) Total petroleum systems of the Paleozoic and Jurassic, Greater Ghawar Uplift and adjoining provinces of central Saudi Arabia and northern Arabian-Persian Gulf

Prothero DR (1994) The late eocene-oligocene extinctions. Annu Rev Earth Planet Sci 22(1):145–165

Reichow MK, Pringle MS, Al'Mukhamedov AI, Allen M, Andreichev VL, Buslov MM, Davies C, Fedoseev GS, Fitton JG, Inger S (2009) The timing and extent of the eruption of the Siberian Traps large igneous province: implications for the end-Permian environmental crisis. Earth Planet Sci Lett 277(1–2):9–20

Ruban DA, Al-Husseini MI, Iwasaki Y (2007) Review of Middle East Paleozoic plate tectonics. GeoArabia 12(3):35–56. https://doi.org/10.2113/geoarabia120335

Sadooni FN, Alsharhan AS (2019) Regional stratigraphy, facies distribution, and hydrocarbons potential of the Oligocene strata across the Arabian Plate and Western Iran. Carbonates Evaporites 34(4):1757–1770. https://doi.org/10.1007/s13146-019-00521-3

Sarkar A, Sarangi S, Bhattacharya SK, Ray AK (2003) Carbon isotopes across the Eocene-Oligocene boundary sequence of Kutch, western India: implications to oceanic productivity and pCO2 change. Geophys Res Lett 30(11). https://doi.org/10.1029/2002GL016541

Schandelmeier H, Reynolds PO, Semtner A-K (1997) Palaeogeographic-palaeotectonic atlas of north-eastern africa, arabia, and adjacent areas: late neoproterozoic to holocene

Sellwood BW, Valdes PJ (2006) Mesozoic climates: general circulation models and the rock record. Sediment Geol 190(1):269–287. https://doi.org/10.1016/j.sedgeo.2006.05.013

şengör AMC (1979) Mid-Mesozoic closure of Permo-Triassic Tethys and its implications. Nature 279(5714):590–593. https://doi.org/10.1038/279590a0

Sengor A, Robertson A, Searle M, Ries A (1990) The geology and tectonics of the Oman region. Geological Society [London]

Sepkoski JJ (1982) Mass extinctions in the Phanerozoic oceans: a review

Sharland PR, Casey DM, Davies RB, Simmons MD, Sutcliffe OE (2001) Arabian Plate sequence stratigraphy, vol 9. Gulf Petrolink

Sheldon ND, Retallack GJ, Tanaka S (2002) Geochemical climofunctions from North American soils and application to paleosols across the Eocene-Oligocene boundary in Oregon. J Geol 110(6):687–696

Sherkati S, Letouzey J (2004) Variation of structural style and basin evolution in the central Zagros (Izeh zone and Dezful Embayment), Iran. Marine Petrol Geol 21(5):535–554

Sherkati S, Molinaro M, de Lamotte DF, Letouzey J (2005) Detachment folding in the Central and Eastern Zagros fold-belt (Iran): salt mobility, multiple detachments and late basement control. J Struct Geol 27(9):1680–1696

Stampfli G, Borel G, Cavazza W, Mosar J, Ziegler P (2001) Palaeotectonic and palaeogeographic evolution of the western Tethys and PeriTethyan domain (IGCP Project 369). Episodes J Int Geosci 24(4):222–228

Stampfli G, Marcoux J, Baud A (1991) Tethyan margins in space and time. Palaeogeogr Palaeoclimatol Palaeoecol 87(1):373–409. https://doi.org/10.1016/0031-0182(91)90142-E

Stampfli GM (2000) Tethyan oceans. Geological Society, London, Special Publications 173(1):1–23

Stampfli GM, Borel G (2002) A plate tectonic model for the Paleozoic and Mesozoic constrained by dynamic plate boundaries and restored synthetic oceanic isochrons. Earth Planet Sci Lett 196(1–2):17–33

Stern RJ, Johnson P (2010) Continental lithosphere of the Arabian Plate: a geologic, petrologic, and geophysical synthesis. Earth-Sci Rev 101(1):29–67. https://doi.org/10.1016/j.earscirev.2010.01.002

Stewart SA (2016) Structural geology of the Rub' Al-Khali Basin, Saudi Arabia. Tectonics 35(10):2417–2438. https://doi.org/10.1002/2016TC004212

Stocklin J (1968) Structural history and tectonics of Iran: a review. AAPG Bull 52:1229–1258

Talbot CJ, Alavi M (1996) The past of a future syntaxis across the Zagros. In: Salt Tectonics (eds) 89109 100 The Geological Society Publishing House, Bath. Geological Society Publishing House

Vannucci G, Pondrelli S, Argnani A, Morelli A, Gasperini P, Boschi E (2004) An atlas of Mediterranean seismicity

Vaslet D (1990) Upper Ordovician glacial deposits in Saudi Arabia . Int Union Geol Sci 13(3):147–161. https://doi.org/10.18814/epiiugs/1990/v13i3/002

von Raumer J, Stampfli G, Borel G, Bussy F (2002) Organizationn of pre-Variscan basement areas at the North-Gondwanan margin. Int J Earth Sci 91:35–52

von Raumer J, Stampfli G, Bussy F (2003) Gondwana-derived microcontinents—the constituents of the Variscan and Alpine collisional orogens. Tectonophysics 365(7–22)

Wade BS, Pearson PN (2008) Planktonic foraminiferal turnover, diversity fluctuations and geochemical signals across the Eocene/Oligocene boundary in Tanzania. Mar Micropaleontol 68(3–4):244–255

Ziegler MA (2001) Late Permian to Holocene paleofacies evolution of the Arabian Plate and its hydrocarbon occurrences

Ziegler A, Eshel G, Rees PM, Rothfus T, Rowley D, Sunderlin D (2003) Tracing the tropics across land and sea: Permian to present. Lethaia 36(3):227–254

4 Ediacaran and Cambrian

Abstract

The Ediacaran and Cambrian geological formations in Arabia provide essential insights into the region's tectonic and sedimentary history. The Infracambrian sequence, encompassing Neoproterozoic to Early Cambrian strata, illustrates a succession of depositional and tectonic processes that influenced the Arabian Plate. The geology of Western Arabia is primarily characterized by Neoproterozoic crust, which is substantially visible due to the rifting of the Red Sea. In contrast, eastern Arabia, including Qatar, does not exhibit exposed Precambrian basement rocks. The eastern portions, characterized by dense Phanerozoic layers, are believed to harbor juvenile Neoproterozoic crust beneath the sedimentary cover, as observed in Oman. Prominent formations, like the Hormuz and Huqf Groups, indicate substantial salt tectonics and hydrocarbon potential throughout the region. The Qatar High separates significant salt basins, with the Hormuz salt mobilization creating elongated formations and salt domes that affect oilfields and subsurface structural traps. The Huqf Group in Oman and the Hormuz Series are geologically interconnected, consisting of clastics, carbonates, and evaporites formed in an arid setting. Islands such as Halul and Sharao, located off the coast of Qatar, exhibit outcrops of the Hormuz Formation, which comprises dolomite, anhydrite, and small igneous intrusions. The dolomitic formations of these islands exhibit the effects of salt diapirism and tectonic brecciation. The Ediacaran-Cambrian sequence includes formations like the Jibalah Group and the Cambrian Siq and Burj Formations, exemplifying depositional conditions from fluvial to tidal systems. Stratigraphic investigations throughout Arabia, particularly in Jordan and Saudi Arabia, facilitate the reconstruction of the regional context of these formations. This chapter offers an extensive overview of the Ediacaran-Cambrian strata of the Arabian Plate, emphasizing its importance in tectonics, sedimentology, and resource exploration.

4.1 Introduction

The oldest rocks in Arabia are assigned to the Infracambrian, an obsolete term that includes parts of the Neoproterozoic (Cryogenian), late Neoproterozoic (Ediacaran), and Early Cambrian. It ended with the appearance of *trilobite*-bearing sediments of the Khabour Quartzite Formation (Late Cambrian-Tremadocian) in the Arabian Platform's northern parts (Baban and Lawa 2016).

Western Arabia's geology is dominated by exposures of the upper continental crust due to the Red Sea's rifting during the Oligo-Miocene. The basement is also visible in sporadic outcrops in western Arabia near the Gulf of Aden and the Indian Ocean (Fig. 4.1). These rocks are Neoproterozoic in age based on radiometric dating. Except for substantial chunks of Paleoproterozoic and Archean crust in the southwest, isotopic data show that the basement is primarily a juvenile continental crust that originated during the Neoproterozoic (Stoeser and Frost 2006; Whitehouse et al. 2001; Windley et al. 1996).

In eastern Arabia, such as Qatar, the setting is different. The Precambrian basement is not exposed and has not been drilled in Qatar. The nature of the continental crust east and north of the Arabian Shield is not fully investigated due to the shortage of outcrops, the thick sediment cover, and little or no publicly available geophysical data. According to some tectonic theories, juvenile "shield-type" rocks reach the Zagros thrust. The most significant impediment to a complete understanding of eastern Arabia's crystalline crust is that it is mainly covered by Phanerozoic strata and only appears in small, dispersed spots in Oman. Despite their small size, basement exposures

F. N. Sadooni and H. A. Al-Kuwari, *Geological Evolution of Qatar and the Arabian Peninsula*, Regional Geology Reviews,
https://doi.org/10.1007/978-3-031-94394-2_4

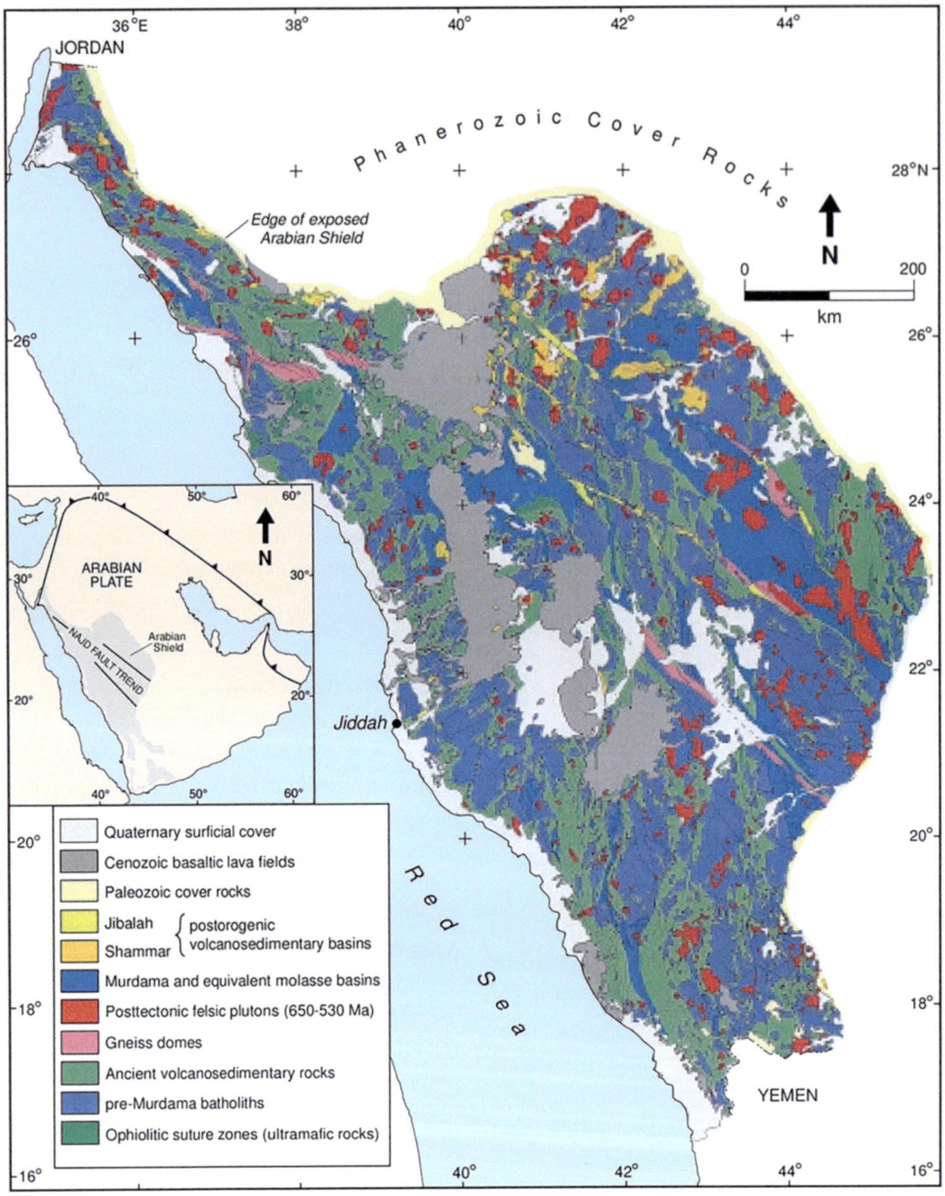

Fig. 4.1 Distribution of the basement rocks in western Saudi Arabia (Nehlig et al. 2002)

in Oman have recently been studied because they are crucial to understanding the crust of eastern Arabia. Most notably, known geochronologic data show that igneous and metamorphic activity occurred during the Tonian (1000–850 Ma) and early Cryogenian periods and stopped by 700 Ma, as opposed to the protracted period of magmatic and metamorphic processes that characterized the Arabian Shield's formation. Subsequent Neoproterozoic activity in Oman became Cryogenian-Ediacaran deposition of unconformable siliciclastic, glaciogenic, carbonate, and evaporite rocks (the Huqf Supergroup) Gass et al. (1990).

Dated Precambrian exposures in Oman using Rb–Sr whole-rock dating, including basement exposures on the Mirbat (Marbat) Peninsula, Al-Hallaniyat (Kuria Muria) Islands, and Jebel Ja'alan yielded dates ranging from 706.40 to 850.27 Ma, with initial 87Sr/86 Sr values ranging from 0.7025 to 0.7041.

The crystalline basement of the eastern parts of the Plate is also Neoproterozoic in age. However, it has a geologic history unique to the Shield, as deduced from the studies mentioned above on the Oman outcrops. There are accreted island arcs and microcontinental terrains in the basement. The Amar Collision is the final collision of the accreted plates, which occurred between 640 and 620 million years ago when the Rayn plate merged with the Afif Terrain (Al-Husseini 2000; Geert et al. 2001). Large-scale horst structures are created

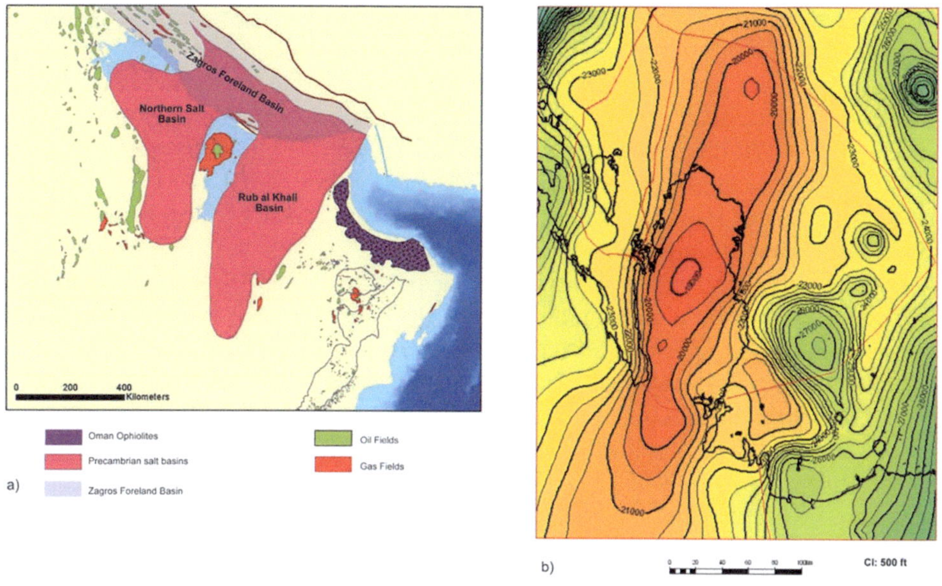

Fig. 4.2 **a** Map of the main structural elements of Qatar and the neighboring areas **b** gravity and magnetic data-based basement structural map of eastern Arabia (Van Buchem et al. 2014)

by this collision, including the Ghawar Horst and the Qatar-Fars High, which were identified using gravity and magnetic data (Fig. 4.2b; Van Buchem et al. 2014). The North Gulf Salt Basin and the Rub Al Khali Basin are two enormous salt basins separated by the Qatar High from the eastern Arabian plate (Fig. 4.2a). Later, salt diapirs were created by mobilizing the heavy Precambrian Hormuz salt. According to Murris (1980), each salt basin has its type of deformation: the northern salt basin is characterized by elongate structures and is controlled by deep, foundation fault block tectonics (for example, the Dukhan field), whereas the southeastern salt basin contains more minor salt domes (e.g., Bul Hanine, Idd El Shargi, Maydan Mahzam, and El Bunduq fields; Fig. 4.2a).

4.2 Depositional Systems

The N–S elongated Qatar High, which reached as far north as coastal Fars, was flanked by Precambrian salt basins in the northwest and southeast (Fig. 4.2a). Because the lower Paleozoic and Pre-Cambrian have yet to be investigated in Qatar (the deepest wells, MK-2 and DK-27, have a TD of 4877 m (16,000 ft) in the Ordovician Qasim Formation), gravity and magnetic studies were employed to analyze this segment of the geologic record (Fig. 4.2b). According to gravity-derived structural maps, there may be thin Pre-Cambrian layers beneath the deepest well penetrations, which indicate that there are around 1219 m (4000 ft) of Cambrian sediments beneath them Geert et al. (2001). Geng et al. (2022) studied the distribution of the Hormuz salt sediments in the United Arab Emirates using the residual gravity anomaly of the sedimentary cover. They suggested that the Hormuz salt sediments are more prevalent in the offshore area in NE–SW to N–S trending basins with an average thickness of 2100 m. They also proposed that the Hormuz, the Ara (in central Oman), and the East Rub Al Khali (in Saudi Arabia) salt basins are probably connected except in the basement high regions.

According to Mattes and Morris (1990), the Huqf Group in Oman is the equivalent of the Hormuz salt group. It is dated as Precambrian-Cambrian and comprises the Abu Mahara Formation, which has a thickness of more than 1400 m and is mainly made up of sandstone and conglomerate. This formation is overlain by the Khufai Formation, which is made up of dolomite and has a thickness of 390 m, and by the Shuram Formation, which is made up of carbonate and clastic mixture and has a thickness of 600 m. The Shuram Formation is followed by the Buah Formation, primarily composed of dolomite. The Ara Formation is the upper formation in the Group, consisting of at least five cycles of alternating carbonates (mostly dolomite) and evaporites with a total thickness of roughly 1700 m.

Stewart (2020) provided a new comprehensive tectonostratigraphic framework of the Ediacaran basins of Saudi Arabia, including the eastern parts neighboring Qatar, based on 2D and 3D seismic reflection data. The following synthesis is based on his work.

4.2.1 The Pre-Hormuz Succession

In the absence of drilled wells or shared domain geophysical data, the pre-Hormuz succession of Qatar is mere speculation based on the available information. As shown in Fig. 4.3, there are two geologic settings as far as Qatar is concerned. The first setting is that of Halul and Sharao islands, where there are some exposures of anhydrite and dolomites belonging to the Hormuz Series; the second setting is that of the mainland, where the Hormuz is missing. In both sets, the oldest strata belong to the Huqf Group, which consists of mixed clastics and carbonates with evaporites, followed by the Badayi andesite, which is overlain by the siliciclastic and carbonates of the Muraykhah Formation (Al-Husseini 2011). At this point, the mainland is distinctive from the Halul Sharao succession. On the mainland, the Muraykhah Formation is overlain by the Kurayshah Group (550–530 Ma), Stewart (2020). This group is formed of a mixed succession of fluvial clastics and basalt and is separated from the Muraykhah Formation by an unconformity surface. According to Al-Husseini (2011) and Stewart (2020), this group is equivalent to the Jifn Formation, which has a thickness of around 2000 m. Both formations are exposed in some fault-cut basins on the Arabian Shield, but the sediments of the Kurayshah Group are prevalent in the Nafud Basin with a thickness of up to 4 km. The Kurayshah and the Jifn strata are nearly equivalent to the Hormuz-Ara Group of Eastern Arabia.

Igneous materials are common in these units. The Kurayshah Group basalts are generally contemporaneous with ash beds in the Ara Group of Oman, while igneous materials occur in Hormuz salt diapirs of the Zagros and Arabian Gulf (Allen 2007; Stewart 2020; Thomas et al. 2015). These igneous materials are part of a more extensive distribution of late Ediacaran to early Cambrian igneous activity recorded along Gondwana's "Cadomian" margin (Moghadam et al. 2017; Stewart 2020).

4.2.2 The Hormuz Series

The Hormuz strata are present only in Halul and Sharao islands offshore the mainland of Qatar. Stewart (2020) summarizes the plate tectonic and depositional context of the Hormuz-Ara evaporite basins from different sources in Fig. 4.4. Radiometric ages of some ash beds and lavas in the Hormuz and Ara basins generated dates ranging from 558 to 542 Ma (Allen 2007; Faramarzi et al. 2015), which means that they are equivalent to Kurayshah Group clastics and volcanic. Correlations have also been established with evaporites from other areas, such as the Indian plate's Bikaner-Nagaur-Potwar basins (Allen 2007; Smith et al. 2012). Evaporite deposition was prevalent west of the Gondwanan margin's Cadomian terranes (Álvaro et al. 2000). This north Gondwanan Ediacaran evaporite basin system necessitates the presence of a long barrier further outboard, to the north in present-day coordinates, separating the evaporite basins from the Panthalassic Ocean. More recent models place South China next to India, with the Arabian sector of the Gondwanan margin separated from the Panthalassic Ocean by a relatively narrow emergent zone (Merdith et al. 2017). Late Ediacaran sedimentary and igneous complexes in Iran (Etemad-Saeed et al. 2015; Moghadam et al. 2017; Moghaddam et al. 2020) and Turkey (Abbo et al. 2015; Gürsu et al. 2015) provide evidence for this belt near the Arabian plate. Late Ediacaran volcanism

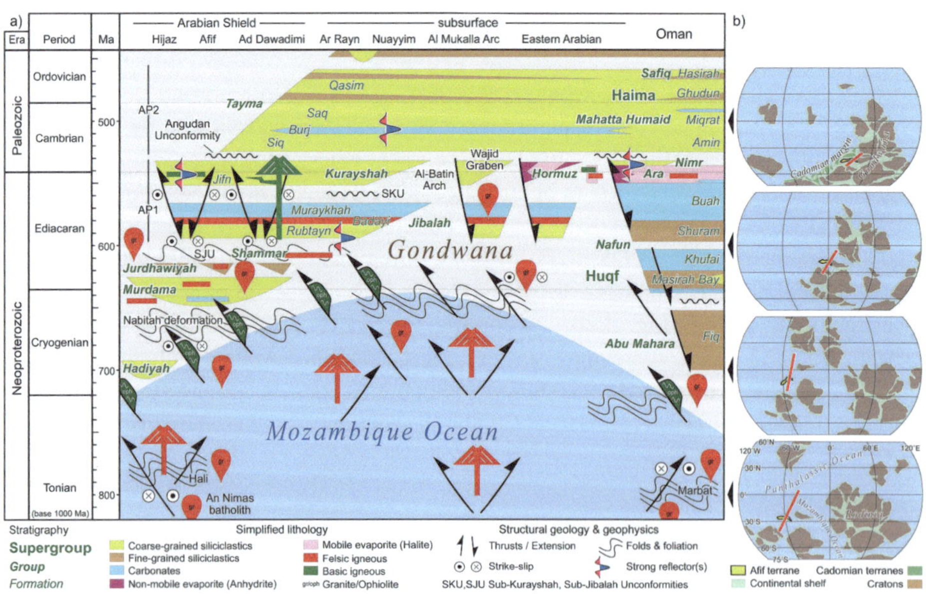

Fig. 4.3 Tectonostratigraphic framework of the Arabian Plate during the Neoproterozoic to the Ordovician. There are two possible settings for Qatar: the mainland, where there are no Hormuz strata, and Halul and Sharao islands, where the Hormuz anhydrite and dolomite were reported (Stewart 2020)

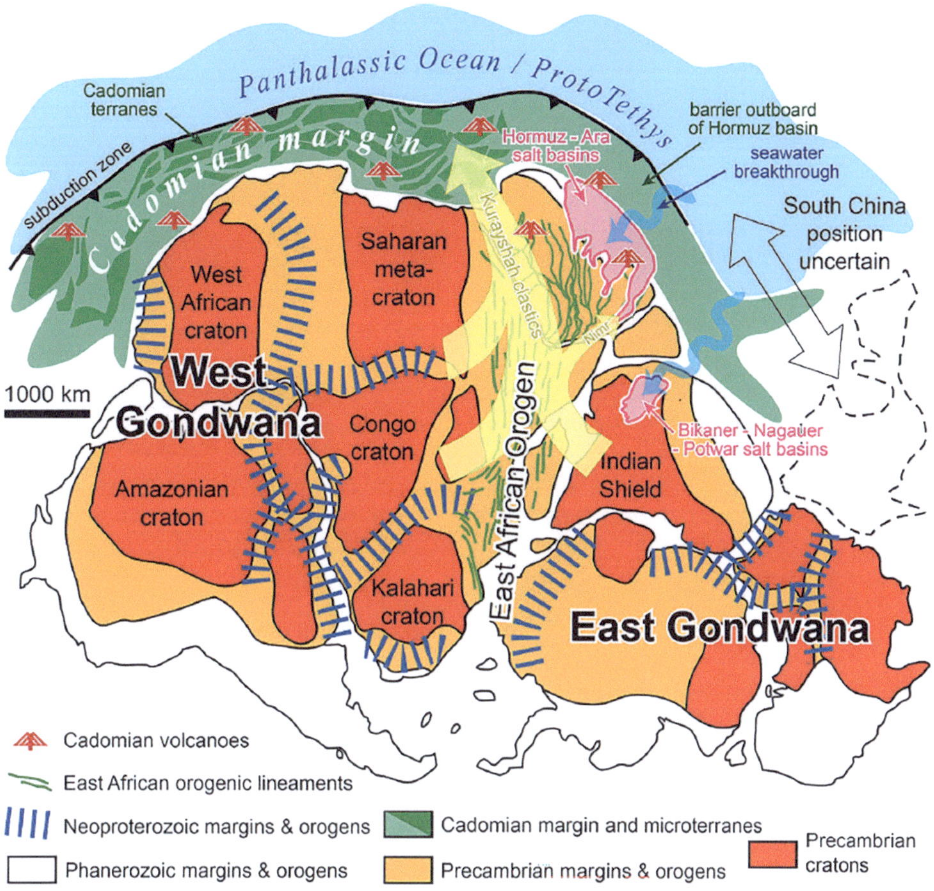

Fig. 4.4 Plate tectonic configuration of the Hormuz-Ara evaporite basins of Gondwana during the end of the Precambrian (Stewart 2020)

is depicted in Fig. 4.4 as a regional belt or arc associated with the active north margin of Gondwana (Cadomian margin), providing context for the volcanic activity. The Kurayshah clastics are depicted as part of a continental-scale superfan funneling northwards along the denuded East African Orogen, a forerunner to the Tayma Group's broadly accepted regional depositional architecture (Meinhold et al. 2013; Morag et al. 2011). The southwest margin of the Hormuz evaporite basin occupies late Ediacaran to early Cambrian subbasins on the eastern side of Saudi Arabia and in Oman, which are investigated in terms of how the evaporites influence the structural style and their relationships with the Kurayshah Group.

A possible stratigraphic configuration of western Qatar can be deduced from a seismic section passing through the Ghawar oilfield in eastern Saudi Arabia (Fig. 4.5). The section shows that the Hormuz salt is concentrated in the structural highs and piercing the strata of the Saq, Burj/Siq undifferentiated and hunching the beds of the Qasim Formation.

The term Hormuz Formation was introduced by Blanford (1872) and Edgell (1991), and the Hormuz Series was renamed by Pilgrim (1908). There is no complete type section of the formation, and strata are deformed by salt tectonics. The age of the formation is assigned as the uppermost Proterozoic to Middle Cambrian by Alavi (2007), which makes it nearly equivalent to the Ara Evaporite group in Oman. Edgell (1991) divided the Hormuz Formation into upper and lower units. The lower unit is formed of alternation of halite, anhydrite, and carbonates with some siltstone and purple or green sandstone. The upper unit of the formation comprises anhydrite and halite with inclusions of gabbro and diabase (Stewart 2016, 2020). The Hormuz sediments are present in many diapiric islands such as the Sir Bin Yas, Dalma, Qarnain, Arzana, Zirku, Das, Dayana, Sir Bu N'Air in the UAE, Halul, and Sharao in Qatar and Jabal Siman hill in southern Iraq. The thickness of the Hormuz Formation ranges between 2000 and 4000 m with an average of 2100 m (Geng et al. 2022).

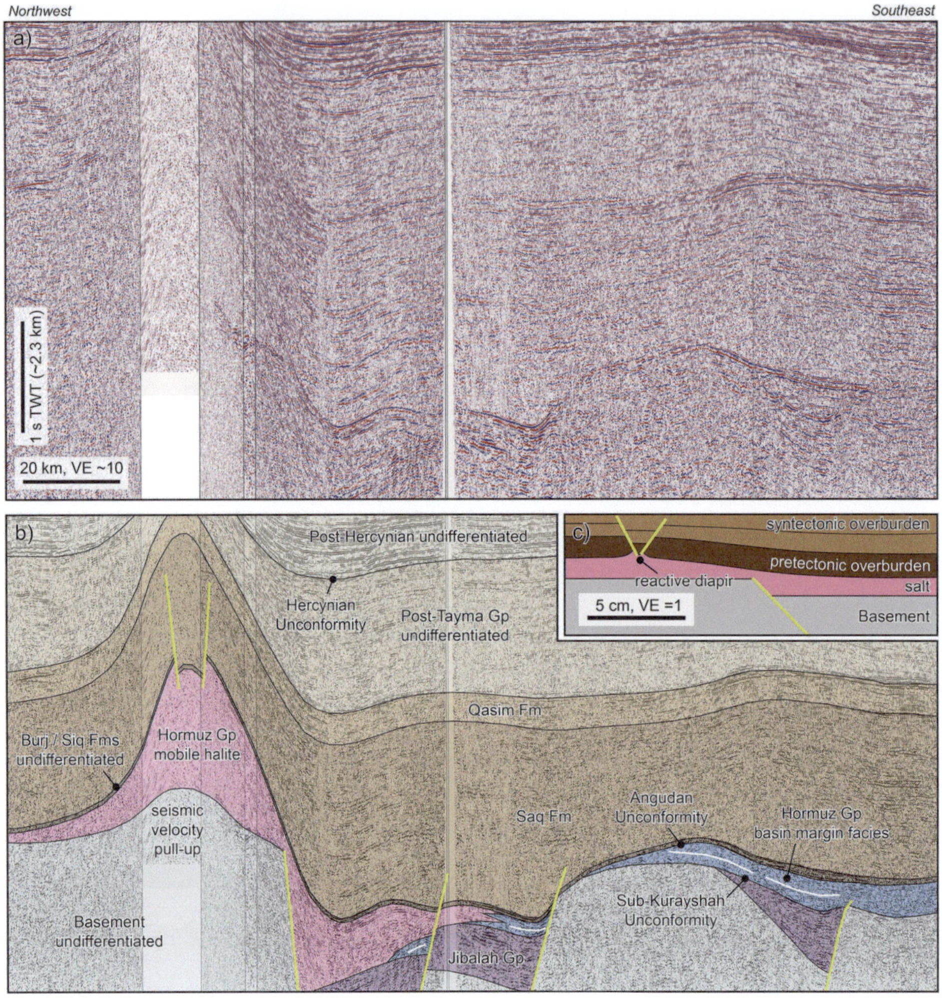

Fig. 4.5 Composite 2D and 3D seismic section through eastern Saudi Arabia passing through the Dammam dome may serve as an approximation of the chronostratigraphic configuration of the mainland of Qatar (Stewart 2020)

The distribution of the Hormuz strata in the United Arab Emirates using residual gravity anomaly supported by well data was documented by Geng et al. (2022). They suggested that many of the major offshore oilfields of the UAE are associated with the salt tectonics of the Hormuz Formation. They indicated that the Hormuz salt is found in N–S to NE–SW basins parallel to the basement highs and that the salt basins are concentrated in topographic lows between the structural highs. They also concluded that the Hormuz salt basins of Arabia are connected to the salt basins of the Ara Formation of Oman except over structural basement highs. Figure 4.6 shows the relation between the development of subsurface structures/oilfields between the UAE and Qatar.

4.2.2.1 The Hormuz Formation in Halul Island

Halul Island is in the eastern part of offshore Qatar, about 100 km from Doha (the capital of Qatar). It has a total area of 1.5 km^2. It comprises small hills that reach a maximum elevation of about 55 m above sea level (Fig. 4.7). The island's sediments consist of anhydrite, carbonates, and clastic deposits, accompanied by igneous rocks. Very few papers have been written on the geology of Halul Island (Pilgrim 1908) provided a general description of Qatar's offshore islands. Shell Company geologists created the first geological map of the island in 1964 (scale 1:10,000). Meneisy (1988) used K/Ar dating to date volcanic rocks on the island, implying that the sediments formed 33 million

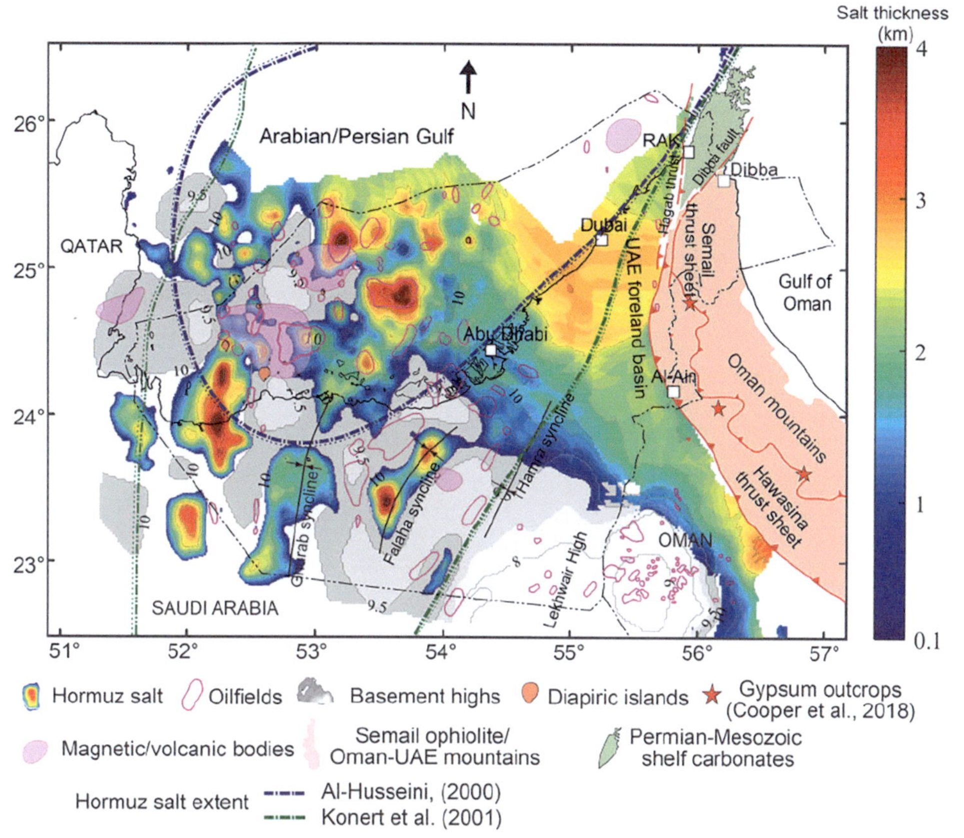

Fig. 4.6 Structural map of the area between UAE and Qatar shows the relationship between the distribution of Hormuz salt and the development of subsurface structures and diapiric islands (Geng et al. 2022)

years ago. Halul Island has a distinctive elongated dome shape, which, like many other similar islands in the Gulf region, results from salt piercement on the surface. The island corresponds to a prominent negative anomaly on the Bouguer map (Shell 1964). The mobility of the Infracambrian Hormuz salt formation is thought to have played an important role in the island's formation. The main structural feature is an angular unconformity between the older Hormuz Formation clastic and carbonate rocks and the younger Miocene dolomitic limestones.

Recent urbanization has obscured the island's geology, and distinguishing the Hormuz suite from the overlying Miocene rocks is no longer possible. Dolomite, composed of thin, friable lamina that may be pseudomorphs after gypsum, and hard, massive pink and purple anhydrite with intense iron staining are among the exposed rocks on the island.

No significant igneous rocks were discovered in the studied sections. However, a composite section of the Island prepared by Shell geologists before the extensive urbanization of the island records igneous rocks. The section descriptions and samples are available in the Qatar Petroleum office on the island. The main lithologies of the island, as sampled by Shell geologists, are (Fig. 4.8):

1. Amygdaloidal andesite and basalt with chlorite and evaporites.
2. Platy, reddish hematite anhydrite with gypsum crystals.
3. Hematitic limey dolomite.
4. Broken, contoured gypsum.
5. Bedded hematitic stained limestone.
6. Interbedded evaporite and grey dolomite.
7. Transparent, crystalline gypsum.
8. Evaporites are slightly reddish stained.

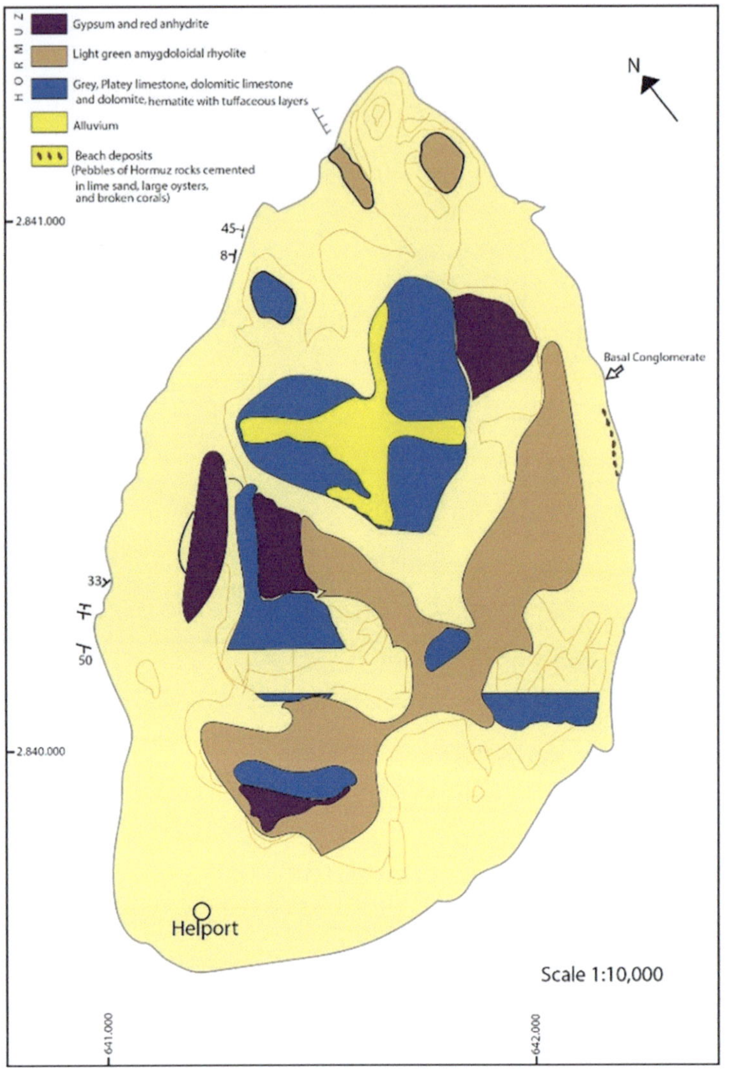

Fig. 4.7 Geologic map of Halul island

Discrepancies between the present stratigraphy of the island and those of the Shell compiled section are due to the disturbance of the surface exposures by urbanization. It can be concluded, however, that the island consists of three main lithologic units:

1. Grey to light yellow hard massive dolomitic limestone and limestone with a minor occurrence of siltstone and mudstone.
2. Pinkish and red, locally coarsely crystalline anhydrite and gypsum.
3. Minor igneous rocks, mostly andesite, with hematitic tuffaceous layers (Fig. 4.9).

The total thickness of the formation in Halul Island has not been penetrated fully, and most of the igneous materials described by early workers are no longer present due to the extensive urbanization of the island.

The correlation of the Hormuz Formation across Halul Island showed considerable variations in the ratio of carbonates to evaporites (Fig. 4.10). The top forty meters of the formation are mainly evaporites in the well Halul-1, changing into predominantly dolomite and dolomitic limestone in Halul 2 and 3. There is a thin bed of reddish-brown clayey silt in the well Halul-1 and a thin cavity at around 11–12 m in the well Halul-3, which may suggest the presence of dissolved halite. These lithological variations are probably attributed to halokinesis (salt tectonics) rather than depositional or diagenetic processes.

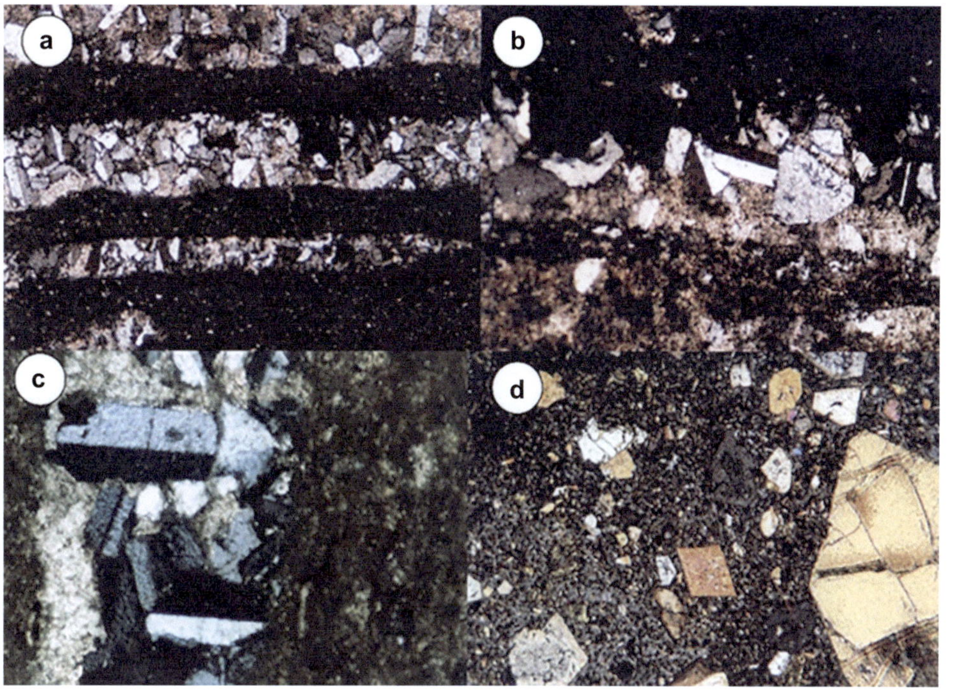

Fig. 4.8 Thin sections of micrographs from the Hormuz Formation in Halul Island **a** fractured dark mudstone and siltstone. The fractures are filled with orthoclase and quartz, indicating igneous intrusion **b** large sanidine crystals filling a fracture in mudstone groundmass, **c** close-up of sanidine crystals, **d** dolomite rhombs replaced by evaporites in a mixed groundmass

Fig. 4.9 Igneous tuffaceous bed from Halul Island

4.2.2.2 The Hormuz Formation in Sharao Island

Sharao Island is a smaller island located approximately 98 km southeast of the Qatar Peninsula (Fig. 4.11). The island is about 2 km long and 1 km wide. Sharao Island is still being researched. It has a relatively high rocky cliff, roughly 12 m high, enclosing a level interior covered with weathered and stained tectonic breccia. The cliff comprises a basal tectonic breccia overlain by a succession of dolomite layers 10–12 cm thick separated by thin weathered-out layers up to 25 cm wide. There are some thin beds of pink gypsum with yellowish siltstone and mudstone (Fig. 4.12).

The dolomite of both Halul and Sharao islands is thinly bedded or contorted. It alternates with pink, ferruginous

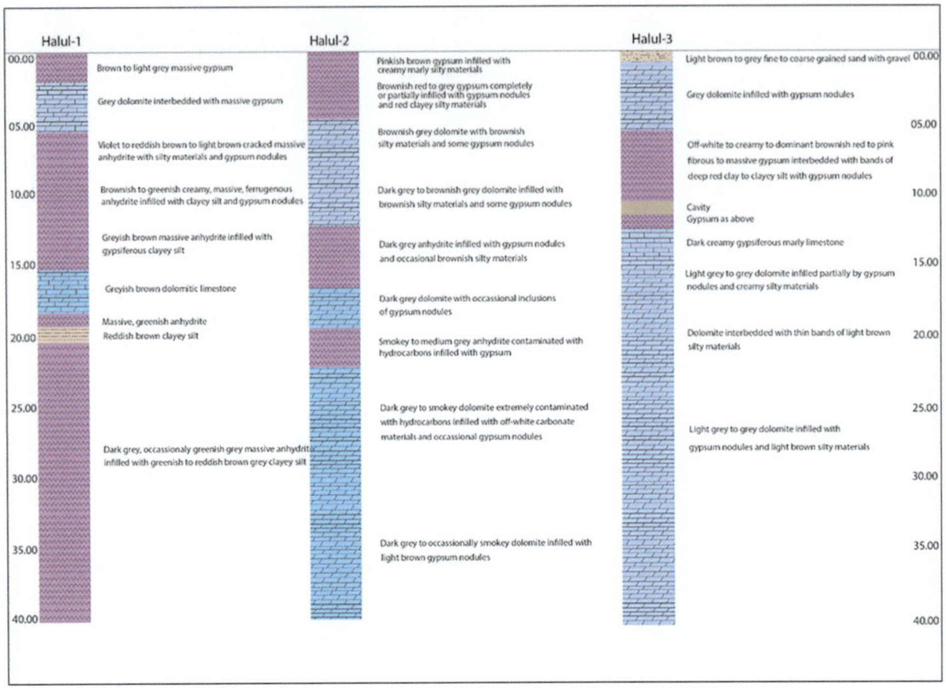

Fig. 4.10 Lithology of the top 40 m of the Hormuz Formation in Halul Island

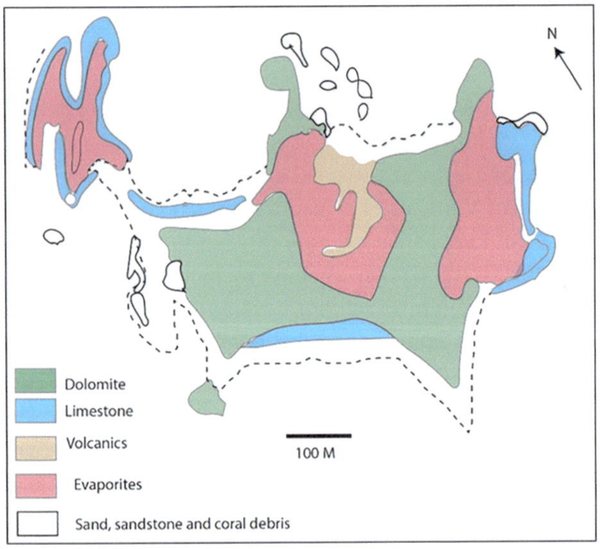

Fig. 4.11 Geologic map of Sharao island

anhydrite and massive gypsum with some thin beds of limestone and reddish-brown clayey silt (Fig. 4.13).

Thin sections of the dolomite assemblages collected from the island show widespread intrusion by igneous materials in the forms of streaks and fractures filled with sanidine, orthoclase, or Quartz. This dolomite is probably hydrothermal because of its close association with igneous intrusions and its strained structures and textures (Fig. 4.8d).

Fig. 4.12 Pink anhydrite overlain by grey anhydritic dolomite and underlain by yellowish siltstone and mudstone, Sharao island

4.2.3 Ediacaran-Cambrian Stratigraphy of the Mainland of Qatar

Without drilled wells or shared domain geophysical data, the Ediacaran-Cambrian stratigraphy of Qatar's mainland can be inferred from the stratigraphy of neighboring countries, such as Saudi Arabia and Jordan. Figure 4.14 shows the Ediacaran to Ordovician stratigraphy of Saudi Arabia and Jordan.

The following synthesis of the Ediacaran- Cambrian stratigraphy of Qatar will be based on the seismic section passing through eastern Saudi Arabia (Fig. 4.15) and the

Fig. 4.13 Thin-bedded Hormuz dolomite from Halul Island

published works on Saudi Arabia and Jordan (e.g., Al-Husseini 2010, 2011, 2014; Khalifa 2015; Powell et al. 2014). The nearest deepest control well is the Khursaniyah-81 well (Fig. 4.17), which is located around 250 km NW of Qatar (Al-Husseini 2010).

4.2.3.1 The Ediacaran Stratigraphy

The Jibalah Group

The Ediacaran strata are found in small, scattered outcrops within the Arabian Shield, western Saudi Arabia. They are still a work in progress, and the following materials reflect the status of our knowledge at the time of writing. They are assigned to the Jibalah Group, described according to Al-Husseini (2011) by a French survey group. Then, several formation names and definitions were proposed to be included in the group by Delfour (1970), Hadley (1974), Hadley (1986). Al-Husseini (2011) provided a comprehensive review of the group, and in another paper, Al-Husseini (2014) correlated the group with the Abu Mahara Supergroup in Oman.

The Jibalah Group strata are exposed in isolated outcrops across the Arabian Shield. These were considered pull-apart basins along the Najd Fault System (Al-Husseini 2011; Delfour 1970). The nearest basins to Qatar that fall on nearly the same latitudes are the Khutayfah, Al Jifn, and Nuqrah basins, which were associated with the Halaban Fault Zone, one of four zones of the Najd Fault System.

In the northern parts of the Arabian Shield, the Jibalah Group is underlain by the Shammar Group. The latter group

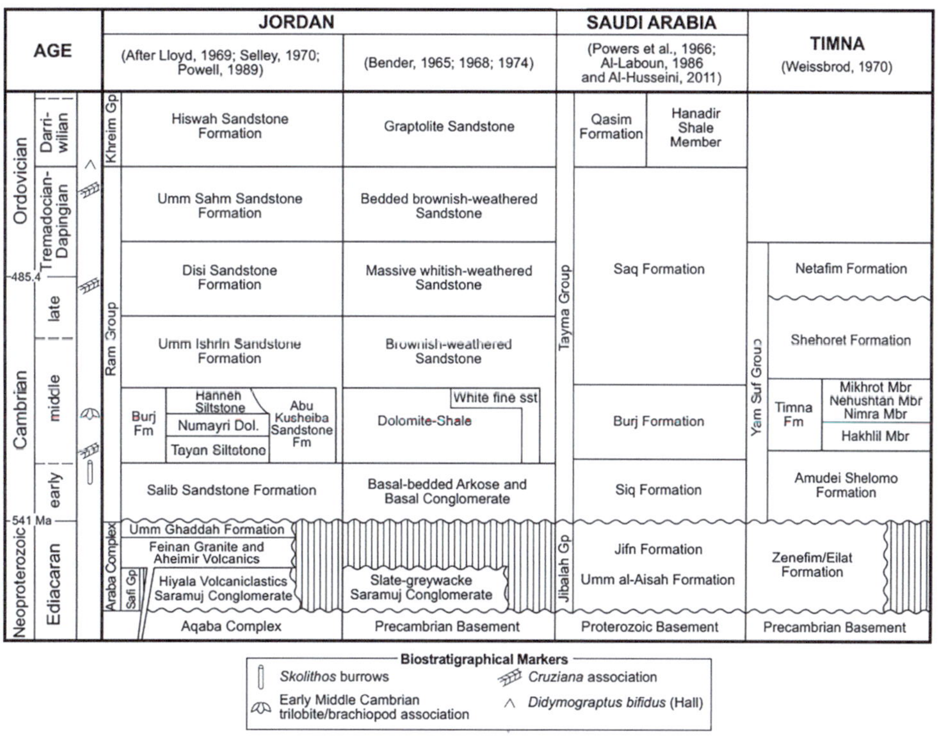

Fig. 4.14 Ediacaran to Ordovician stratigraphy of Saudi Arabia and Jordan (Powell et al. 2014)

Fig. 4.15 Ediacaran to Cambrian rock outcrops in Ghour Al-Safi, southern Jordan, show the grey-colored Saramuj Conglomerate Formation at the bottom, the brown above is the Salib Formation, and the greenish at the top is the Burj Formation (Courtesy of Dr. Issa Makhlouf)

is formed of igneous suits, including rhyolites and granite. The basal part of the Jibalah Group consists of basal conglomerates derived from the Shammar Group. The Jibalah is overlain by the Siq Sandstone Formation of the lower Cambrian. The Siq Sandstone is a widespread clastic front assigned to the Asfar Sequence (Al-Husseini 2010) and is equivalent to the Salib Arkosic Sandstone of Jordan (Powell et al. 2014), Fig. 4.15.

The Jibalah Group includes many formations, the most important of which are the Umm Al ‘Aisah and the Jifn Formations. The Umm Al ‘Aisah Formation consists of two units. The lower conglomerate bed is roughly 500 M thick of boulders and cobbles of granite and rhyolite with volcanic sandstone overlain by the cherty limestone unit, which is formed of a mixture of cherty limestone, calcareous shale and chert beds, inclusions, and lenses within the limestone strata. The total thickness of this unit is around 320 M. It is considered equivalent to the Muraykhah Formation.

The type locality of the Jifn Formation is located at Umm Al ‘Aisah mountain, between the SE part of the Al Jifn Basin and the NW part of the Nuqrah Basin (Al-Husseini 2011). The Jifn Formation consists of the “Upper Conglomerate” unit followed by alternations of sandstone, shale, and minor carbonates. Red to brown sandstone is the most common lithology, but it contains some carbonates and andesite-basalt flows. The average thickness of the formation is around 2000 m.

In Jordan, the Jibalah Group is equivalent to the Safi Group, which consists of the Saramuj Conglomerate Formation and the overlying Haiyala Volcaniclastic Formation (Jarrar et al. 1991; Powell 1988; Powell et al. 2015). The Saramuj Formation is exposed along some valleys SW of Karak town, adjacent to the Dead Sea. Powell et al. (2015) selected the well Ajlun-1 as a subsurface reference section of the formation. The Saramuj Formation comprises a polymict conglomerate of pebbles, cobbles, and boulders with arkosic sandstone (Fig. 4.16).

Al-Husseini (2014) correlated the Jibalah Group in some of the isolated basins in which its rocks are exposed with the Nafun formations of Oman based on new geochronological data. He suggested that the younger parts of the Lower Nafun Supersequence are correlated with the Lower Jibalah Supersequence as represented by the Umm Al-‘Aisha Formation in the Jifn Basin. The Upper Nafun Supersequence is correlated with the Upper Jibalah Supersequence, which overlies the Umm Al-‘Aisha Limestone with its basal unit, the Jifn Conglomerate.

4.2.3.2 The Cambrian Stratigraphy

The early Cambrian strata of the Arabian Plate were assigned to the Asfar Sequence by Al-Husseini (2010). The Asfar Sequence comprises quartz-rich arkosic sandstone and conglomerate, shale, and siltstone, attending a total thickness of around 750 m in Jordan and more than 700 m in Oman. These sediments were deposited in continental

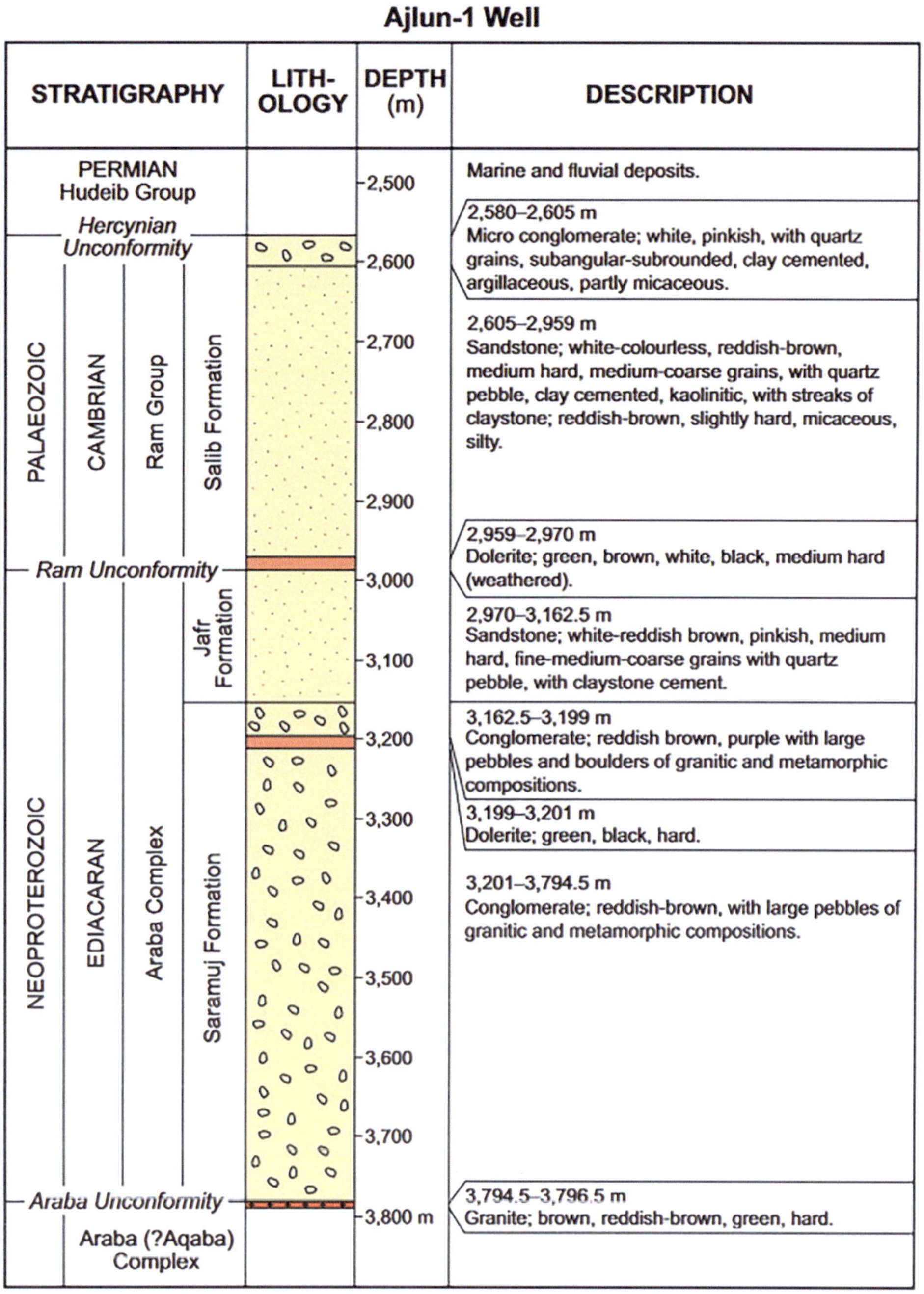

Fig. 4.16 Stratigraphic section of the Well Ajlun-1 shows the thickness and lithology of the Saramuj Conglomerate Formation (Powell et al. 2015)

settings all over the region. The Sequence includes four formations. These include the Salib Arkosic sandstone of Jordan, the Siq Formation of Saudi Arabia, the Amin Formation of Oman, and the Lalun Sandstone Formation of Iran.

The Siq Formation

The Siq Formation is the oldest formation of the Tayma Group of Saudi Arabia. It has been encountered in the well of Khursaniyah-81 in western Saudi Arabia, and hence it is of relevance to the Cambrian stratigraphy of Qatar. Its rocks are exposed in a narrow belt along the Arabian Shield from the southeast to the east. It is also reported in outliers above the Shield and the subsurface, as in the Khursaniyah well. The Tayma Group includes four formations. These are the Zig, the Burj, the Saq, and the Qasim Formations.

The Siq Formation was described for the first time by Bramkamp et al. (1963) and was formally defined by Powers et al. (1966). The Siq Formation is overlain by the Saq Sandstone Formation. The reference section of the Formation

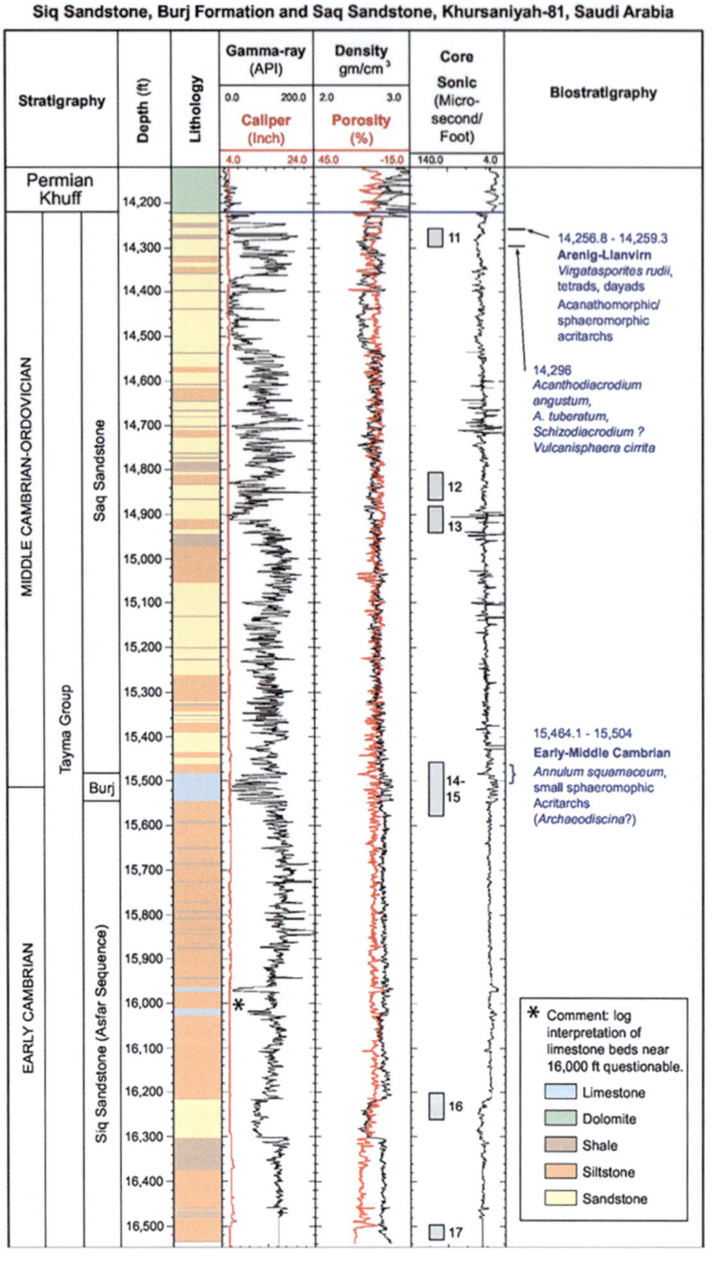

Fig. 4.17 Lithostratigraphic section of the Khursaniyah-81 well, the nearest deep well to Qatar (Al-Husseini 2010)

is taken at the well Khursaniyah-81 (Fig. 4.17). The Saq Formation, with a total thickness of 384 m, was encountered below the Permian-Lower Triassic Khuff Formation. The Saq Formation is overlain by around 18 m of carbonate rocks assigned to the Burj Limestone Formation. (Al-Husseini 2010) noted, based on written communication with D. Janou, that although both formations are formed of sandstone, the Siq and Saq Formations are distinguishable from each other by the sharp contact between them, which contains lag deposits. Furthermore, the Siq sandstone is dark red to pink, while the Saq has a light brown to yellow sandstone.

The outcrops of both the Siq and the Burj Limestone Formations were described from areas where it overlies the Arabian Shield and overlain unconformably by the Saq Sandstone Formation. The thickness of this formation ranges from a few meters to 97 m. It consists of dark to red-pink or yellow sandstone with iron oxides. The section is divided into seven units (Fig. 4.18). These units range in thickness from 7 to 50 m and consist mainly of sandstone with white pebbles and sandstone and shield rock lithoclasts with siltstone and claystone. The sediments of Units 1 and 2 were deposited in a braided stream environment, while the other units, except Unit 7, were

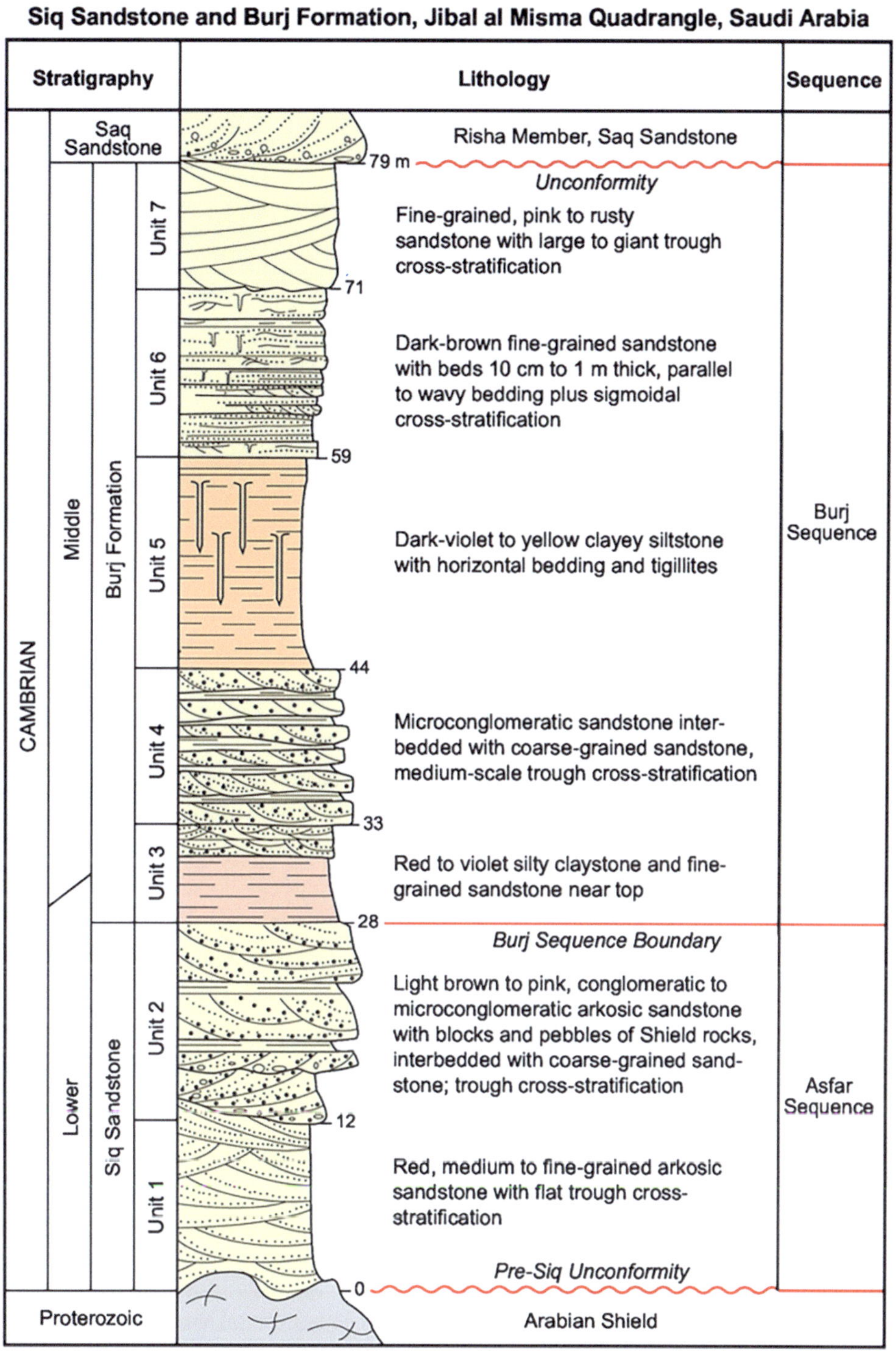

Fig. 4.18 Lithologic section of the Siq Sandstone and the Burj Limestone Formations in Jabal al Misa, Saudi Arabia (Al Husseini 2010)

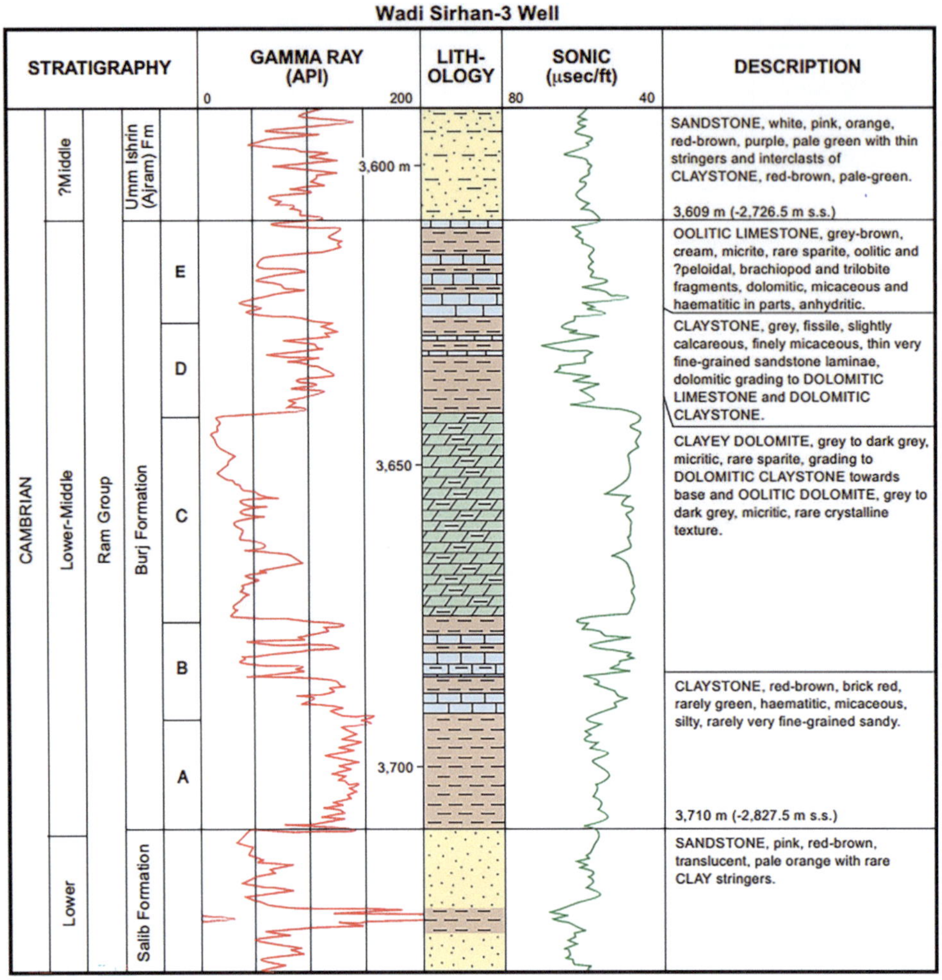

Fig. 4.19 Lithostratigraphy of the Burj Limestone Formation in the well Wadi Sirhan-3, Jordan (Powell et al. 2014)

formed in a coastal to tidal flat environment. Unit 7 sediments are probably of eolian origin.

The Burj Limestone Formation

The Burj Limestone Formation is part of the Ram Group in Jordan, which encompasses the siliciclastic formations and the marine carbonates/siliciclastics that overlain uncomfortably the Neoproterozoic basement (Aqaba and Arab Complexes) (Powell et al. 2014). The group sediments range in age from early Cambrian to the Ordovician. The main formations in the group include from bottom: Salib Sandstone, Burj Formation, Umm Ishrin Sandstone, Disi Sandstone, and Umm Sahm Sandstone. The Burj Limestone Formation has been encountered in the well Khursaniyah-8, so it is relevant to the stratigraphy of Qatar.

The first mention of the Burj Formation was by Blanckenhorn (1912) (cited by Richter and Richter 1941). The type section of the formation is in Khirbet El Burj, and because it is not fully exposed there, a new reference section has been selected in the same area by Rushton and Powell (1998). The Burj Formation was also encountered in some wells. According to Powell et al. (2014), the best sections are found in the wells of Wadi Sirham-3 and 10 and Al Jafr-1 in southern Jordan, the Wadi Ghada-2 in central Jordan, and the North Highland-1 in the north. The most representative section of the formation is found in the well Wadi Sirhan-3 and appears in Fig. 4.19. The following synthesis is based on the work of Powell et al. (2014).

At the outcrops and subsurface, the Burj Formation has a maximum thickness of 135 m. It is made up of two siliciclastic members (Tayan and Hanneh) separated by a carbonate member (Numayri) (Fig. 4.19). Red-brown, mauve, green, and buff micaceous siltstone and sandstone with ripple cross-lamination, ripple marks, and sparse burrows make up the siliciclastic members; thin beds of buff-brown sandstone with bimodal trough cross-bedding are also present. Numerous ichnofossils exist. The carbonates are yellow–brown or grey, with dolomitic limestone, dolomite, and sandy dolomite lenses; cross-lamination is common in the siliciclastic lenses; and oncolites, ooids, peloids, and bioclasts are present in the cleaner carbonates. Packstones, grainstones, and floatstones-rudstones with shelly bioclasts such as brachiopods, trilobites, echinoids, ooids, and pisoliths are common in some beds. The upper part of the carbonate member near the Safi area contains microbial mats and thin beds of siltstone containing shelly fossils and glauconite peloids. Siliciclastic beds above the carbonate member exhibit mega-ripples and bimodal cross-bedding, including tidal bundles.

The formation has been assigned to the Lower-Middle Cambrian based on an association of trilobites, brachiopods, hyolithids, and the calcareous algae *Girvanella* sp.

The Burj Formation shows typical transgressive and regressive system tracts as indicated by the present fauna, the distinctive sedimentary structures, and the facies distribution. The *Tayan Member*, formed of siltstone, represents a transgressive system tract (TST) that overlays the fluvial sandstone of the Lower Cambrian Salib Formation. The sediments of the Tayan member shows wave ripples, oscillation ripples, cross-beddings, intraclasts, and traces of *Cruziana* and *Planolites* burrow. All these sedimentary features suggest a southward marine incursion over an alluvial plain with the predominance of a shallow subtidal to an intertidal depositional system (Powell et al. 2014). The continuation of this TST led to the deposition of the *Numayri Dolomite Member*, where the Maximum Flooding Surface was positioned at the top of the relatively thick carbonate unit. The presence of a mixture of ooids, broken brachiopods, cross-beddings, cross-lamination, and the abundance of siliciclastic materials in the carbonate sediments indicate the prevalence of coastal marine lagoon with periodic storm events. There are some domal microbial mats at the top of the unit, indicating intertidal marine conditions. The Regressive System Tract (RST) is reflected in the renewal of the siliciclastic influx from the recently uplifted land and the termination of the carbonate depositional system, which led to the deposition of the *Hanneh Siltstone Member*. The sedimentary structures and the trace fossils in this member are the same as that of the Tayan Member. They indicate that the sediments were deposited under a tidal-dominated marine shoreline environment. These sediments were capped with coarse-grained sandstone that was deposited by braided to meandering rivers and belong to the fluvial deposits of Umm Ishrine Formation during the whole Cambrian Period. There are different variants of these depositional settings in different parts of Jordan.

The Saq Formation

The Saq Formation was initially defined in an unpublished report by Burchfiel and Hoover in 1935 from Jabal Saq as part of the now obsolete Uyun Formation and formally upgraded to the status of a formation by Steineke et al. (1958) and later described by Powers et al. (1966). They defined its type section at Gebel Saq, representing the most exposed section in Al Qasim Province's central region, northwest of Gebel Isbaa. Vaslet et al. (1985) classified the Saq Formation as having two formal members: the Risha Member at the bottom and the Sajir Member at the top. The Saq Formation has an incongruent relationship with the underlying Precambrian Basement Complex. The contact between the Saq Formation and the underlying Precambrian Basement Complex can be seen in a few places and is hidden beneath some sand dunes in others. The upper contact of the Saq Formation with the overlying Anz Formation is sharp between the uppermost Saq Formation's coarse-grained pebbly reddish buff sandstone and the basal Anz Formation's white, thin-bedded micaceous sandstone with siltstone interbeds. Glaciers removed the Anz Formation and the top part of the Saq Formation in some locations (northeast Gebel Saq) (Khalifa, 2015). In Khursaniyah-81 well, western Saudi Arabia, the Saq Formation is directly overlain by the Permian–Triassic Khuff Formation (Fig. 4.17).

According to Powers et al. (1966), the lithology of the Saq Formation is uniform in both horizontal and vertical directions. It is formed mainly of cross-bedded puff to grey and white quartz sandstone with grains that range from poorly to well sorted, often friable, but may be consolidated with a black ferruginous quartzite surface. There are some local occurrences of siltstone. Khalifa (2015) described the Saq Formation from the Al Qasim region, Saudi Arabia, and found that the formation is formed of polymictic conglomerates that occur locally in channels formed by the Najd fault system. The Saq Formation is exposed along the western margin of the Arabian Basin in the Al Qasim region. The Saq Formation is composed of two major lithologic units. The lower unit comprises massive, cross-bedded sandstones and siltstone–mudstone channel layers. These strata are interpreted as channel facies probably deposited in a fluvial or estuarine environment. The upper unit contains sandy siltstone and bioturbated thin-bedded and rippled sandstone with trace fossils of *Cruziana* and vertical burrows such as *Skolithos* (Fig. 4.20). The Saq was removed in some areas by the glacial movement of the overlying Lower Ordovician–Lower Silurian Zarqa and Sarah Formations. Sedimentation occurred on a broad shelf on a passive margin with low accommodation.

The Saq Formation is equivalent to the lower and middle units of the Araba Formation in Egypt, the Salib, Umm Ishrine, and Disi Formations in Jordan, the Zabuk, Burj, and the Sosink Formations in Syria, the Barut and the Zagun Formations in Iran, and the Karim, Haradh, and Amin and the Miqrat and Mahwis Formations in Oman (Khalifa 2015) (Fig. 4.21).

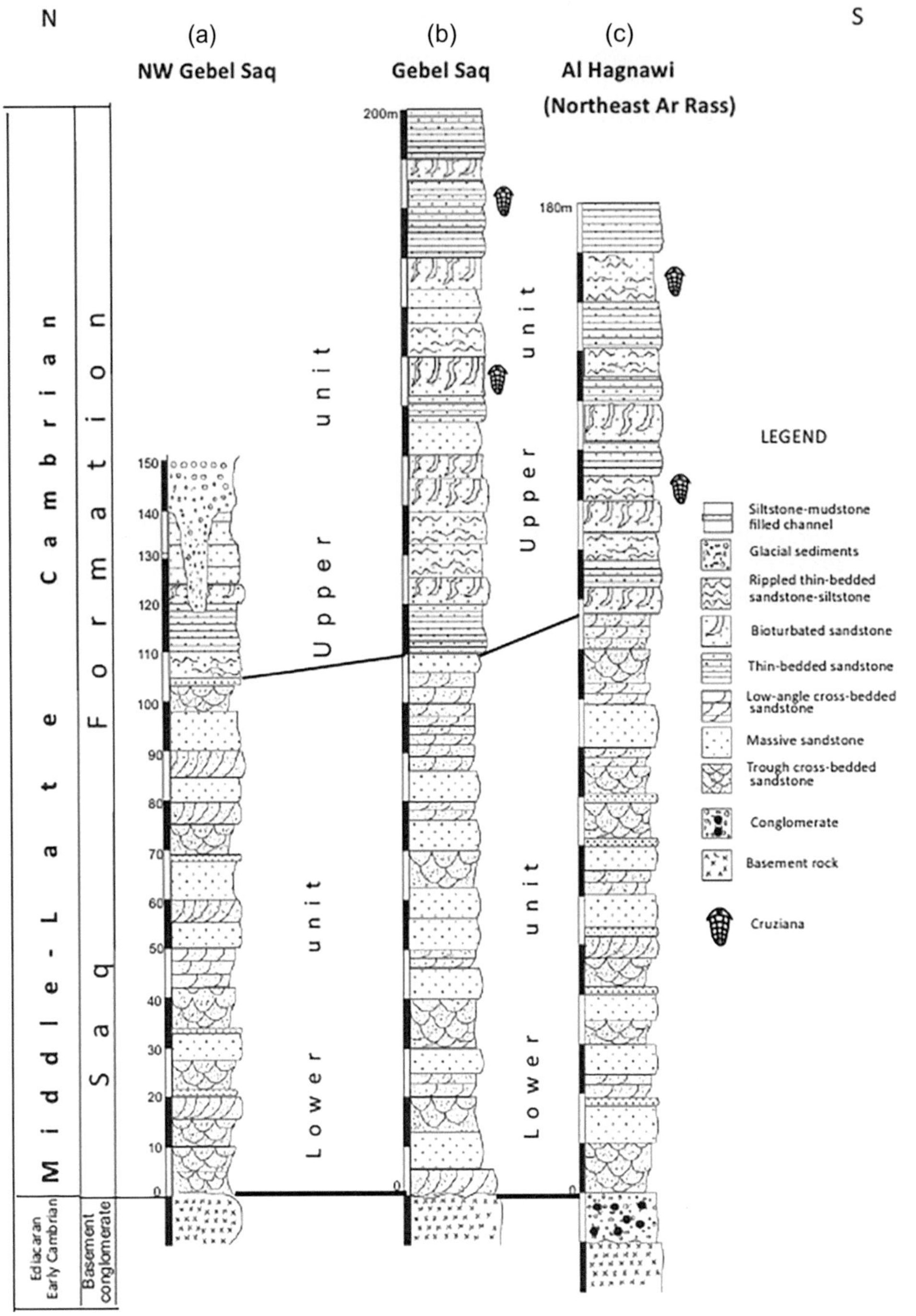

Fig. 4.20 Different stratigraphic units of the Saq Formation in Gebel Saq and Al Hagnawi areas, Saudi Arabia (Khalifa 2015)

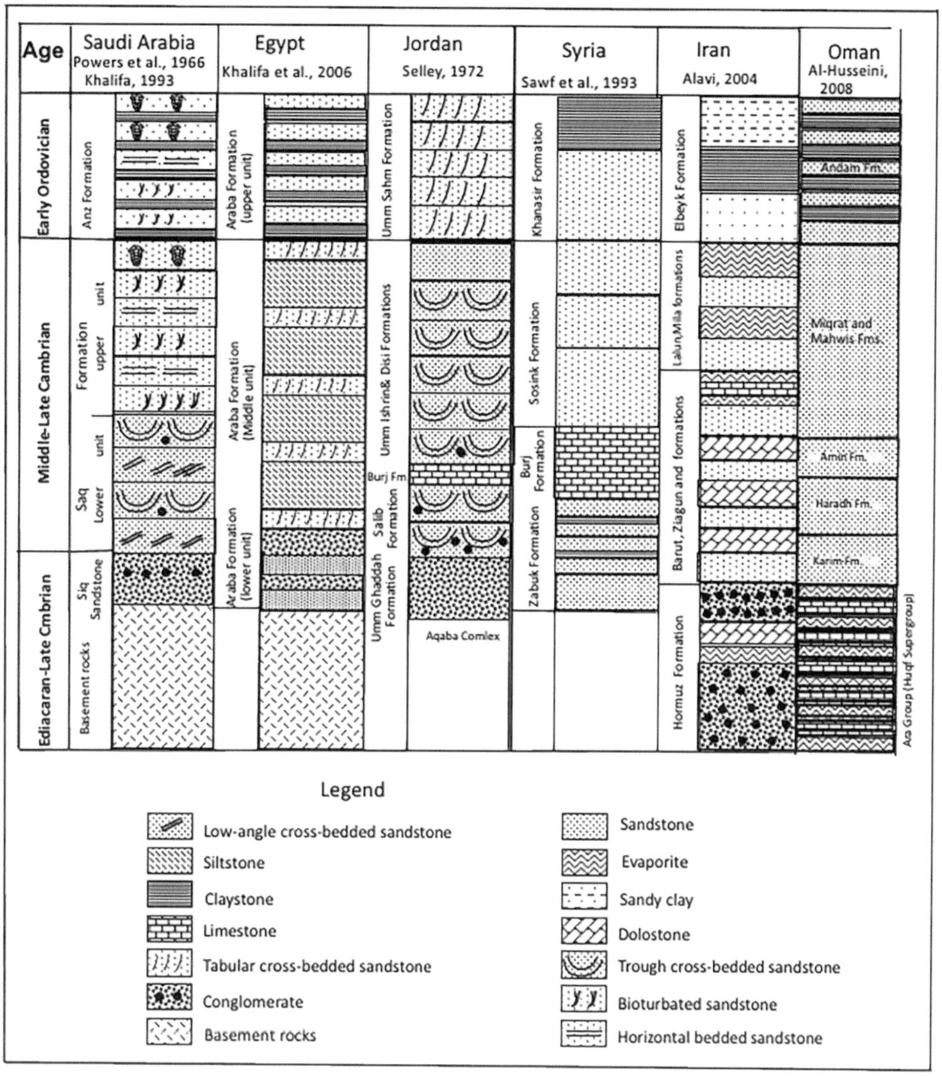

Fig. 4.21 Correlation of the Saq Formation over Eastern Arabia and Iran (Khalifa 2015)

References

Abbo A, Avigad D, Gerdes A, Güngör T (2015) Cadomian basement and Paleozoic to Triassic siliciclastics of the Taurides (Karacahisar dome, south-central Turkey): Paleogeographic constraints from U–Pb–Hf in zircons. Lithos 227:122–139. https://doi.org/10.1016/j.lithos.2015.03.023

Al-Husseini MI (2000) Origin of the Arabian plate structures: Amar collision and Najd Rift. GeoArabia 5(4):527–542. https://doi.org/10.2113/geoarabia0504527

Al-Husseini MI (2010) Middle east geologic time scale 2010: Early Cambrian Asfar sequence. GeoArabia 15(1):137–160. https://doi.org/10.2113/geoarabia1501137

Al-Husseini M (2011) Late Ediacaran to early Cambrian (Infracambrian) Jibalah group of Saudi Arabia. GeoArabia 16(3):69–90. https://doi.org/10.2113/geoarabia160369

Al-Husseini MI (2014) Proposed correlation of Oman's Abu Mahara Supergroup and Saudi Arabia's Jibalah Group. GeoArabia 19(2):17–48. https://doi.org/10.2113/geoarabia190217

Alavi M (2007) Structures of the Zagros fold-thrust belt in Iran. Am J Sci 306(9):1064–1095. https://doi.org/10.2475/09.2007.02

Allen PA (2007) The Huqf Supergroup of Oman: basin development and context for Neoproterozoic glaciation. Earth-Sci Rev 84(3):139–185. https://doi.org/10.1016/j.earscirev.2007.06.005

Álvaro JJ, Vennin E, Moreno-Eiris E, Perejón A, Bechstädt T (2000) Sedimentary patterns across the Lower–Middle Cambrian transition in the Esla nappe (Cantabrian Mountains, northern Spain). Sedimentary Geol 137(1):43–61. https://doi.org/10.1016/S0037-0738(00)00134-2

Baban D, Lawa F (2016) Palynological and stratigraphical evidences on the Age of the outcropped Khabour Formation near Chalki Nasara Village, Kurdistan Region, northern Iraq. J Zankoy Sulaimani, Part-A-(Pure Appl Sci) Special Issue, GeoKurdistan II:293–318

Blanford W (1872) Note on the geological formations seen along the coasts of Bilúchístán and Persia from Karáchí to the head of the Persian Gulf, and on some of the Gulf Islands. Records of the Geological Survey of India 5(2):41–45

Bramkamp RA, Brown GF, Holm DA, Layne Jr NM (1963) Geologic map of the Wadi As Sirhan quadrangle, Kingdom of Saudi Arabia

[Report] (200A). IMAP, Issue. U. S. G. Survey. https://pubs.usgs.gov/publication/i200A

Delfour J (1970) Le Groupe de J'Balah, une nouvelle unite du Bouclier Arabe. Bureau de Recherche Geologique et Minieres Bull (Ser. 2) 4(4):19–32

Edgell HS (1991) Proterozoic salt basins of the Persian Gulf area and their role in hydrocarbon generation. Precambrian Res 54(1):1–14. https://doi.org/10.1016/0301-9268(91)90065-I

Etemad-Saeed N, Hosseini-Barzi M, Adabi MH, Sadeghi A, Houshmandzadeh A (2015) Provenance of Neoproterozoic sedimentary basement of northern Iran, Kahar Formation. J Afr Earth Sci 111:54–75. https://doi.org/10.1016/j.jafrearsci.2015.07.003

Faramarzi NS, Amini S, Schmitt AK, Hassanzadeh J, Borg G, McKeegan K, Razavi SMH, Mortazavi SM (2015) Geochronology and geochemistry of rhyolites from Hormuz Island, southern Iran: a new record of Cadomian arc magmatism in the Hormuz Formation. Lithos 236–237:203–211. https://doi.org/10.1016/j.lithos.2015.08.017

Gass IG, Ries AC, Shackleton RM, Smewing JD (1990) Tectonics, geochronology and geochemistry of the Precambrian rocks of Oman. Geol Soc London, Special Publ 49:585–599

Geert K, Afifi AM, Al-Hajri SS, Droste HJ (2001) Paleozoic stratigraphy and hydrocarbon habitat of the Arabian plate. GeoArabia

Geng M, Ali MY, Fairhead JD, Pilia S, Bouzidi Y, Barkat B (2022) Hormuz salt distribution in the United Arab Emirates: Implications for the location of hydrocarbon fields. Marine Petrol Geol 143:105797. https://doi.org/10.1016/j.marpetgeo.2022.105797

Gürsu S, Moeller A, Göncüoglu MC, Köksal S, Demircan H, Köksal FT, Kozlu H, Sunal G (2015) Neoproterozoic continental arc volcanism at the northern edge of the Arabian Plate, SE Turkey. Precambr Res 258:208–233

Hadley DG (1974) The Taphrogeosynclinal Jubaylah Group in the Mashhad Area, Northwestern Hijaz, Kingdom of Saudi Arabia. Min Resour Bull 10:18

Hadley DG (1986) Explanatory notes to the geologic map of the Sahl Al Matran Quadrangle, Kingdom of Saudi Arabia. Geoscience Map GM-86 C, scale 1:250,000, sheet 26C

Jarrar G, Wachendorf H, Zellmer H (1991) The Saramuj Conglomerate: evolution of a Pan-African molasse sequence from southwest Jordan. N Jb Geol Paläont 6:335–356

Khalifa MA (2015) The Ediacaran-Cambrian and Ordovician rocks of Al Qasim Province, Saudi Arabia: facies, depositional history and regional correlation. J Afr Earth Sc 109:263–282. https://doi.org/10.1016/j.jafrearsci.2014.12.010

Mattes BW, Morris SC (1990) Carbonate/evaporite deposition in the Late Precambrian—Early Cambrian Ara formation of Southern Oman. Geol Soc London, Special Publ 49:617–636

Meinhold G, Morton AC, Avigad D (2013) New insights into peri-Gondwana paleogeography and the Gondwana super-fan system from detrital zircon U-Pb ages. Gondwana Res 23(2):661–665

Meneisy M (1988) New potassium/argon ages for the volcanic rocks of Halul Island (Qatar). Qatar Univ Sci Bull 8:287–295

Merdith AS, Collins AS, Williams SE, Pisarevsky S, Foden JD, Archibald DB, Blades ML, Alessio BL, Armistead S, Plavsa D, Clark C, Müller RD (2017) A full-plate global reconstruction of the Neoproterozoic. Gondwana Res 50:84–134. https://doi.org/10.1016/j.gr.2017.04.001

Moghadam HS, Li X-H, Santos JF, Stern RJ, Griffin WL, Ghorbani G, Sarebani N (2017) Neoproterozoic magmatic flare-up along the N. margin of Gondwana: the Taknar complex, NE Iran. Earth Planet Sci Lett 474:83–96. https://doi.org/10.1016/j.epsl.2017.06.028

Moghaddam PS, Salehi MA, Jafarzadeh M, Zohdi A (2020) Provenance, palaeoweathering and tectonic setting of the Ediacaran Bayandor Formation in NW Iran: implications for the northern Gondwana continental margin during the late Neoproterozoic. J Afr Earth Sci 161:103670. https://doi.org/10.1016/j.jafrearsci.2019.103670

Morag N, Avigad D, Gerdes A, Belousova E, Harlavan Y (2011) Detrital zircon Hf isotopic composition indicates long-distance transport of North Gondwana Cambrian-Ordovician sandstones. Geology 39(10):955–958

Murris RJ (1980) Middle East: stratigraphic evolution and oil habitat. AAPG Bull 64(5):597–618. https://doi.org/10.1306/2f918a8b-16ce-11d7-8645000102c1865d

Nehlig P, Genna A, Asfirane F, BRGM F, Guerrot C, Eberlé JM, Kluyver HM, Lasserre JL, Le Goff E, Nicol N, BRGM F, Salpeteur N, Shanti M, Thiéblemont D, Truffert C, BRGM F (2002) A review of the Pan-African evolution of the Arabian shield. GeoArabia 7(1):103–124. https://doi.org/10.2113/geoarabia0701103

Pilgrim HE (1908) The geology of the Persian Gulf and the adjoining portions of Persia and Arabia. Office the Geological Survey. https://books.google.com.qa/books?id=3BRgGwAACAAJ

Powell JH (1988) The geology of the Karak area

Powell JH, Abed AM, Le Nindre Y-M (2014) Cambrian stratigraphy of Jordan. GeoArabia 19(3):81–134. https://doi.org/10.2113/geoarabia190381

Powell JH, Abed A, Jarrar GH (2015) Ediacaran Araba complex of Jordan. GeoArabia 20(1):99–156. https://doi.org/10.2113/geoarabia200199

Powers RW, Ramirez LF, Redmond CD, Elberg Jr EL (1966) Geology of the Arabian Peninsula: sedimentary geology of Saudi Arabia [Report] (560D). (Professional Paper, Issue. U. S. G. Survey. https://pubs.usgs.gov/publication/pp560D

Richter R, Richter E (1941) Die Fauna des Unter-Kambriums von Cala in Andalusien

Rushton AWA, Powell JH (1998) A review of the stratigraphy and trilobite faunas from the Cambrian Burj Formation in Jordan. Bull Natural History Museum. Geology Ser 54:131–146. https://www.biodiversitylibrary.org/part/314062

Shell CQ (1964) Geologic notes on Halul Island (Unpublished report)

Smith AG, Bhat GM, Craig J, Thurow JW, Thusu B, Cozzi A (2012) A review of the Ediacaran to Early Cambrian ('Infra-Cambrian') evaporites and associated sediments of the Middle East. In Geology and hydrocarbon potential of Neoproterozoic–Cambrian basins in Asia, Vol 366. Geological Society of London. https://doi.org/10.1144/sp366.12

Steineke M, Bramkamp RA, Sander NJ (1958) Stratigraphic relations of Arabian Jurassic oil. American Association of Petroleum Geologists Symposium, Tulsa, U.S.A.

Stewart SA (2016) Structural geology of the Rub' Al-Khali Basin, Saudi Arabia. Tectonics 35(10):2417–2438. https://doi.org/10.1002/2016TC004212

Stewart SA (2020) Structural style of buried Ediacaran basins in Saudi Arabia on seismic reflection data. Precambr Res 351:105954. https://doi.org/10.1016/j.precamres.2020.105954

Stoeser DB, Frost CD (2006) Nd, Pb, Sr, and O isotopic characterization of Saudi Arabian Shield terranes. Chem Geol 226(3):163–188. https://doi.org/10.1016/j.chemgeo.2005.09.019

Thomas RJ, Ellison RA, Goodenough KM, Roberts NMW, Allen PA (2015) Salt domes of the UAE and Oman: probing eastern Arabia. Precambr Res 256:1–16. https://doi.org/10.1016/j.precamres.2014.10.011

Van Buchem F, Svendsen N, Hoch E, Pedersen-Tatalovic R, Habib K (2014) Depositional history and petroleum habitat of Qatar

Vaslet D, Beurrier M, Villey M, Manivit J, Le Strat P, Le Nindre Y, Berthiaux A, Brosse J, Fourniguet J (1985) Explanatory notes to the geologic map of the Al Faydah Quadrangle, Kingdom of Saudi Arabia. Geoscience Map GM-102C, scale 1: 250,000, sheet 25G

Whitehouse MJ, Windley BF, Stoeser DB, Al-Khirbash S, Ba-Bttat MAO, Haider A (2001) Precambrian basement character of Yemen and correlations with Saudi Arabia and Somalia. Precambrian Res 105(2):357–369. https://doi.org/10.1016/S0301-9268(00)00120-0

Windley BF, Whitehouse MJ, Ba-Bttat MAO (1996) Early Precambrian gneiss terranes and Pan-African island arcs in Yemen: crustal accretion of the eastern Arabian Shield. Geology 24(2):131–134. https://doi.org/10.1130/0091-7613(1996)024%3c0131:Epgtap%3e2.3.Co;2

The Paleozoic

5

Abstract

This chapter offers a detailed examination of the Paleozoic stratigraphy and depositional history of the Arabian Plate, focusing on Qatar and adjacent areas. The Paleozoic rocks of Arabia are categorized into two principal exposure zones: one in the west, from Jordan to Yemen, and another in the east, encompassing Turkey to Oman. Only a few wells penetrated the deeper Paleozoic strata. Contemporary stratigraphic classifications categorize these rocks into groups, including Tayma, Tabuk, Qaliba, and Huj, with most studies concentrated on Saudi Arabia and Oman. The Paleozoic succession in Qatar has been inadequately studied, yet it is crucial for regional correlations and hydrocarbon potential assessment. The Qasim Formation (Middle to Late Ordovician) signifies shallow marine and deltaic settings, but the Silurian Qusaiba and Sharawra formations encompass organic-rich shales essential for oil generation. The Devonian Tawil Formation in Qatar comprises cross-bedded sandstones with minimal marine influence, reflecting river and coastal depositional environments. This chapter examines the most important formations' regional lithofacies and sedimentary environments, correlating them with their Saudi Arabian equivalents. Such studies improve the comprehension of the Arabian Plate's stratigraphy and depositional history, especially its Paleozoic hydrocarbon reservoirs, providing significant regional exploration and production insights.

5.1 Introduction

Two longitudinal zones of Paleozoic rocks running north to south in the Arabian Plate. The first is located to the east and extends from the Turkish mountains the "Amanus Mountains" down to the Huqf region and Oman Mountains in Oman passing via the Zagros and Taurus Mountains in Iran and Iraq (Beydoun 1986). The second zone lies to the west and stretches between Asir in Yemen and southern Jordan. Between these two zones, Paleozoic rocks are covered with a thick succession of rocks that may reach around 16 km. Only a few wells have penetrated this succession (Fig. 5.1).

Compared to the Mesozoic strata, Paleozoic rocks have been investigated only recently after discovering mineral deposits and hydrocarbons in Oman and Saudi Arabia (McGillivray and Husseini 1991). The Permian Khuff Formation received high attention because of its high hydrocarbon potential and hence most of the published work has been focused on it. For the first time, Beydoun (1986) comprehensively reviewed and showed the economic significance of the layers older than the Khuff. Alsharhan and Kendall (1986) illustrated the sedimentary sequence in the Arabian Gulf region at different geological timescales from the Precambrian to the Jurassic period. They observed that the first economically viable oil reserve found in the region was in southern Oman in the Haushi and Haima groups of the Marmul Field, which were characterized as fluvial Ordovician to Permian sands. They also suggested that oil was sourced from the underlying and deeply buried Huqf Group.

Alsharhan et al. (1993) reported the existence of the Upper Carboniferous to Lower Permian in Oman, Yemen, and Saudi Arabia. These strata are formed of sandstones, varvites with dropstones, diamictites, and shale. They suggested that these sediments represent grounded tillites and glaciolacustrine to marine deposits. They showed that the sands of the Haushi Group host oil accumulations found in the Late Precambrian marine algae-bearing limestones are the potential source of these accumulations. De la Grandville (1982) mentioned that producing oilfields in south Oman include Permian sediments, Permo-Carboniferous periglacial clastics, and Ordovician clastic. The geochemical model built by Abu-Ali et al. (1999) revealed that the hot shale Silurian "Qusaiba" is

F. N. Sadooni and H. A. Al-Kuwari, *Geological Evolution of Qatar and the Arabian Peninsula*, Regional Geology Reviews,
https://doi.org/10.1007/978-3-031-94394-2_5

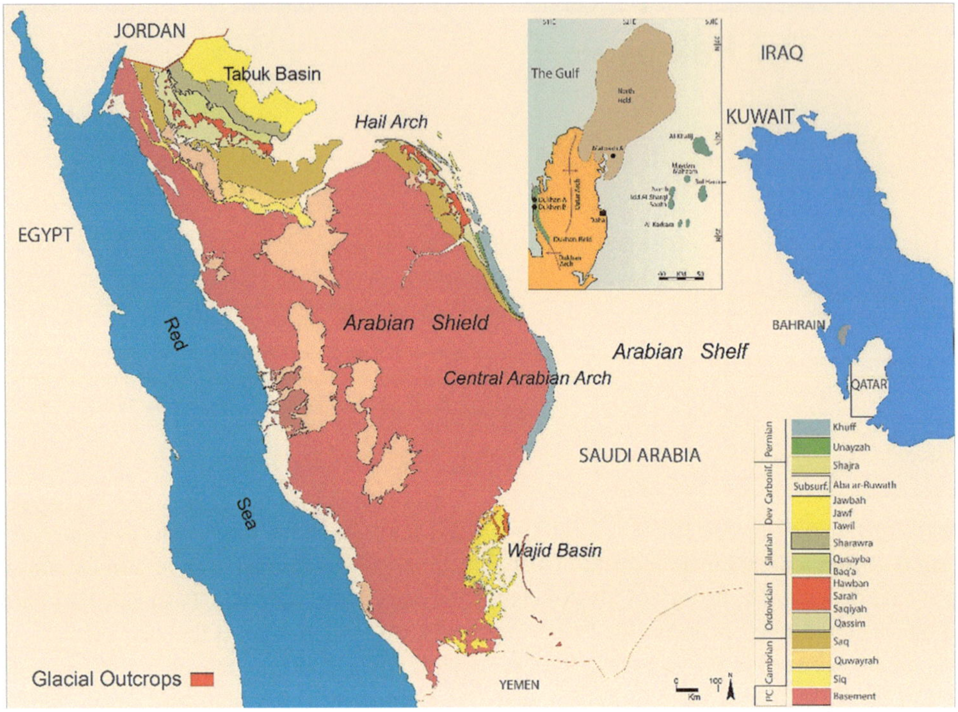

Fig. 5.1 Distribution of the Paleozoic rocks of the Arabian Peninsula (Al-Saad and Sadooni 2016)

the main source of hydrocarbons of the Paleozoic in the Permian Unayzah Formation in the center of Saudi Arabia. Senalp and Al-Duaiji (2001b) analyzed the geological features of the Ordovician Qasim Formation in the center of Saudi Arabia. Senalp and Al-Laboun (2000) portrayed a vivid picture of the Upper Ordovician glacial sediments of central Saudi Arabia based on subsurface and surface materials. It was mentioned that the Upper Ordovician stratigraphic column in the central part of the country showed glacially formed unconformities and a mixture of fluvial, marine, and glacial sediments that filled the deeply incised valleys of the region. They suggested that these sediments are derived from the southern polar sheet of the Gondwana continent that covered present-day Africa, eastern South America, and major parts of western Arabia.

The first stratigraphic scheme for the Paleozoic layers of Saudi Arabia was presented by Powers (1968). He proposed the name Tabuk Formation encompasses the whole Ordovician-Silurian strata. Powers (1968) classified the Tabuk Formation to six members, including two anonymous members. The Gondwanan continental glaciation (Hirnantian) from the Ordovician strata of Saudi Arabia was first described by McClure (1978) (Beuf et al. 1970; Miller and Al-Ruwaili 2007). Such a finding resulted in a considerable review of Powers's (1968) stratigraphy (Vaslet 1987, 1990; Vaslet et al. 1987). Accordingly, the stratigraphic sequence was classified into glacial and non-glacial sediments. Two phases of ice sheet advancement characterized the glaciation event in Arabia. The Zarqa Formation and Sarah Formation represented these phases. Clark-Lowes (2005) proposed that potential reservoir rocks might be represented by the glacial sediments of the Sarah Formation based on the development of paleo valleys in this formation. Le Heron et al. (2009) suggested that the Lower Ordovician (Hirnantian) glacial period was relatively short, lasting less than 0.5 Myr. They mentioned that the sedimentary regime resulted from fast ice streams during glacial maxima. Also, local tectonics may have affected the product of meltwater that followed the glacial recession and sediment geometry.

5.2 Stratigraphy

The Saudi Stratigraphic Committee (2013) introduced the most recent succession for the Paleozoic strata in Saudi Arabia. The new scheme divided these strata into four groups: Tayma, Tabuk, Qaliba, and Huj. The Tayma Group comprises all the deposited sediments prior to the first glacial period during the Late Ordovician. The Tabuk Group encompasses solely the glacial deposits of the Late Ordovician, including the glacial and periglacial deposits that filled the paleo-valleys incised into the Qasim Formation or older strata. The Qaliba Group contains the regressive lower Silurian clastics that formed the primary hydrocarbon source in this stratigraphy. The Huj Group includes all the Lower Carboniferous and Upper Silurian strata (Fig. 5.2).

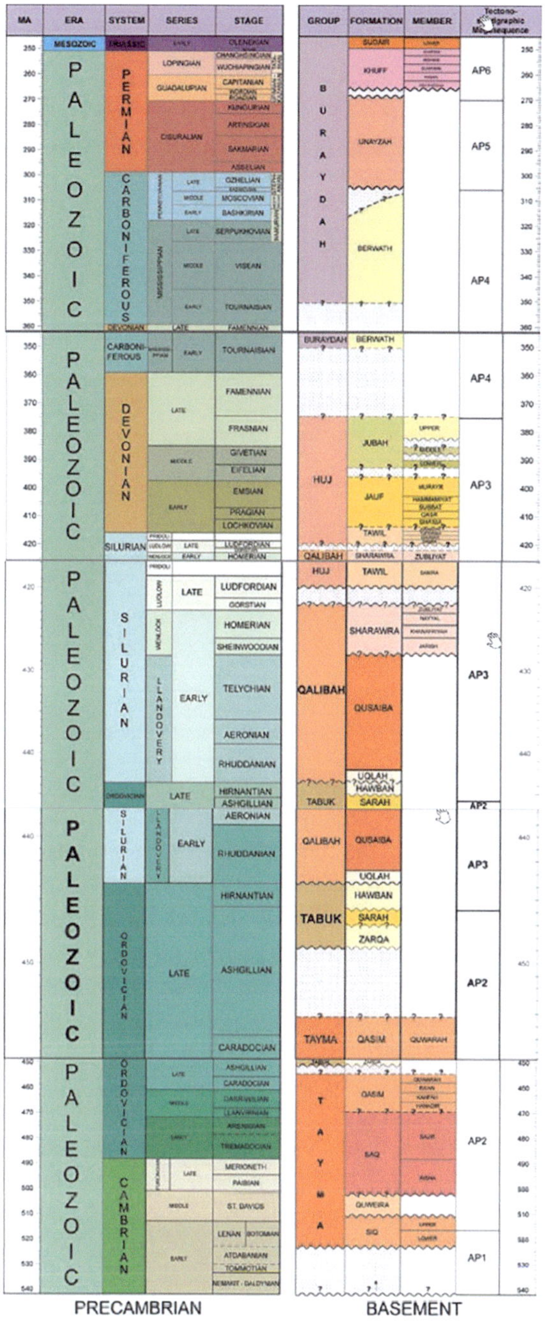

Fig. 5.2 Stratigraphic subdivisions of the Paleozoic strata of Saudi Arabia (Saudi Stratigraphic Committee 2013)

Drilling the well Matbakh-2, which penetrated 1500 m of pre-khuff strata, on Qatar Arch provided primary data on the pre-Khuff stratigraphic sequence in Qatar, Dukhan DKG-27 was drilled subsequently. Focke et al. (1986) used the two wells to describe the whole section in some detail. This semi-internally published data of Qatar Petroleum is not widely recognized. Thus, the Qatari section had never been contained within any regional work on the pre-Khuff stratigraphy of the Greater Arabian Basin, including that of Sharland et al. (2001). Focke et al. (1986) followed Powers (1968) scheme with some revisions and classified the stratigraphic sequence in Qatar to Haushi, Tawil, Sharawra, and Tabuk formations.

Al-Saad and Sadooni (2016) changed some formation names to align them with the Saudi nomenclature by using the regional correlation of the Paleozoic succession of Qatar with its Saudi equivalent. This practice improved the regional correlation and knowledge of these layers across the Arabian gulf region.

The lack of sufficient data makes it difficult to prove or deny the presence of the Tabuk Group strata, which represent the glacial and periglacial sediments in Qatar. Therefore, the section is classified into three groups revealed by specific formations, namely the Tawil formation to demonstrate the Huj Group, the Sharawra and Qusaiba formations to illustrate the Qalibah Group, and the Qasim Formation represent the Tayma Group. This nomenclature makes it possible to insert the Tabuk Group strata if their existence is verified in Qatar in future. Figure 5.3 compares Al-Saad and Sadooni's (2016) stratigraphy and Focke et al. (1986) scheme.

According to Konert et al. (2001), the Late Cambrian and Early Ordovician periods in northern Arabia saw increased clastic influx, which ended carbonate deposition. Over the Middle Cambrian carbonates, a prograding clastic sheet was deposited. The Arabian Plate's Cambrian-Ordovician boundary is poorly defined in the rock record. The platform was inundated again during the Early Ordovician, with deeper marine environments establishing basinward in northern and central Iran. The central Iran microplates had mixed clastic and carbonate settings, and various environments were present along the basin margin. The transitional boundary between the Disi and Umm Sahm formations in Jordan was difficult to identify. Late Cambrian to Early Ordovician palynomorphs were found in the upper Saq Formation of Saudi Arabia, which covered most of the country. In the Late Cambrian of Oman, a marine-influenced deposition environment was established in the Ghaba and Fahud salt basins. Regression accompanied the remainder of the Arenig stage, and a prograding braid-delta system was deposited.

During the Middle Ordovician, a transgression led to the deposition of middle to outer neritic shales over most of the Arabian Plate, including Qatar. This primary maximum flooding surface is dated at 465 Ma and is of Llanvirn age, corresponding to MFS O30 (Konert et al. 2001; Sharland et al. 2001). In certain areas, these shales contain high amounts of organic matter, indicating limited water circulation. The stippled regions in Fig. 5.4 denote the outbuilding deltas during the subsequent regressive stage identified in Oman and northern Saudi Arabia.

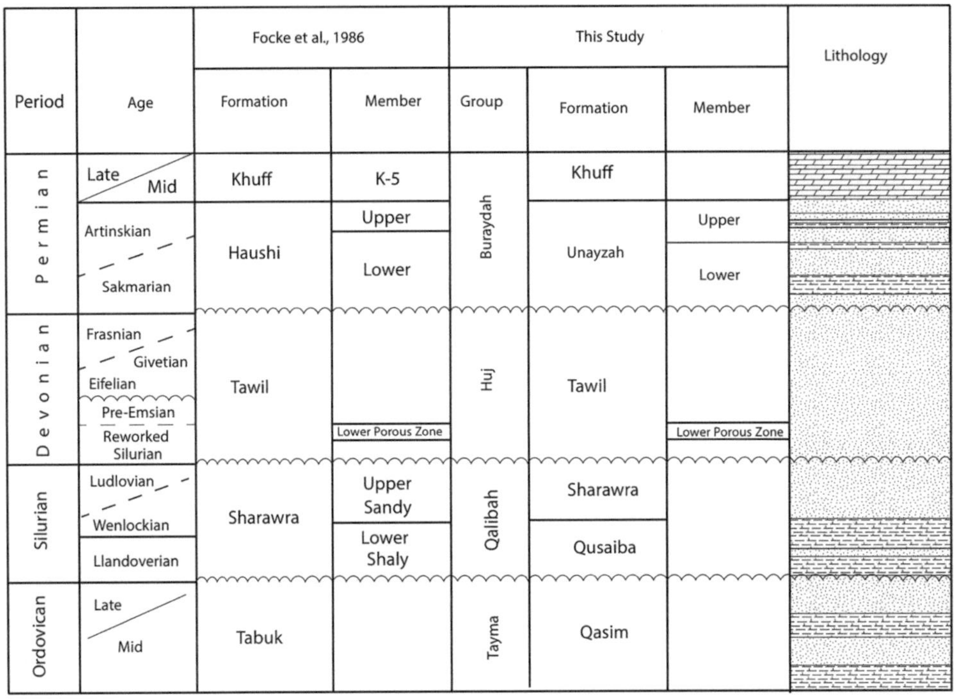

Fig. 5.3 Comparison of the Paleozoic stratigraphic system of Qatar as suggested by Focke et al. (1986) and Al-Saad and Sadooni (2016)

5.3 Depositional History

5.3.1 Tayma Group (Late Early Cambrian—Early Late Ordovician)

The Tayma Group was introduced by Vaslet et al. (1987) to include all the lower Paleozoic sedimentary sequences that were deposited prior to the first Late Ordovician glaciation in Saudi Arabia. It encompasses the formations of Siq, Quweira, Saq, and Qasim.

5.3.1.1 The Qasim Formation

The Qasim Formation was deposited during the Middle and Late Ordovician periods on a stable marine shelf adjacent to the Paleo-Tethys ocean, gently sloping toward the northeast. This shelf provided space for depositing a siliciclastic wedge measuring over 1220 m on the edge of Gondwana. The Qasim Formation and the subsequent Zarqa and Sarah formations are visible in two outcrop belts running approximately 900 km along the northern and northeastern margins of the Arabian Shield in Saudi Arabia (Senalp and Al-Duaiji 2001b).

The Qasim Formation, which comprises units 1 to 3 of the obsolete Tabuk Formation, as defined by Powers et al. (1966) and subsequently amended by Powers (1968), represents the lower portion of the formation. The Ordovician section of the Tabuk Formation was examined by Davis et al. (1981) in the Ha'il region and by Clark-Lowes (1980) and McClure (1988) in the Qasim region. Through these studies and a compilation by Wolfard (1981), the lithostratigraphy and sedimentology of the formation were defined.

The Formation was initially defined by Williams et al. (1986) from the Jabal Habashi quadrangle in the Qasim area and corresponds to the lower part of the Tabuk Formation1 as defined by Powers et al. (1966) and amended by Powers (1968). Other researchers, such as Manivit et al. (1986) and Vaslet et al. (1987), have also described the Qasim Formation. Williams et al. (1986) noticed that the reference section for the Formation corresponds to the reference sections of its four members: the Hanadir, Kahfah, Ra'an, and Quwarah, but no specific type section has been designated for the Qasim Formation as a whole.

In Saudi Arabia, the Qasim Formation comprises two siliciclastic coarsening-upward sequences, each with a basal shale/siltstone member and an overlying sandstone member. In its central outcrop belt in the Baq'a quadrangle, the siliciclastic sequence attends a thickness of 194 m. In the subsurface, the Qasim thickens toward the north and northeast, measuring more than 1220 m in Iraq and Syria. The incised glacial paelovalleys featured by the drop in sea level throughout the Late Ordovician glaciation cut the upper beds in some regions (Ghazwani 2012).

The basal argillaceous member of the Qasim, the Hanadir, conforms with the arenaceous Saq Formation. In the subsurface, the lower contact is intersected in the Qasim-801 water well. The upper boundary is marked by a pronounced unconformity, with significant parts of the formation removed by the

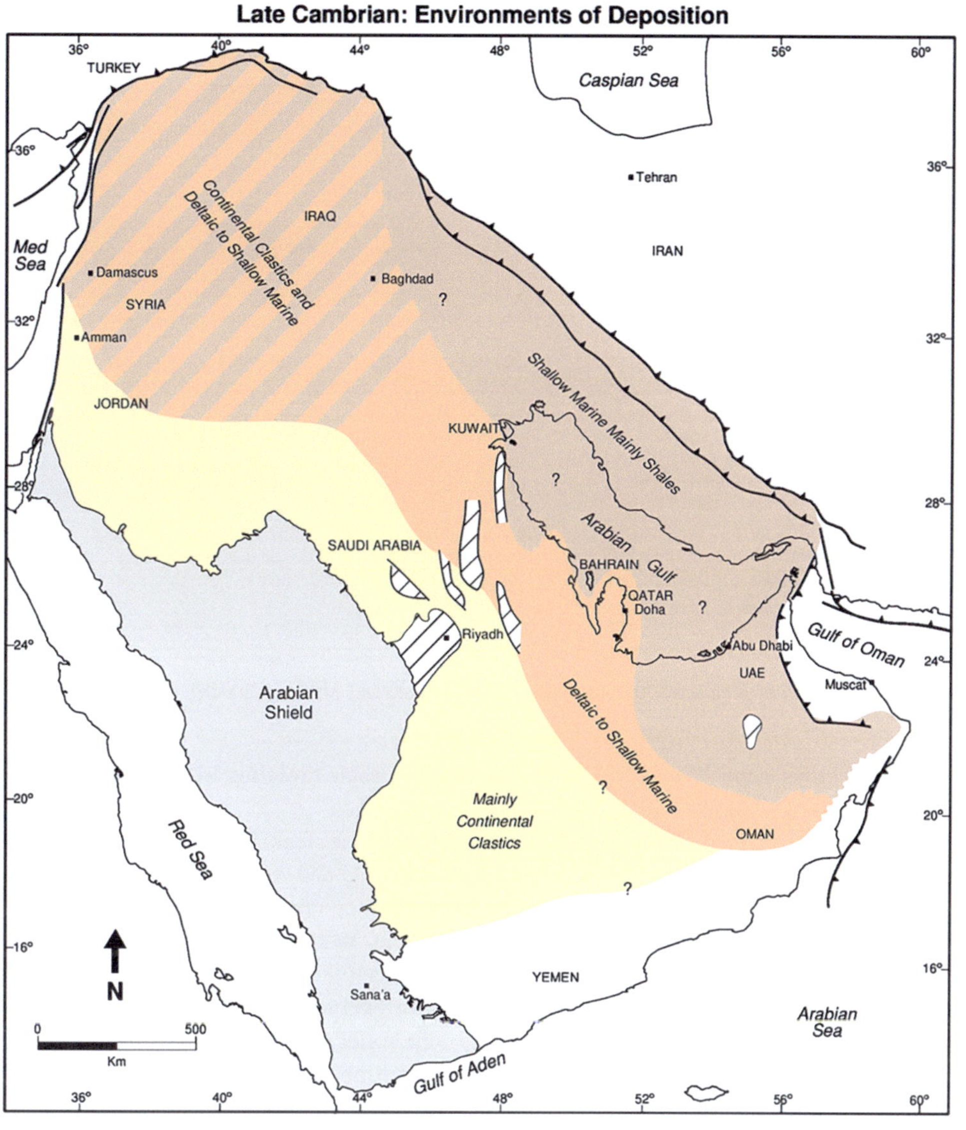

Fig. 5.4 Depositional environments of Eastern Arabia during the middle Ordovician (Konert et al. 2001)

effects of the Ashgillian (Late Ordovician) glacial activity. In the subsurface, the upper contact of the Qasim changes basinward as the magnitude of erosion at the base of the Zarqa, and Sarah diminishes toward the north and northeast.

Dating of the Ordovician Qasim Formation ranges from late Llanvirnian to late Caradocian (possibly Ashgillian), with Graptolites, trilobites, and chitinozoa commonly used to determine the age at outcrops. The Hanadir and Ra'an members contain many of these age-diagnostic fossils.

The Qasim Formation has a combined thickness of 261 m at the type localities of its four members. At the same time, in reference sections in northwestern Saudi Arabia, it measures 358 m thick (Fig. 5.5). It is unconformably overlain by the glacial and periglacial deposits of the Zarqa or Sarah formations of the Tabuk Group. In general, its relationship with the underlying Saq Formation is conformable. In some cases, it may be disconformable, as evidenced by conglomerates or a thin, red ferruginous breccia at the base of the Qasim Formation.

In northwestern Saudi Arabia, the Qasim Formation has been assigned to Early to Late Ordovician depending on trilobite and graptolite faunas by Vaslet et al. (1994) and Janjou et al. (1996c), Janjou et al. (1997). Senalp and Al-Duaiji (2001b) classified the formation as Late Llanvirnian to Late Caradocian, and probably even Ashgillian. The outcrop samples were dated primarily depending on the recognition of chitinozoans, trilobites, and graptolites that are commonly found in the Ra'an and Hanadir shale members. Figure 5.6 provides a detailed description of the formation's major lithologies.

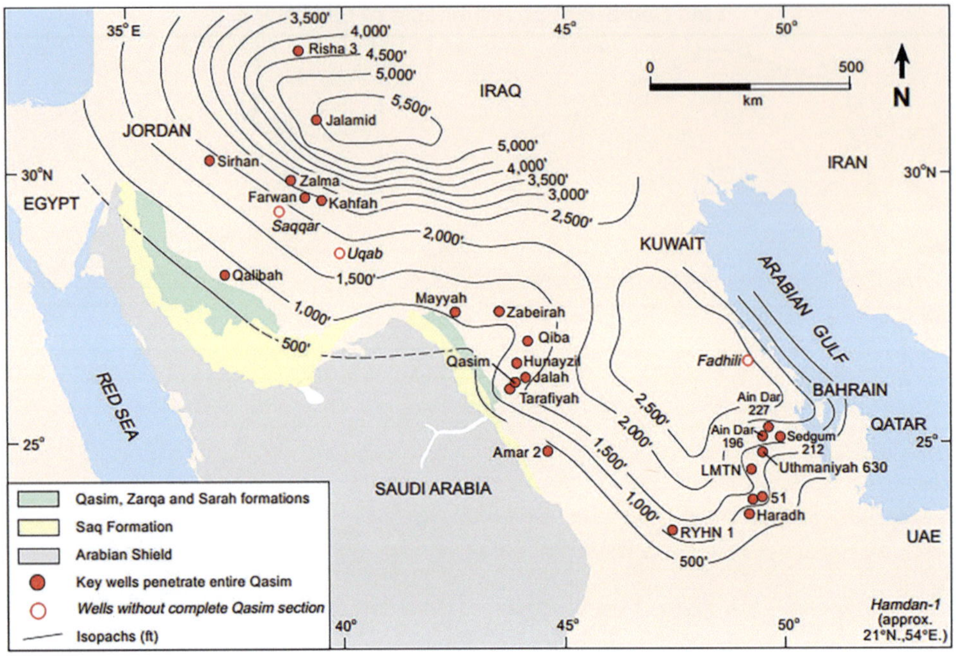

Fig. 5.5 Isopach map of the Qasim Formation across the Arabian platform (Senalp and Al-Duaiji 2001b)

Focke et al. (1986) gave the name "Tabuk" to describe these layers in Qatar. The term Qasim will replace that outdated term following the more recent Saudi stratigraphic nomenclature scheme. Consequently, some palynological assemblages and the well-known trace fossils of *Cruziana* and *Scolithos* obtained from the Matbakh MK-2 well in Qatar indicated that the Qasim Formation belongs to the Middle to Late Ordovician age. In the well Dukhan DKG-27, the formation was dated to the Early Silurian (Fig. 5.7). The current level of data cannot solve this age controversy. The formation consists of arkosic and micaceous sandstone with siltstone and shale. The sandstone contains quartz, up to 34% plagioclase, and less than 15 cm of thin laminated to bedded sandstone viewing fining upward sequences. Sometimes, the sandstone changes into shale or is completely removed (Focke et al. 1986).

In the Dukhan DKG-27 well, larger units with a thickness of about 15 m were contained in the formation with upward coarsening trends. Focke et al. (1986) proposed that the water in the deposition environment of these units may have been affected by traction currents or storm-agitation. Bioturbation significantly affected the formation part that contains fine-grains. *Scolithos* and *Cruziana* were the most common ichnofacies showing vertical or oblique burrows. Figure 5.8 shows samples of lithologies present in core materials of the formation recovered from the well Mk-2. The present lithologies contain:

1. Silty shale revealing load structures and dewatering characteristics.
2. Sandstone featured fine grains presenting contortion, scouring, and well-developed cross-lamination.
3. Claystone interbedded with contorted and thin sandstone layers, with numerous root-like deformations extending between the claystone and sandstone, likely representing cast-load structures from sandy materials passing through the softer clay unit.
4. Sandstone displaying unidirectional ripple marks, interspersed with bioturbated shale and some sandy nodules, possibly indicating local reworking.

The detailed lithologies of the formation in MK-2 and Dukhan DKG-27 show that finer materials dominate the formation at MK-2 with higher gamma ray values. This section can be correlated with a fair level of confidence with the Ra'an Shale Member of the Qasim Formation in Saudi Arabia.

Reports from some regions of the Arabian Basin such as Iraq, Saudi Arabia, and Oman showed the trace fossil *Cruziana* (Seilacher 2007; van Bellen et al. 1959). The whole Qasim Formation is supposed to be formed in locally deltaic, shallow marine environments. The fluvial/glacial sediments of the Sahar Formation, recognized in many areas in Saudi Arabia, cannot be identified in the Qatari section. The currently available data is not sufficient to support that the glacial sediments extended to Qatar at that geologic time.

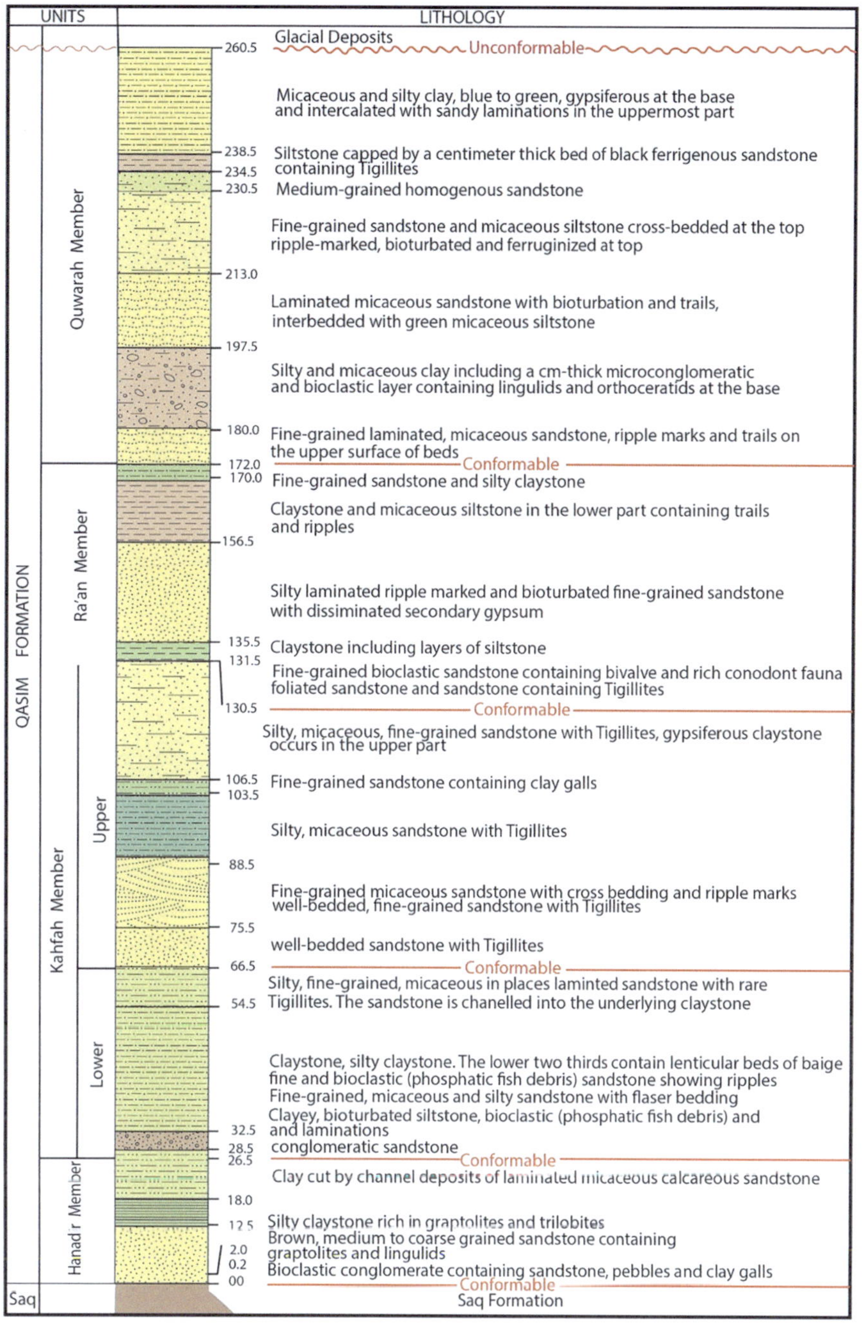

Fig. 5.6 Detailed lithologic description of the composite-type section of the Qasim Formation at the Jabal Habashi quadrangle (Saudi Stratigraphic Committee 2013)

5.3.2 The Silurian

The Silurian period brought a significant paleogeographical shift to Arabia, including Qatar, due to various tectonic events that affected the southern margin of Gondwana, such as the closure of the Rheic Ocean and the formation of the Variscan orogeny. These events resulted in the development of a foreland basin in Arabia filled with sediments eroded from the Arabian Shield. During the Llandovery, a major phase of global warming led to a rapid rise in sea level, causing the flooding of the Arabian Platform, including S10, dated at 440 Ma. Shallow to open marine environments were established in marginal areas, and deeper marine environments covered the inundated platform and extended southward along a subsiding intrashelf basin in central Saudi Arabia (Husseini 1991; Jones and Stump 1999; Mahmoud et al. 1992). Anoxic water bottom conditions in the sediment-starved basin resulted in preserving organic-rich shales, known as the prolific Silurian "hot shale." These shales include the Qusaiba shale in Saudi Arabia, Mudawwara Formation in Jordan, Sahmah Formation in Oman, Abba Formation in Syria,

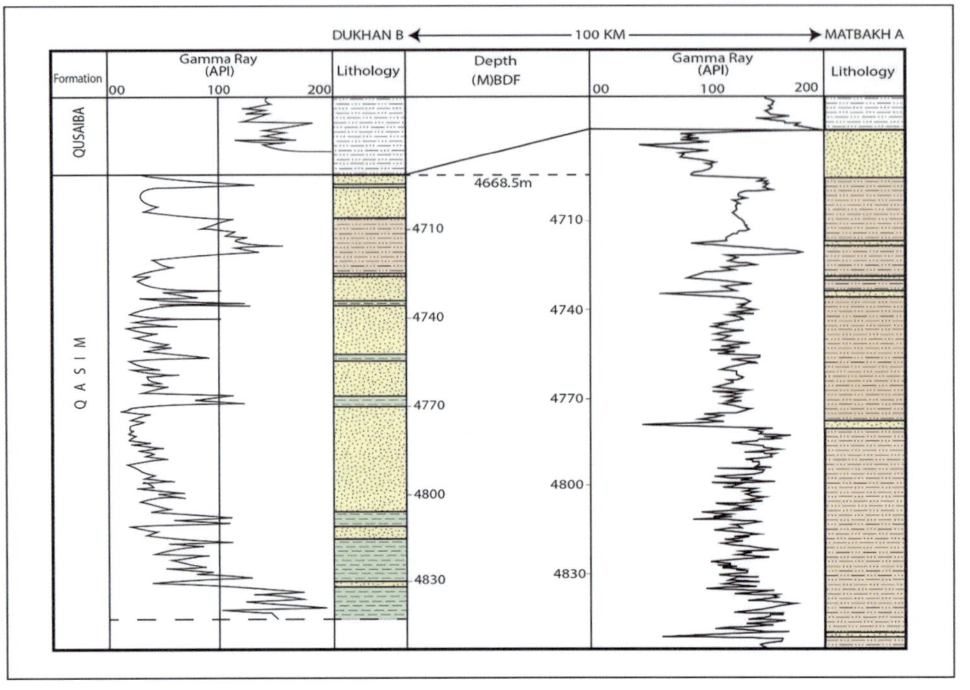

Fig. 5.7 The lithology of the Qasim Formation in Matbakh MK-2 and Dukhan DKG-27 wells (Focke et al. 1986, ESC Archives)

Fig. 5.8 Core samples from the Qasim Formation in the well Matbakh Mk-2. To the right, trace fossils of Cruziana and Scolithos (Focke et al. 1986, The ESC Archives)

Dadas Formation in southeast Turkey, and Ghakum Formation in Iran. The Qusaiba shale is the primary source of Paleozoic hydrocarbons in Saudi Arabia (Abu-Ali et al. 1991; Mahmoud et al. 1992). In the northern parts of the basin, a second, younger source rock of possibly Wenlock age occurs (Aqrawi 1998). Following the initial transgression, there was a thick (> 1000 m) coarsening-upward sequence of shales and sandstone of Llandovery to Pridoli age, such as the Qalibah in Saudi Arabia, which prograded basinward (Mahmoud et al. 1992). However, middle to outer neritic environments persisted in the north, such as the Abba Formation in Syria and the Dadas Formation in southeast Turkey, and in the east, such as the Gahkum Formation in Iran, throughout the remainder of the Silurian (Konert et al. 2001; Sharland et al. 2001).

5.3.2.1 The Qalibah Group (Silurian)

Janjou et al. (1996b) proposed the Qalibah Group to consider the Silurian formations of Sharawra, Qusaiba, and Uqlah. Only Qusaiba and Sharawra formations were recognized in Qatar's subsurface.

The Qusaiba Formation

The Qusaiba was a member of the so-called Qalibah Formation and was recently upgraded into a formation status. Janjou et al. (1996b) described a new type section from the Ad Dahkiyah region in Saudi Arabia. In the central and northwestern parts of Saudi Arabia, the formation was recognized as Llandoverian (Early Silurian) depending on chitinozoans, graptolites, and trilobites (Clark-Lowes 1980).

The formation is classified into five units in its type section. From the bottom, Unit 1 is formed of siltstone and fine-grained sandstone, Unit 2 consists of alternations of fine-grained, micaceous sandstone characterized by hummocky cross-stratification with black clayey siltstone. Unit 3 comprises two beds; the lower bed is formed of micaceous silty claystone, and the upper bed is formed of fine to medium sandstone grains showing hummocky cross-stratification. Unit 4 has the same lithologies and arrangement as Unit 3. In contrast, Unit 5 consists of claystone interbedded with a fine-grained coarsening upward sequence of sandstone with hummocky parallel to low-angle cross-stratification. Bioturbation is common in some of the horizons of this unit (Saudi Stratigraphic Committee 2013) (Fig. 5.9). Sections of the Qusaiba formation correlate with the Saudi "hot shale beds," which are marked as the primary hydrocarbon source of the Paleozoic play (Jones and Stump 1999).

Sharawra Formation

The Sharawra Formation has an average thickness of 433 m in its new type locality at the Al Qalibah quadrangle (Saudi Stratigraphic Committee 2013). The formation gross lithology contains upward coarsening sequences of siltstone and sandstone with dark-red phosphatic grains featured with pillow and ball structures and red micro-conglomeratic clasts. Al-Ruwaili (2000) dated the formation to a Ludlovian (Late Silurian) age.

Janjou et al. (1996a) divided the formation into four formal members. From the bottom, these are the Jarish Member, which disconformably overlies the Qusaiba Formation. It comprises around 97 m of sandstone, siltstone, and fossil debris of bivalves, trilobites, ortho cones, and small gastropods. The Khanafriyah Member comprises around 110 m of sandstone and siltstone showing pillow and ball structures and contains traces of *Diplocraterion* and *Cruziana*. The sandstone coarsens upward, forming a tabular surface capped by hardground. The Nayyal Member comprises around 125 m of sandstone and siltstone that becomes thicker and coarser upward. The fine grains of sandstone form thin beds with micro-conglomeratic clasts and phosphatic grains at the base. A rusty-brown fine-grained sandstone marks the top of the member. In some horizons, the sandstone has pillow and ball structures with a size range from one meter and several meters. The top member is the Zubliyat, which comprises around 119 m of upward-coarsening siltstone and sandstone sections. The top of the member is a zone of claystone pebbles and a composite bench of fine grains of sandstone featuring by rusty-brown color with erosional contact at its base. This section is channeled and overlain by a conglomerate at the base of the Tawil Formation (Fig. 5.10).

The Sharawra Formation in Qatar is comprised of thinly bedded shales, siltstones, and sandstones in the upper part, known as the Upper Sandy Member, and a monotonous sequence of grey, locally blackish, micaceous siltstone and claystone in the lower part, known as the Lower Shaly Member. This formation rests unconformably on the albite-rich arkosic sandstone of the Qasim Formation and is overlain by Devonian Tawil sandstone. A noticeable decrease in the Thorium content on the Gamma Ray marks the transition from the Sharawra to the Tawil. The sandstones of the Upper Sandy Member are typically grey-to-brownish or whitish and contain micas such as sericite and biotite. The individual sandstone beds often amalgamate into sand units up to 10 m thick, and cross-bedded flasers and ripples are common sedimentary features. There are also occasional slump/traction features that resemble turbidite.

Sediments and palynomorphs substantiate the marine environment. The Lower Shaly Member has a correlation with the Ra'an Shale Members and Qusaiba in Saudi Arabia, and it contains source-rock-prone intervals that predominantly consist of Sapropelic organic matter, making it the primary source of the gas present in the tight streaks of the pre-Khuff section and the gas in the North Field Khuff. The Upper Sandy Member is composed of well-sorted quartz arenite, with up to 80% solid rock volume quartz and varying amounts of Potassium feldspar (2–13%), volcanic rock fragments, micas,

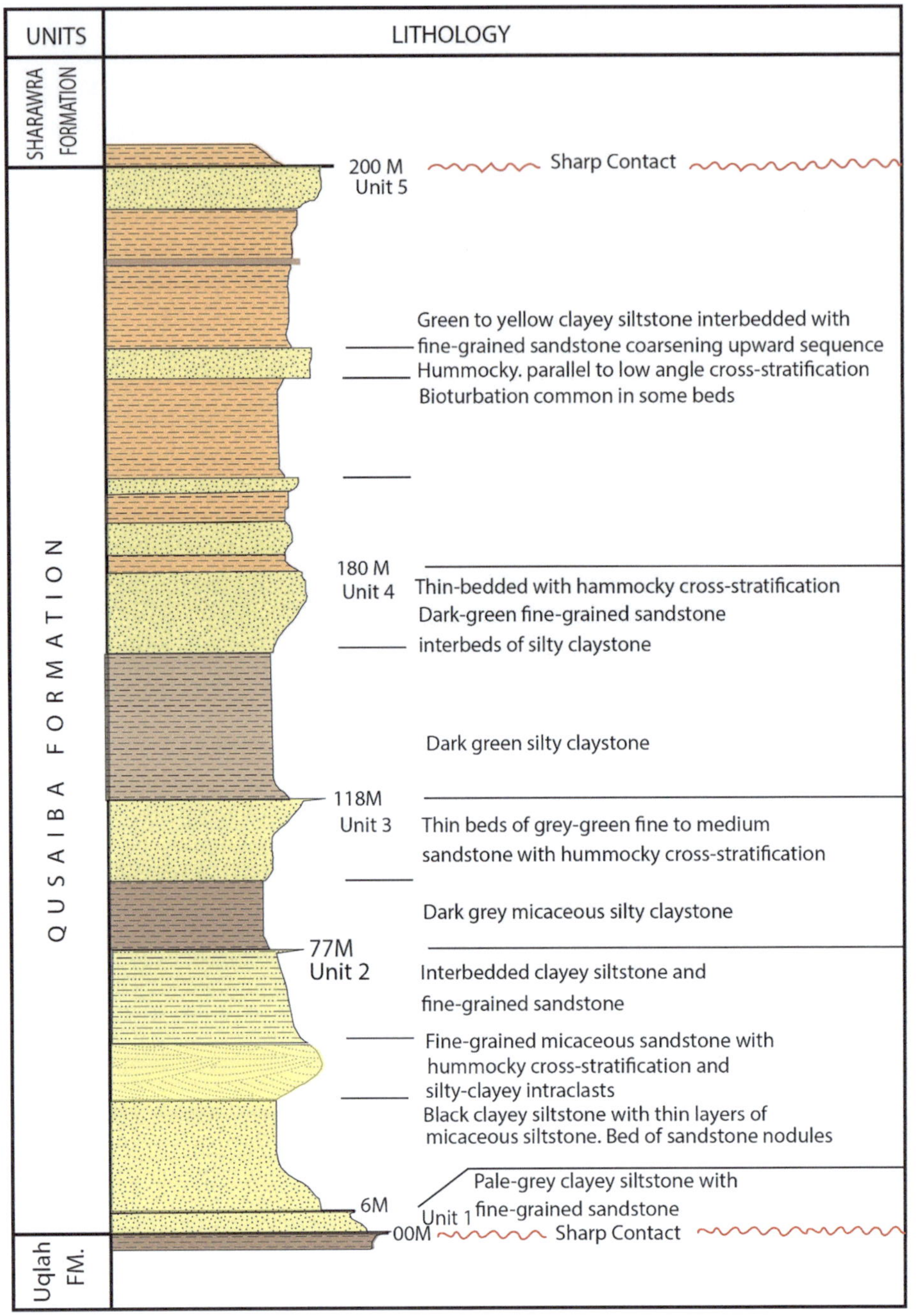

Fig. 5.9 Composite stratigraphic section of the Qusaiba Formation in Al Qalibah quadrangle, Saudi Arabia (Saudi Stratigraphic Committee 2013)

and clay-sized particles such as illite, smectite, and chlorite. Extensive quartz overgrowth and the replacive carbonate have significantly reduced primary porosity. However, the early leaching of diagenetically unstable components and the late leaching of some replacive carbonate have created moldic porosity. The secondary pores are poorly interconnected or filled with twisted illite crystals, significantly reducing permeabilities. The Lower Shaly Member consists of grey and blackish siltstones and claystone (Focke et al. 1986) (Fig. 5.11).

5.3.3 The Late Silurian-Early Carboniferous

Late Silurian and Devonian sediments are not well represented in the geological record of eastern Arabia due to Hercynian tectonic activity, which caused widespread uplift and erosion. In Saudi Arabia, the Tawil Formation, which comprises conti-

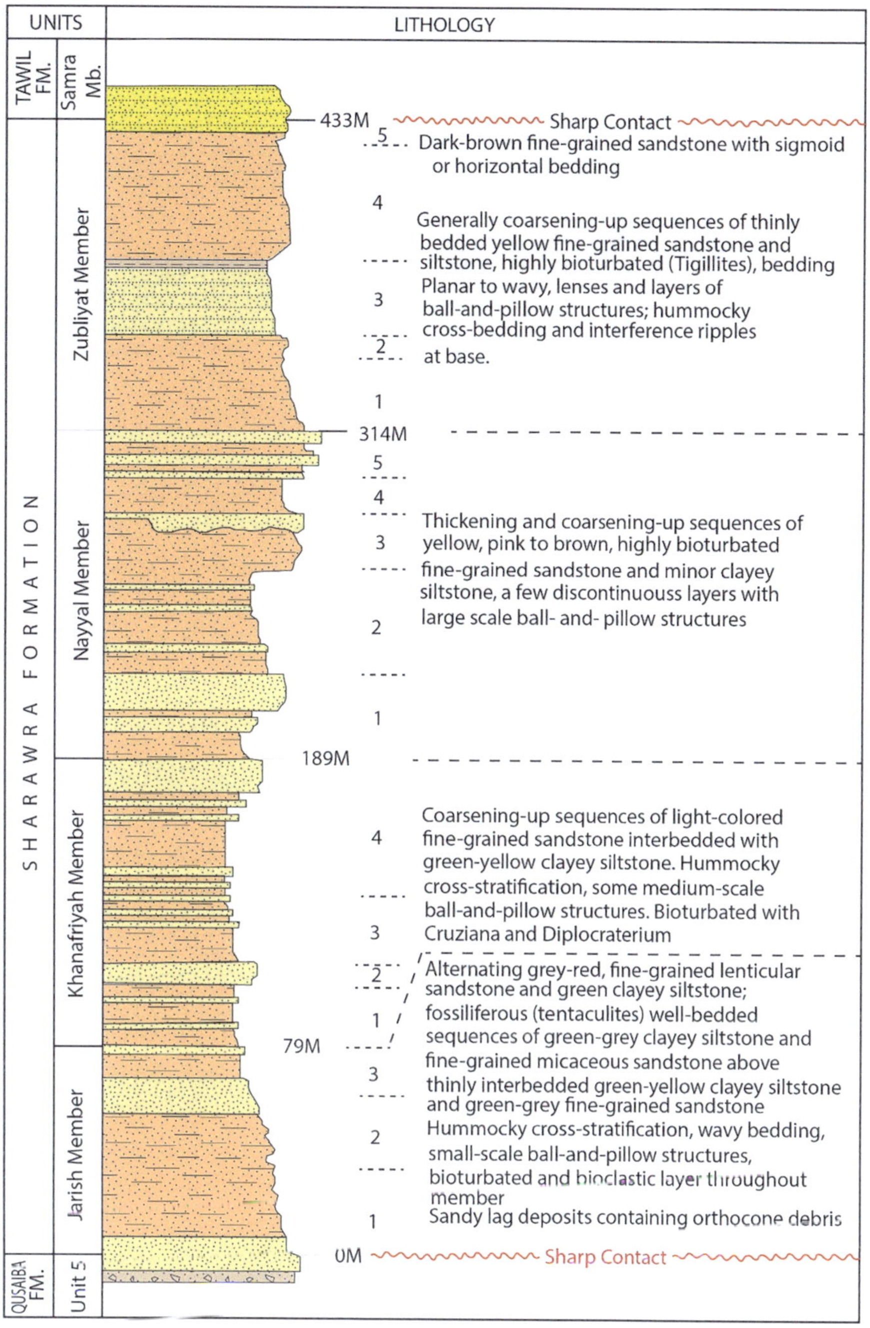

Fig. 5.10 A composite stratigraphic section of the Sharawra Formation in Al Qalibah quadrangle, Saudi Arabia (Saudi Stratigraphic Committee 2013)

nental clastics, is found above the pre-Tawil unconformity and is followed by the marine Jauf Formation, deposited during the Pragian to Emsian (Al-Hajri et al. 1999). The Misfar Formation in Oman also records marine incursions during this period, with the deposition of anoxic mudstones in lower coastal plain environments. The lack of pre-Emsian deposits in Syria, Turkey, and Iraq suggests a structurally high position with respect to the depocenter in Saudi Arabia.

In the Middle and Late Devonian, continental clastics were deposited in Saudi Arabia's Jubah Formation, replaced by mixed marine siliciclastics and carbonates in the north (Hazro and Kayayolu formations in southeast Turkey). While continental environments were established in the center of Saudi Arabia, Syria, and Iraq, marginal marine environments persisted in Turkey and Oman. However, the age determination of these deposits in Syria and Iraq is highly uncertain and may represent the latest Devonian (Aqrawi 1998).

The resumption of marine environments, particularly during the latest Devonian in the northern region, suggests a differential downwarp of the northern margin of Gondwana, which is also observed in interior Iran. The Uppermost

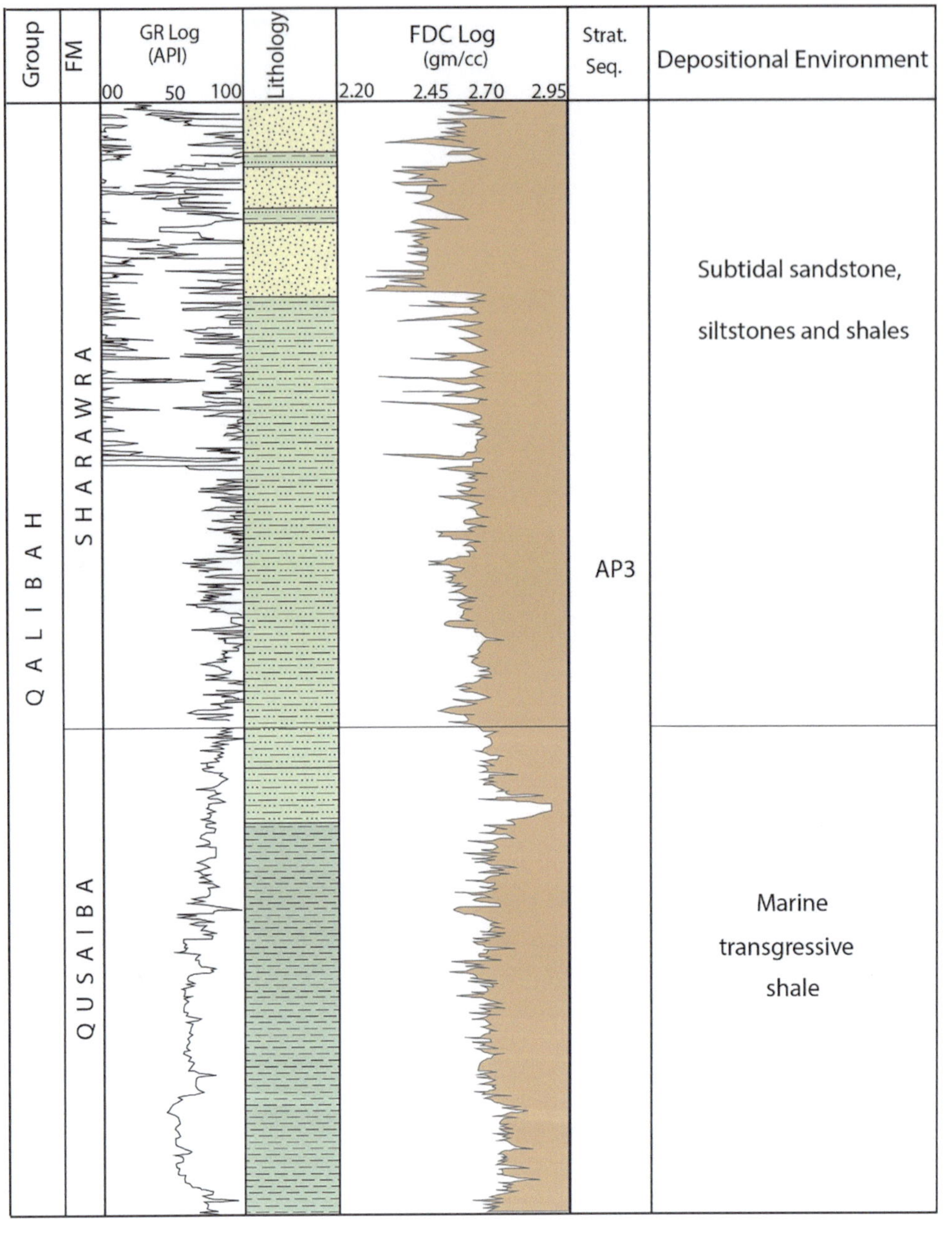

Fig. 5.11 Detailed lithologic section of the Qusaiba and Sharawra formations in Qatar (Al-Saad and Sadooni 2016)

Devonian strata in the Alborz Mountains in Iran rest directly on Cambrian or Lower Ordovician sequences, while a more continuous Paleozoic section, including older Devonian, is preserved in the basin south of the mountains (Wensink 1991). These relationships suggest that the northern margin of Gondwana became tectonically unstable, marking the beginning of the Hercynian Orogeny (Konert et al. 2001).

5.3.3.1 The Huj Group (Late Silurian to Early Carboniferous)

Janjou et al. (1996c) introduced the Huj Group to encompass the Upper Silurian-Lower Carboniferous rocks in the northwestern parts of Saudi Arabia. The group consists of Jubah, Jauf, and Tawil Formations. The strata of the Tawil Formation were only recognized in the subsurface sections of Qatar.

The Tawil Formation

Steineke et al. (1958) published for the first time the Tawil Formation. The Al Jawf area contains the type locality of the formation, which is 200 m thick at the type section. The Tawil Formation was reclassified as "Tabuk Formation" member by Bramkamp et al. (1963), Powers et al. (1966), and Powers (1968). The Tawil Sandstone member was identified as the lowermost member of the enlarged Jauf Formation by Bahafzallah et al. (1981) and Al-Laboun (1982). Vaslet et al. (1987) proposed that these Early Devonian rocks are a separate formation, owing to a possible disconformity at the base of the Jauf Formation as defined by Powers (1968).

In its type locality, the formation contains medium to coarse sandstone interbedded with red shale and large-scale cross-bedded, particularly at its lower part. The formation base contains beds of micro-conglomeratic and coarse-grained sandstone with quartz gravels forming a lenticular band at the lower part (Fig. 5.12). The formation age is

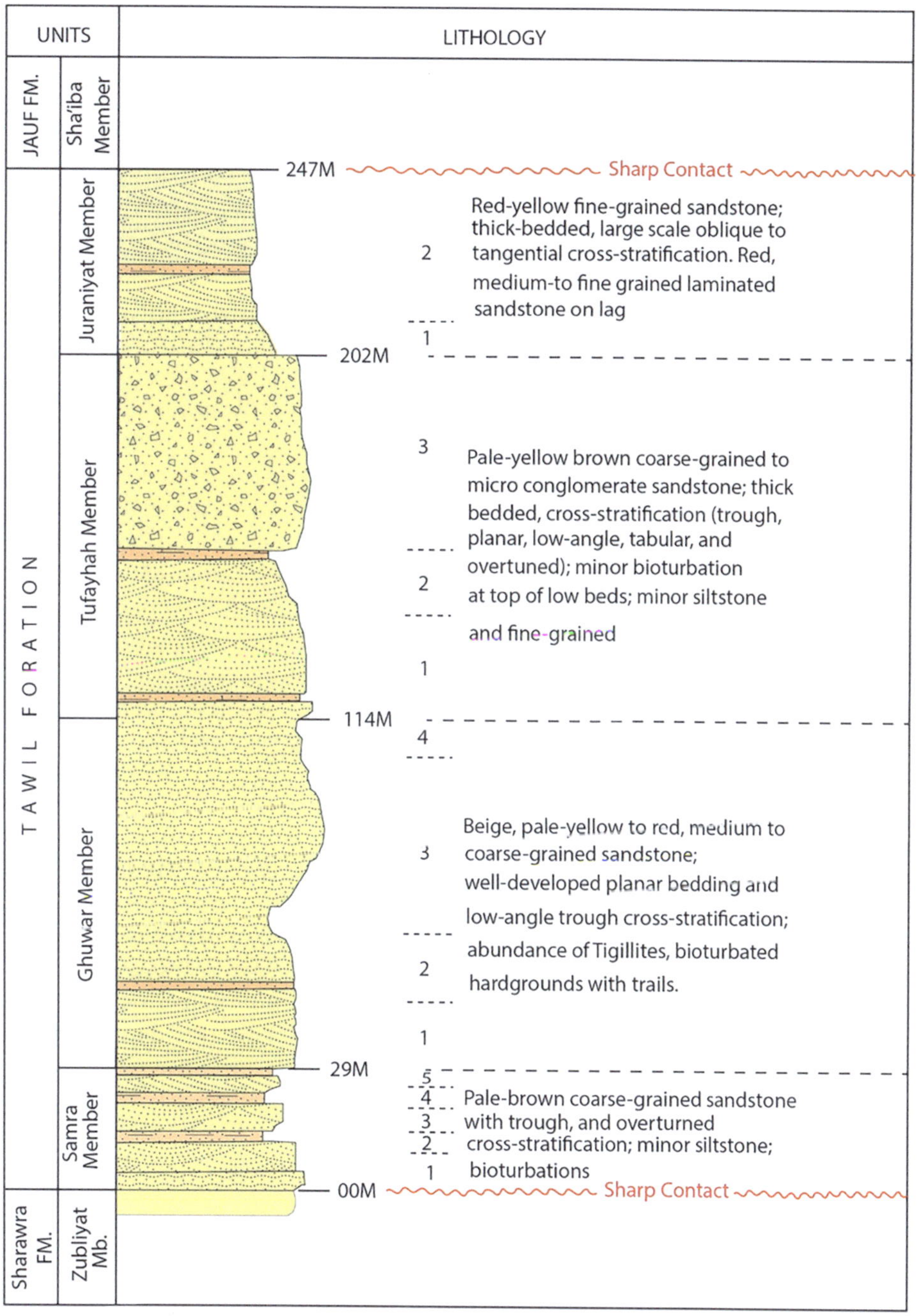

Fig. 5.12 A detailed stratigraphic composite-type section of the Tawil Formation in the Al Qalibah quadrangle (Saudi Stratigraphic Committee 2013)

controversial because it has no major fossils. Laboun (1982) assigned the formation to an Early Devonian (Gedinnian) age because the overlying Jauf Formation was assigned to the Middle Devonian and well samples that contained some microflora. Janjou et al. (1996a) designated the formation age to a Late Silurian to Early Devonian since the basal member of the underlying Jauf Formation was assigned to the Lochkovian-Pragian age. The Formation is subdivided into four members, from the top downwards: Juraniyat, Tufayhah, Ghuwar, and Samra members.

The Samra Member comprises medium—to coarse-grained sandstone that displays approximately 29 m of fining and thinning upward successions. Some layers of fine-grained sandstone within red siltstone exhibit current ripples.

The Ghuwar Member's basal part comprises pink clayey-silty sandstone intercalated with tabular layers of coarse-grained sandstone. A microconglomerate containing quartz gravel makes up its base, which is overlain by a thick layer of medium-grained bioturbated sandstone that has Tigillites in some areas. It has an overall thickness of 85 m.

The Tufayhah Member is 88 m thick and is divided into two units. The upper unit is composed of medium to coarse grains of sandstone with microconglomeratic beds toward the top and a coarse-grained basal section. The lower unit contains medium to coarse grains of sandstone with a microconglomerate containing quartz gravel at its base. Its top portion is several meters thick and comprises brown or red silty-clayey sandstone with Tigillites.

The Juraniyat Member has a thickness of 45 m and is made up of fine-grained purplish-red sandstone and siltstone. A microconglomerate with small quartz pebbles is situated at the bottom, followed by a fine-grained sandstone.

Al-Harbi and Khan (2008) studied the Tawil Formation strata in central Saudi Arabia. They discovered that the formation is formed exclusively of mature Quartz arenite, with quartz forming up to 96% of the studied samples. They suggested that the sediments were of a stable cratonic source, and their accumulation occurred in a passive margin basin.

More detailed depositional settings were proposed by the Saudi Stratigraphic Committee (2013). It is suggested that the depositional environments shifted from a coastal plain to shallow channels in distal braided streams, leading to the formation of sand bars. While the Samra Member was deposited in a sandy riverbed that was braided to straight as migrating linguoid and transverse bars formed. The lower part of the Ghuwar Member marks a significant transgressive event, while the remaining portion shows a cycle of regression and transgression. The sandstones of the Tufayhah Member indicate a resumption of an alluvial environment of the braided-stream type. The sedimentary structures near the top of the unit show a renewed transgression. The Juraniyat Member is characterized by tidal sand bars that are embedded in a deltaic to estuarine complex, suggesting an environment of mixed tidal and fluvial influence within a marginal-marine depositional setting.

5.3.4 The Early Devonian Transgression

Dossary et al. (2017) studied the Tawil Formation in Ghawar and neighboring fields. They suggested that the Tawil Formation consists mostly of siliciclastics, primarily fluvial sandstones in the southwest and western regions, and more distal facies in the north and northeastern Arabian Gulf area. The depositional environment is mostly fluvial distributary channels that are influenced by tides within an overall deltaic setting. Petrographic data reveal that the key factors that control reservoir properties are the lithofacies of the Tawil Formation, which consist of well-sorted, fine- to medium-grained quartz arenites dominated by monocrystalline quartz grains and a few mica flakes.

In Qatar, The Tawil Formation contains a repetitive sequence of thinly bedded sandstones, shaly sandstones, and siltstones that alternate with shale bands that are a few meters thick. It is of Devonian age and underlain by the Silurian Sharawra Formation, while its top is truncated by the Hercynian unconformity. The sandstones in the upper part are typically cross-bedded and contain plant debris and mica flakes concentrated along bedding planes. Quartz and siderite predominantly cement the sandstones, while salt and gypsum cement occur to a lesser extent. The Lower Porous Zone, near the bottom of the Tawil Formation, is up to 100 m thick and is the primary target for pre-Khuff gas exploration. It comprises dark shaly sandstones and subordinate carbonates alternating with white to brown, medium- to coarse-grained, rather friable quartzose to arkosic sandstones. Parallel bedding is prominent, and the poikilotopic siderite cement determines the friability and reservoir quality of the cleaner sandstones, as authigenic quartz overgrowth is limited.

The lack of specific environmental indicators prevents a detailed depositional interpretation. Still, the upper part of the Tawil is thought to be deposited in an alluvial system and most likely on a fluvial plain of braided streams. At the same time, sedimentary structures and marine acritarchs suggest a shallow-marine, sublittoral environment for the lowermost Tawil. Powers et al. (1966) described Devonian sediments of marine, transitional, and continental origin from the Arabian Peninsula.

The Tawil Formation comprises two rock types: rock-type 1 (tight upper Tawil sandstone or quartz psammite) and rock-type 2 (porous lower Tawil sandstone or arkosic psammite). Rock-type 1 is predominantly quartzose to slightly arkosic sandstone with feldspar admixtures, illitized and leached feldspars, and quartz cementation resulting in poor reservoir

Fig. 5.13 Cores from the Tawil Formation in the well DK-6

properties (Fig. 5.13). Rock-type 2 represents good reservoir quality in the Lower Porous Zone, with intergranular and secondarily enlarged porosity, thin illite coatings on framework grains, and late diagenetic carbonate cement occurring patchily or in layers, replacing quartz and feldspar. Vuggy porosity created by leaching of feldspar and other microscopic indeterminate rock fragments is present, but later-stage authigenic clay pore filling occurs on clay coats. While some intervals within the Tawil Lower Porous Zone are tight, rock-type 5 represents the ideal end-member of good reservoir quality intervals (Focke et al. 1986) (Fig. 5.14).

5.3.5 The Mississippian-Pennsylvanian (Buraydah Group)

Vaslet et al. (1988) introduced the Buraydah Group in the Saudi stratigraphic system to contain the layers passing through the Paleozoic–Mesozoic boundary. The Paleozoic section of the Buraydah includes the Khuff, Unayzah, and Berwath formations.

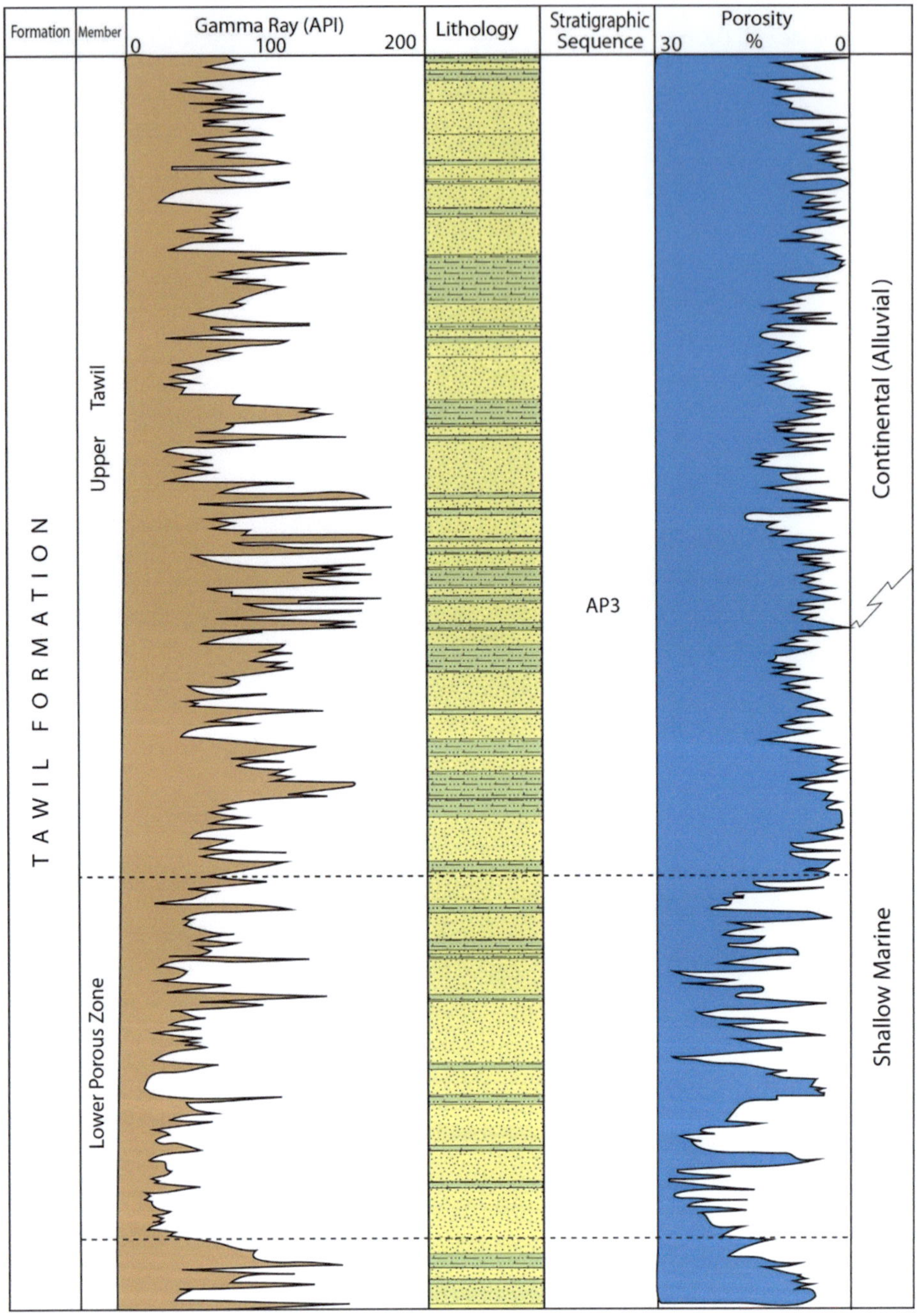

Fig. 5.14 Detailed stratigraphic section of the Tawil Formation in the well Matbakh-2, Qatar (Al-Saad and Sadooni 2016)

5.3.5.1 Unayzah Formation

The Unayzah Formation is an important hydrocarbon reservoir in Saudi Arabia and has the same potential in other countries such as Kuwait and Qatar. The Unayzah Formation was presented to the stratigraphy scheme of Saudi Arabia by Al-Laboun (1982) to accommodate the clastic strata (initially called the fifth lower member of the Khuf Formation or the pre-Khuff clastics) subjacent to the Khuff carbonates in Al Qasim (Saudi Stratigraphic Committee 2013; Delfour et al. 1982; Al-Laboun 1982). It was named after the town of Unayzah in the Qasim area of eastern Saudi Arabia. A complete section was reported from the Qusayba area, north of Unayzah town, which is around 34 m thick.

The Unayzah Formation has a Late Carboniferous (Pennsylvanian) to Early Permian age, based on various fossil evidence. El-Khayal et al. (1980) described fossil plants from the lower part of the Unayzah Formation, which include species of genera *Marattiopsis, Validopteris, Cordites, Annularia, Lobatannularia, and Plecopteris* indicating an age of Late Carboniferous (Pennsylvanian) to Early Permian. The formation was subsequently assigned to Late Carboniferous (Pennsylvanian) to Middle Permian (Al-Laboun 1984, 1986) and later modified to Late Carboniferous (Pennsylvanian) to Early Permian age (Al-Laboun 1987). A plant bed in the formation has been dated as Early Late Permian (Kazanian) by El-Khayal and Wagner (1985). The same authors proposed a later Permian (Tatarian) for the mixed sandstone, shale, and carbonate that formed the basal part of the Khuff Formation.

A composite-type section of the formation was described by the Saudi Stratigraphic Committee (2013). The selected section is located on the way between Buraydah and Unayzah towns. It consists of three sections along the road and in an escarpment at that road. They also selected a reference section at Wadi Shaqra, Qusayba depression in the Al Qasim region. Ferguson and Chambers (1991) suggested a subsurface reference section in the well Hawtah-1 (Saudi Stratigraphic Committee 2013). The thickness of the formation is 32.2 m in the type section and around 34 m in the reference section.

In the type locality, the Unayzah Formation consists of mixed lithologies, including thin-bedded sandstone, siltstone, shale, and marl with subordinate limestone and anhydrite. In some sections, there is a cross-bedded, fine-grained sandstone bed in the middle part of the formation. Some slumped and tinted red dolomite beds can be seen at the top of the section (Fig. 5.15).

The Formation is disconformable at the top with the As Shiqqah Member of the Khuff Formation. El-Khayal and Wagner (1985) reported the presence of a break that they used as the boundary between the Unayzah and the Khuff formations. The formation is separated by an angular unconformity from the Carboniferous to Cambrian rocks in the northern parts of Arabia and with a nonconformity from the Precambrian rocks in the central parts of Saudi Arabia. The base of the formation is associated with the Hercynian unconformity, and it is not exposed in the outcrops.

Senalp and Al-Duaiji (2001b) studied the Unayzah Formation in surface and subsurface sections. The outcrops of the formation in the towns of Quway'iyah, Ash-Shiqqah, and Unayzah (the original type section) and also in Wadi Al-Shajara (the original reference section of the formation) were the subject of that study. The Unayzah outcrops in the Quway'iyah section rest between the basement and the Khuff Formation. They also remeasured the type section of the Unayzah Formation and the overlying Ash-Shiqqah Formation in light of new subsurface data obtained from exploration wells. The total thickness of the two formations is 31 m, but the lower contact of the formation was not exposed. They provided a detailed new description of the type section with more sedimentological details. They also documented a thicker section of the formation in the downtown area in a long north–south trending escarpment where the formation is 100 m thick. The lower 1.5 m of the section consists of a massive red mudstone with laminae of iron-stained siltstone that contains root structures (rhizoliths) and abundant plant fossils. This is the "Unayzah plant bed" that dates as early Late Permian (Kazanian). This unit is overlain by 4 m of red sandstone that sometimes cuts into the mudstone. The sandstone ranges from fine to coarse-grained (becoming pebbly at the base) with moderate soring and cross-bedding. Senalp and Al-Duaiji (2001b) suggested that this sandstone was deposited by a meandering river system as a part of a coastal plain environment. This unit is overlain by a red to light brown mudstone around 3 m thick. This mudstone was deposited in a swamp or marsh environment.

The contact between the clastic sediments and the overlying carbonate section is sharp. This contact is formed of a weathered paleosol and represents the "stratigraphic break" proposed by El-Khayal and Wagner (1985), and it is associated with the pre-Ash-Shiqqah unconformity.

According to Senalp and Al-Duaiji (2001a, b), the original reference section of the Unayzah Formation in Wadi Al-Shajarah, the Qusaiba Depression, represents the thickest exposed section of the formation. The lower part of the section is underlain by the pre-Unayzah unconformity that cuts either the Tawil or Qusiaba formations and is covered by thick caliche sediments that represent the pre-Ash-Shiqqah unconformity. The total thickness of the measured section is 36.5 m and is made up of mudstone interbedded with four sandstone units. The mudstone beds are between 4.2 and 10.9 m thick and are usually intersected by gypsum veins. The sandstone beds are 1.5–6.5 m thick; they represent point bars of low sinuosity meandering rivers. The mudstone sediments were formed in a flood plain between the meander belts.

Unayzah Formation in the subsurface

In the subsurface, the Unayzah formation encompasses the strata between the pre-Unayzah and pre-Ash-Shiqqa unconformities. It comprises a complex succession of the thick, red-colored continental clastic sequence formed of conglomerate, sandstone, siltstone, mudstone, nodular anhydrite, and caliche. This sequence was deposited as an alluvial fan that graded down-slope into braided river system plains and playas. There are some sediments of eolian origin also. This succession indicates deposition under arid climatic conditions (Senalp and Al-Duaiji 1995, 2001a) The Unayzah strata show wide lateral variations; in some areas, the Unayzah Formation is represented by well-sorted sandstone beds, while massive mudstone becomes dominant in the same succession. These rapid lateral facies variations probably resulted from the geometry of the underlying old topographic surface of the Pre-Unayzah unconformity or relatively rapid climatic changes.

Senalp and Al-Duaiji (2001a) suggested that the Unayzah sediments were deposited under two major depositional settings. These are the alluvial fan and playa and the eolian systems. They divided the alluvial fan deposits into proximal and distal facies. The coarse materials were deposited in the

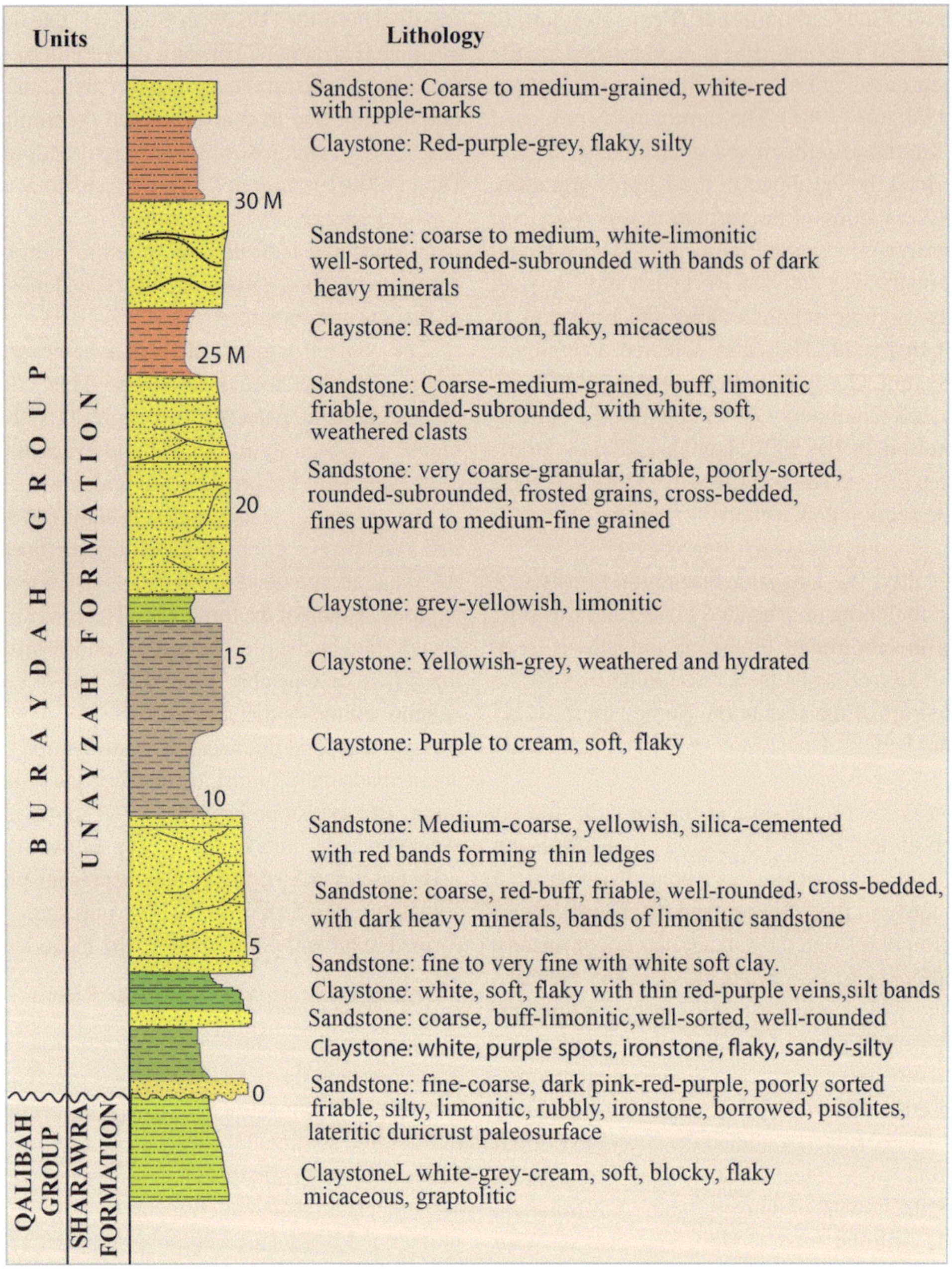

Fig. 5.15 Detailed lithostratigraphic section of the Unayzah Formation in Unayzah town, the Qasim region (Saudi Stratigraphic Committee 2013)

upper or inner part of the fan, while the finer materials were in the lower parts of the fan. The interbedded mudstones in the upper part of the formation were deposited in playas (inland sabkhas) and flood plains. The playa sediments are closely associated with eolian sediments. The eolian system encompasses sandstone interbedded with fluvial red sandstone and the playas mudstones.

The Unayzah Formation in Kuwait

The Unayzah Formation has been drilled in two wells in northern Kuwait, where considerable coring was conducted. The Unayzah Formation unconformably overlies a thick carbonate succession of the Hormuz Group in northern Kuwait. It is unconformably overlain by a thick carbonate sequence of the Basal Khuff Clastics (Abd el-aal et al. 2023; Tanoli et al. 2008). The formation's thickness varies in value and tends to increase westward, reaching a maximum depth of 6334 m before gradually decreasing toward the NE and SE, reaching a minimum depth of 5197 m (Fig. 5.16).

According to Tanoli et al. (2008), the Unayzah Formation is distinguished by two lithologies: pebbly sandstones, red mudstones, and siltstones. The pebbly sandstones constitute

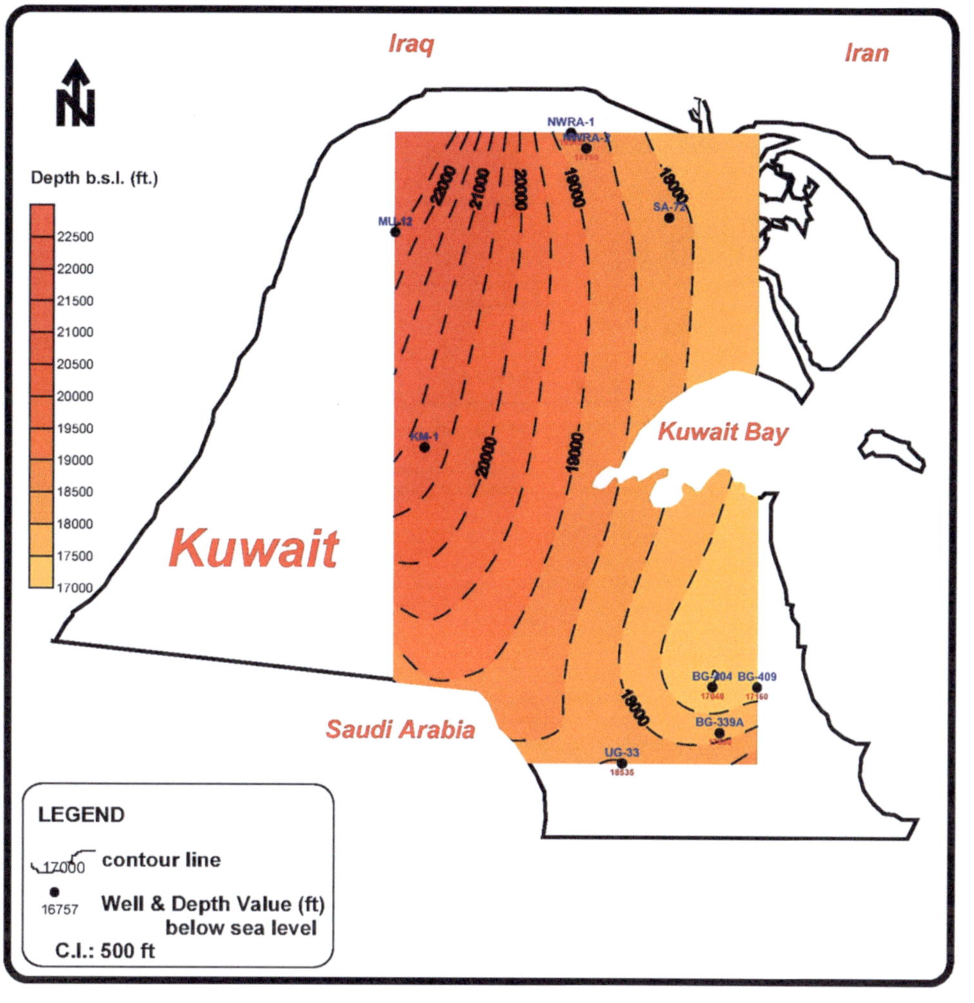

Fig. 5.16 Depth map of the Unayzah Formation in the subsurface of Kuwait (Abd el-aal et al. 2023; Abdullah et al. 2017)

the main body of fining-upward cycles with a conglomeratic base and fine-grained sandstone, siltstone, and shale in the upper part. Within the cored interval, the red oxidized mudstone and siltstone zone are repeated at least three times. It contains distinct pedogenic features such as root casts, mottling, reduction spots, and fissure fills. There is also the development of locally well-sorted sandstone and black shale lithologies (Fig. 5.17).

The cored intervals of the Unayzah Formation have been divided into six major lithofacies based on their prominent features, including lithology, color, sedimentary structures, and bioturbation (Tanoli et al. 2008). Some of these facies were repeated many times within the studied interval. The first facies (UN1) is formed of well-sorted, cross-bedded sandstone, occurring only at the formation's lower part. It consists of finding-upward cycles with an average thickness of around 1 m. It consists of coarse-medium-grained, well-sorted sandstone with greenish clay laminae that increase upward. The second facies (UN2) is the medium to dark grey fine-grained sandstone and overlies a 1.5 m red siltstone. It comprises grey muddy sandstone or siltstone at bas and fines upward into carbonaceous siltstone and shale at the top. The third facies (UN3) consists of upward-fining coarse, pebbly sandstone that developed in the lower and middle parts of the formation. It comprises pebbly sandstone or conglomerate at the base that fines upward into fine-grained sandstone. The pebbles are formed of quartz and poorly sorted and angular to subangular. The fourth facies (UN4) is made up of fine-grained sandstone with siltstone and can be found in two associations, either as fining upward sequences starting with the conglomerate to coarse–fine-grained sandstone or acting as a base of a small fining-upward cycle.

The facies are composed of fine-grained sandstone changing upward into carbonaceous siltstone and shale. Facies five (Un-5) is a distinctive facies formed of red mudstone and siltstone. It is formed of red silty mudstones alternating with thin local cream and grey mudstones. In some cases, think, fining-upward cycles may be found in these facies. They are mm-thick red and formed of darker parallel laminations. These facies contain some rootlet casts and

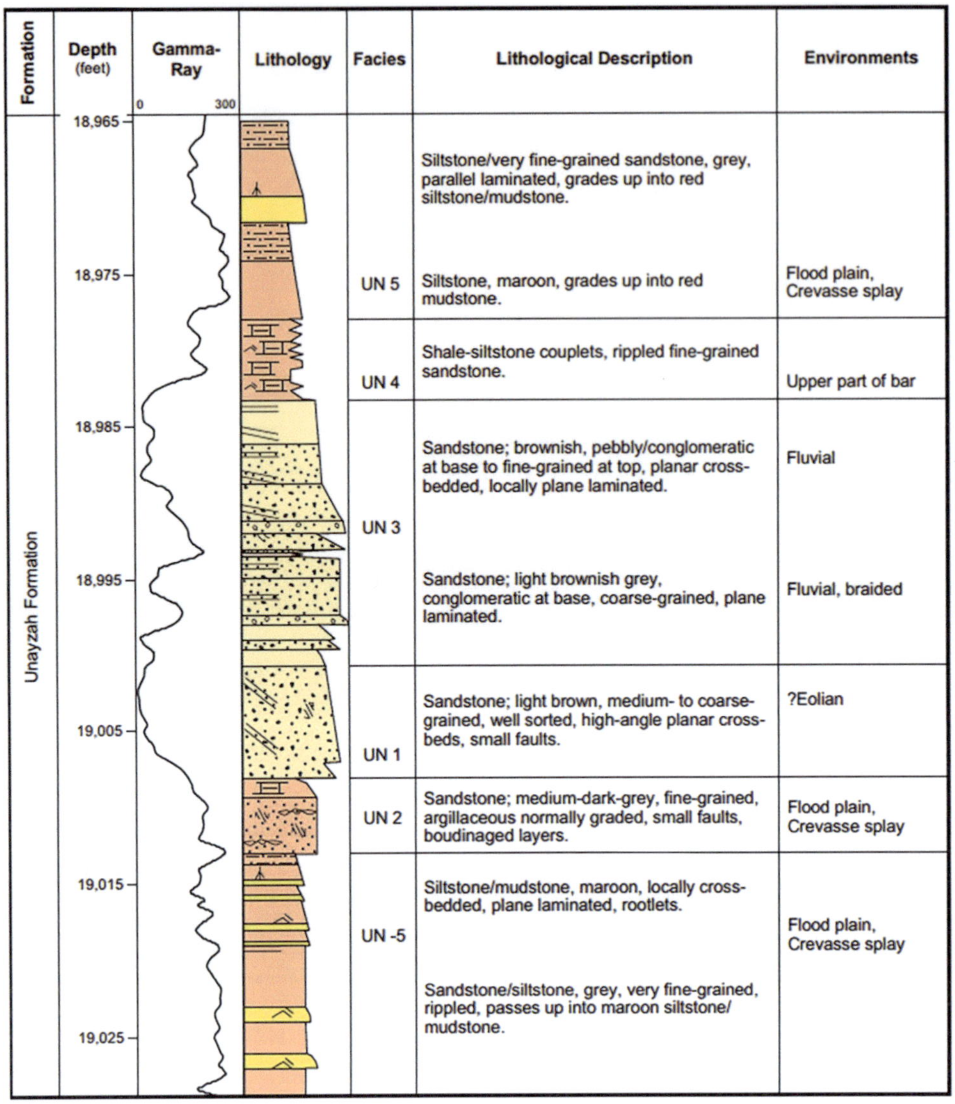

Fig. 5.17 The different sedimentary facies forming the Unayzah formation in the subsurface of Kuwait with their depositional environment (Tanoli et al. 2008)

other pedogenic structures. The sixth facies (UN-6) comprises around 4 m of black, highly carbonaceous shale with thin laminations of fine-grained sandstone and siltstone alternating beds. Leaf imprints are reported from the black shale. The black shale changes to around 2 m of clean, well-sorted, medium to fine-grained sandstone (Fig. 5.17).

These lithofacies are classified into four major paleoenvironments: eolian, braided fluvial, flood plain, and coastal plain. High-angle, cross-bedded sandstones with a probable dune affinity represent the possible eolian facies. Pebbly sandstones make up the braided fluvial facies and can be progressively more fine-grained. The red mudstone, shale, and siltstone are interpreted as having formed in a flood plain because they display characteristics typical of these deposits, such as abundant root casts and other indicators of subaerial exposure. Three occurrences of these facies are particularly notable within the succession under study. Tidal channels contain fine-grained and clean sandstone, crevasse splay and interfluve sediments contain bioturbated sandstone, siltstone, and shale, and dark grey to black shale of coastal marsh origin represent the coastal plain environments (Tanoli et al. 2008).

The Ga'ara Formation in Western Iraq

The Ga'ara Formation represents the oldest exposed unit in the west of Iraq. The formation is Upper Carboniferous to Lower Permian, equivalent to the Unayzah Formation of eastern Arabia. It was deposited under a humid (with interludes of the dry), hot climate. The formation thickness thins out toward the northern and northwestern parts. It consists of sandstone and mudstones deposited from the bottom by a delta, meandering river, lake, and braided river sequences. It comprises mature quartz arenites and kaolinite beds. It also contains iron deposits such as pisolites, ooliths, and sandy iron ores.

In its type locality, the formation is formed of varicolored sandstones with sandy marls in the upper part of the unit. The marls and shales contain plant remains. Palynological studies on the subsurface section of the formation assigned a Late Carboniferous to Early Permian age to the formation (Nadir et al. 1993).

The Ga'ara Formation represents the last clastic deposition of the Paleozoic in Iraq. Its deposition was associated with the global regressive phase affecting most of the Arabian Platform (Al-Laboun 1986; Alsharhan et al. 1993). The lower contact of the formation is unconformable with the Lower Carboniferous (Toumisian–Visean) Harur Formation. The upper contact is also unconformable with the Permian Chia Zairi Formation (Khuff-equivalent).

Al-Juboury and Al-Hadidy (2005) divided the sediments of the Ga'ara Formation into four facies. These include the *deltaic sequence*, which forms the lower part of the formation. It comprises siltstone, sandstone, and shale forming a coarsening-upward sequence. The sandstone is formed of well-sorted quartz arenite with some mica debris. The sandstone is cemented by carbonate (calcite and dolomite) or silica. The second facies is the *meandering fluvial system,* which consists of alternating fine pebbly sandstone, sandstone, and siltstone, forming a fining-upward sequence. The sandstone is dross-bedded quartz arenite with some mica fragments. The third facies is of a *lacustrine depositional system.* These sediments are formed of varved featureless mudstone. The fourth lithofacies is the *braided river system,* which consists of fine to coarse-grained quartz arenite and fine pebbly sandstone. The sandstone is well-sorted and has some plant remains.

The Ga'ara formation had been deposited in a series of transitional depositional systems. The first phase commenced with subsidence controlled by deep-seated faults that led to the development of an intracratonic faulted basin. This was followed by the clastic influx, which represents the second phase. This may result from a marine regression accompanied by an uplift that affected the deposition area. This deposition was generally controlled by the subsidence and old topographic surface of the basin and the succession of deltaic, fluvial, and lacustrine depositional regimes. These transitions are probably developed as a result of combined tectonics and climatic conditions (Al-Juboury and Al-Hadidy 2005).

The Unayzah Formation in Abu Dhabi

In Abu Dhabi, the Unayzah Formation is bounded by the Hercynian and the pre-Khuff unconformities at the bottom and top. The formation is divided into Unayzah C, B, and A, and the contact between the two subdivisions signals a major shift in depositional and tectonic configuration (Aktas et al. 2000; Heine 2000; Taher et al. 2012). Unayzah A is of wider distribution than the other subdivisions of the formation. According to Taher et al. (2012), It may represent regional flooding during the Sakmarian age and be related to the Rahab Shale of Saudi Arabia.

Unit Unayzah A represents a coarsening-upward sequence composed of multiple upward thinner, cleaner, and coarser parasequences, forming a Highstand System Tract (HST). In Abu Dhabi, Unayzah A is divided further into three members. Member Unayzah A1 comprises well-rounded coarse sand, whereas Unayzah A2 comprises dune sand deposits. Unayzah A3 consists of siltstone and shale sediments.

Hu et al. (2017) suggested that Unayzah B and Unayzah C were deposited as fluvial and continental clastics sediments of arid to semi-arid characteristics. After the Late Carboniferous glaciation, the fluvial and eolian conditions resumed again, as reflected by the sedimentary features of the Unayzah A sediments. They suggested the Unayzah sequence in Abu Dhabi represents a cyclical assemblage of the eolian dune, ephemeral fluvial, sabkha, and lacustrine deposits. The melted glacial water caused the formation of floodplains in the south and lacustrine conditions in the offshore area of Abu Dhabi in the Unayzah A3 sequence. This flooding continued during the deposition of Unayzah A2 shale and siltstone sediments. During the deposition of the Unayzah A2, the climate started to get drier gradually, which led to the deposition of eolian sediments in the Unayzah A1 sequence. These eolian conditions were neighboring a fluvial regime (Fig. 5.18). Further to the north, the dune and fluvial systems mixed with the advancement of the fluvial system. In the eolian areas, ephemeral fluvial sands accumulated locally. The fluvial sediments reflect the oscillation of dry and wet climatic conditions. Under a wet climate, floodplains of fine sediments were developed in the fluvial setting. The mud was probably deposited from suspension in the ephemeral lakes, whereas the fluvial system was dominated by channels and sheet flood sands during dry climatic conditions.

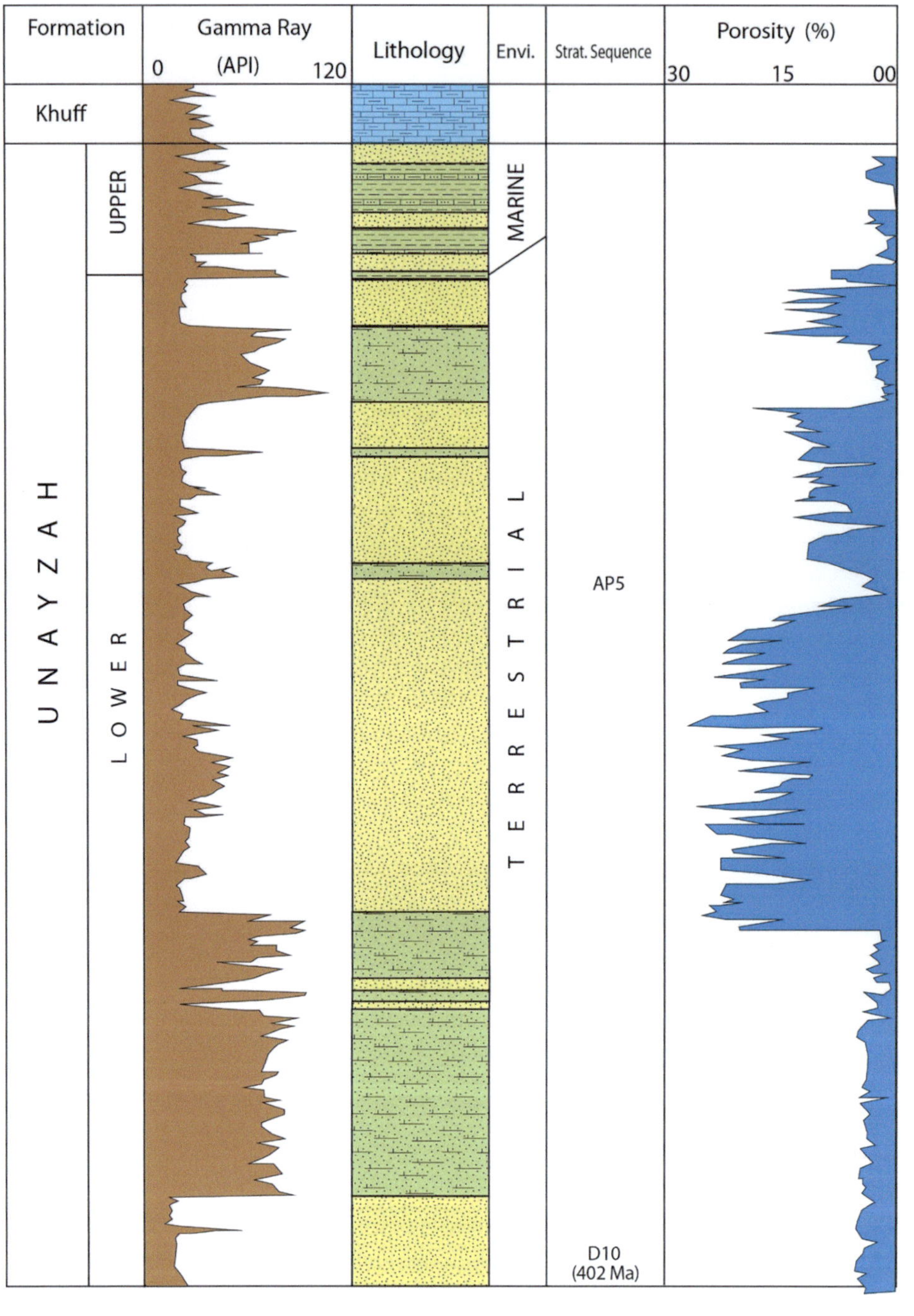

Fig. 5.18 Stratigraphy and lithologic subdivisions of the Unayzah Formation in the well Matbakh-2 (Al-Saad and Sadooni (2016))

The Unayzah Formation in Qatar

Focke et al. (1986) called these strata Haushi Formation in Qatar. The term Haushi has already been given as a group to the Carboniferous-Permian sediments of Oman. Al-Saad and Sadooni (2016) suggested replacing it with Unayzah in the Qatari stratigraphic section to solve this confusion and align it with the Saudi stratigraphic nomenclature. The lower contact is unconformable with the Tawil Formation; the upper contact is conformable with the Khuff Formation. Focke et al. (1986) suggested that this Formation is of the Early Permian age (Artinskian-Sakmarian). The Unayzah Formation has been penetrated in a few wells in Dukhan and Matbakh (Fig. 5.19). This synthesis is based on cores recovered from these few wells and the work published by Focke et al. (1986). The formation generally consists of pebbly sandstone, siltstone, conglomerates and breccias, and shale with minor lignite, anhydrite, wood remains, and some coal beds.

At the upper part of the formation, the sandstone rarely exceeds 3 m thick. The sandstone is very mature and cemented with silica. The high maturity of the sand, existence of trace fossils, and cross-bedding indicate that the depositional environment of these sandstones was shallow marine.

Fig. 5.19 The two contrasting lithologies of the Unayzah Formation; the light-colored is sandstone, and the dark-colored is shale

At the lower part of the formation, the sandstone is formed of thicker beds, ranging in thickness from 6 to 12 m, and contains fine-grained to pebbly sandstone. There are breccias and conglomerates less than 30 cm thick occur in fining-upward sequences. Sedimentary structures include parallel to low-angle cross beddings and plant debris that forms discontinuous laminations or along bedding planes. These sandstone bodies are interbedded with red, silty to argillaceous layers up to 40 m thick. This succession was formed in the braided fluvial system. Zig-zag structures were reported from this sandstone. These are formed of streaks of darker grains taking a zig-zag shape. Such features are characteristics of alluvial depositional systems under arid climatic conditions. They are believed to form due to the infiltration of clay minerals into the sand during desiccation periods. The clay is accumulated along the coarse cross-beds of the sandstone or deposited parallel to the fluctuating water table (Fig. 5.20f).

Focke et al. (1986) divided the sediments of this formation into three major types. These are well-sorted *quartz arenite* which is found in the upper part of the formation. It consists of around 90% quartz with a minor amount of feldspar, and it is cemented by carbonate minerals. The second type is fine-grained to pebbly sandstone. This rock type is formed of detrital quartz (up to 70% of the rock volume), feldspar, metamorphic and volcanic rocks, and mica (less than 10% in total). This facies represents the coarser fraction of the continental part of the formation. It shows many indications of reworking and transportation. The third type is shale, which constitutes the lower part of the formation. Al-Saad et al. (2019) investigated the potentiality of this section as a hydrocarbon source rock in Dukhan Field. They reported that the total organic carbon (TOC) contents indicate gas-prone-kerogen types III and IV. They suggested that many of these materials were recycled from old sediments. Most organic matter was subjected to oxidation, leaching, biodegradation, and recycling due to the changes in the depositional environments ranging from river systems to marine settings.

Fig. 5.20 **a** The contact between shale and sandstone **b–e** plant remains and wood pieces **f** zig-zag pattern **g** graded bedding

References

Abd el-aal Ae-aK, Al-Awadhi JM, Al-Dousari A (2023) The geology of Kuwait. Springer International Publishing. https://doi.org/10.1007/978-3-031-16727-0

Abdullah F, Shaaban F, Khalaf F, Bahaman F, Akbar B, Al-Khamiss A (2017) Petrophysics and hydrocarbon potential of Paleozoic rocks in Kuwait. J Asian Earth Sci 148:105–120. https://doi.org/10.1016/j.jseaes.2017.07.057

Abu-Ali MA, Franz UA, Shen J, Monnier F, Mahmoud MD, Chambers TM (1991) Hydrocarbon generation and migration in the paleozoic sequence of Saudi Arabia. Middle East Oil show

Abu-Ali MA, Rudkiewicz J-LL, McGillivray JG, Behar F (1999) Paleozoic petroleum system of Central Saudi Arabia. GeoArabia 4(3):321–336. https://doi.org/10.2113/geoarabia0403321

Aktas GG, Al-Khalifa M, Al-Qassab H (2000) Sequence stratigraphy and reservoir architecture of Unayzah continental deposits, Hawtah trend, Saudi Arabia (Abstract). GeoArabia 5(1):11

Al-Hajri SiA, Filatoff J, Wender LE, Norton AK (1999) Stratigraphy and operational palynology of the Devonian System in Saudi Arabia. GeoArabia 4(1):53–68

Al-Harbi OA, Khan MM (2008) Provenance, diagenesis, tectonic setting and geochemistry of Tawil Sandstone (Lower Devonian) in Central Saudi Arabia. J Asian Earth Sci 33(3):278–287. https://doi.org/10.1016/j.jseaes.2008.01.004

Al-Juboury A, Al-Hadidy A (2005) Sedimentology of the Permo-Carboniferous Ga'ara Formation, Iraq. The Nonmarine Permian. In: Lucas SG, Zieglar KE (eds) New Mexico museum of natural history and science bulletin, 30, pp 9–14

Al-Laboun A (1982) The subsurface stratigraphy of the pre-Khuff formations in central and northwestern Arabia. Unpublished PhD thesis, Faculty of Earth Sciences, King Abdulaziz University, Jiddah, Saudi Arabia, 102

Al-Laboun AA (1984) Paleozoic oil and gas potential of Arabian Basin. AAPG Bull 68(9):1199–1199

Al-Laboun AA (1986) Stratigraphy and hydrocarbon potential of the Paleozoic succession in both Tabuk and Widyan basins, Arabia

Al-Laboun AA (1987) Unayzah formation: a new Permian-Carboniferous unit in Saudi Arabia. AAPG Bull 71(1):29–38

Al-Ruwaili MH (2000) New Silurian acritarchs from the sub-surface of northwestern Saudi Arabia

Al-Saad H, Sadooni FN (2016) Stratigraphy and petroleum systems of the palaeozoic (pre-khuff) succession, Qatar. J Petrol Geol 39(4):357–373. https://doi.org/10.1111/jpg.12657

Al-Saad H, Al-Khafaji AJ, Sadooni FN (2019) Evaluation of the source rock potential of the Unyazah Formation (late Carboniferous-Early Permian) in Dukhan Field, Qatar. Petrol Sci Technol 37(14):1655–1664. https://doi.org/10.1080/10916466.2019.1602636

Alsharhan AS, Kendall CGSC (1986) Precambrian to jurassic rocks of Arabian Gulf and adjacent areas: their facies, depositional setting, and hydrocarbon habitat1. AAPG Bull 70(8):977–1002. https://doi.org/10.1306/94886650-1704-11d7-8645000102c1865d

Alsharhan AS, Nairn AEM, Mohammed AA (1993) Late Palaeozoic glacial sediments of the southern Arabian Peninsula: their lithofacies and hydrocarbon potential. Mar Petrol Geol 10(1):71–78. https://doi.org/10.1016/0264-8172(93)90101-W

Aqrawi AA (1998) Paleozoic stratigraphy and petroleum systems of the western and southwestern deserts of Iraq. GeoArabia 3(2):229–248

Bahafzallah A, Jux U, Omara S (1981) Stratigraphy and facies of the Devonian Jauf Formation, Saudi Arabia

Beuf S, Biju Duval B, Charpal OD, Rognon P, Gariel O, Bennacef A (1970) Les gres du paleozoique inferieur au sahara. sedimentation et discontinuites, evolution structurale d'un craton

Beydoun ZR (1986) The petroleum resources of the middle east: a review. J Petrol Geol 9(1):5–27. https://doi.org/10.1111/j.1747-5457.1986.tb00370.x

Bramkamp RA, Brown GF, Holm DA, Layne Jr NM (1963) Geologic map of the Wadi As Sirhan quadrangle, Kingdom of Saudi Arabia [Report] (200A). IMAP, Issue. U. S. G. Survey. https://pubs.usgs.gov/publication/i200A

Clark-Lowes D (1980) Sedimentology and mineralization potential of the Saq and Tabuk formations, al-Qasim District. Imperial College of Science and Technology, Cover Rock Contract, Open-File Report CRC/Ic-7, Saudi Arabia, Ministry of Petroleum and Mineral Resources, Deputy Ministry for Mineral Resources, Jiddah, Open File Report DGMR-767, 88

Clark-Lowes DD (2005) Arabian glacial deposits: recognition of palaeovalleys within the Upper Ordovician Sarah Formation, Al Qasim district, Saudi Arabia. Proc Geol Assoc 116(3):331–347. https://doi.org/10.1016/S0016-7878(05)80051-3

Davis GR, Shearman DJ, Fergusson J, Davies RG (1981) Reconnaissance of the area of Quaternary volcanic influence on cover rocks east of Ha'il. Imperial College of Science and Technology Cover Rock Contract Open-file Report CRC/IC 18, Open-file Report, Issue

De la Grandville BF (1982) Appraisal and development of a structural and Stratigraphic trap oil field with reservoirs in glacial to periglacial elastics. The deliberate search for the subtle trap

Delfour J, Dhellemmes R, Elsass P, Vaslet D, Brosse J, LeNindre Y, Dottin O (1982) Explanatory notes to the geologic map of the ad-Dawadimi quadrangle, sheet 24G. Minist Pet Miner Resour 36:1–85

Dossary N, Tourqui H, Breuer P, Soua M, Lacsamana E (2017) Tawil formation, Eastern Saudi Arabia: sequence stratigraphic analysis and implications for better understanding of reservoir heterogeneities geosciences technology workshop, Amman, Jordan

El-Khayal A, Wagner R (1985) Upper Permian stratigraphy and megafloras of Saudi Arabia: palaeogeographic and climatic implications. Compte Rendu X Congress International Stratigraphie Geologie Carbonifere, Madrid

El-Khayal AA, Chaloner WG, Hill CR (1980) Palaeozoic plants from Saudi Arabia. Nature 285(5759):33–34. https://doi.org/10.1038/285033a0

Ferguson GS, Chambers TM (1991) Subsurface stratigraphy, depositional history, and reservoir development of the early-to-late Permian Unayzah formation in Central Saudi Arabia. Middle East Oil show

Focke J, Munn D, Al Kuwari S, Frikken H, Frei H (1986) Petrographic atlas of rock types, common in the subsurface of Qatar and some recent equivalents. General Petroleum Corporation (Offshore Operations), Qatar, pp 110–189

Ghazwani A (2012) Investigation of source rock potential and micropaleontology of the middle Ordovician Hanadir member of the Qasim Formation, Northwestern Saudi Arabia King Fahd University of Petroleum and Minerals. http://eprints.kfupm.edu.sa/id/eprint/138787

Heine CJ (2000) The Permo-Carboniferous Unayzah depositional system of Saudi Arabia. AAPG. Abstracts 84:1436–1437

Hu J, Al Blooshi A, Mavromatidis A, Witte J, Neves F, Caeiro MD (2017) Regional characterization and stochastic object modeling of the Pre-Khuff clastics in Abu Dhabi. In: Abu Dhabi international petroleum exhibition & conference

Husseini MI (1991) Tectonic and depositional model of the Arabian and adjoining plates during the Silurian–Devonian1. AAPG Bull 75(1):108–120. https://doi.org/10.1306/0c9b274f-1710-11d7-8645000102c1865d

Janjou D, Halawani M, Rasin P (1996a) The Tawil Formation (Late Silurian to Early Devonian) in the Tabouk basin, northwestern Saudi Arabia: correlation with the Wajid Sandstone, Early Paleozoic evolution IN NW Gondwana. Ser Corr Geologica 12:107–120

Janjou D, Halawani M, Al-Muallem M, Robelin C, Brosse J, Courbouleix S, Dagain J, Genna A, Razin P, Roobol M (1996b) Geologic map of the Al Qalibah quadrangle, sheet 28C. Kingdome of Saudi Arabia (with text): Saudi Arabian Deputy Ministry for Mineral Resources, Jiddah, Geoscience Map GM-135

Janjou D, Halawani MA, Al-Muallem MS, Brosse JM, Becq-Giraudon JF, Dagain J, Genna A, Razin P, Roobol MJ, Shorbaji H, Wyns R (1996c) Geologic map of the Tabuk quadrangle, sheet 28B, (with text). Kingdom of Saudi Arabia, Saudi Arabian Deputy Ministry for Mineral Resources

Janjou D, Halawani MA, Brosse J-M, Al-Muallem MS, Becq-Giraudon JF, Dagain J, Genna A, Razin P, Roobol MJ, Shorbaji H, Wyns R (1997) Explanatory notes to the geologic map of the Tabuk Quadrangle, Kingdom of Saudi Arabia. Geoscience map; GM-137, scale 1:250,000, sheet 28B., Issue

Jones PJ, Stump TE (1999) Depositional and tectonic setting of the lower Silurian hydrocarbon source rock facies, Central Saudi Arabia1. AAPG Bull 83(2):314–332. https://doi.org/10.1306/00aa9a7a-1730-11d7-8645000102c1865d

Konert G, Afifi AM, Al-Hajri SiA, Droste HJ (2001) Paleozoic stratigraphy and hydrocarbon habitat of the Arabian plate. GeoArabia 6(3):407–442. https://doi.org/10.2113/geoarabia0603407

Laboun AA (1982) The subsurface stratigraphy of the pre-Khuff formations in central and northwestern Arabia King Abdulaziz University

Le Heron DP, Craig J, Etienne JL (2009) Ancient glaciations and hydrocarbon accumulations in North Africa and the Middle East. Earth-Sci Rev 93(3):47–76. https://doi.org/10.1016/j.earscirev.2009.02.001

Mahmoud MD, Vaslet D, Husseini MI (1992) The lower Silurian Qalibah formation of Saudi Arabia: an important hydrocarbon source rock 1. AAPG Bull 76(10):1491–1506. https://doi.org/10.1306/bdff8a24-1718-11d7-8645000102c1865d

Manivit J, Vaslet D, Le Strat APB, Fourniguet J (1986) Explanatory notes to the geologic map of the Buraydah quadrangle, Sheet 26G

McClure HA (1978) Early paleozoic glaciation in Arabia. Palaeogeography, Palaeoclimatology, Palaeoecology, 25(4):315–326. https://doi.org/10.1016/0031-0182(78)90047-0

McClure HA (1988) Chitinozoan and acritarch assemblages, stratigraphy and biogeography of the early palaeozoic of Northwest Arabia. Rev Palaeobot Palynol 56(1):41–60. https://doi.org/10.1016/0034-6667(88)90073-5

McGillivray JG, Husseini MI (1991) The Paleozoic petroleum geology of central Arabia Journal Name: Geobyte; (United States) 75(8); Conference: International conference and exhibition of the American Association of Petroleum Geologists (AAPG): the way ahead-hydrocarbons for the 1990s, London (United Kingdom), 29 Sep–3 Oct 1991, United States. https://www.osti.gov/biblio/5741494

Miller MA, Al-Ruwaili MH (2007) Preliminary palynological investigation of Saudi Arabian Upper Ordovician glacial sediments. Revue de Micropaléontologie 50(1):17–26. https://doi.org/10.1016/j.revmic.2007.01.002

Nadir AD, Khalaf FH, Yusuf RA (1993) Palynology of the upper part of the gaara formation in the western Iraqi Desert. Mu'tah J Res Stud: Nat Appl Sci Ser 8(4):77–137

Powers RW (1968) Asie: Arabie Saoudite. Lexique Stratigraphique Internationale, Paris

Powers RW, Ramirez LF, Redmond CD, Elberg Jr EL (1966) Geology of the Arabian Peninsula: sedimentary geology of Saudi Arabia [Report] (560D). Professional Paper, Issue. U. S. G. Survey. https://pubs.usgs.gov/publication/pp560D

Saudi Stratigraphic Committee (2013) Phanerozoic Stratigraphy of Saudi Arabia, Part 1. Paleozoic successions of the Arabian Shelf (cover rocks). Special Publication SGSSP-2012–1, First edition, 69 p

Seilacher A (2007) Cruziana stratigraphy. In: Trace fossil analysis. Springer Berlin Heidelberg, pp 187–200. https://doi.org/10.1007/978-3-540-47226-1_14

Senalp M, Al-Duaiji A (1995) Stratigraphy and sedimentation of the Unayzah reservoir, central Saudi Arabia. AAPG Bull 79(CONF-950995-)

Senalp M, Al-Duaiji A (2001a) Sequence stratigraphy of the Unayzah reservoir in central Saudi Arabia. Saudi Aramco J Technol 20:20–43

Senalp M, Al-Duaiji AA (2001b) Qasim Formation: Ordovician storm- and tide-dominated shallow-marine siliciclastic sequences, Central Saudi Arabia. Geoarabia 6(2):233–268. https://doi.org/10.2113/geoarabia0602233

Senalp M, Al-Laboun A (2000) New evidence on the Late Ordovician glaciation in central Saudi Arabia. Saudi Aramco J Technol 2000:11–40

Sharland PR, Casey DM, Davies RB, Simmons MD, Sutcliffe OE (2001) Arabian Plate sequence stratigraphy, vol 9. Gulf Petrolink

Steineke M, Bramkamp RA, Sander NJ (1958) Stratigraphic relations of Arabian Jurassic oil. In: American association of petroleum geologists symposium, Tulsa, U.S.A.

Taher A, Al Shateri A, Al Mehsin K, Witte J, Al Zaabi M (2012) Hydrocarbon habitat of the early Permian Unayzah Formation in Abu Dhabi. In: Abu Dhabi international petroleum conference and exhibition

Tanoli SK, Husain R, Sajer AA (2008) Facies in the Unayzah Formation and the Basal Khuff Clastics in subsurface, northern Kuwait. GeoArabia 13(4):15–40. https://doi.org/10.2113/geoarabia130415

van Bellen RC, Dunnington HV, Wetzel R, Morton D (1959) Lexique stratigraphique international: Asie. Iraq, vol 3C

Vaslet D (1987) Early Paleozoic glacial deposits in Saudi Arabia: a lithostratigraphic revision

Vaslet D (1990) Upper Ordovician glacial deposits in Saudi Arabia. Int Union Geol Sci 13(3):147–161. https://doi.org/10.18814/epiiugs/1990/v13i3/002

Vaslet D, Kellogg K, Berthiaux A, Strat PL, Vincent P (1987) Explanatory notes to the geologic map of the Baq'a quadrangle, Sheet 27F, Kingdom of Saudi Arabia, sheet 27F. Kingdom of Saudi Arabia, Kingdom of Saudi Arabia

Vaslet D, Borsse J-M, Breton J-P, Manivit J, Le Stat P, Fourniguet J, Shorbaji H (1988) Geologic map of the Shaqra quadrangle, sheet 25 H Kingdom of Saudi Arabia, Saudi Arabian Deputy Ministry for Mineral Resources

Vaslet D, Christian R, Al-Muallem MS, Halawani MA, Brosse J, Breton JP, Courbouleix S, Roobol MJ, Dagain J (1994) Explanatory notes to the geologic map of the Tayma Quadrangle, Kingdom of Saudi Arabia. Geoscience map GM-134, scale 1:250,000, sheet 27C

Wensink H (1991) Late Precambrian and Paleozoic rocks of Iran and Afghanistan. In: The Phanerozoic geology of the world, I, The Palaeozoic, A, pp 147–218

Williams PL, Vaslet D, Johnson PR, Berthiaux A, Le Strat P, Fourniguet J. (1986) Geological map of the Jabal Habashi quadrangle, Sheet 26F

Wolfard R (1981) Lower Paleozoic of the Middle East, 6–130. Lower Paleozoic of the Middle East, Eastern and Southern Africa and Antarctica. Wiley, New York

6 The Late Permian-Early Triassic

Abstract

This chapter examines the Late Permian to Early Triassic Khuff Formation, an important carbonate-evaporite reservoir throughout the Arabian Plate. During this epoch, the Arabian Plate, encompassing Qatar, constituted a segment of the Pangean supercontinent, with tectonic processes influencing its stratigraphy. Rifting along the Neo-Tethys coast approximately 300 million years ago commenced crustal thinning and sinking, creating an extensive epeiric shelf sea spanning from Iran to Saudi Arabia. This transgression altered the depositional environment from siliciclastic to carbonate-evaporite as the Arabian Plate migrated toward warmer, dry climates. The Pre-Khuff Unconformity, roughly 268 million years old, signifies the origin of the Neo-Tethys Ocean and a notable marine transgression. The Khuff Formation, thoroughly examined for its hydrocarbon reservoirs, comprises stratified carbonates and evaporites, extensively found throughout Saudi Arabia, Oman, Iran, and Qatar. Stratigraphic analyses have divided the Khuff into various units, each indicative of distinct depositional habitats, from tidal flats to open sea conditions. Prominent subunits comprise oolitic and bioclastic grainstones, dolomitized mudstones, and microbial facies, signifying a high-energy, shallow marine carbonate platform affected by tidal and storm dynamics. The Khuff Formation in Qatar has increased thickness in the northeast, indicative of regional tectonics and structural development influenced possibly by underlying Hormuz salt. The Khuff depositional system is recognized as one of the earliest extensive carbonate ramps in the Arabian Plate, setting a precedent for carbonate-dominated sedimentation. This carbonate ramp model encompasses a progression from high-energy oolitic shoals to low-energy tidal flats, highlighting the formation's considerable hydrocarbon potential. This chapter offers a comprehensive analysis of the Khuff formation lithofacies, stratigraphy, and depositional conditions in the Gulf region, emphasizing its significance as a principal oil reservoir.

6.1 Introduction

During the Permo-Triassic period, the Arabian plate, including Qatar, was an essential part of the supercontinent Pangea, and there was minimal motion between Africa and Arabia. Rifting started along the Neo-Tethys margin 300 million years ago, resulting in lithospheric thinning and stretching, which led to subsidence (Stampfli and Borel 2002). Additionally, from 300 to 260 million years ago, the Paleo-Tethys Ocean was subducted under the Eurasian Plate and intracontinental rifting occurred between the Cimmerian and Arabian terranes (Konert et al. 2001; Pöppelreiter and Marshall 2013). During the Late Permian, increased accommodation space due to crust stretching accompanied the formation of the Neo-Tethys Ocean along the Oman-Zagros suture. The pre-Khuff unconformity marks the birth of this new ocean. Sea-floor spreading, thermal subsidence, and the associated transgression led to the evolution of a large epeiric shelf sea that extended from southern Iran to Saudi Arabia (Pöppelreiter and Marshall 2013; Sharland et al. 2001).

The separation of the Arabian Plate and the Iran-Turkey Plate began in the early Late Devonian and was completed by the early Late Permian, allowing the Paleo-Tethys Sea to advance from the south onto the Arabian Plate. During the Permo-Carboniferous ice age, the Arabian Plate was between 30° S and 55° S, moving northward with Gondwana (Al-Dukhayyil 2012; Konert et al. 2001; Sharief 1983; Ziegler et al. 1997). By the Late Permian onset of greenhouse conditions (Fielding et al. 2008; Frakes et al. 1992), the Arabian Plate lay at 10° S to 35° S, and by the end of Khuff deposition in the Early Triassic, northernmost Arabia had reached 5° S (Konert et al. 2001; Sharland et al. 2001).

F. N. Sadooni and H. A. Al-Kuwari, *Geological Evolution of Qatar and the Arabian Peninsula*, Regional Geology Reviews,
https://doi.org/10.1007/978-3-031-94394-2_6

A major unconformity lies beneath the Khuff Formation, known as the Middle Permian break-up unconformity or the "Pre-Khuff Unconformity." Its age is estimated to be around 268 million years old (according to Ogg et al.'s 2008-time chart). It signifies the beginning of the Neo-Tethys Ocean along the existing Zagros-Oman suture and the significant Permian transgression. This led to the expansion of the Neo-Tethys Ocean (Al-Aswad 1997; Konert et al. 2001; Sharland et al. 2001). During this major transgressive event, shallow seas covered a large portion of the Arabian platform, marking a transition from primarily siliciclastic to carbonate-evaporite deposition as the Arabian Plate moved toward warmer, drier, and lower latitudes in the north. Consequently, shallow marine carbonates and evaporites were deposited across much of the Arabian Plate (Fig. 6.1) (Al-Jallal 1995, 1989; Ziegler 2001). Throughout the Early Triassic, sea-floor spreading persisted in the Neo-Tethys ocean basin (Al-Aswad 1997).

The Middle Permian climate is thought to be transitional between the Upper Carboniferous icehouse and the early Triassic greenhouse (Ziegler 2001). During the Permian Triassic, global greenhouse and ice-free conditions led to a complex array of meter and submeter-scale parasequences. The mineralogy and dolomitization potential of deposited carbonates were influenced by aragonite-high Mg seas (Al-Dukhayyil 2012; Scotese et al. 1999; Stanley and Hardie 1998). During the Middle Permian, the Khuff deposition started near the developing Neo-Tethys Ocean (Sharland et al. 2001). The Khuff platform was exposed to storms moving

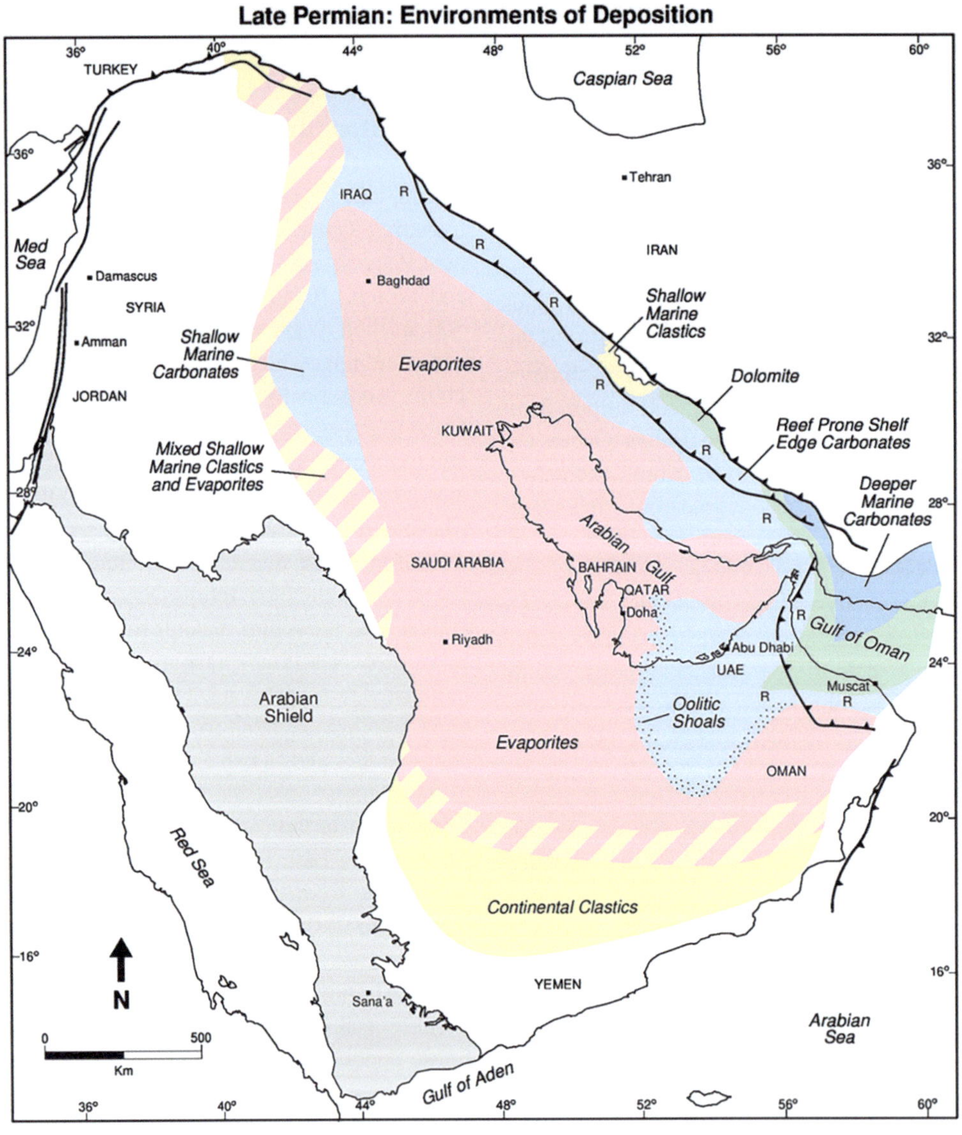

Fig. 6.1 Paleogeographic map of the late Permian-early Triassic Khuff Basin of eastern Arabia (Konert et al. 2001)

westward due to the Neo-Tethys Ocean's constant expansion, which was open to the east and situated in the path of a hurricane. In a paleogeographic position around 30° S of the equator, hurricanes may have swept the shallow platform across Arabia, with the wave base moving grains down to a depth of a few tens of meters (Oosterhout and Pöppelreiter 2013; Stampfli and Borel 2002).

The resulting megasequence, with the base of AP6 dated at 255 Ma, is composed of continental to marine sandstones and shales (Senalp and Al-Duaiji 1995), which were supplied from the west. The continental deposits northward included coal deposits, indicating wetter tropical environments, such as the Kas Formation in southeast Turkey. This was followed by an extensive deposition of carbonates and evaporites of the Khuff Formation in Saudi Arabia and Oman and Dalan and Kangan formations in Iran (Al-Jallal 1995), covering the entire Arabian shelf in shallow marine to tidal flat environments. These carbonates and anhydrite were deposited after the continental deposits and formed the upper part of the Khuff sequence (Konert et al. 2001; Senalp and Al-Duaiji 1995).

The Khuff basin during Permo-Triassic times is among the largest epicontinental seas in geologic times. It occupied some 2500 km in the NW–SE direction and some 1500 km in the SW-NE direction, with a total area of 3.7 million km^2 (Aigner and Dott 1990; Al-Jallal 1995).

During the Permo-Triassic period, the Arabian basin experienced gradual subsidence. Over roughly 20 million years, a maximum thickness of 1000 m of Khuff sediment accumulated (Al-Husseini and Matthews 2005). This deposition rate is less than 50 m/m.y., at the lower end of the range observed in epicontinental basins, where subsidence generally varies between 20 and 250 m/m.y. (Leighton and Kolata 1995).

The Khuff Formation is found in Saudi Arabia, Oman, Iran, and the United Arab Emirates. The isopach map of the Khuff Formation in eastern Arabia is shown in Fig. 6.2. The thickness of the Khuff Formation ranges between 200 feet (61 m) and 4800 feet (1463 m). The average thickness of the formation is around 400 m in Saudi Arabia up to 1667 m in Iran and Oman. The average formation thickness in the offshore area is around 3000 feet (914 m).

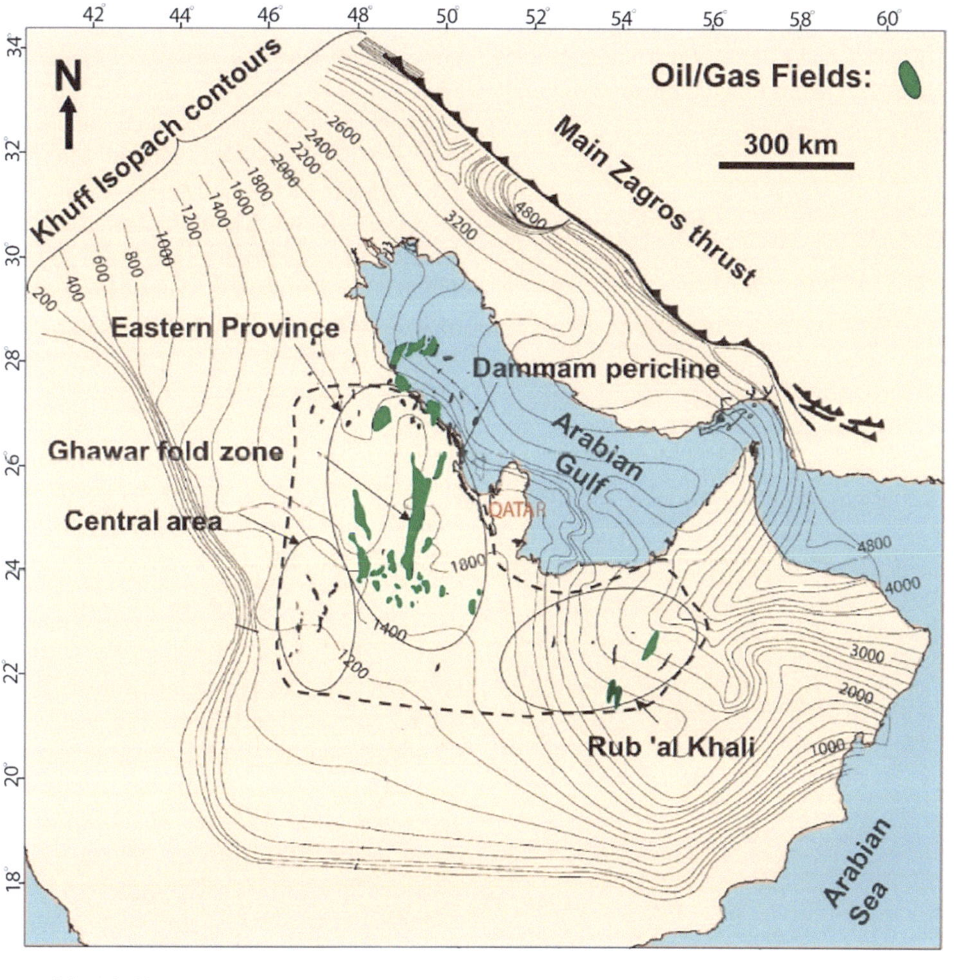

Fig. 6.2 Isopach map of the Khuff Formation in Eastern Arabia (Al-Jallal 1995; Ameen et al. 2010)

This chapter is dedicated to the Khuff Formation, the only formation reported in Qatar from this cycle.

6.2 Stratigraphy

6.2.1 The Khuff Formation

The Khuff Formation, as a major hydrocarbon reservoir in the whole region, has been the subject of extensive studies since the middle of the last century. The formation was described for the first time by Max Steinke (Fig. 6.3) in an unpublished report in 1937 to cover the exposed lowermost carbonate succession of the obsolete Mustawi Group (Al-Aswad 1997; Powers et al. 1966). The formation was named after the town of Khuff (Lat. 24° 55′ N, long. 44° 43′ E). The first formal description of the formation was by Steineke and Bramkamp (1952). The formal designation of the type section was then published by Steineke et al. (1958). Powers et al. (1966) attempted to divide the formation into three units. In this classification, the upper and lower units are formed of limestone, while the middle unit is composed of shale. The subsurface section of the formation was divided into four units, namely Khuff D, Khuff C, Khuff B, and Khuff A. Each unit represents a transgressive system tract. Units C and B are reservoir units also (Al-Jallal 1987).

Fig. 6.3 Max Steineke, the geologist who described the Khuff and other Saudi Arabian formations (Wikipedia 2024)

Delfour et al. (1982) divided the Khuff Formation into five informal members. These are Unayzah, Huqayl, Duhaysan, Midhnab, and Khartam. The Unayzah member has been upgraded to a formation status by Al-Laboun (1987) (see Chap. 5).

Saltworks (2024) (internet source) provided a more detailed classification of the Khuff Formation in both the outcrops and the subsurface (Fig. 6.4). Accordingly, the formation is divided into two carbonate units: the lower and upper Khuff with a middle anhydrite unit.

In the outcrops, the Khuff is divided into three members from the bottom; the Bih, the Hagil, and the lower Ghail (with the Upper Ghail assigned to the overlying Sudair Formation). The Upper Khuff is divided into four units (K1–K4), and the lower unit is divided into three units (K5–K7). The Upper Khuff is divided into four reservoir units (K1-K4) separated by four barrier or dense units (D1–D4) in the subsurface. The Lower Khuff is divided into three reservoir units (K5–K7) separated by three barrier or dense units (D5–D7). So, the subsurface units, although they have the same names, are not fully correlatable with the outcrops.

The Upper Khuff has four maximum flooding surfaces, MFS1-MFS-4 at 249 Ma, 252 Ma, 255 Ma, and 258 Ma. The Lower Khuff has three maximum flooding surfaces, MFS5-MFS7, at 263 Ma, 265 Ma, and 268 Ma (Fig. 6.4).

Vaslet et al. (2005) provided detailed age assignments to the different members of the Khuff Formation based on a combination of different fossil assemblages. They suggested that the Khuff Formation's base is identified as a regional unconformity (Pre-Khuff Unconformity), while its top is marked by the conformable contact with the overlying Sudair Shale. The Ash Shiqqah Member is tentatively dated to the Middle Permian (Capitanian) due to the presence of fusulinids *Monodiexodina kattaensis* and *Reichelina* sp. The Huqayl Member is estimated to be from the Late Permian Wuchiapingian (Dzhulfian) era, as evidenced by an assemblage of smaller foraminifers, mainly *Pseudomidiella* cf. *labensis, Earlandia* spp., and *Neodiscus* aff. *qinglongensis*. The Duhaysan Member is assigned to the Late Permian (Wuchiapingian-Changhsingian) period based on *Hemigordius baoqingensis*, *Graecodiscus* cf. *kotlyarae*, "*Dentalina*" *hoi*, and *Colaniella* cf. *minuta*. The Midhnab Member is identified as Changhsingian (Dorashamian) due to the presence of *Paradagmarita* sp. and "*Glomospirella spirillinoides*" and the disappearance of the genera *Nankinella* and *Globivalvulina*. Within the continental facies in the upper part of the Midhnab Member, incised channel facies revealed the Late Permian Midhnab Flora. The Lower Khartam Member is also considered as latest Permian (Dorashamian) based on the presence of several species of *Paradagmarita* and "*Nodosaria*" *dzhulfensis* in the member's lowest beds, as well as ostracods *Paraparchites* spp., *Knoxiella* spp., and *Kloedenellitina* sp. throughout the Lower

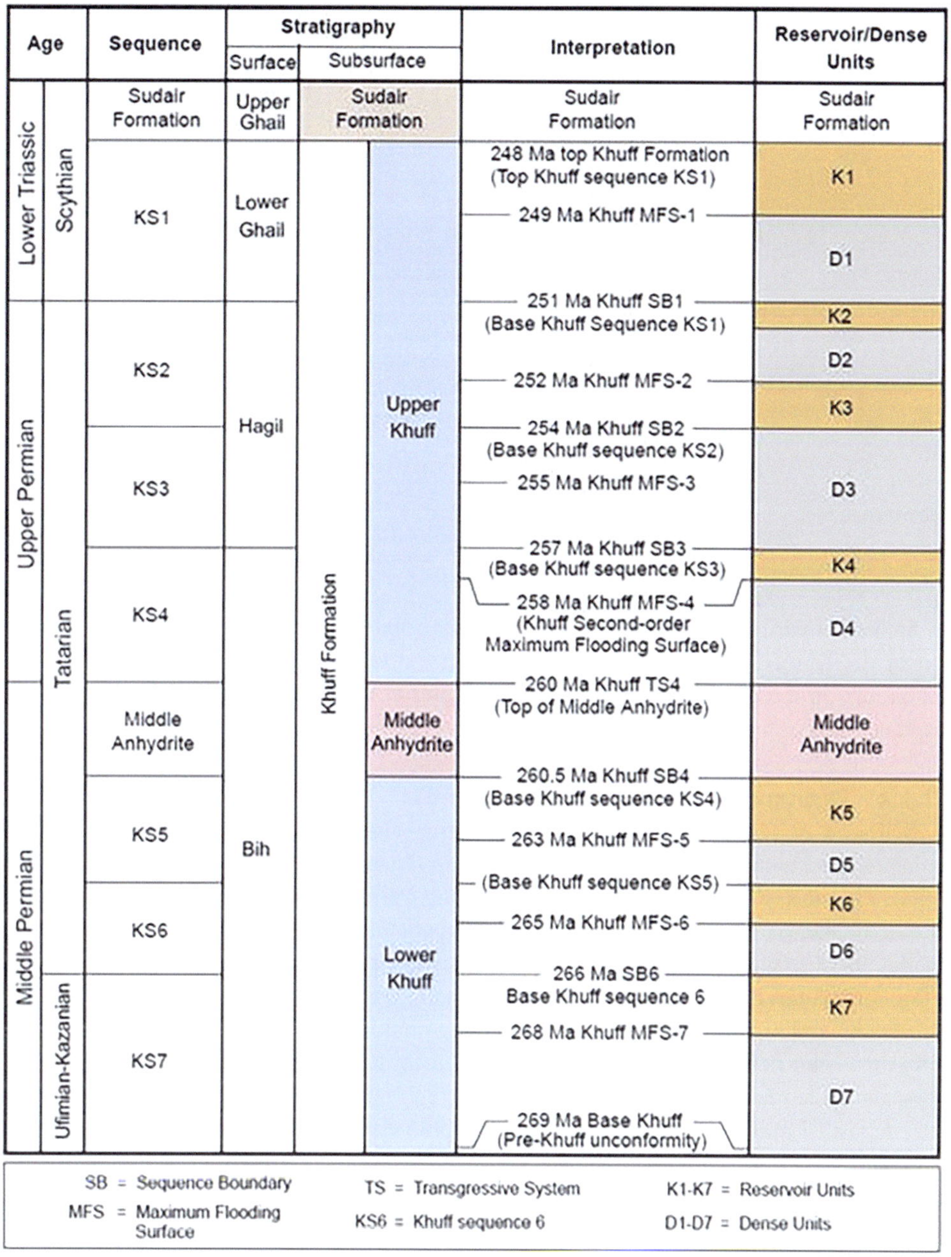

Fig. 6.4 Detailed vertical subdivision of the Khuff Formation (Alsharhan 2006)

Khartam Member. The Upper Khartam Member is assigned to the Early Triassic ('scythian') age due to the occurrence of the characteristic annelid *Spirorbis phlyctaena* (Table 6.1).

6.2.1.1 Regional Distribution

The stratigraphic subdivisions of the Khuff Formation across the region remain similar, with some local variations. In Saudi Arabia, there is a bifold classification, where the formation is divided into members in the outcrops (namely As Shiqqa, Huqal, Midhnab, and Khartam). In contrast, it is divided into units (KA-KD) in the subsurface.

They are divided into four units (KS1–KS4) but are not fully correlated with the Saudi subsurface units. In the United Arab Emirates, the Khuff Formation is divided into seven units (KS1–KS7) in the subsurface and four units (KS1–KS4) in the outcrops. The surface and the subsurface units are not correlated. In Oman, the formation is divided into five units in the subsurface (K1–K5), with K5 subdivided into K5A, B, and C. In the outcrops, the Khuff is divided into six units (KS1–KS6). The two subdivisions are not correlated. The Khuff is divided into four units (K0–K3) in Bahrain, partially correlated with the Saudi subdivisions. In Iraq, the upper Khartam member is equivalent to the lower part of Mirga Mir Formation, and the rest of the Khuff Formation is equivalent to the Chia Zairi Formation. The Chia Zairi is divided into upper and lower carbonate units separated by

Table 6.1 Diagnostic fossils and possible age of the different members of the Khuff Formation (Vaslet et al. (2005)

Lithostratigraphy		Biostratigraphy		Chronostratigraphy	
		Foraminifers-annelids	Ostracods	Stages	Subsystems
Khartam	Upper	*Spirorbis phlyctaena*	*Langdaia* cf. *suboblong*	Scythian	Early Triassic
	Lower	"Dentalina" hoi	Argoviella khartamensis Argoviella permiana? Knoxiella infirma	Changhsingian (= Dorashamian)	Late Permian
	Lowermost	"Nodosaria" dzhulfensis			
Midhnab Mbr		Paradagmarita–"Glomospirella spirillinoides"			
Duhaysan Mbr		"Dentalina" hoi–Hemigordius baoqingensis–*Graecodiscus* cf. kotlyarae *Colaniella* sp.		Wuchiapingian/Changhsingian	
Hukayl Mbr		*Aeolisaccus* spp.–*Pseudomidiella* cf. labensis–Neodiscus aff. Qinglongensis		Wuchiapingian (= Dzhulfian)	
Ash Shiqqah Mbr		Monodiexodina kattaensis–Reichelina sp.		Capitanian (= Midian)	Middle Permian

the Satina anhydrite member in a similar tripartite subdivision of the Khuff Formation. More details about the local stratigraphy in these countries will be provided later in this chapter.

6.2.1.2 Regional Depositional Environments

Al Marhoon et al. (2014) identified three major global geological events during the east Arabian Permian–Triassic period: the most significant mass extinction at the Permo-Triassic boundary, the extended prevalence of superanoxia, and the initial fragmentation of Pangea. The passive margin in this context lies on top of a multi-terrain basement, which underwent varying tectonic subsidence, creating multiple subbasins. Mud-rich facies primarily accumulated in these subbasins (paleo-lows), while grain-rich facies (reservoirs) were deposited on the neighboring paleotopographic elevations. They developed a series of Isopach maps and regional composite cross-sections for the area, using seismic data and well data calibration. These maps revealed thickness variations corresponding to the irregular basin topography caused by the differential subsidence of basement blocks. The researchers concluded that the Basal Khuff clastics were likely deposited in rifted basins consisting of horsts, grabens, and slanted normal fault blocks. These features were observed in flattened seismic sections, which employed the top of the Khuff formation as a reference point. This rift system quickly transitioned into a passive margin.

The Khuff depositional system is the first carbonate shelf that started the dominance of carbonates as the main sediments in eastern Arabia till our present. The same configuration would be repeated at different parts of the Arabian Plate during the geologic time afterward. The major components of such a system are formed of shallow-water carbonates alternate or interbed with evaporites with clastics at the outer fringes of the basin and reef at the shelf break. Figure 6.5 illustrates the main depositional environments that led to the deposition of the Khuff Formation.

The Khuff carbonate system is underlain by the clastics of the Unayzah Formation (see Chap. 5). The contact is unconformable and marks the start of the Early Permian regional transgression on the old clastic front that dominated large parts of Arabia during most of the Paleozoic. There is a clastic wedge at the western extension of the shelf in Saudi Arabia, derived from the nearby land. The shelf is formed of normal shallow marine carbonates of microbial mudstone, skeletal wackestone, and packstone and punctuated with zones of oolitic grainstone. The Middle Anhydrite bed represents a temporary desiccation period that interrupted carbonate deposition, but it resumed again. There are also lenses of anhydrite across the shelf. Coral reefs are formed along the shelf break and adjacent to the Zagros suture zone.

The different local depositional environments in the different countries of eastern Arabia and Iran will be dealt with in more detail later in this chapter.

6.2.2 The Khuff Formation in Saudi Arabia

The Khuff Formation is usually overlain unconformably by the Sudair Formation. This formation is formed of maroon or green shale. In central Saudi Arabia, Al-Aswad (1997) mentioned that the Sudair shale is present only in isolated patches, which suggests that the upper part of the Khuff is preserved there. The formation is underlain by the Unayzah Formation, and where the Unayzah is missing, the formation is underlain by various formations of ages ranging

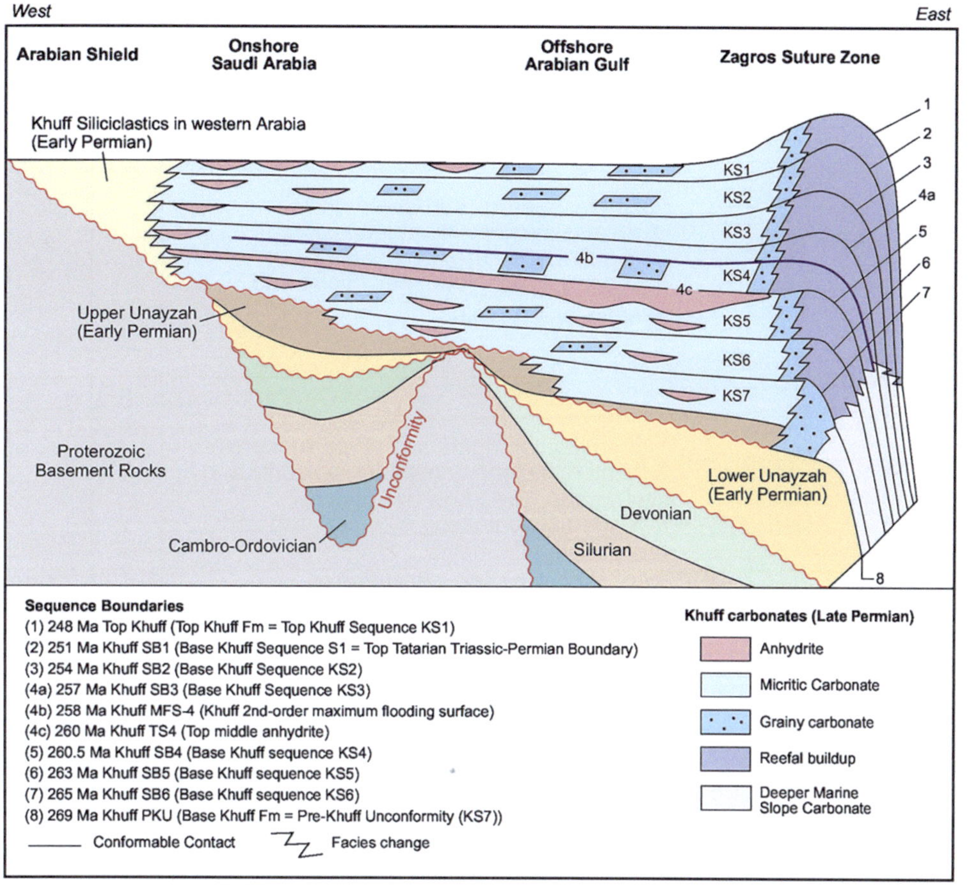

Fig. 6.5 Regional depositional environment of the Khuff Formation in eastern Arabia and Iran (Alsharhan 2006)

from the Proterozoic to the Carboniferous. The relationship between the Khuff and the Unayzah formation is problematic and has witnessed several changes (see (Al-Laboun 1987). Generally, the boundary between the two formations is placed at the base of mixed siliciclastic and carbonate strata.

6.2.2.1 Stratigraphic Subdivisions

Following Delfour et al. (1982), the Khuff Formation is subdivided into four members, excluding the Unayzah. These members are briefly described below, with their characteristics summarized in Fig. 6.6 (Al-Aswad 1997).

1. Huqayl Member: Comprising mudstone, wackestone, heavily cemented, and dolomitized occasionally with shales, the Huqayl Member also contains lithoclastic wackestone-packstone with pelecypods and fish remains in its basal part. Gypsiferous mudstone is present at some levels, and a massive gypsum unit (three meters thick) is found in the lower part of the member. The uppermost part consists of mudstone and fine dolomite sequence, transitioning to the Duhaysan Member through pelletal wackestone containing ooids and rare skeletal remains.
2. Duhaysan Member: This member is characterized by alternating shales and limestones with minor dolomites. It is generally rich in fossils, displaying a diverse range of fauna. Dominant skeletal wackestone and packstone containing foraminifera, fish remains, brachiopods, ostracods, bryozoans, and algae. The shales within this member contain foraminifera, gastropods, pelecypods, worm tubes, and ostracods. Dolomite and mudstone are found near the top of the member. The transition to the Midhnab Member occurs from shale or claystone, lime mudstone, or fine dolomite below to skeletal wackestone containing bactritoid ammonoids above.
3. Midhnab Member: The Midhnab Member exhibits the highest faunal diversity and density in the Khuff Formation. It consists of alternating carbonates and shale sequences, with some sections containing shale and gypsum. Dolomitization is pervasive at certain levels. The lower half of the member includes gypsum beds, and its basal part begins with a skeletal wackestone and packstone heavily cemented, sometimes containing *Bactrites* sp. The transition to the Khartam Member happens from shale to skeletal packstone containing brachiopods.

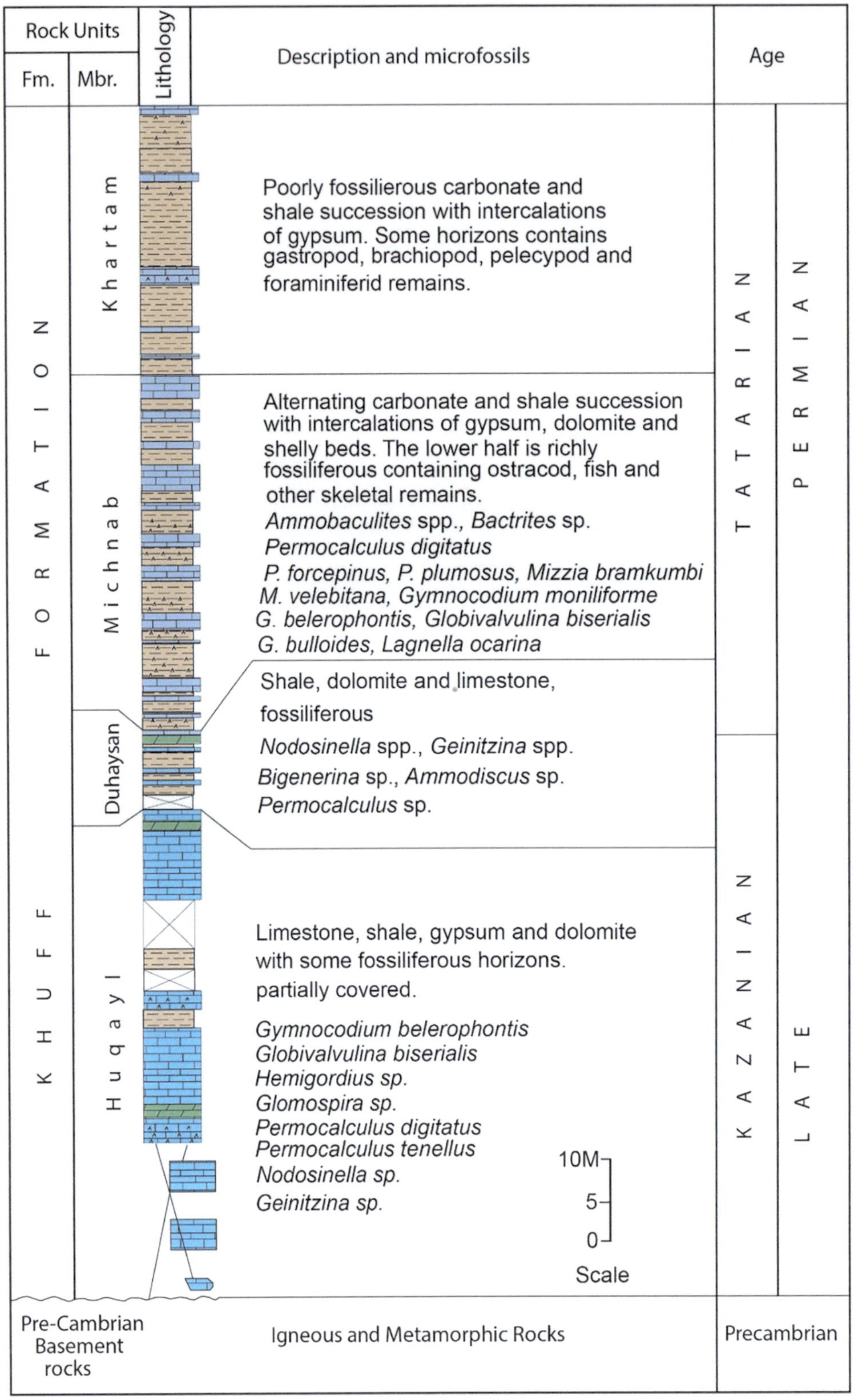

Fig. 6.6 Lithology and fossil contents of the Khuff Formation members in central Saudi Arabia (Al-Aswad 1997)

4. Khartam Member: The Khartam Member contains the highest proportion of shale in the Khuff Formation. This member is generally less fossiliferous, and its shales are highly gypsiferous.

Figure 6.6 shows a gradual increase in the amount of shale in the Khuff succession toward the top. The lower part is composed of limestone and dolomite with algae and a benthonic *foraminifera* population. The middle part shows the appearance of shale horizons, but the carbonate units are still thicker. Shale becomes the dominant lithology in the upper part, with thin limestone horizons with rare or no fossils.

Vaslet et al. (2005) revisited the Khuff outcrops in central Saudi Arabia and suggested some modifications to the stratigraphic subdivisions of the formation. The major change involves naming what was called the Unayzah Member to the Ash Shiqqah Member. This member comprised the lower part of the formation. It consists of claystone interbedded with sandstone beds changing gradually to bioclastic argillaceous dolomite.

The Huqayl Member has a homogenous lithology in terms of facies and thickness. Two major units can be recognized in this member. These are packstone and gypsum-bearing claystone with evaporites in the subsurface. The base of the lower unit is characterized by the presence of marine fauna (foraminifera and algae), which indicates that this unit was deposited under transgressive marine conditions.

The dolomitic limestone facies first appears within the Duhaysan Member. Carbonates dominate the sediments of this member, gradually changing to argillaceous facies to the south. The clastic content of this facies is minor.

The Midhnab Member has a relatively vertical homogenous facies distribution that can be easily recognized. At the base is a reworked limestone or dolomitic limestone with many marine fossils. It is overlain by a fossiliferous limestone changing upward to gypsiferous and dolomitic claystone. Locally, the upper part of the member contains plant-bearing claystone with channels of sandstone containing fossil driftwood or lacustrine limestone.

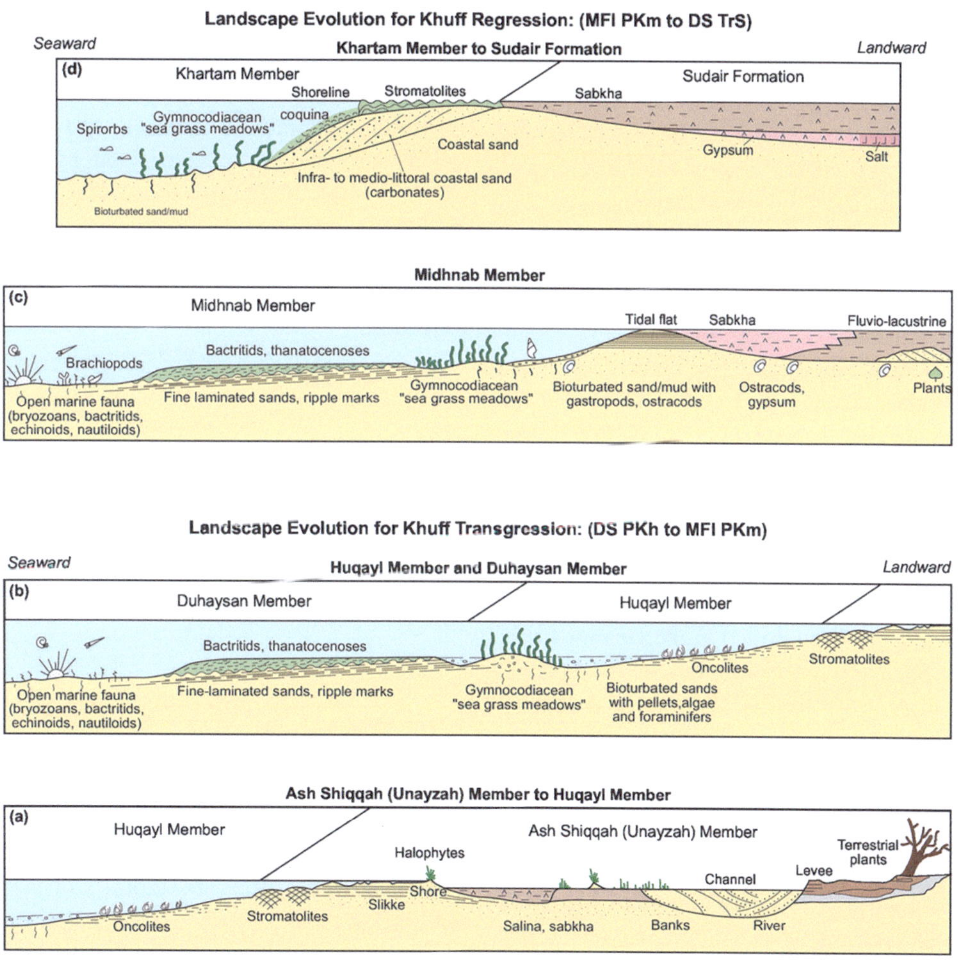

Fig. 6.7 Depositional environments of the different Khuff members (compiled from (Vaslet et al. 2005)

The lower part of the Khartam Member is mainly argillaceous and dolomitic or sandy. It overlies the detrital gypsiferous or lacustrine sediments at the upper part of the Midhnab Member. The upper part comprises constant, relatively thick oolitic, peloidal, and bioclastic limestone.

6.2.2.2 Depositional Environments

The depositional environment of the Huqayl Member is a continuation of the Unayzah Formation (see Chapter Six). It is dominated by a combination of fluvial conditions with some inland and coastal sabkhas. The fluvial conditions are reflected by plant remains and levee deposits with gypsiferous and argillaceous sediments. The marine conditions are of limited distribution, and the limestone is microbial with stromatolites and oncolites, suggesting shallow, protected coastal lagoonal conditions (Fig. 6.7a).

The Duhaysan Member represents the start of full marine conditions throughout the area. The carbonate ramp is formed of normal marine conditions with foraminifera, fish remains, brachiopods, ostracods, bryozoans, and algae, and restricted coastal lagoons with pellets, algae, and benthonic foraminifera. Seagrass meadows are developed locally on elevated areas (Fig. 6.7b).

The deposition of the Midhnab Member was associated with a relative retreat of the sea level as reflected by the development of fluvial and lacustrine conditions and wide coastal sabkhas. Tidal flat sediments of bioturbated wackestone and packstone with gastropods and ostracods became a prominent part of the shelf. Marine conditions remained the same with open marine fauna such as Bryozoa, echinoids, nautiloids, and bactritids (Fig. 6.7c).

That sea retreat continued during the deposition of the Khartam Member with the development of gypsiferous clastic sediments of the Sudair Formation. The marine sediments occupy a smaller area, consisting of carbonate sand at the coastal parts with coquina, capped with stromatolitic limestone. Bioturbated carbonate mud and sand continued to be deposited in the deeper parts of the basin with spirorbs and seagrass (Fig. 6.7d).

6.2.2.3 Khuff Formation in the Subsurface

The Khuff Formation has been drilled in the Ghawar Field, eastern Saudi Arabia. The formation was encountered at around 3000 m deep. It overlies the siliciclastic sediments of the Unayzah Formation and is overlain by the clastic sediments of the Sudair Formation. The total drilled thickness is around 450 m. The bearing units are located within the formation's upper part, totaling around 270 m thick (Al-Dukhayyil 2012).

The formation has been divided into four units. The lower unit is Khuff D which also includes the Basal Khuff Clastics (BKC) or the Khuff E (Al-Jallal 1995), then Khuff C, B, and A. The age of units D and C is Permian, and units B and A is Triassic (Al-Dukhayyil 2012; Al-Dukhayyil and Tawil 2006; Hughes 2005; Hughes et al. 2009). Units D (including the Basal Khuff Clastics) are probably late Wordian to Capitanian in age. The lower part of the Khuff C is Wuchiapingian-Changhsingian, while its upper part is Changhsingian (Al-Dukhayyil 2012; Alsharhan 2006; Hughes et al. 2009). Units B and A are early Triassic (Induan) (Al-Jallal 1995; Hughes et al. 2009).

Al-Dukhayyil (2012) placed the Wuchiapingian-Changhsingian boundary at the upper part of a series of parasequences capped by emergence breccias (Fig. 6.8) which he considered as a major sequence boundary. The brecciated zone represents a major sea-level fall. This boundary seems to separate a barren section below from a foraminifera-rich section at the top.

Main Lithofacies

A set of cores recovered from the formation in some wells in the Ghawar field were examined by Al-Dukhayyil (2012). He recognized the following lithofacies in the studied sediments (Fig. 6.9).

Paleosol Beccias with rhizoliths

These units range in thickness between 0.15 and 2.0 m. They consist of breccia with rhizoliths and collapse featured and infilled with sand to gravel intraclasts formed of anhydritic dolomitic mudstone and wackestone. They usually overlie units characterized by desiccation cracks and microbial lamination. These units are associated with sequence boundaries and even some parasequences. They are products of subaerial exposure where alternations of desiccation, dissolution, and brecciation would lead to their formation.

Nodular and massive anhydrite

Anhydrite is found in units ranging in thickness between 5 cm and 1.5 m. They usually occur below exposure surfaces. The anhydrite is whitish to grey, massive, nodular, or bedded, with some occasional fine laminations. These are some nodules or patches within the dolomite. These sediments are deposited in a sabkha system (Al-Jallal 1995, 1989).

Laminated anhydrite

These units are formed of laminations or beds ranging in thickness between 0.1 and 4.0 m. They are overlain and underlain by dolomite with sharp or gradational contacts. The anhydrite shows a color gradation from block to grey or white. They are occasionally interbedded with massive or nodular anhydrite. The dark color is probably due to organic-rich argillaceous materials. These evaporites are subaqueous and were deposited in salinas or coastal lagoons. Some rare

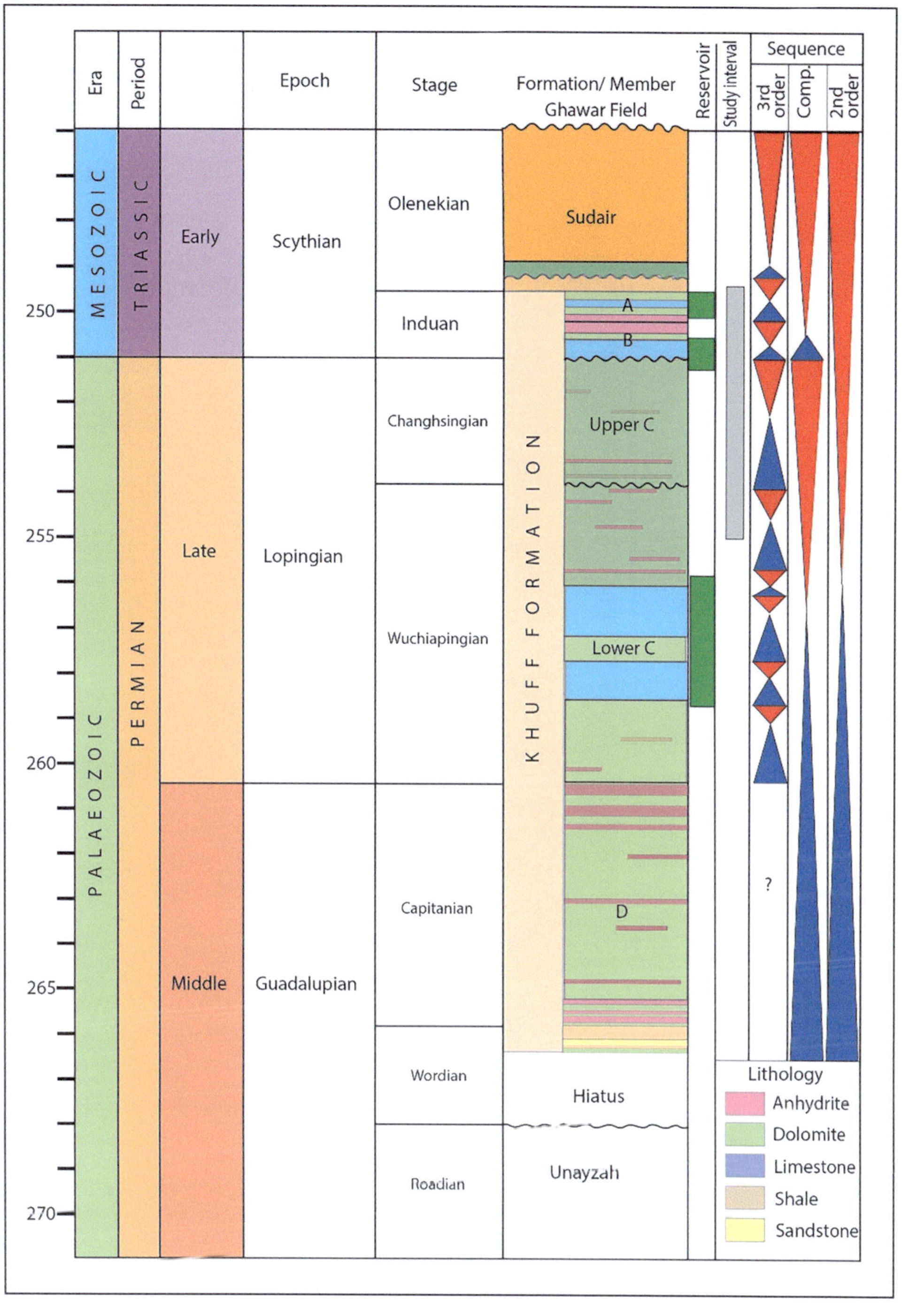

Fig. 6.8 Stratigraphic subdivisions and cycles of the Khuff Formation in the Ghawar Field, Saudi Arabia (Al-Dukhayyil 2012)

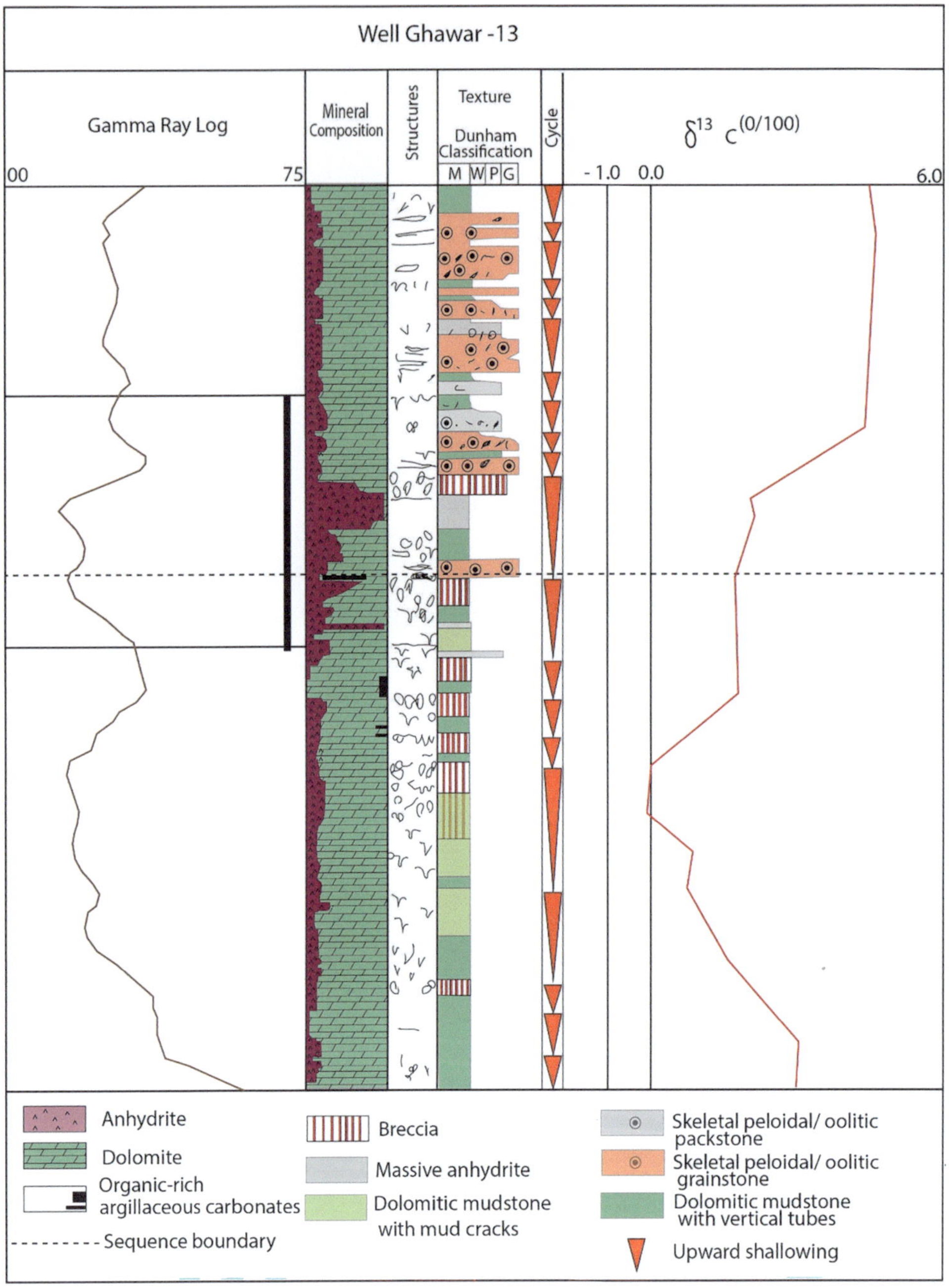

Fig. 6.9 Detailed lithologic section of the Khuff Formation in the Ghawar Field, Saudi Arabia (Al-Dukhayyil 2012)

burrow mottling is generated by salt-tolerant shrimps (Al-Dukhayyil 2012).

Laminated dolomitic mudstone

These sediments represent the final sedimentation products since they cap the parasequences and are overlain by subaerial exposure surfaces. They range in thickness from 5 cm to 2 m thick. They are characterized by thin microbial laminations, fenestral porosity, burrows, and mud cracks. They are formed in intertidal flats similar to the present-day intertidal flats of the sabkha ecosystem in the Arabian Gulf (Al-Dukhayyil 2012; Sadooni et al. 2010).

Burrowed dolomite

This lithofacies is restricted to the Late Permian part of the formation. It consists of beds ranging in thickness between 0.5 and 3.5 m. Dolomite is aphanitic and dark grey in color. There is also some peloidal wackestone with anhydrite nodules. The characteristic feature of this lithofacies is the presence of vertical burrows. There are up to 25 cm in length and 1 cm in diameter. They are usually filled with anhydrite. This facies was deposited in restricted lagoons with active microbial activity.

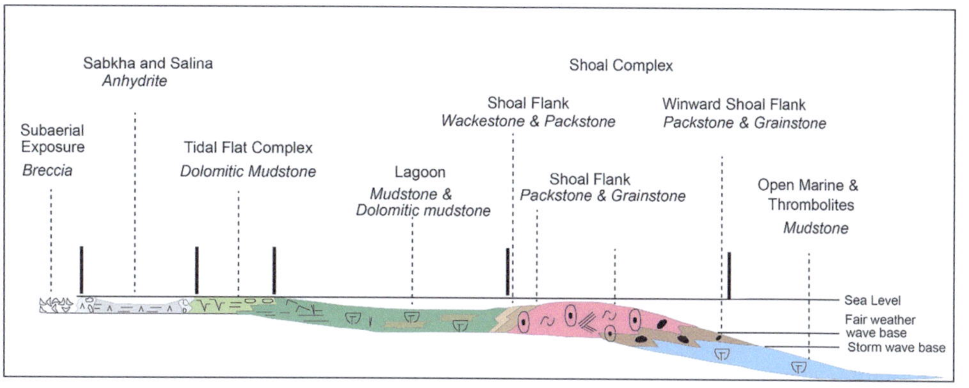

Fig. 6.10 Depositional model of the Khuff Formation in Ghawar Oilfield (Al-Dukhayyil 2012)

Lime mudstone and wackestone

This facies is formed of laminated and burrowed lime mudstone and fine peloidal and gastropod-bearing wackestone. The sediments are grey to dark grey and thin to medium bedded. This facies was deposited in protected lagoons and shallow tidal flats with limited water circulations (probably behind oolitic or skeletal shoals). The scarcity of bioturbation is probably related to the high salinity of the depositing environment (Al-Dukhayyil 2012; Al-Jallal 1989).

Oolitic grainstone

This lithofacies represents the major porous parts of the formation. It forms units 0.2–12 m thick with a wide distribution across the basin. They are light to medium brown with cross and planar bedding. Some of them are extensively dolomitized. The ooliths are well-sorted, fine to medium. In other variants of these facies, ooliths are replaced by peloids (which may be micritized ooliths) or even skeletal grains such as gastropods, foraminifera, green algae, brachiopods, bryozoans, and crinoid debris. These sediments were deposited in high-energy, oolitic shoal as indicated by the well-sorting of the grains, cross-bedding presence, and the scarcity of fossil contents (Al-Dukhayyil 2012; Rankey et al. 2006).

Intraclastic packstone/grainstone

This facies is up to 2 m thick of intraclastic limestone, forming local layers of flat pebbles conglomerate with some ooliths and peloids. Al-Dukhayyil (2012) suggested that this facies was formed on wind-facing oolitic shoal flank, and these sediments resulted from storm action.

Basinal mudstone

This is the deep-water sediments of the Khuff ramp. It formed around 8 m of dark grey to dark brown mudstone with decimeter-scale cycles of microbial laminations, thrombolites, and occasional horizontal burrows. This lithofacies represents the most distal sediments of the Khuff succession. They were deposited in distal marine, low energy, and oxygen settings below the storm-wave base (Al-Dukhayyil 2012).

6.2.2.4 Depositional Environments

The Khuff Formation in the Ghawar Oilfield represents a typical carbonate ramp system under arid climatic conditions, as reflected by the presence of sabkhas and salinas. The ramp also comprises a relatively wide tidal flat complex that includes coastal and inner lagoons with limited water circulation. The ramp's gentle flank is represented by an oolitic shoal where the main reservoir unit (i.e., the oolitic grainstone) formed. The windward flank of the shoal was subjected to occasional storms leaving lithoclastic packstone and grainstone. The distal and deeper part of the ramp is characterized by mudstone with thrombolites. This depositional system will be repeated at different parts of the Arabian Plate through geologic time (Fig. 6.10).

6.2.3 The Khuff Formation in Iran

The Khuff Formation has been reported from the Pars Oilfield in Iran, which is the continuation of the North Gas field of Qatar, and hence, reviewing the Khuff sediments in Iran is relevant in this context. The Khuff strata in Iran is divided among many formations and members. From the top, they include the Kangan Formation, and Upper Dalan Formation, followed by an anhydrite unit, then the Nan Member and the Lower Dalan Formation (Fig. 6.11).

The Lower Dalan Formation overlays the Faraghan Sandstone Formation. The Lower Dalan comprises oolitic and peloidal limestone (Fig. 6.11).

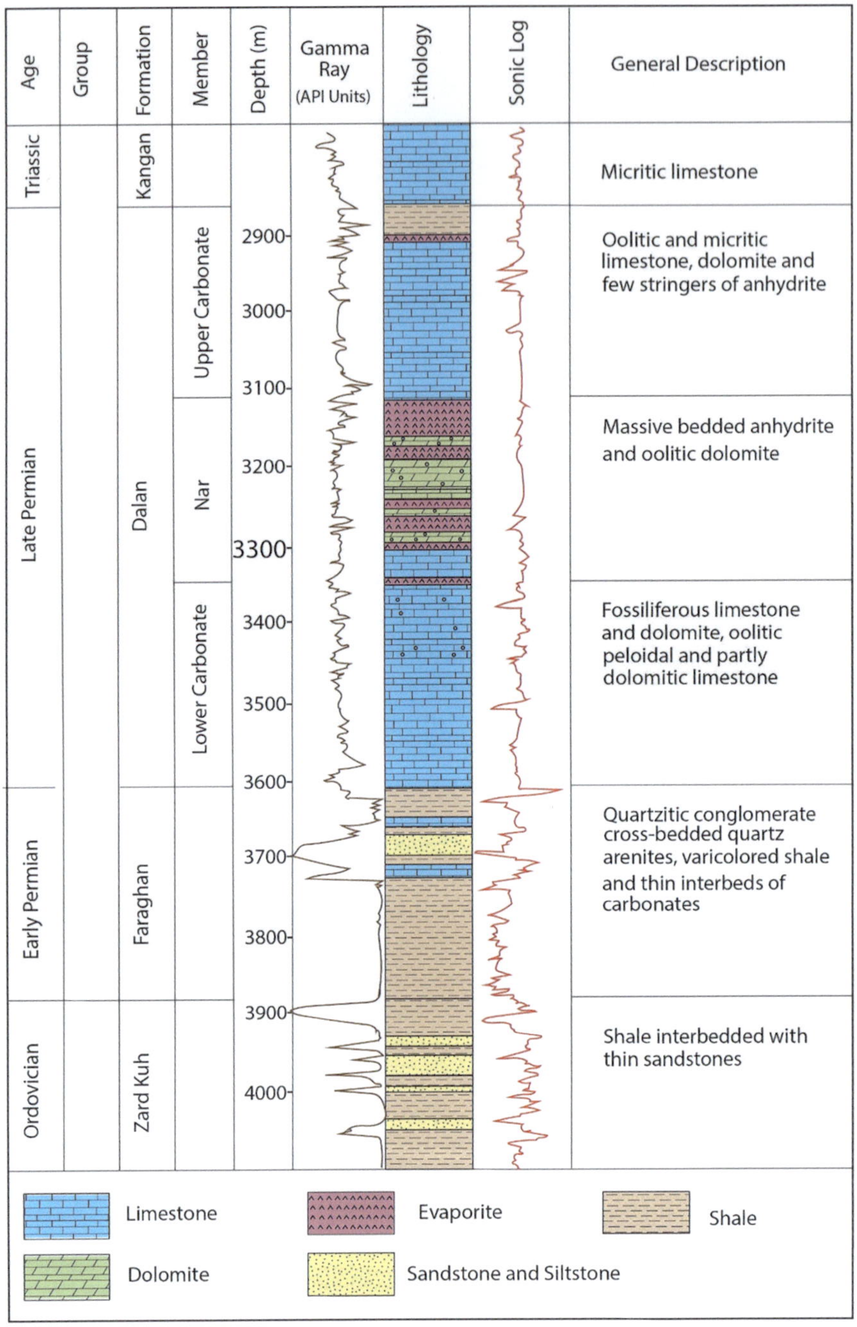

Fig. 6.11 Stratigraphic section of the Permian-Early Triassic strata of southwestern Iran (Alsharhan and Nairn 1997)

6.2.3.1 The Kangan Formation (Early Triassic, Scythian)

The formation name is derived from the name of a village in the Fars Province, southern Iran, near the Arabian Gulf. It overlies the Permian Dalan Formation disconformably, but it gradually passes to the Aghar Shale Member of the Dashtak Formation. The Kangan Formation is 178 m thick in the subsurface and 140 m in the outcrops. The formation comprises three main lithofacies (Fig. 6.11). The lower part is formed of pelletal and oolitic grainstone and dolomitic mudstone with some minor evaporites and sand and silt grains. The middle division consists of shale, argillaceous limestone, and dolomite. This section is characterized by the presence of abundant pelecypod *Claraia*. The top part is made up of evaporites, dolomitic limestone, dolomite, and anhydrite (Alsharhan and Nairn 1997).

6.2.3.2 The Dalan Formation (Late Permian)

The name of the formation is after the well Dalan-1, located around 110 KM southwest of Shiraz, southwestern Iran.

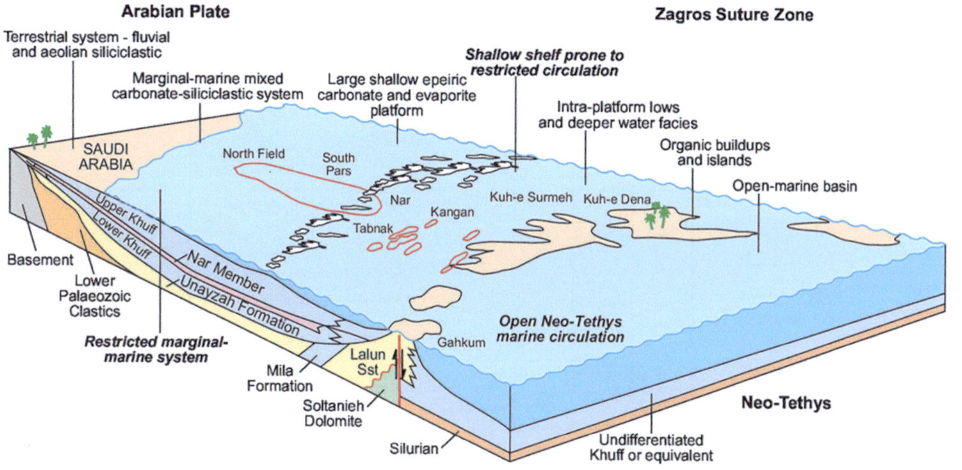

Fig. 6.12 Depositional model of the Kangan and the Dalan formation in southwestern Iran (Insalaco et al. 2006)

The maximum thickness of the formation is around 748 m. The formation was divided by Edgell (1977) into three units. These are two informal carbonate units, Upper and Lower Dalan, separated by a formal anhydrite unit, the Nar Member. This member consists of massive and bedded anhydrite, anhydritic dolomite, and oolitic dolomite. The upper and lower carbonate units are formed of skeletal limestone and dolomite, grading upward into oolitic and peloidal limestone with some dolomitized horizons. Microfossils suggest that the Dalan Formation is of the Late Permian age (Kazanian-Kungurian and possibly early Dzulfian; (Szabo and Kheradpir 1978).

The Upper Dalan and Kangan formation are collectively called the "Upper Dalan-Kangan reservoirs." This combined unit is equivalent to the Upper Khuff in Qatar (Insalaco et al. 2006), and it is the main gas reservoir in the South Pars Field, a continuation of the North Field in Qatar (Fig. 6.12). In the South Pars Field, the Upper Dalan Member comprises an oolitic shoal grading upward into a basinal bioturbated mudstone and crinoid-bearing wackestone.

The lower part of the Triassic Kangan Formation is formed of laminated mudstone changing upward into thick bioturbated mudstone. The upper part of the formation is formed of oolitic and bioclastic grainstone.

Insalaco et al. (2006) suggested that the Upper Khuff depositional system changed from a more arid system during the Late Permian to a subtropical climate during the Early Triassic as the system shifted northward. This change is reflected in the abundance of evaporites and hypersaline lagoons in the Permian, while the Early Triassic strata were affected by freshwater. Furthermore, Angiolini et al. (2003) suggested that the Khuff ramp witnessed high temperatures due to the global warming that followed the Carboniferous-Early Permian glaciation.

The sediments of the Kagan Upper Dalan reservoirs were deposited in a wide range of depositional environments, including (Insalaco et al. 2006).

(1) Evaporitic flats (supratidal/intertidal to subtidal setting).
(2) Tidal flats (intertidal to supratidal setting, with various local environments such as beach ridges, intertidal flats, tidal channels, microbial mud, and sand flats).
(3) Subtidal lagoon (subtidal setting, including microbial buildups).
(4) Leeward shoals (subtidal to intertidal setting).
(5) Oolitic to oolitic bioclastic shoal belts (subtidal to intertidal setting).
(6) Composite sand wave constructions (subtidal setting).
(7) Middle shelf (subtidal setting) and outer-ramp settings.

6.2.4 The Khuff Formation in Qatar

The reference section of the Khuff Formation in Qatar was described from the well Dukhan-65. The formation was encountered between drilled depths of 2895–3394 m with an average thickness of 499 m. In this well, the Khuff Formation is formed from the top of 81 m of dolomite with ghosts and relics of pellets and ooliths. The dolomite is generally fine, but it is saccharoidal in some places. This is followed by 31 m of dolomitic limestone and dolomite with preserved pelletal and oolitic texture. The third unit consists of around 134 m of dolomite with nodular and thin-bedded anhydrite at the lower part. The fourth unit is made up of interbedded dolomite and anhydrite 15 m thick. The final unit comprises 238 m of grey dolomite with relics of limestone texture and nodular anhydrite (Alsharhan and Nairn 1997; Sugden and Standring 1975) (Fig. 6.13).

The formation was encountered in Dukhan and North fields and the well Musaiymir-1 in eastern Qatar.

In the North Field, Focke et al. (1986) reported a conspicuous marker on the Gamma-Ray log at the top of the K3 unit. This "log kick" is correlated through Oman and Iran, where it is identified as the possible Permian–Triassic boundary based

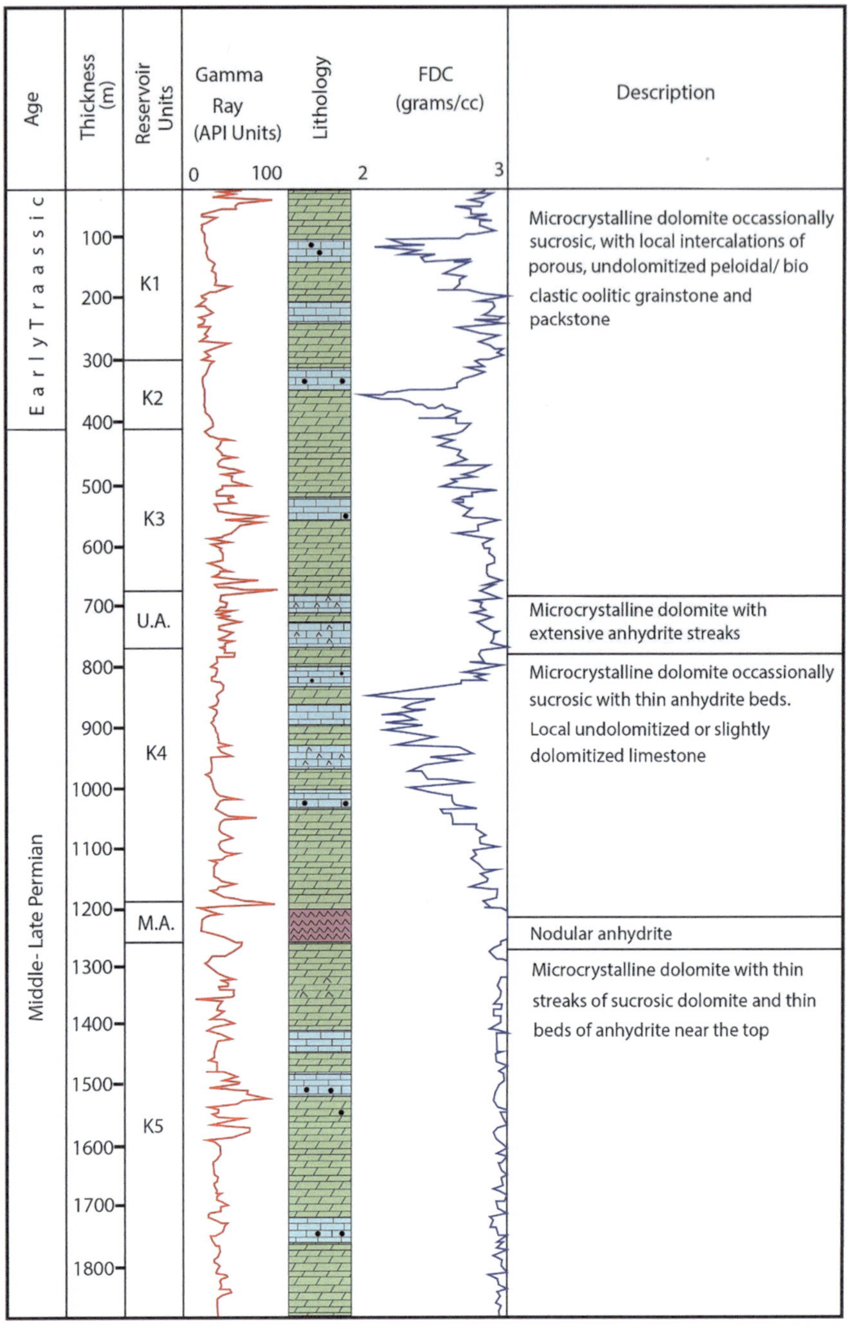

Fig. 6.13 Lithologic succession of the Khuff Formation in Qatar (Alsharhan and Nairn 1997)

on paleontological data. This horizon, in Qatar, is characterized by the presence of a subaerial exposure surface. It is characterized by calcrete, burrows, and intraclasts (Fig. 6.14).

The high Gamma-Ray reading is probably due to paleosoil (caliche) resulting from the local weathering of rocks during the exposure period.

The Khuff Formation shows a clear thickening northeast from around 515 m (1700 feet) in Dukhan Field to more than 900 m (3000 ft) in the extreme northeastern parts of the North Field. This, according to Focke et al. (1986), is aligned with the general trend of the thickness distribution of the Khuff Formation across eastern Arabia (see Fig. 6.2). However, the thickness of the formation in the subsurface of Qatar is impacted by other factors including Qatar Arch and the Hormuz Salt vertical tectonics which may induce structural growth in the area. This can be inferred from the seismic structural map of the area (Hamam and Nasrulla 1989).

Fig. 6.14 Possible Permian–Triassic boundary in the subsurface of Qatar (Focke et al. 1986, The ESC Archives)

There is a good correlation between the main facies of the Khuff Formation across Saudi Arabia, Qatar, Iran, and the United Arab Emirates. The formation shows the same oolitic packstone and grainstone facies all over these countries, suggesting a giant trans-countries oolitic shoal. Variation in the thickness of this facies is probably the result of salt tectonics.

6.2.4.1 Lithofacies

The Khuff Formation in Qatar, with a thickness of approximately 550 m (or 1800 ft), comprises lithologically oolitic, peloidal, dolomitic grainstone, and packstone, which form cycles that end with the formation of subtidal and supratidal, dolomitized mudstone, wackestone, and anhydrite in rather shallow, semi-restricted to open marine environments (Alsharhan and Nairn 1997; Schlumberger 1981). These sedimentary facies display a transition from current onshore toward offshore, where broader marine conditions supplant the dominant supratidal environment. Given the region's low relief, minor sea-level fluctuations led to significant shifts in the ancient shoreline's location.

In the Khuff carbonate reservoirs, five megacycles (K1–5) have been identified (Fig. 6.13), two of which are situated in the Triassic period, in contrast to the four described in Saudi Arabia. The lowest cycle (K-5) is composed of dense, slightly argillaceous dolomites, occasionally interspersed with anhydrite, and is topped by the thick middle anhydrite, serving as a useful marker horizon.

Fig. 6.15 Well-sorted, oolitic grainstone. Some oolith centers are dissolved, creating moldic porosity, which is subsequently isolated by cementation (Focke et al. 1986, The ESC Archives)

Each subsequent four cycles (K1–4) can be divided into subcycles and shares the same lithologies: dense, dolomitic mudstone or wackestone; thinly stratified, dolomitic packstone or grainstone; and secondary, porous, dolomitic grainstone. Anhydrite beds are not uncommon, while undolomitized limestone is seldom found.

Focke et al. (1986) and Holail et al. (2006) described the following lithofacies from the subsurface of Qatar. These are:

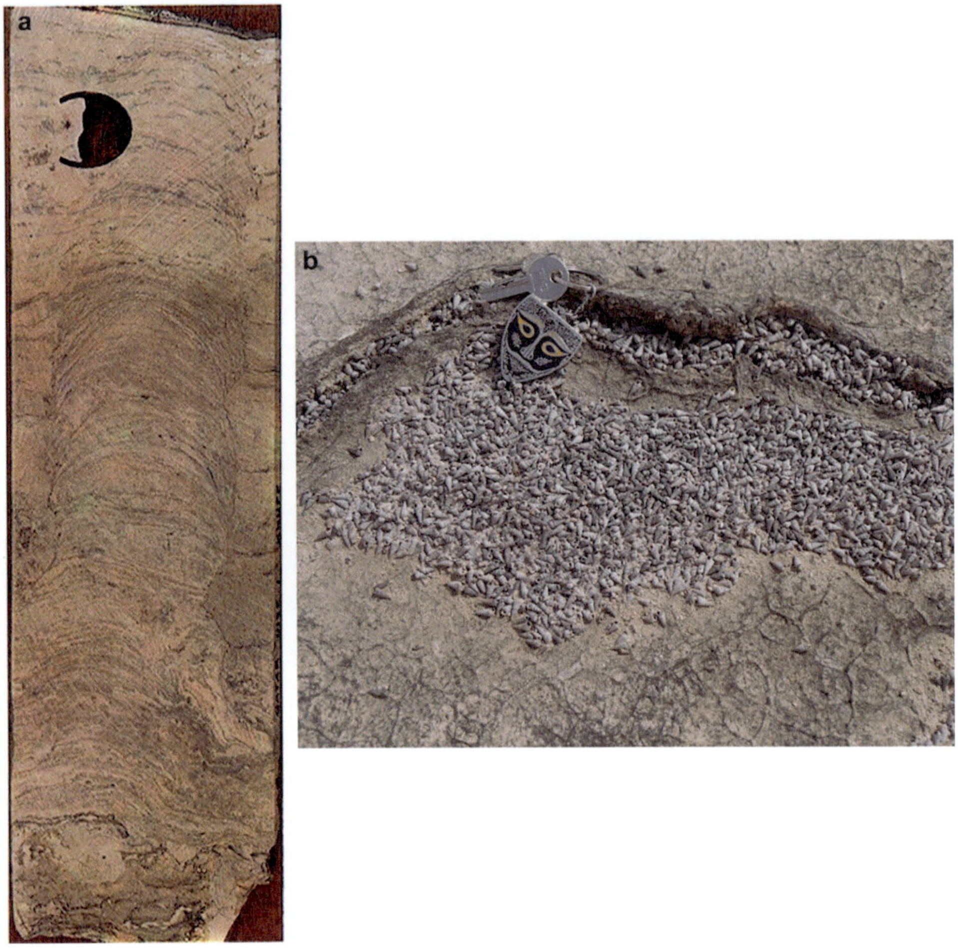

Fig. 6.16 **a** Well-preserved microbial mat in a slabbed core from the subsurface of the Khuff Formation in Qatar **b** Gastropods feeding on the microbial mats in the recent intertidal zone of Khor Al-Adaid (Focke et al. 1986, The ESC Archives)

Dolomitic Oolitic packstone and grainstone

This facies is the most dominant and characteristic of the Khuff Formation across the area. It is also the main reservoir unit. It consists of ooliths ranging in diameter between 0.2 and 0.25 mm (Fig. 6.15). This facies also contain some peloids and intraclasts. Some of the ooliths are extensively dolomitized, while others have quartz nuclei. The matrix is extensively dolomitized. The dolomite is medium-grained, ranging in size between 100 and 250 microns.

The well-rounded ooliths suggest that this facies was deposited in a high-energy agitated shoal under normal shallow marine conditions. Most of the ooliths are dissolved, creating extensive oomoldic porosity. Extensive cementation was also reported from this facies isolating these molds and reducing permeability. This may indicate a phase of dissolution by meteoric water followed by cementation.

Bioclastic Mudstone and Wackestone

This facies comprises grey to light grey mudstone and wackestone with skeletal debris of fossils, mainly benthic foraminifera, echinoderms, and shell fragments with some peloids and unidentified heavily micritized grains. The matrix is formed of micrite and microspar that are extensively leached locally, creating vugs and molds. This facies is transitory between the oolitic facies and the lagoonal microbial mudstone. It was formed in a shallow marine shelf. Extensive micritization may be due to early diagenetic microbial activity.

Microbial Mudstone

Microbial mudstone is a common facies in the Khuff sediments. It has been described from Iran also as consisting of mudstone or wackestone with microbial laminations a few mm to several cm in scale. They may be planar, wavy, domal, or wrinkled. They are associated with peloids or fine bioclastic materials. Insalaco et al. (2006) suggested that sediments were deposited in the intertidal zone as microbial mats. In Qatar, the microbial structures are well-preserved, probably due to rapid burial (Fig. 6.16) because recent microbial mats in the Arabian Gulf coastal sabkhas tend to

Fig. 6.17 Dolomitic anhydritic mudstone facies with bioturbation (Focke et al. 1986, The ESC Archives)

disappear if buried to a depth of around 60 cm (Sadooni et al. 2010). However, Focke et al. (1986) suggested that the preservation of these materials may be due to the absence of predators or competitors, such as gastropods that usually feed on recent microbial mats in the intertidal zone. Thus, they were capable of microbial buildups or (mud mounts) in some cases.

The microbial mats of the Khuff Formation in the subsurface of Qatar are usually laminated, curved upwards, and extensively dolomitized. In some cases, they are leached, generating some dissolution porosity.

Dolomitic anhydritic mudstone

This facies is the more evaporitic part of the above facies. It is usually dolomitized and characterized by either anhydrite nodules or dispersed, detrital quartz grains. The dolomite has an aphanitic texture (with a crystal size of 20–80 microns). The quartz grains are sub-rounded light grey and form between 8 and 10% of the total volume of this facies. Anhydrite laths were also found displacing or replacing some parts of the matrix. Some framboidal pyrite crystals are also present.

This facies is formed in an inland or coastal lagoon with restricted water circulation. The presence of anhydrite laths may suggest supratidal conditions. Detrital quartz grains may indicate clastic influx from nearby land to the lagoon. Framboidal pyrite suggests some microbial activity, probably during the early diagenetic processes. Such activity is reflected in the presence of well-developed burrowing (Fig. 6.17).

6.2.4.2 Depositional Environments

The above facies association indicates that the Khuff Formation in the subsurface of Qatar has been deposited in a carbonate ramp under arid climatic conditions. The ramp had a wide and well-developed oolitic shoal that extended all over the area. Shallow-water carbonates with benthonic foraminifera were deposited in the tidal flats of the ramp. The shallower parts were characterized by coastal and inland lagoons grading to a sabkha system, where thick microbial mats were forming. These mats may be able to form some buildups due to the absence of predators and the low-oxygen levels of the associated sediments.

Figure 6.18 illustrates a typical regressive sedimentary cycle within the Khuff Formation in Qatar. The cycle starts with a cross-bedded, oolitic grainstone that was formed under normal marine conditions in an oolitic shoal with good water agitation and probably bidirectional currents that generated the cross-bedding and the rounded nature of the ooliths. These ooliths are well-sorted and have some interplanar porosity that may have resulted from selective leaching of the more soluble components. This facies is overlain by mudstone with organic-rich bands and birds-eye porosity indicating gas bubbling from organic-rich sediments. These sediments were probably formed in the intertidal zone. The cycle ends upward with a subaerial exposure surface. This hiatus is characterized by the presence of cracked and cemented mudstone and topped up by a relatively thick clay seam which may represent a relatively long weathering phase. This is followed by a new sedimentary cycle characterized by subtidal marine sediments with some cement patches.

The Khuff Formation in Qatar was deposited in a carbonate ramp similar to the depositional system reported from other parts of the basin, which has the same major facies.

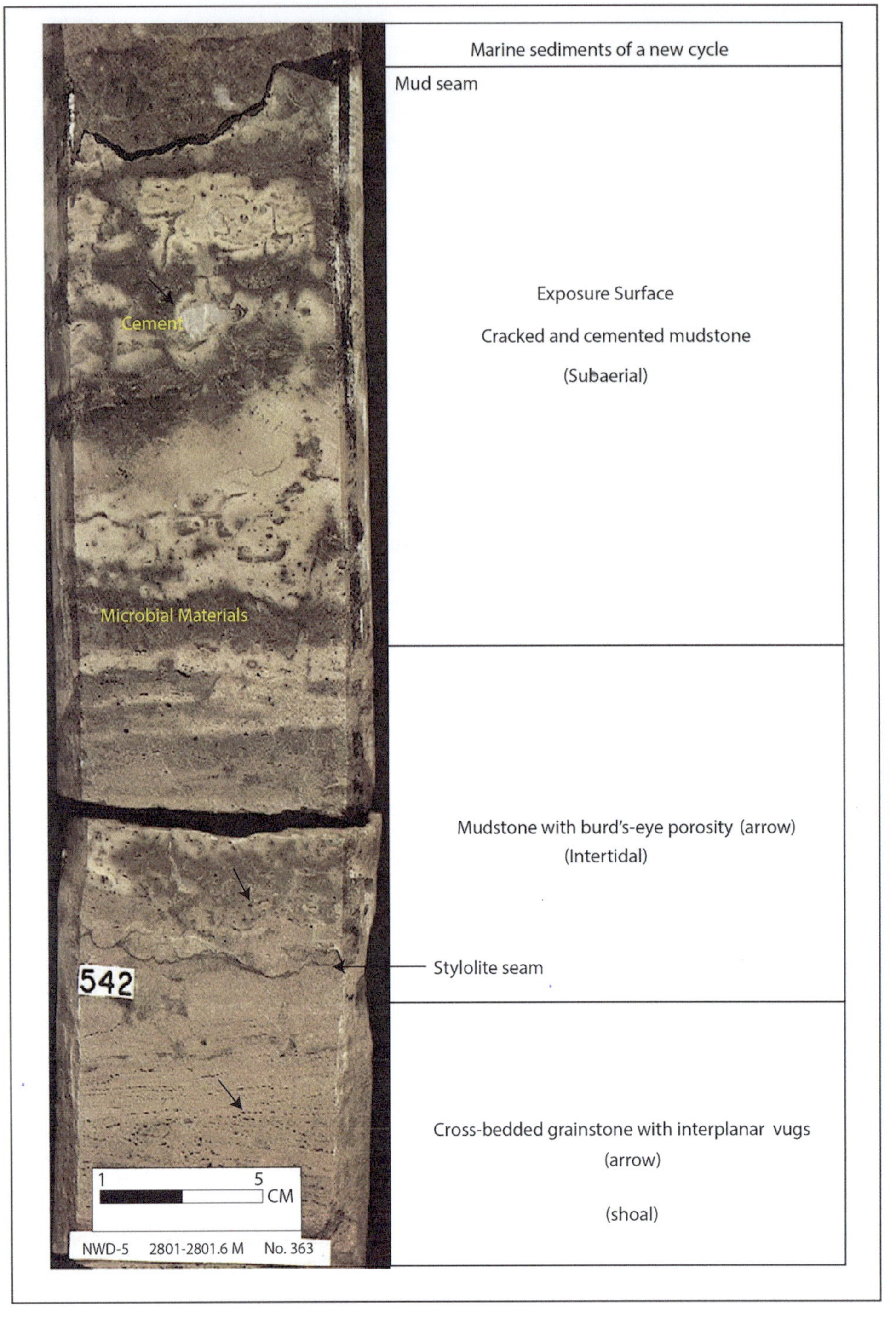

Fig. 6.18 Typical sedimentary cycle in the Khuff Formation in Qatar (modified from Focke et al. 1986)

References

Aigner T, Dott RH (1990) Processes and patterns in Epeiric basins: special issues. Elsevier

Al Marhoon MM, Khalil M, Al Sheikh MO (2014) Tectonic controls on the stratigraphic architecture of the Permo-Triassic Khuff Formation, Eastern Saudi Arabia In: AAPG international conference and exhibition, Istanbul, Turkey

Al-Aswad A (1997) Stratigraphy, sedimentary environment and depositional evolution of the Khuff Formation in south-central Saudi Arabia. J Pet Geol 20(3):307–326

Al-Dukhayyil RK (2012) Sequence stratigraphy and chemostratigraphy across the Permo-Triassic extinction event, Upper Khuff carbonates, Ghawar Field, Saudi Arabia

Al-Dukhayyil RK, Tawil AAA (2006) Reservoir architecture of the Triassic Khartam carbonate sequence, Khuff outcrop analogue in Al-Qassim, central Saudi Arabia. In: 7th Middle East geosciences conference, Manama, Bahrain

Al-Husseini M, Matthews RK (2005) Arabian orbital stratigraphy: periodic second-order sequence boundaries? GeoArabia 10(2):165–184. https://doi.org/10.2113/geoarabia1002165

Al-Jallal I (1995) The Khuff Formation: its regional reservoir potential in Saudi Arabia and other Gulf countries; depositional and stratigraphic approach. In: Middle east petroleum geosciences conference, GEO

Al-Jallal IA (1987) Diagenetic effects on reservoir properties of the Permian Khuff Formation in Eastern Saudi Arabia. Middle East oil show

Al-Jallal IA (1989) Depositional environments, diagenesis and reservoir characteristics of the Permiun Khuff formation in Eastern Saudi Arabia

Al-Laboun AA (1987) Unayzah formation: a new Permian-Carboniferous unit in Saudi Arabia. AAPG Bull 71(1):29–38

Alsharhan AS (2006) Sedimentological character and hydrocarbon parameters of the Middle Permian to Early Triassic Khuff Formation, United Arab Emirates. GeoArabia 11(3):121–158. https://doi.org/10.2113/geoarabia1103121

Alsharhan AS, Nairn AEM (1997) Sedimentary basins and petroleum geology of the Middle East. Elsevier

Ameen MS, Buhidma IM, Rahim Z (2010) The function of fractures and in-situ stresses in the Khuff reservoir performance, onshore fields, Saudi Arabia. AAPG Bull 94(1):27–60. https://doi.org/10.1306/06160909012

Angiolini L, Balini M, Garzanti E, Nicora A, Tintori A (2003) Gondwanan deglaciation and opening of Neotethys: the Al Khlata and Saiwan Formations of Interior Oman. Palaeogeogr Palaeoclimatol Palaeoecol 196(1):99–123. https://doi.org/10.1016/S0031-0182(03)00315-8

Delfour J, Dhellemmes R, Elsass P, Vaslet D, Brosse J, LeNindre Y, Dottin O (1982) Explanatory notes to the geologic map of the ad-Dawadimi quadrangle, sheet 24G. Minist Pet Miner Resour 36:1–85

Edgell HS (1977) The Permian system as an oil and gas reservoir in Iran, Iraq and Arabia. In: Second Iranian geological symposium (Tehran), Tehran, pp 161–201

Fielding CR, Frank TD, Birgenheier LP, Rygel MC, Jones AT, Roberts J (2008) Stratigraphic imprint of the Late Palaeozoic Ice Age in eastern Australia: a record of alternating glacial and nonglacial climate regime. J Geol Soc 165(1):129–140. https://doi.org/10.1144/0016-76492007-036

Focke J, Munn D, Al Kuwari S, Frikken H, Frei H (1986) Petrographic atlas of rock types, common in the subsurface of Qatar and some recent equivalents. In: General petroleum corporation (offshore operations), Qatar, pp 110–189

Frakes LA, Francis JE, et al (1992) Climate modes of the Phanerozoic: the history of the Earth's climate over the past 600 million years. Cambridge University Press

Hamam K, Nasrulla I (1989) Drilling operations and evaluation of Matbakh-2, a deep formation test in Qatar offshore. Country reports and case studies presented at the seminar on deep formations in the Arab countries: hydrocarbon potential and exploration techniques. Organisation of Arab Petroleum Exporting Countries, E229-E241

Holail HM, Kolkas MM, Friedman GM (2006) Facies analysis and petrophysical properties of the lithologies of the north gas field, Qatar. Carbonates Evaporites 21(1):40–50. https://doi.org/10.1007/BF03175467

Hughes GW (2005) Saudi Arabian Permo-Triassic biostratigraphy, micropalaeontology and palaeoenvironment. In: Powell AJ, Riding JB (eds) Recent developments in applied biostratigraphy, vol 1. Geological Society of London. https://doi.org/10.1144/tms001.7

Hughes GWAG, Demchuk TD, Gary AC (2009) Micropaleontology and paleoenvironments of Saudi Arabian Upper Permian carbonates and reservoirs. In: Geologic problem solving with microfossils: a volume in honor of Garry D. Jones, vol 93. SEPM Society for Sedimentary Geology. https://doi.org/10.2110/sepmsp.093.111

Insalaco E, Virgone A, Courme B, Gaillot J, Kamali M, Moallemi A, Lotfpour M, Monibi S (2006) Upper Dalan Member and Kangan Formation between the Zagros Mountains and offshore Fars, Iran: depositional system, biostratigraphy and stratigraphic architecture. GeoArabia-Manama 11(2):75

Konert G, Afifi AM, Al-Hajri SiA, Droste HJ (2001) Paleozoic stratigraphy and hydrocarbon habitat of the Arabian Plate. GeoArabia 6(3):407–442. https://doi.org/10.2113/geoarabia0603407

Leighton MW, Kolata DR (1995). Selected Interior Cratonic basins and their place in the scheme of global tectonics. In: Interior cratonic basins, vol 51. American Association of Petroleum Geologists

Oosterhout C, Pöppelreiter vM (2013) Plate Reconstructions and distribution of sedimentary facies during Permo-Triassic opening of Neo-Tethys Ocean EAGE Special Volume Permo-Triassic Sequence of the Arabian Plate

Pöppelreiter M, Marshall E (2013) The Khuff Formation: play elements and development history of an epicontinental carbonate platform. In: Permo-Triassic sequence of the Arabian Plate, pp 387–400

Powers RW, Ramirez LF, Redmond CD, Elberg Jr EL (1966) Geology of the Arabian Peninsula: sedimentary geology of Saudi Arabia [Report] (560D). Professional Paper, Issue. U. S. G. Survey. https://pubs.usgs.gov/publication/pp560D

Rankey EC, Riegl B, Steffen K (2006) Form, function and feedbacks in a tidally dominated ooid shoal, Bahamas. Sedimentology 53(6):1191–1210

Sadooni FN, Howari F, El-Saiy A (2010) Microbial dolomites from carbonate-evaporite sediments of the coastal Sabkha of Abu Dhabi and their exploration implications. J Petrol Geol 33(4):289–298. https://doi.org/10.1111/j.1747-5457.2010.00480.x

Saltworks CPL (2024) Khuff Platform—no modern same-scale counterpart

Schlumberger SL (1981) Well evaluation conference-United Arab Emirates/Qatar. In: Schlumberger Services Tech. Paris, France

Scotese CR, Boucot AJ, McKerrow WS (1999) Gondwanan palaeogeography and paloclimatology. J Afr Earth Sci 28(1):99–114. https://doi.org/10.1016/S0899-5362(98)00084-0

Senalp M, Al-Duaiji A (1995) Stratigraphy and sedimentation of the Unayzah reservoir, central Saudi Arabia. AAPG Bull 79(CONF-950995-)

Sharief FA (1983) Permian and Triassic geological history and tectonics of the middle east. J Petrol Geol 6(1):95–102. https://doi.org/10.1111/j.1747-5457.1983.tb00266.x

Sharland PR, Casey DM, Davies RB, Simmons MD, Sutcliffe OE (2001) Arabian Plate sequence stratigraphy, vol 9. Gulf Petrolink

Stampfli GM, Borel G (2002) A plate tectonic model for the Paleozoic and Mesozoic constrained by dynamic plate boundaries and restored synthetic oceanic isochrons. Earth Planet Sci Lett 196(1–2):17–33

Stanley SM, Hardie LA (1998) Secular oscillations in the carbonate mineralogy of reef-building and sediment-producing organisms driven by tectonically forced shifts in seawater chemistry. Palaeogeogr Palaeoclimatol Palaeoecol 144(1):3–19. https://doi.org/10.1016/S0031-0182(98)00109-6

Steineke M, Bramkamp R (1952) Stratigraphical introduction. Jurassic Ammonites from Jebel Tuwaiq, Central Arabia Philos Trans Ser B 236:241–256

Steineke M, Bramkamp RA, Sander NJ (1958) Stratigraphic relations of Arabian Jurassic oil. In: American association of petroleum geologists symposium, Tulsa, U.S.A.

Sugden W, Standring AJ (1975) Qatar Peninsula. Lexique Stratigraphique Int III:1–85

Szabo F, Kheradpir A (1978) Permian and Triassic stratigraphy, Zagros basin, south-west Iran. J Petrol Geol 1(2):57–82. https://doi.org/10.1111/j.1747-5457.1978.tb00611.x

Vaslet D, Le Nindre Y-M, Vachard D, Broutin J, Crasquin-Soleau S, Berthelin M, Gaillot J, Halawani M, Al-Husseini M (2005) The Permian-Triassic Khuff Formation of central Saudi Arabia. GeoArabia 10(4):77–134. https://doi.org/10.2113/geoarabia100477

Wikipedia (2024) Max Steineke. https://en.wikipedia.org/wiki/Max_Steineke

Ziegler MA (2001) Late Permian to Holocene paleofacies evolution of the Arabian Plate and its hydrocarbon occurrences

Ziegler AM, Hulver ML, Roeley DB (1997) Permian world topography and climate. In: Martini IP (ed) Late glacial and postglacial environmental changes-Quaternary, Carboniferous-Permian and Proterozoic. Oxford University Press, pp 111–146

7 The Jurassic

Abstract

The Jurassic succession of the Arabian Plate is renowned for its rich petroleum systems, with Upper Jurassic reservoir rocks producing a substantial share of global oil and gas. This chapter examines the stratigraphy, tectonic development, and depositional conditions of Jurassic sediments throughout Arabia, encompassing Saudi Arabia, Qatar, Kuwait, Iraq, and the United Arab Emirates. The chapter opens with an overview of the Jurassic stratigraphic framework, highlighting the difficulties in regional connections across various outcrops and borehole data. The tectonic evolution is examined through the division of the Gondwana and Laurasia supercontinents and the formation of the Neo-Tethys Sea. This tectonic activity shaped the formation of large shallow carbonate platforms throughout Arabia, resulting in diverse depositional regimes governed by regional arches and basins, including the Qatar-Fars Arch and the Gotnia Basin. The chapter elucidates the impacts of climatic and sea-level fluctuations on sediment deposition, forming organic rich source rocks, carbonate reservoirs, and evaporite seals. The chapter examines the stratigraphic units, particularly the productive Middle and Upper Jurassic formations (Hanifa, Jubaila, and Arab) that contain substantial hydrocarbon reserves. These units exhibit alternating carbonate and evaporite cycles, with a depositional model predicated on intrashelf basins partitioned by reefal barriers. Comprehensive facies investigation indicates ecosystems varying from high-energy oolitic shoals to confined lagoons and anoxic basinal conditions. Diagenetic processes, such as dolomitization and anhydrite cementation, substantially affected reservoir quality. The Jurassic system in Arabia constitutes a cohesive hydrocarbon system characterized by intricate stratigraphy shaped by tectonic activity, eustatic sea-level fluctuations, and depositional mechanisms.

7.1 Introduction

The Jurassic section in Arabia is among the richest petroleum systems in the world (Fig. 7.1) (Beydoun 1988). Upper Jurassic source rocks have probably generated one-quarter of the World's discovered oil and gas (Klemme 1993, 1994). This system comprises several regional organic-rich source rocks, stacked reservoirs, multiple seals, and numerous super-giant oil fields. Most of the Saudi oil reserve is hosted in the Jurassic sediments. Although these strata were deposited on a broad platform showing some "layer cake" stratigraphy, but as Al-Husseini (1997) had noticed that their chronological and stratigraphic correlations across outcrops, wells, and even between countries remain inconclusive. In Qatar, Upper Jurassic reservoir rocks are essential in several oilfields. Their distribution is controlled mainly by the NNE-SSW trending Qatar-Fars Arch. This structure was positive during most of the Paleozoic and started to subside gradually during the Jurassic, thus controlling the depositional regime during that period (Saint Marc 1978; Sugden and Standring 1975).

The first Jurassic rocks documented in the region were in Saudi Arabia and Iraq. Marine Jurassic rocks outcrop approximately 320 km northwest of Riyadh, the capital of Saudi Arabia. They form a prominent limestone ridge known as the Jebel Tuwaiq. These strata are nearly flat, with minor normal faulting, but the eastern side gradually dips under the Cretaceous and Tertiary layers of the coastal plain and the Rub al-Khali. The limestone and shale components of these strata wedge north and south (Arkell 1956). Philby (1922) was the first to describe the sediments at Jebel Tuwaiq, but the formal description and nomenclature of the Jurassic sediments in Saudi Arabia were accomplished by the Arabian-American Oil Company geologists (Arkell 1956; Bramkamp and Steineke 1952; Steineke et al. 1958).

The primary escarpment rocks belong to the Tuwaiq Mountain Limestone, which is 215 m thick and dates back

F. N. Sadooni and H. A. Al-Kuwari, *Geological Evolution of Qatar and the Arabian Peninsula*, Regional Geology Reviews,
https://doi.org/10.1007/978-3-031-94394-2_7

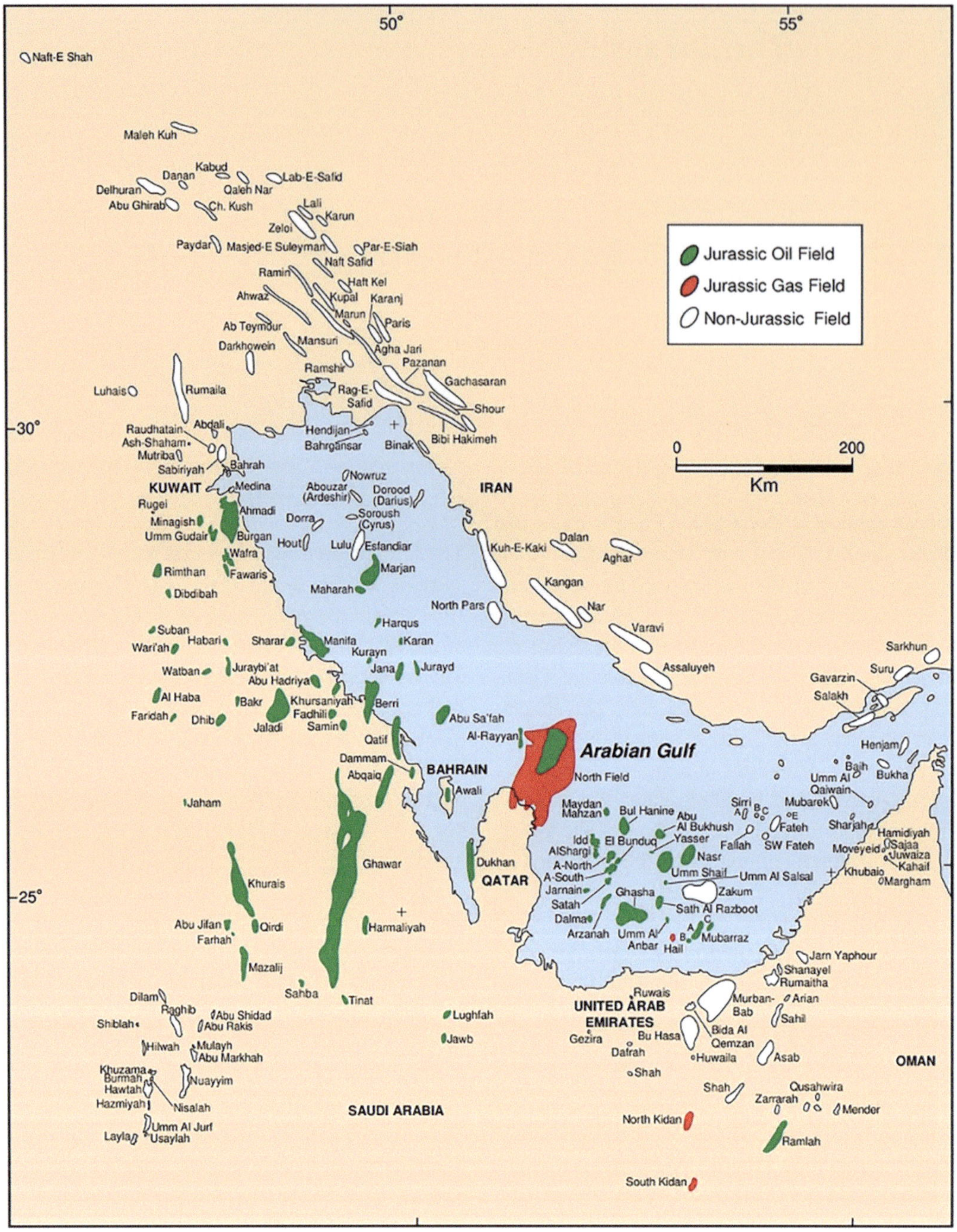

Fig. 7.1 Main Jurassic oil and gas fields of the Gulf region (Al-Husseini 1997)

to the Middle Callovian period. This limestone bed forms the barren plateau that Philby traversed, where he discovered the first ammonites (Erymnoceras) and other fossils. Other limestones lower in the succession of shales and sandstones form smaller escarpments. The total thickness of the Jurassic layers is over 1000 m, even though the range spans only from the Lower Toarcian to Lower Kimmeridgian periods (Arkell 1956).

In Iraq, the Jurassic strata were drilled by Iraq Petroleum Company and Basra Petroleum Company. The stratigraphy and regional correlation of these rocks were summarized in (van Bellen et al. 1959).

Most of the formation names and ages were based on the work conducted on the Jurassic outcrops of Saudi Arabia by the Arabian-American Oil Company (Aramco) and the subsurface data from Iraq and Iran. The age attributions

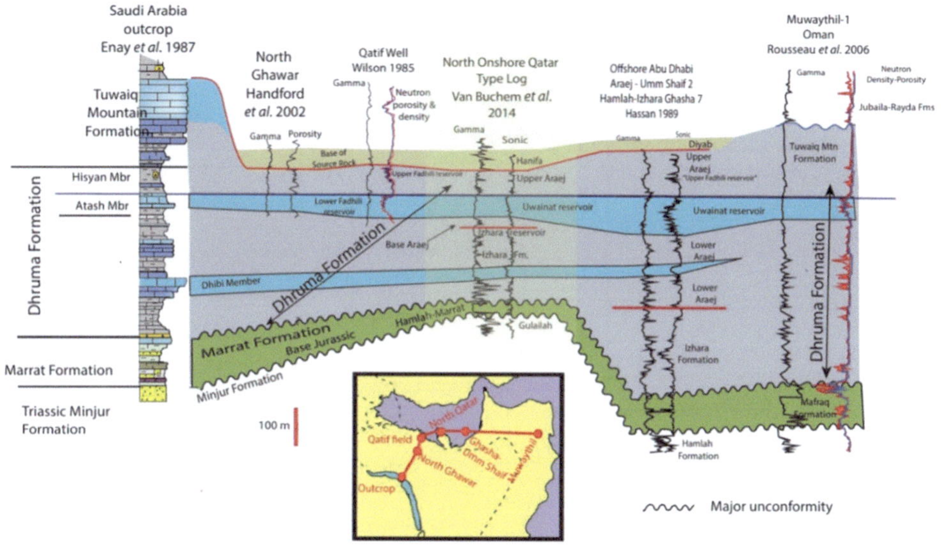

Fig. 7.2 Correlation of the Jurassic strata across the Arabian Plate showing similar formation names (Wilson 2020)

of these formations were based on fossil evidence and stratigraphic regional correlation, as documented in the Lexicons of Iraq (van Bellen et al. 1959), Saudi Arabia (Powers 1968), and Qatar (Sugden and Standring 1975). In Bahrain, the characterization and age designation of the Jurassic formations were described by Chaube et al. (1995). Similarly, in Kuwait, they were defined by Yousif and Nouman (1997) and in the United Arab Emirates (UAE), they were outlined by Hassan (1989). In Oman, the Jurassic formations were detailed in a lexicon-style format by Clarke (1988).

The same terms are used to describe the Jurassic strata across the Arabian Plate with minor variations. This is due mainly to the cake-layered stratigraphy of the Jurassic strata and the contemporaneous exploration efforts conducted during the last century by the Arabian-American Oil Company (ARAMCO) and the Iraq Petroleum Company geologists. Figure 7.2 shows a regional correlation of the Jurassic sediments in the Gulf region.

Le Nindre et al. (1990) and Vaslet et al. (1991) pioneered the study of sequence stratigraphy for the Jurassic strata in Saudi Arabia. Their research leaned on surface data, aligning the Mesozoic formations of Central Arabia with the global eustatic cycles presented by Haq et al. (1988). The criteria they used for their assignments incorporated lithology, depositional settings, and biostratigraphic dating methods.

Later, Grabowski and Norton (1995) introduced a streamlined regional tectonic and sequence stratigraphic outline for the Arabian Plate. Their classification of most rock units, from the Permian to the recent periods, relied on the global eustatic cycles detailed by Haq et al. (1988). They underscored the ability to align these sequences with well logs across vast stretches of the Arabian Platform. The global eustatic sequence model they showcased offers a robust template for the chronostratigraphic alignment of Middle Eastern rock units. Interestingly, their classification for Saudi Arabia echoed the findings of McGuire et al. (1993) for the Upper Jurassic. Yet, these findings diverged notably from those by Le Nindre et al. (1990) and Vaslet et al. (1991). In Fig. 7.3, there's a comparative analysis of the results from Le Nindre et al. (1990), Vaslet et al. (1991), McGuire et al. (1993), Grabowski and Norton (1995), and the categorizations by Powers (1968).

Adding to the narrative, Al-Husseini (1997) proposed a provisional Jurassic sequence stratigraphic structure for the region. This structure is based on the investigations of Le Nindre et al. (1990) and Vaslet et al. (1991) concerning Central Saudi Arabia. Al-Husseini then expanded this framework to other countries, grounded in regional insights and established stratigraphic connections.

Wilson (2020) authored and edited a full memoir on the Jurassic intrashelf basin. The work comprehensively analyzes the geologic evolution and petroleum potential of the basin and the Jurassic strata in the Arabian Plate.

7.2 Tectonic Evolution

The separation of the Gondwana-Laurasia supercontinent and the opening of the Neo-Tethys Sea (the eastern and northeastern parts of the present-day Arabian Plate) started during the early Triassic (see Chap. 6). Tethyan rifting during the Jurassic was characterized by the westward drifting of Gondwana and the opening of the central part of the Atlantic Ocean. The westward drift of Gondwana was the major megatectonic event affecting the Arabian Intrashelf Basin during the Callovian–Tithonian period. This was followed up by the development of both the Somali and Mozambique Basins that separated the old Gondwana into two major parts. The advent of the Jurassic time was associated with global sea

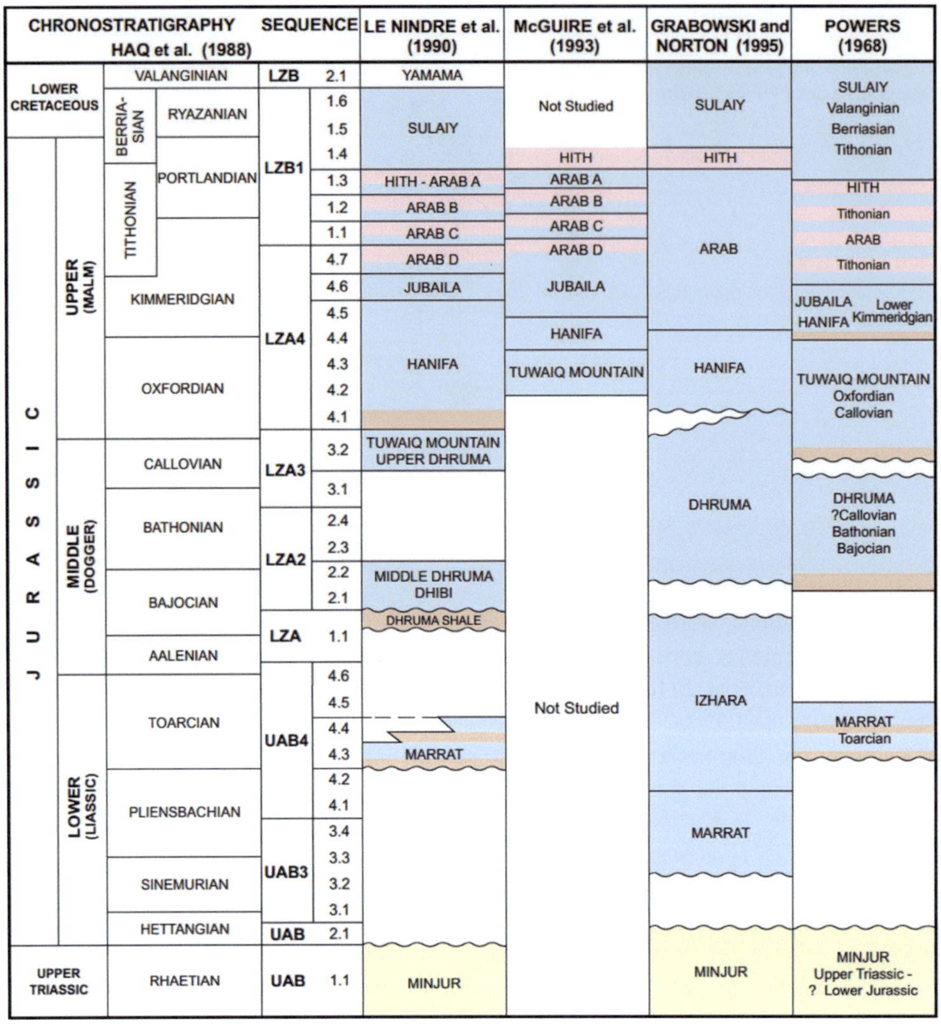

Fig. 7.3 Sequence stratigraphy of Central Saudi Arabia by Le Nindre et al. (1990), Vaslet et al. (1991), McGuire et al. (1993), and Grabowski and Norton (1995). The stratigraphic units of Powers (1968) are also shown (Al-Husseini 1997)

rise. Many parts of the Middle East and Europe exposed during the Triassic period became submerged, and the Jurassic Sea covered more frontiers. The continental configuration during the Early (Hettangian), Middle (the Toarcian), and Late (Oxfordian) Jurassic is shown in Fig. 7.4.

During the Early Jurassic period, the Arabian plate, including Qatar, was part of the passive margin of the newly opened Neotethys. It was tectonically stable and situated on the Equator. This facilitated the establishment of a broad, shallow shelf, where carbonates began accumulating on the shelf and inner platform.

Towards the end of the Early Jurassic, the climate grew more humid, considerably reducing evaporite deposition, prevalent over most of eastern Arabia during the Triassic. Tectonic differentiation, basin-rim buildups, and sea-level rise led to the development of intershelfal basins such as the Lurestan Basin in the north. By the late Middle Jurassic, the Rimthan Arch started segregating the Lurestan Basin into the Gotnia Basin in the northern present-day Arabian Gulf (southern Iraq and Kuwait) and the Arabian Basin in the west and south of the current Arabian Gulf (Al-Husseini 1997). The Qatar arch running NE–SW separated the Arabian intrashelf basin from the Rub' Al-Khali intrashelf basin (known as the Southern Gulf Basin of Qatar and the United Arab Emirates) towards the southeast (McGuire et al. 1993).

The climate gradually turned hot and dry, with high eustatic sea levels reflected by the scarcity of clastic sediment and the high carbonate production rates. During that time (lower Early Jurassic: Late Oxfordian to early Kimmeridgian), the Hanifa source rocks were deposited in the intrashelf basin of the Arabian Basin, primarily being low-energy, laminated, dark, organic-rich lime muds collected under anoxic bottom-water conditions. To the north of this intrashelf basin, high-energy, shallow-water grainstones and evaporitic peritidal sediments were deposited across the Rimthan Arch (McGuire et al. 1993). To the east, the continental margin separated the platform from an open ocean where minimal deposition occurred (Murris 1980).

During the Middle and lower Late Jurassic (the Callovian to early Oxfordian), lowstand conditions within the

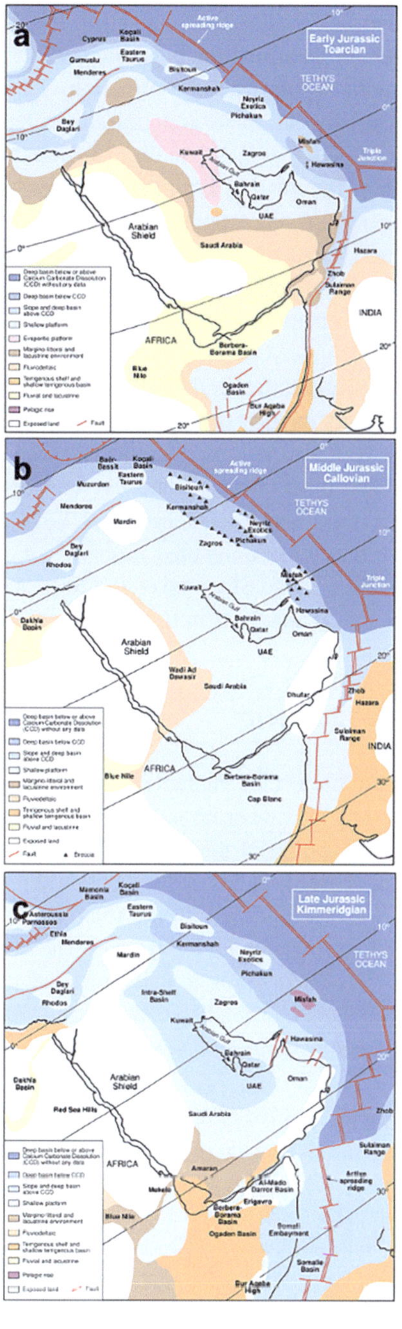

Fig. 7.4 Plate tectonic configuration of the Arabian Plate during the **a** early; **b** middle, and **c** late Jurassic (Torsvik and Cocks 2017)

intrashelf basin became restricted, creating organic-rich rocks in the anoxic intrashelf basins with favorable conditions for depositing microbial carbonate. This deposition resulted in developing of an exceptionally organic-rich and petrographically unique laminated carbonate source rock within the basin (e.g., the Gotnia, Arabian, and Rub' Al-Khali Basins), which constitute the primary source formation. Some interpretations attribute the onset of source rock deposition to anoxic conditions that developed alongside rising sea levels.

During the Late Jurassic, eastern Arabia was part of a relatively open marine Tethyan shelf characterized by rising sea levels and a greenhouse climate. Within this setting, a highly productive shallow-water carbonate factory deposited a carbonate rim (The Tuwaiq Mountain Formation in Saudi Arabia). This carbonate rim acted as an intrashelf basin, surrounded by a vast basinal area as the shallow-water carbonate progradation ceased due to an end-Callovian to early Oxfordian lowstand. Consequently, the Tuwaiq Mountain Formation rim became subaerially exposed.

Towards the late Jurassic period, the deposition of the Arab-Hith carbonate-evaporite sequences was influenced by a combination of eustatic sea-level fluctuations, a land barrier on the broad Tethys shelf, and a subtle westward tilt. It is worth noting that previous interpretations needed to fully recognize the land barrier's significance in shaping these sequences' deposition.

The total thickness of the Triassic strata all over the Arabian Plate and Iran ranges between 450 and 900 m, formed mainly of carbonates and evaporites. The average thickness of the Jurassic sediments in Qatar is around 900 m.

Wilson (2020) suggested that the following merits characterize the Jurassic Arabian Basin:

- The Arabian Intrashelf Basin represents an integral hydrocarbon system that includes rich source rocks, variable porous and permeable shallow-water carbonate reservoirs, and a regional seal.
- The Jurassic basin was found as a single basin in Abu Dhabi, Qatar, Bahrain, Oman's edge, and Saudi Arabia during the Callovian. It was situated on the expansive marine Tethyan shelf yet distanced 200–300 km from its boundary. Its formation was influenced by significant sea-level rise and minor land subsidence, with the latter being tied to Tethys's passive margin movement and rifting.
- Within this intrashelf basin, there are two zones with slightly thicker Mid and Upper Jurassic strata: the Rub' al Khali Basin to the east and the Arabian Basin to the west, as indicated by Ayres et al. (1982) and Ziegler (2001). Although numerous sources like Cole et al. (1994), Ziegler (2001), and Lindsay et al. (2006) suggest the Qatar Arch as a divider, there was only minor ridging and thinning during that period between the two zones. Both Van Buchem et al. (2014) and De Keyser and Kendall (2014) have posited that the Qatar Arch wasn't a prominent structural high and thus wasn't pivotal during the Mid- to Upper Jurassic sedimentation.
- On the marine Tethyan shelf, in a scenario of increasing sea levels and a greenhouse climate, the Tuwaiq Mountain Formation's carbonate edge was developed by a highly efficient shallow-water carbonate mechanism,

forming the intrashelf basin. The progression of shallow-water carbonate ceased at the end of the Callovian and beginning of the Oxfordian period, revealing a vast basin area encircled by the exposed Tuwaiq Mountain Formation rim.

- The global transition from the end of the Callovian to the early Oxfordian period created conditions conducive to cyanobacteria-microbial carbonate deposition within the basin. This led to the formation of a distinct laminated carbonate source rock that is rich in organic matter. In contrast, some interpretations attribute this source rock formation to the anoxic conditions arising from increased sea levels. Models by Dromart et al. (2003) and Donnadieu et al. (2011) suggested a glacial origin for this period.
- During the latter part of the Jurassic period, the deposition of the Arab-Hith carbonate-evaporite strata was impacted by sea-level variations, the formation of a terrestrial barrier on the vast Tethys shelf, and a gentle westward inclination. Wilson (2020) concluded that previous studies might have overlooked the significant role played by the land barrier.

7.3 Jurassic Stratigraphy

The Saudi stratigraphic nomenclature, as formulated by Powers (1968) is the basis for the Jurassic stratigraphy of the Gulf region. This system was modified later with the accumulation of data from the surface geologic survey and the drilling activities in the eastern part of Saudi Arabia. Since the same nomenclature system is applied in Qatar, it seems imperative to cover its major units in some detail.

In Saudi Arabia, Jurassic rocks can be categorized into two main groups: the Buraydah Group and the Shaqra Group. Within the Buraydah Group is the Minjur Sandstone, recognized as the sole Triassic-Jurassic clastic formation in the country. This was formed in a continental setting during the Late Triassic to Early Jurassic periods when tectonic activities elevated and eroded the Arabian Plate. Following the Minjur Sandstone, an unconformity signifies a major sedimentation gap during the Early Jurassic, particularly from the Hettangian to Pliensbachian phases.

On the other hand, the Shaqra Group encompasses seven sequential formations that illustrate the primary marine ebb and flow patterns from the Middle to Late Jurassic. In order of their age, these formations are: Marrat, Dhruma, Tuwaiq Mountain, Hanifa, Jubaila, Arab, and Hith. The foundational Jurassic formation, marking the beginning of marine settings in Saudi Arabia, is the Marrat Formation. This formation is characterized by its shale and limestone deposits originating from a shallow-shelf setting during the Toarcian phase. Notably, the Marrat Formation houses organic-rich black shales, which could potentially serve as sources for hydrocarbons.

The same stratigraphic nomenclature is applied in Qatar with minor modifications (Fig. 7.5). The Early Jurassic is designated as Hamlah/Marrat Formation, the Izhara and Areej Formations represent the middle Jurassic, and the upper Jurassic sediments include the Hanifa, the Jubaila/Darb, the Arab, and the Hith Formations. The Jurassic section is underlain by the Gulailah Formation and overlain by the Cretaceous Sulaiy Formation (van Buchem et al. 2014). This is the nomenclature that will be adapted in this chapter, although Ibrahim et al. (2002) and Sharland et al. (2001) and provided different schemes that appear in Figs. 7.6 and 7.7.

7.3.1 Lower Jurassic

Two major tectonic events characterized the early rifting of Gondwana during the Lower Jurassic. The first was associated with the first separation of the Indian Plate from Gondwana Africa/Arabia (Farouk et al. 2018; Grabowski and Norton 1995). This was concluded with the total separation of the Arabian Plate during the Late Jurassic. The second event began in the late-early Jurassic and impacted the northern parts of the Arabian Plate, leading to the creation of accommodation along the subsiding shelf (Farouk et al. 2018; Sharland et al. 2001).

A widespread marine transgression during the Early Jurassic (Toarcian) led to the development of shallow marine carbonate platforms in the north and eastern parts of the Arabian Plate. The southern parts of the plate were exposed and eroded, while clastic sediments accumulated in the western parts (Fig. 7.8).

7.3.1.1 Bottom of Form

The Marrat Formation (Toarcian)

The Marrat Formation was deposited on the early Jurassic epeiric sea, covering most of eastern Arabia under tropical climatic conditions (Al-Mojel et al. 2020; Murris 1980; Powers et al. 1966). The formation outcrops in the central part of Saudi Arabia formed the base of the Jabal Tuwaiq escarpments. It was described for the first time in

Fig. 7.5 Lithostratigraphic section of the Jurassic units in a typical well from the northern onshore area of Qatar (van Buchem et al. 2002)

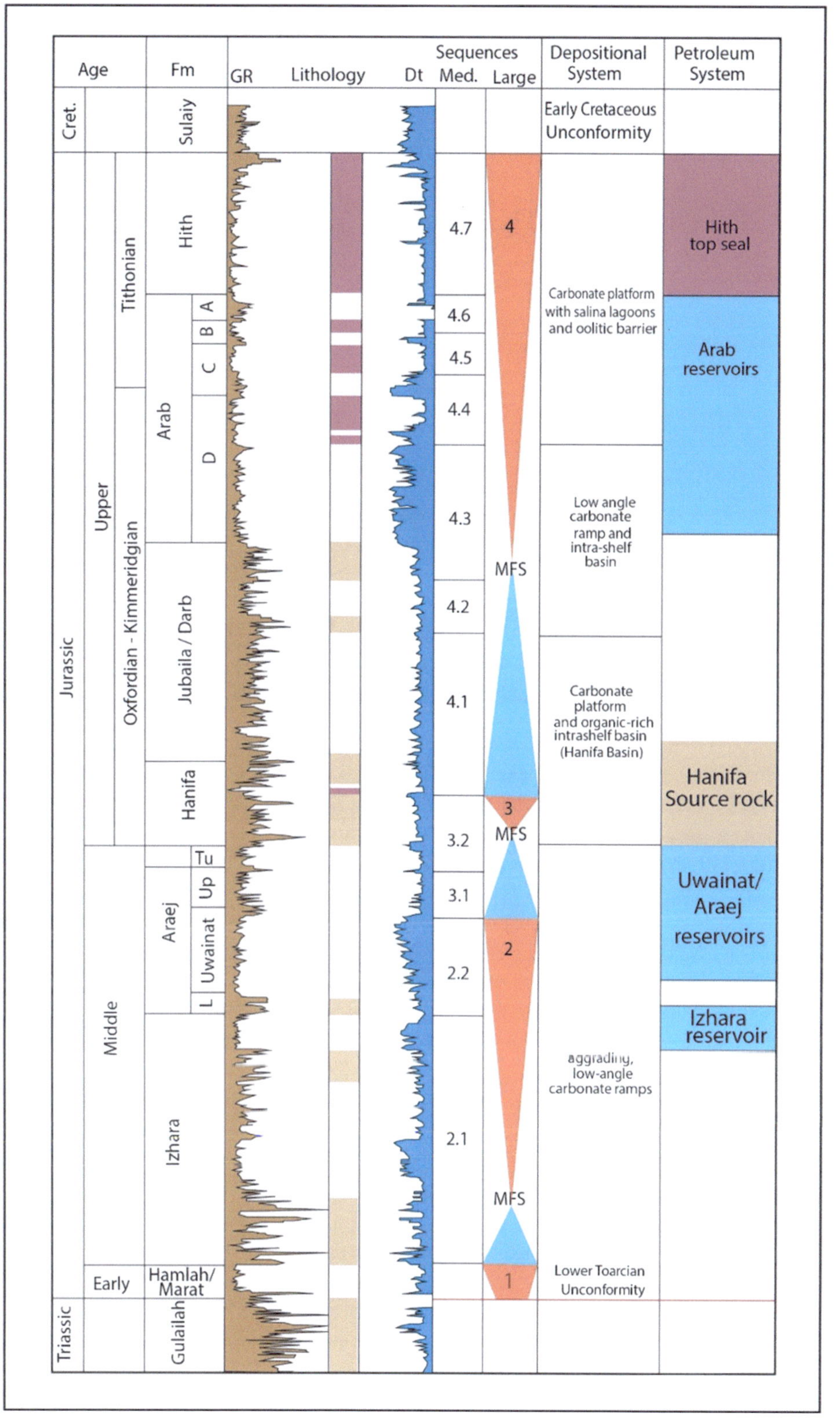

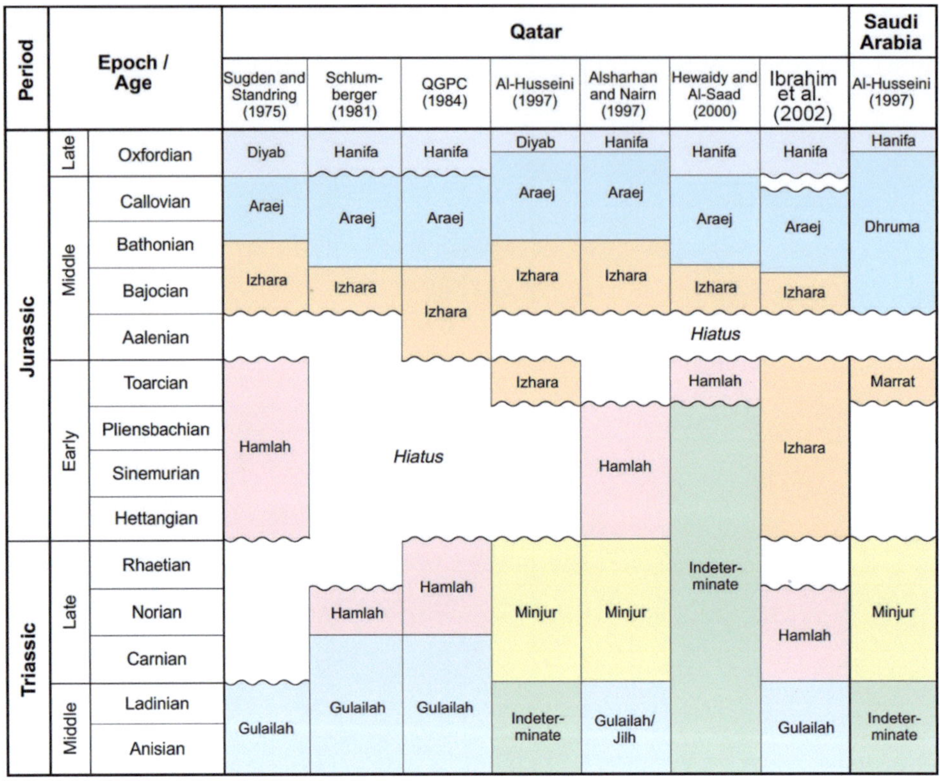

Fig. 7.6 Another suggested nomenclature for the Jurassic stratigraphy of Qatar based on pollen analysis (Ibrahim et al. 2002)

an unpublished report by Steineke in 1937 and formally published by Bramkamp and Steineke (1952) from its type section near the town of Marah. The Marrat Formation comprises 111 m of red shale between two limestone units in this area. The definition was updated by Powers et al. (1966) and Powers (1968) and a better reference section was described by McClure (in Powers et al. 1966) and Manivit et al. (1990) from Khashm Ad Dhibi area.

In central Saudi Arabia, the Marrat Formation outcrops in discontinuous areas that extend for more than 650 km in an arc-shaped belt. The formation ranges in thickness between 111 and 142 cm. It is overlain by the Middle Jurassic Dhruma Formation and underlain by the Late Triassic Minjur Formation. The contact between the above and below formations has been reported to be unconformable (Fig. 7.9).

The Marrat Formation has been divided into three informal units, the Lower, Middle, and Upper Marrat Formation, by Powers et al. (1966) and modified by Manivit et al. (1990). The boundaries between these units are mostly geomorphic and lithologic features visible in the field as three cuestas (Al-Mojel et al. 2020).

Lower Marrat Unit (47 m)

This unit is marked at the lower part by an iron-stained unconformity surface separating the Marrat Formation from the underlying Late Triassic Minjur Formation. This unit comprises shale, sandstone, and thin dolomitic beds. The rare fauna indicates deposition under continental to intertidal-subtidal depositional systems.

Middle Marrat Unit (40 m)

The lower part shows an increase in the carbonate components. The unit is formed of fossiliferous and dolomitic limestone followed by red shale and siltstone free of fossils. The upper part is made up of calcareous shale. The rocks were deposited under intertidal to subtidal environments.

Upper Marrat Unit (39 m)

This unit comprises bioturbated, lagoonal limestone containing ammonites overlain by a gypsum bed (7 m thick). Powers (1968) considered this gypsum bed as the base of the Middle Jurassic Dhruma Formation. In the subsurface, the Upper Marrat is the hydrocarbon reservoir unit of the formation.

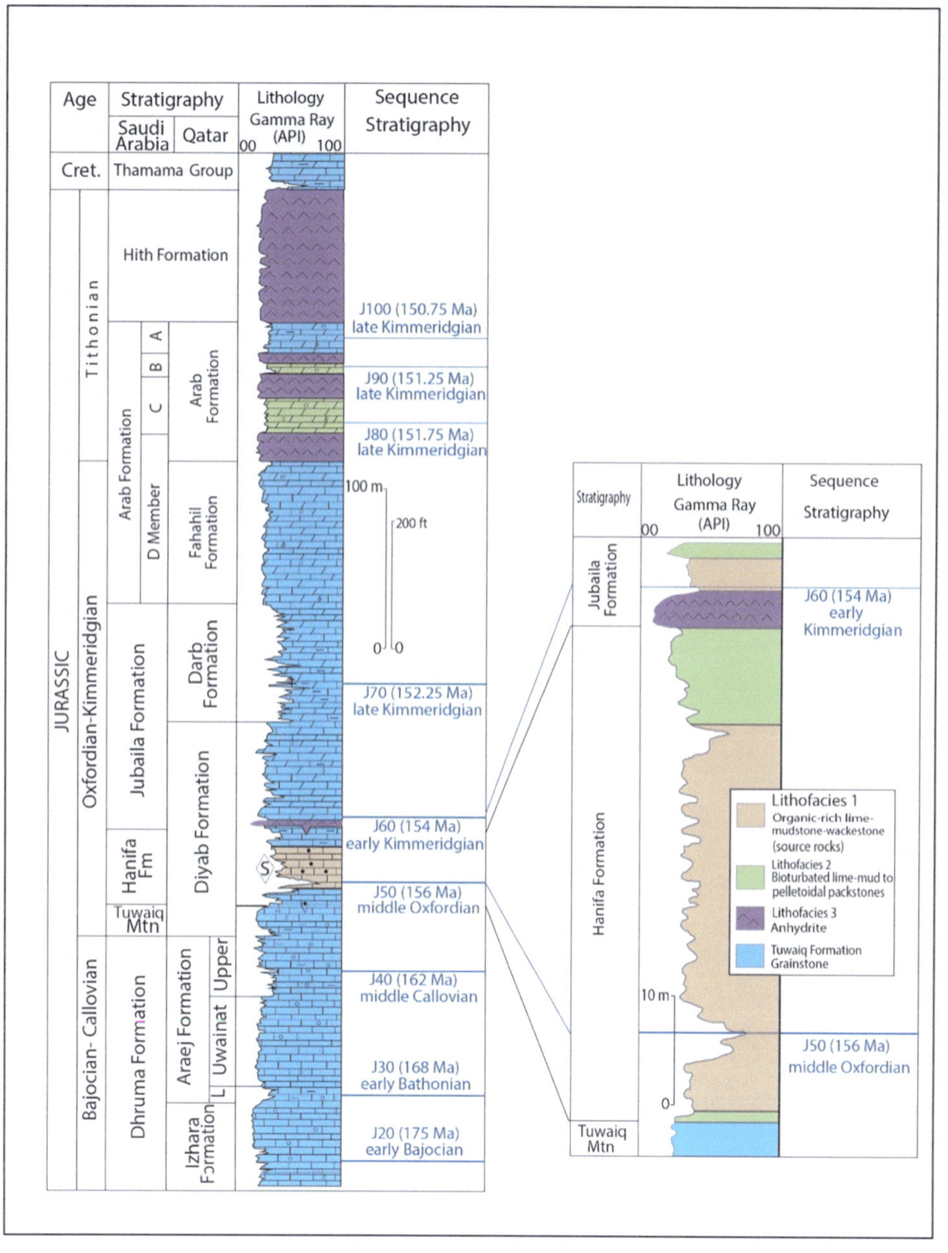

Fig. 7.7 Sequence stratigraphy of the Jurassic system in Qatar, modified from Droste (1990)

A detailed sequence stratigraphy and facies analysis of the formation was carried out by Farouk et al. (2018) and Al-Mojel et al. (2020). Accordingly, the formation has been divided into a set of facies starting from the bottom and including floodplain paleosol facies that consist of red claystone with hematite and rare goethite, supratidal to upper intertidal facies formed of dolomitic quartz arenite and bird's eye dolomicrite, restricted lagoon gypsum-carbonate facies which comprise bedded gypsum intercalated with dolomitic limestone beds, restricted clastic/carbonate lagoonal facies that consist of quartz lime mud and mudstone, calcareous ferruginous quartz arenite which is formed of fine to medium-grained quartz cemented in calcareous and iron-oxide materials, shallow subtidal-lagoonal facies formed of thin-bedded limestone with abundant trace fossils, and shallow subtidal carbonate ramp which comprises of thin to medium-bedded fossiliferous limestone with some ammonites. The detailed stratigraphic subdivisions and lithologies are shown in Fig. 7.10.

The Marrat sediments resulted from two interacting depositional systems: a siliciclastic-prone proximal domain and a carbonate-prone distal domain (Al-Mojel et al. 2020; Farouk et al. 2018). The siliciclastic system consists of a range of alluvial depositional environments with wide coastal plains and inner platforms where shale sediments were deposited. Low-energy tidal flats with occasional wave-dominated environments characterized these clastic-dominated environments. In the inner platforms, high-energy sandstone

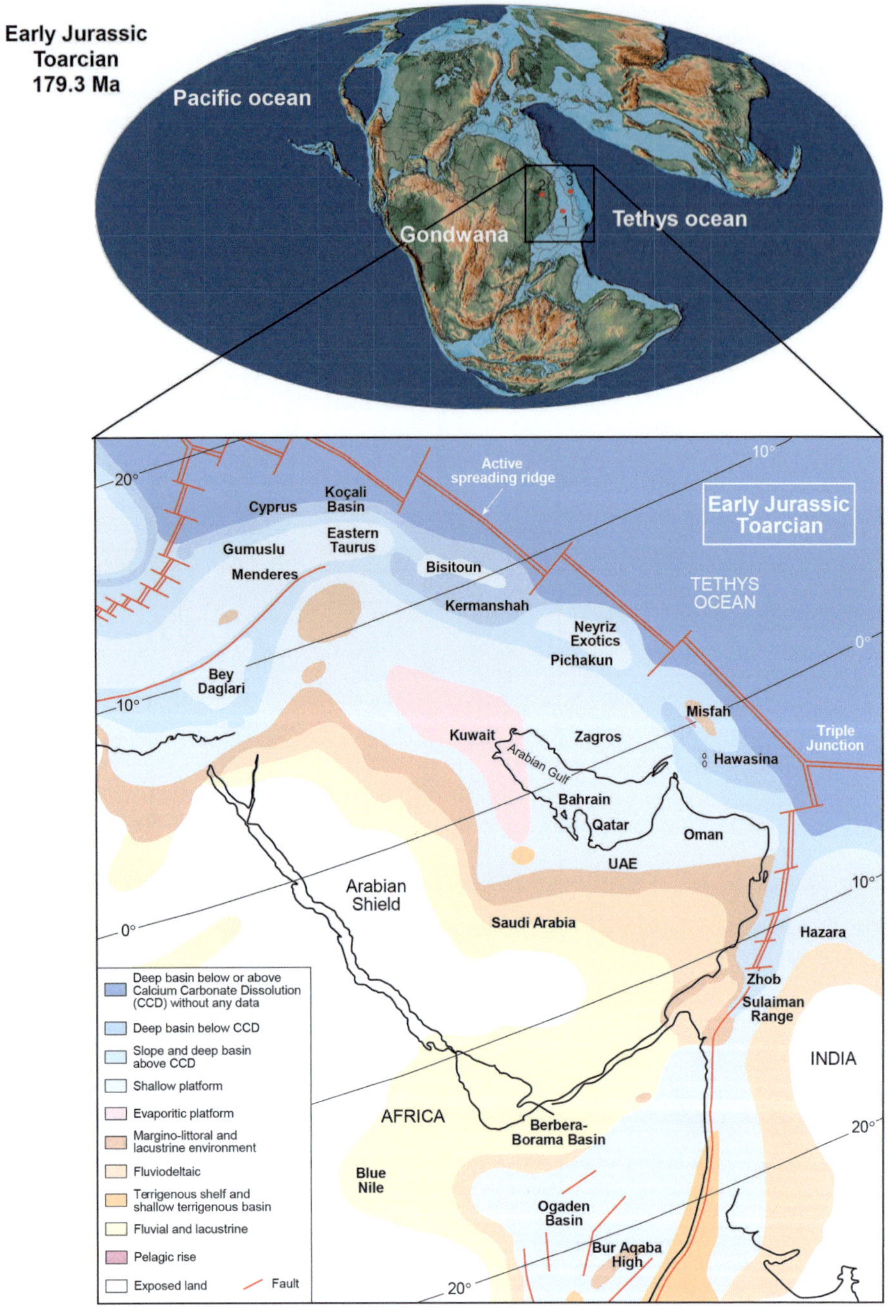

Fig. 7.8 Paleogeographic map of the Arabian Plate during the Early Jurassic (Toarcian) showing the primary depositional environment (Farouk et al. 2018)

belts formed due to repeated transgressive cycles (mostly TST and MFS).

The carbonate system comprises inner platforms depositing shale and lime mud carbonates. The shale sediments were occasionally influenced by siliciclastic input, whereas the lime mud carbonates represent the normal, open marine environment. The sediment nature and the present fossils suggest that the depth of the Marrat Formation's depositional environments is just a few meters. A tentative depositional model of the formation appears in Fig. 7.11.

7.3.1.2 The Marrat Formation in Qatar

The Marrat Formation has been described as the Hamla Formation (Sugden and Standring 1975) from the well Dukhan 65 between the drilled depths 2367 and 2436.5 m with a total thickness of 70 m. It consists, from the bottom, of 8 m of sandy marls with beds of shale; dolomite interbedded with dense limestone and anhydrite (41 m thick); dolomite with anhydrite streaks interbedded with green shale and grey marl

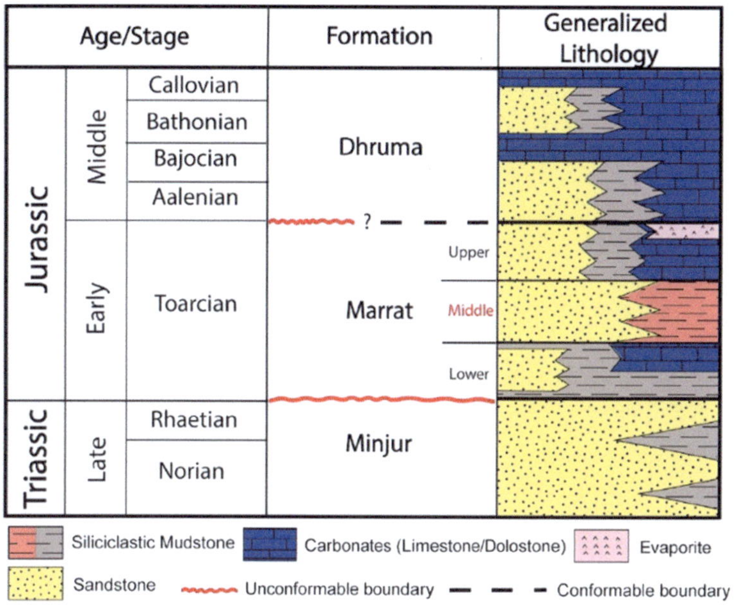

Fig. 7.9 Stratigraphic status and age of the Marrat Formation in central Saudi Arabia (Al-Hussaini et al. 2021)

(9 m thick); shale unit with marl, glauconitic sandstone, and dolomite (6 m thick) at the top unit is 5 m of dolomite, dense to porous, saccharoidal with stringers and nodules of anhydrite becoming more argillaceous downward. The lower boundary is unconformable with the underlying Gulailah Formation and conformable with the overlying Izhara Formation. Sugden and Standring (1975) suggested that the Hamlah Formation is encountered in Dukhan Field only and may be represented in part or all by the Gulailah Formation in the offshore area of Qatar. From the offshore area, van Buchem et al. (2002) reported that the Early Jurassic succession consists of the Marrat or Hamlah Formations and the section consists of Toarcian sandy marl, shale, glauconitic sandstone, and dolomite with stringers of anhydrite. They suggested that this section is bounded at the bottom by the Late Triassic/Early Jurassic unconformity and at the top by the Late Turonian unconformity (Fig. 7.12).

Based on pollen analysis, Ibrahim et al. (2002) suggested that the Marrat Formation of Saudi Arabia is equivalent to the lower part of the Izhara Formation (Fig. 7.6). This conclusion is aligned with the nomenclature suggested by de Matos et al. (1995) for the same succession in Abu Dhabi. However, the section assigned to the Marrat Formation in Qatar by Ibrahim et al. (2002) is formed mainly of clastics with some anhydrite, which does not represent the entire lithologic components of the Marrat Formation as described earlier in this chapter from Saudi Arabia. A detailed correlation between the Jurassic succession in Qatar and Saudi Arabia requires further investigation. Achieving regional correlation with the Saudi nomenclature (where the Jurassic is far more studied) will bring many benefits to understanding better the Jurassic strata of Qatar. In this work, the nomenclature suggested by van Bucham et al. (2002) will be adapted, and the Early Jurassic sediments of Qatar will be assigned to the Marrat or Hamla Formation.

7.3.2 The Middle Jurassic

The early Jurassic (Toarcian) transgression was followed by the Aalenian Hiatus that continued from the late Toarcian to the Aalenian. The gypsum layer at the top of the Marrat Formation was related to this break by Le Nindre et al. (1990).

A new transgression occurred during the Middle Jurassic (Bajocian-Bathonian), and carbonate resumed deposition over most parts of the Arabian Plate in the shallow marine shelf. The clastic deposition continued to the western and southern parts of the plate. In the north, the Gotnia Basin developed in Kuwait and south Iraq, where a mixed siliciclastic and carbonate formed (Fig. 7.13).

As a result of this transgression, the Dhruma Formation was deposited in Saudi Arabia and Bahrain (Al-Hussaini et al. 2021). A mixture of calcareous shale with limestone laminations was described as the Dhruma Formation in Kuwait. According to Al-Hussaini et al. (2021), this section is probably equivalent to the lower part of the Saudi Dhruma Formation. In Qatar, these sediments were assigned to the Izhara Formation. This formation is partially correlatable with the Saudi Dhruma Formation. The Araej Formation is of Bathonian-Callovian age and represents deeper water conditions (Sugden and Standring 1975).

7.3.2.1 The Izhara Formation

The middle Jurassic Izhara Formation was described by Sugden and Standring (1975) from the well Kharaib 1,

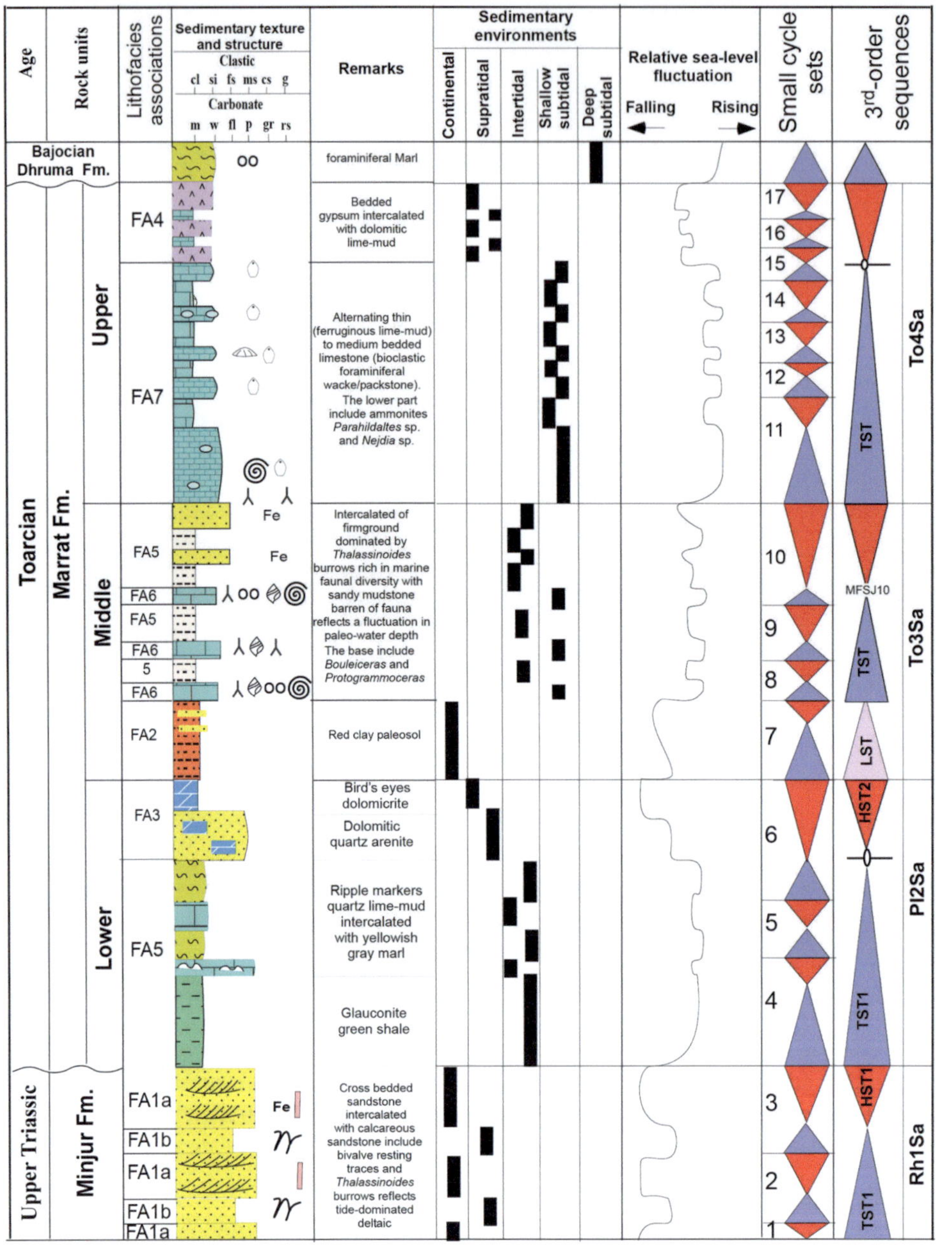

Fig. 7.10 Main lithologic subdivisions and depositional environments of the Marrat Formation in its type locality (Farouk et al. 2018)

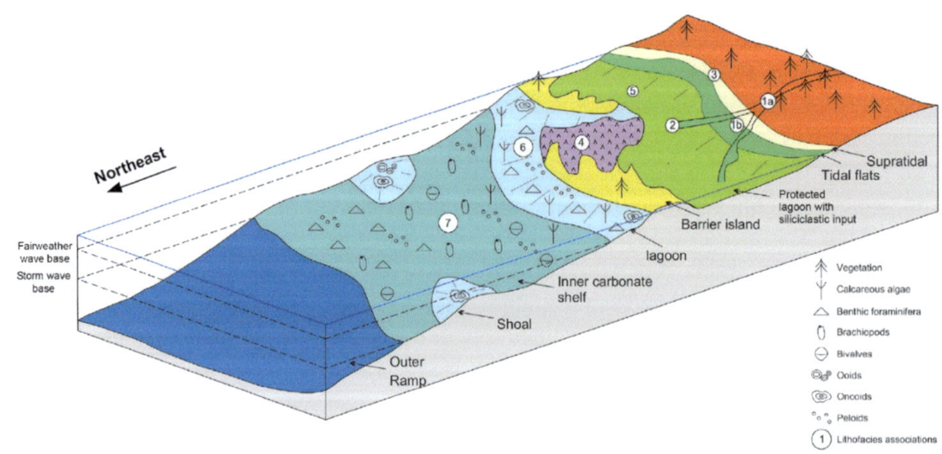

Fig. 7.11 Depositional environments of the Marrat Formation in central Saudi Arabia (Farouk et al. 2018)

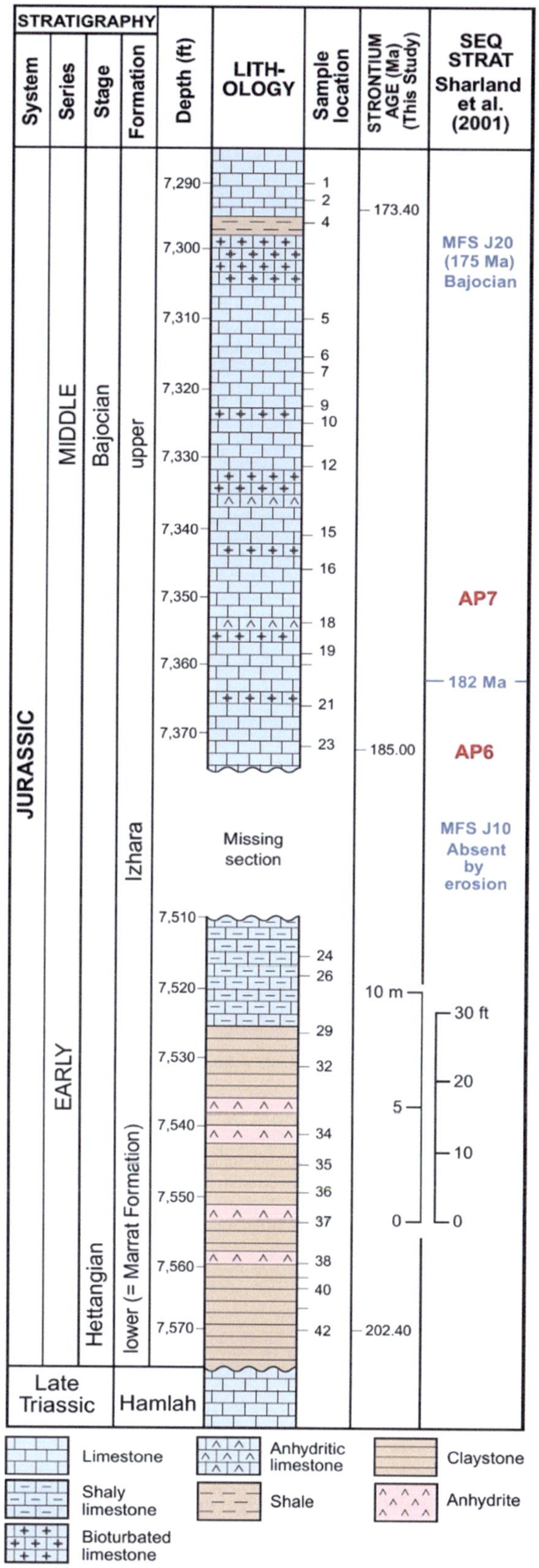

Fig. 7.12 Lithostratigraphy of the Early-Middle Jurassic sediments from a well in Qatar (Ibrahim et al. 2002)

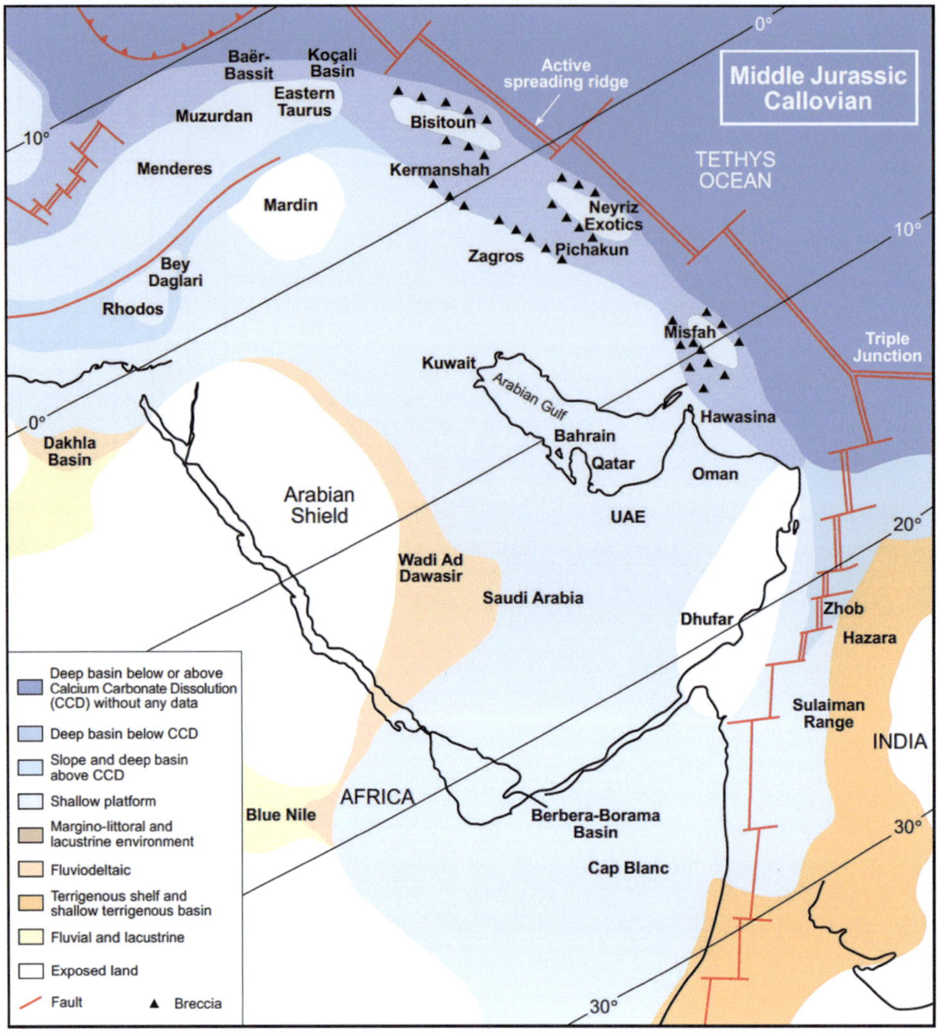

Fig. 7.13 Paleogeographic map of the Arabian Plate during the Middle Jurassic (Callovian) showing the primary depositional environment (Al-Husseini 1997)

between the drilled depths of 2317 and 2454 m, with a total thickness of 137 m. It consists of five different lithologies, from bottom, limestone, and dolomite with some argillaceous beds at the upper and lower parts (40 m thick); dolomite, becoming more porous and saccharoidal toward the top (10.6 m); limestone, dolomitic (6 m); limestone with streaks formed of pelletal limestone, pyrite-stained, argillaceous occasionally becoming dolomitic at the bottom with marl and shale (56 m).

Ibrahim et al. (2002) studied the formation from some onshore wells and reported that it comprises argillaceous limestone at the top and bottom with a middle unit of dolomitic limestone. The limestone is intercalated with laminations of shale, marl, claystone, and anhydrite at the lower part. The limestone is grain-supported and formed of wackestone/packstone with oncoidal grainstone unit in the middle. They suggested that the Izhara Formation extends from the Early to Middle Jurassic, with a major hiatus occurring in the middle of the formation (Al-Husseini 1997). The upper contact with the Gulailah Formation is unconformable, but its lower contact with the Araej Formation is conformable (Fig. 7.14).

The Izhara Formation is equivalent to the Lower Dhruma Formation in Saudi Arabia. The Middle Jurassic Dhruma Formation outcrops in belts parallel to the Marrat Formation in central Saudi Arabia. It has conformable contact with the underlying Marrat Formation and is overlain by the Tuwaiq Mountain Limestone Formation. It represents a good reservoir and source rocks in some parts of the country (Ayres et al. 1982; El-Sorogy et al. 2017; Murris 1980). The formation is assigned to the Bajocian-Callovian based on paleontological data. The correlation between the Dhruma Formation in Saudi Arabia with similar deposits in Qatar and Oman is shown in Fig. 7.15.

The formation has been described from the United Arab Emirates by Alsharhan (1989). It consists of argillaceous lime mudstone with bioclastic, peloidal limestone deposited under shallow marine conditions.

Fig. 7.14 Lithostratigraphy of the Early-Middle Jurassic Izhara Formation in the subsurface of Qatar (Ibrahim et al. 2002)

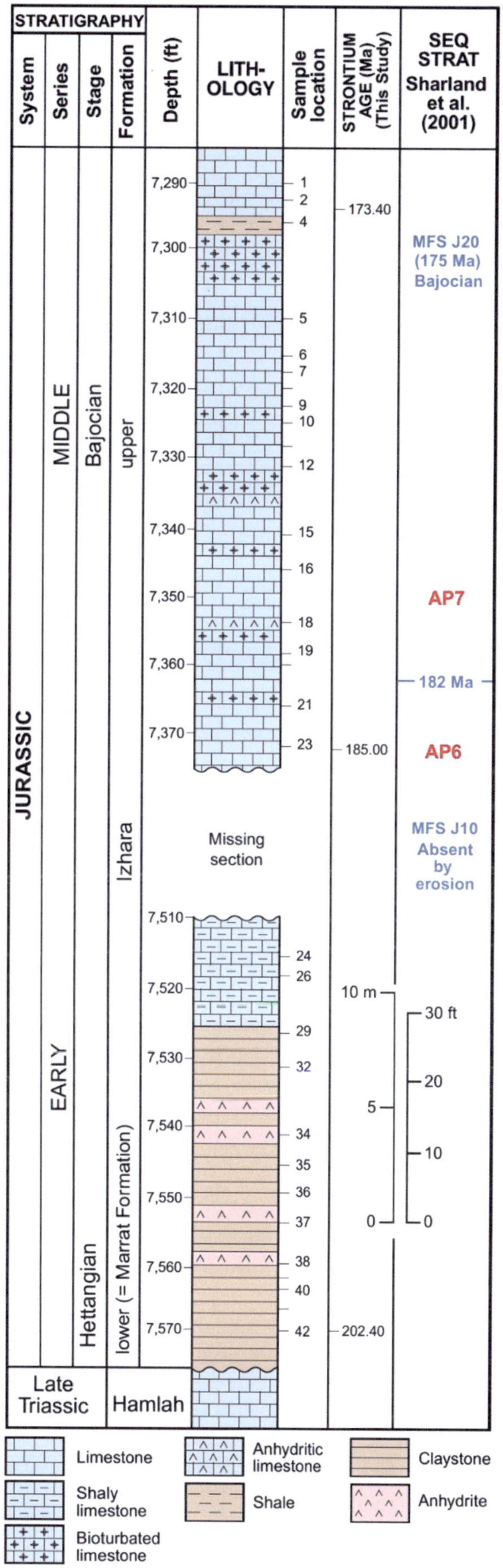

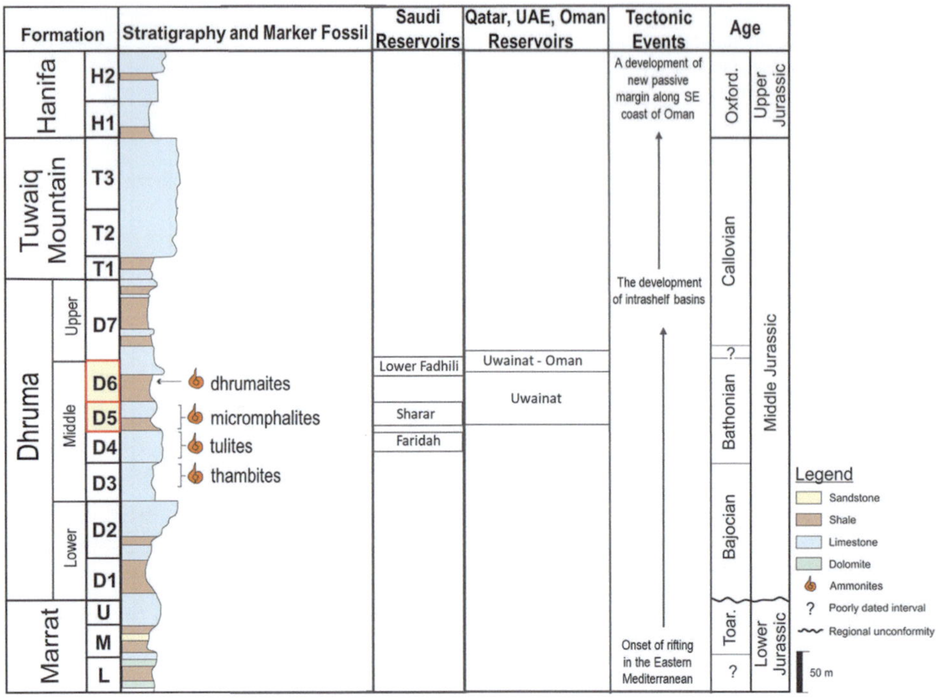

Fig. 7.15 Correlation of the Dhruma Formation in Saudi Arabia with similar strata in Qatar and Oman (AlZoukani et al. 2022)

Fig. 7.16 Correlation of the Upper and Middle Jurassic strata in the Arabian Plate (Alsharhan and Whittle 1995)

Age		Upper Jurassic		Middle Jurassic			
		Oxfordian		Bajocian	Bathonian		Callovian
QATAR	Formation	Hanifa		Izhara	Aarej		
	Member				Lower Aarej	Uwainat	Upper Aarej
SAUDI ARABIA	Formation	Hanifa	Tuwaiq Mountain	Dhruma			
	Member			Lower Dhruma	Middle Dhruma	Upper Dhruma	
						Atash	Hisyan
UAE	Formation	Diyab		Izhara	Aarej		
	Member				Lower Aarej	Uwainat	Upper Aarej

7.3.2.2 The Araej Formation

The Araej Formation was introduced to the Jurassic stratigraphy of Qatar by Walter Sugden in an unpublished report transcribed by Sugden and Standring (1975) from Qatar Petroleum Company well Kharaib No. 1. The formation was encountered between drilled depths of 2137 and 2316 m, with a total thickness of 180 m. It is equivalent to the Upper and Middle Dhruma Formation in Saudi Arabia (Powers et al. 1966). Based on this correlation, the formation was assigned to the Callovian-Bathonian. The underlying formation is the Izhara Formation with a conformable contact (see above). The overlying formation is the Diyab Formation, with a probably disconformable contact represented by a "sharp junction between a coarse, pyritic, *trocholina*-bearing grainstone of the Araej Formation with argillaceous, lime mudstone and fine pelletal limestone of the basal Diyab" (Sugden and Standring 1975).

In its type locality, the Araej Formation comprises wackestone to mudstone interbedded with pelletal packstone and grainstone, where the pellets are stained with pyrite (28.6

Fig. 7.17 Detailed lithology of the Araej Formation in Qatar (Al-Saad et al. 1992)

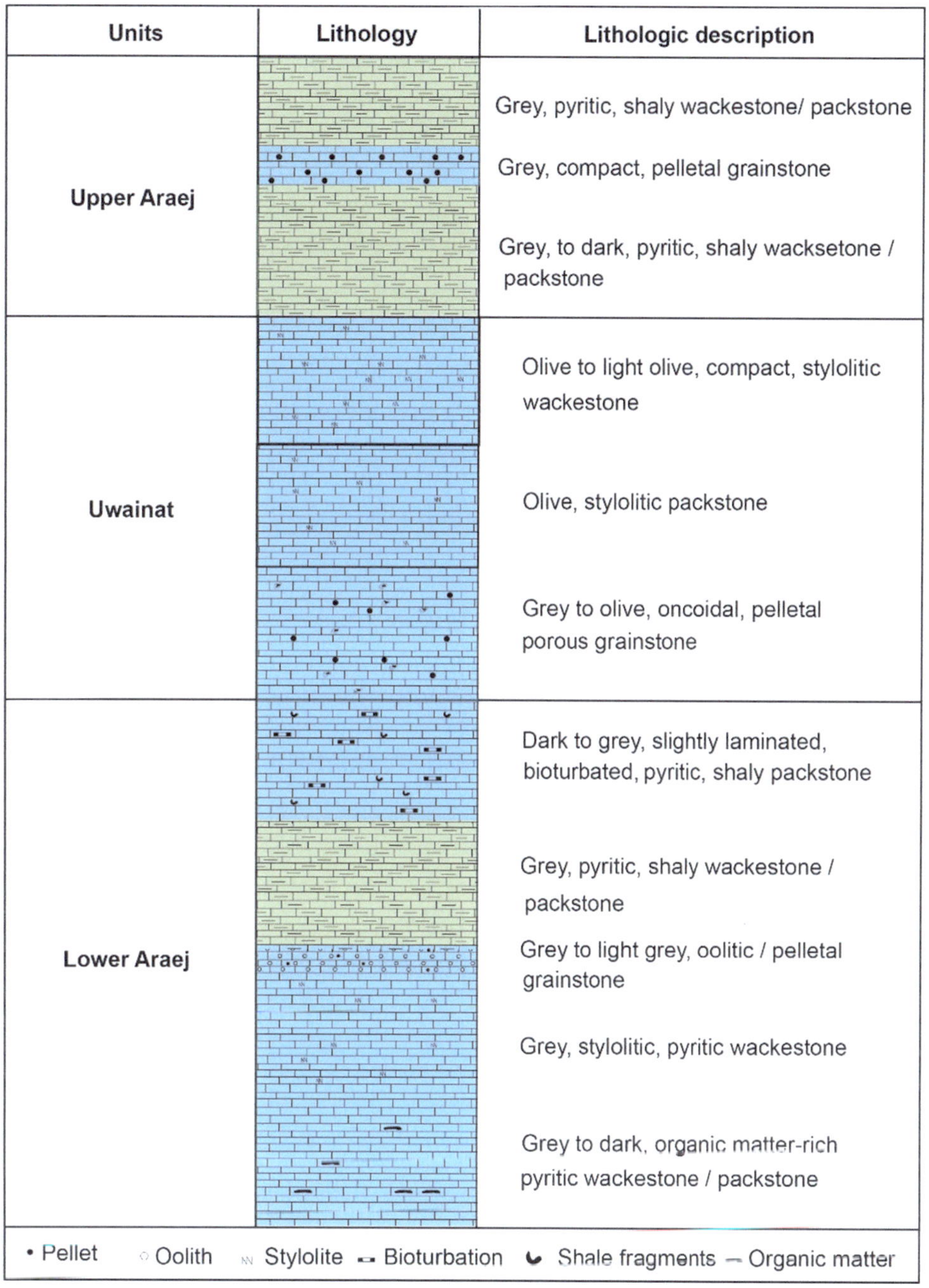

m thick) followed by interbedded pelletal grainstone and packstone, wackestone, and mudstone with some cemented oolitic limestone (59 m thick). The third unit consists of coarse, bioclastic, and pelletal grainstone and packstone with pyrite staining (61 m thick). The top unit is formed of foraminifera-bearing, pelletal grainstone, and packstone interbedded with wackestone and mudstone with some argillaceous horizons (31 m thick).

Both the Izhara and Araej formations are equivalent to the Dhruma Formation in Saudi Arabia (Fig. 7.16). The Dhruma Formation is divided into three parts: lower, middle, and upper. The Upper Dhruma is divided into two members Atash and Hisyan. In the United Arab Emirates, the same Qatari stratigraphic nomenclature is used (Alsharhan and Whittle 1995).

In Qatar, the Araej Formation has been divided into the following members:

The Lower Araej Member

This is the thickest member of the formation, ranging in thickness between 94 and 117 m. The lithologies of the Lower Araej Member have been grouped into three suits from top to bottom. The first lithology consists of fractured packstone with low fossil diversity. It contains rare

Fig. 7.18 Log signature, cycles, units, depositional environments, and insoluble residues of the Lower Araej Member (Al-Saad et al. 1992)

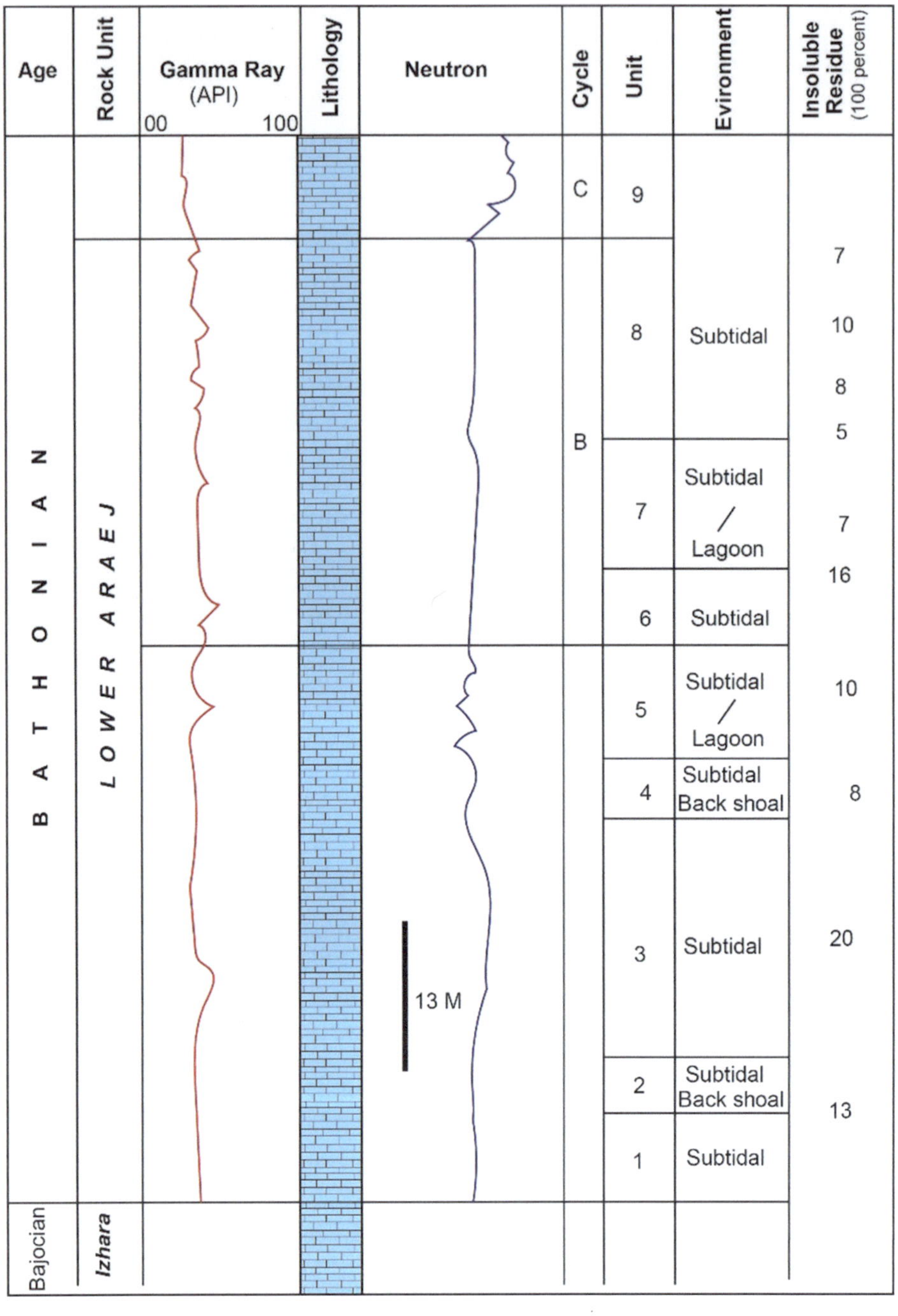

foraminifera and fine quartz grains with some anhydrite that has been dissolved and replaced by calcite cement. It also has some bitumen deposits. The second lithology comprises argillaceous wackestone with abundant dolomite and equant calcite cement. Most of the skeletal grains are extensively micritized or partially replaced by anhydrite cement. The third lithology is formed of coarse, peloidal, and oncoidal dolomitized grainstone. Fossils include Dasycladacean and blue-green algae. There is some intergranular saddle dolomite.

This member is generally a repetitive succession of wackestone to packstone facies alternating with thin horizons of packstone and grainstone. The presence of terrigenous, fine quartz grains, and anhydrite suggest deposition under restricted subtidal to lagoonal conditions (Al-Saad et al. 2002). The same lithologies have been described in the United Arab Emirates, (Alsharhan and Whittle 1995). This member is separated with clear contact from the overlying Uwainat Member (Figs. 7.17 and 7.18).

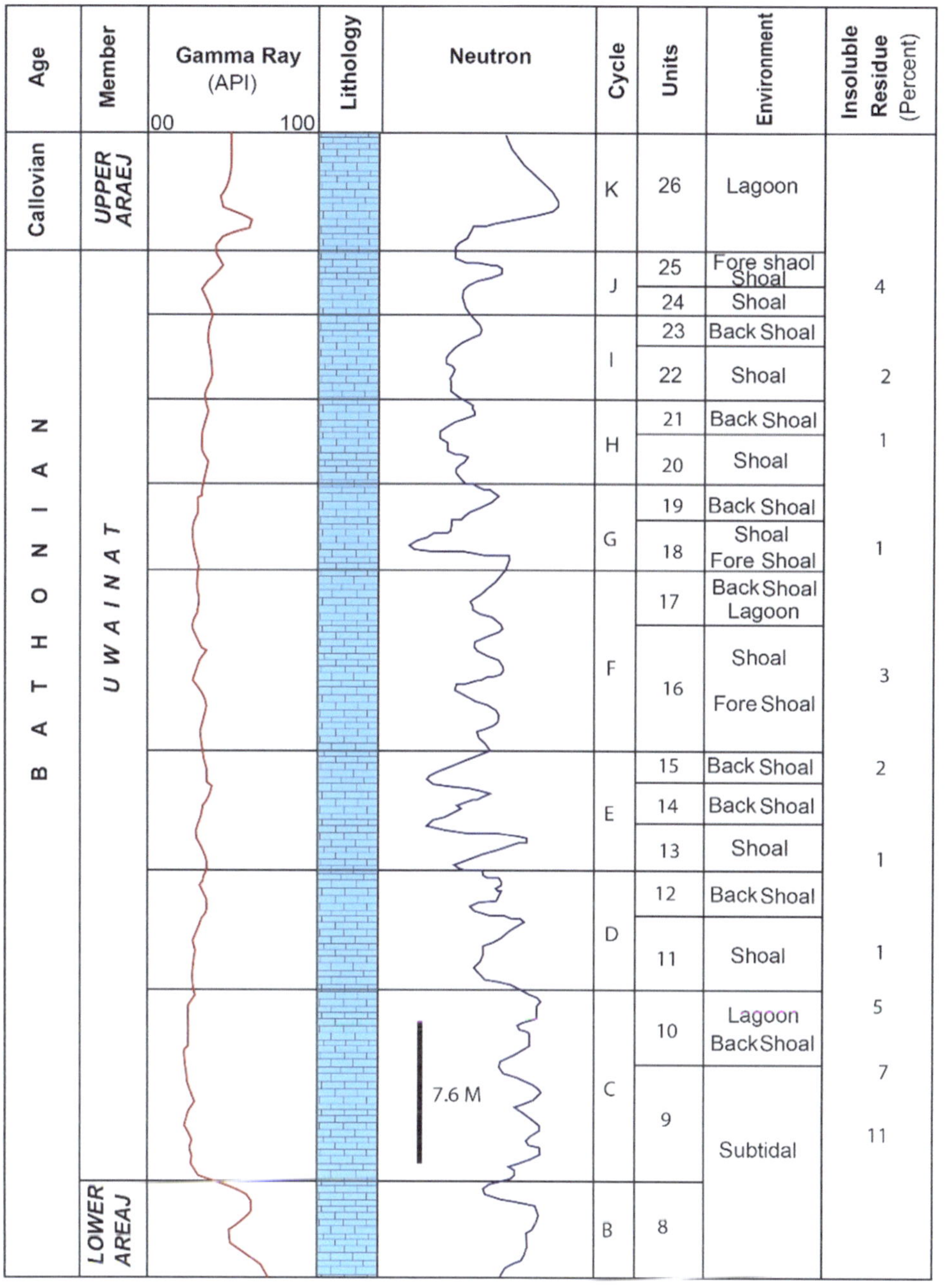

Fig. 7.19 Log signature, cycles, units, depositional environments, and insoluble residues of the Uwainat Member (Al-Saad et al. 1992)

The Uwainat Member

This is the middle member of the Araej Formation and the most important due to its significant oil and gas accumulations. The name is derived from that of an area near the Dukhan oil field. The thickness of this unit ranges between 50 and 55 m, which is fairly uniform. It is separated from the Upper and Lower Araej Members by relatively consolidated rocks that act as a seal for this reservoir unit. The Uwainat Member is generally formed of grey and olive lime mudstone alternating with grainstone stained by bitumen. The contacts between these cycles are reflected as gradual changes in color and texture (grain size). There are some sharp changes, particularly in the grainstone units. These units are rich in skeletal grains such as benthic foraminifera. Other skeletal grains include mollusks, echinoderms, bryozoans, corals, and algae. Some of the grains are strongly micritized. There are some veins and fissures filled with ferroan calcite and anhydrite cement. Al-Saad et al. (1992) reported that insoluble residue analyses suggest that at the top and lower parts of the Uwainat Member, clay minerals are the most abundant insoluble components with few scattered quartz grains. The insoluble residues range between 1 and 4% at the upper part and 7%-11% at the lower part. Quartz grains are probably detrital and

Fig. 7.20 Log signature, cycles, units, depositional environments, and insoluble residues of the Upper Araej Member Jurassic (Al-Saad et al. 1992)

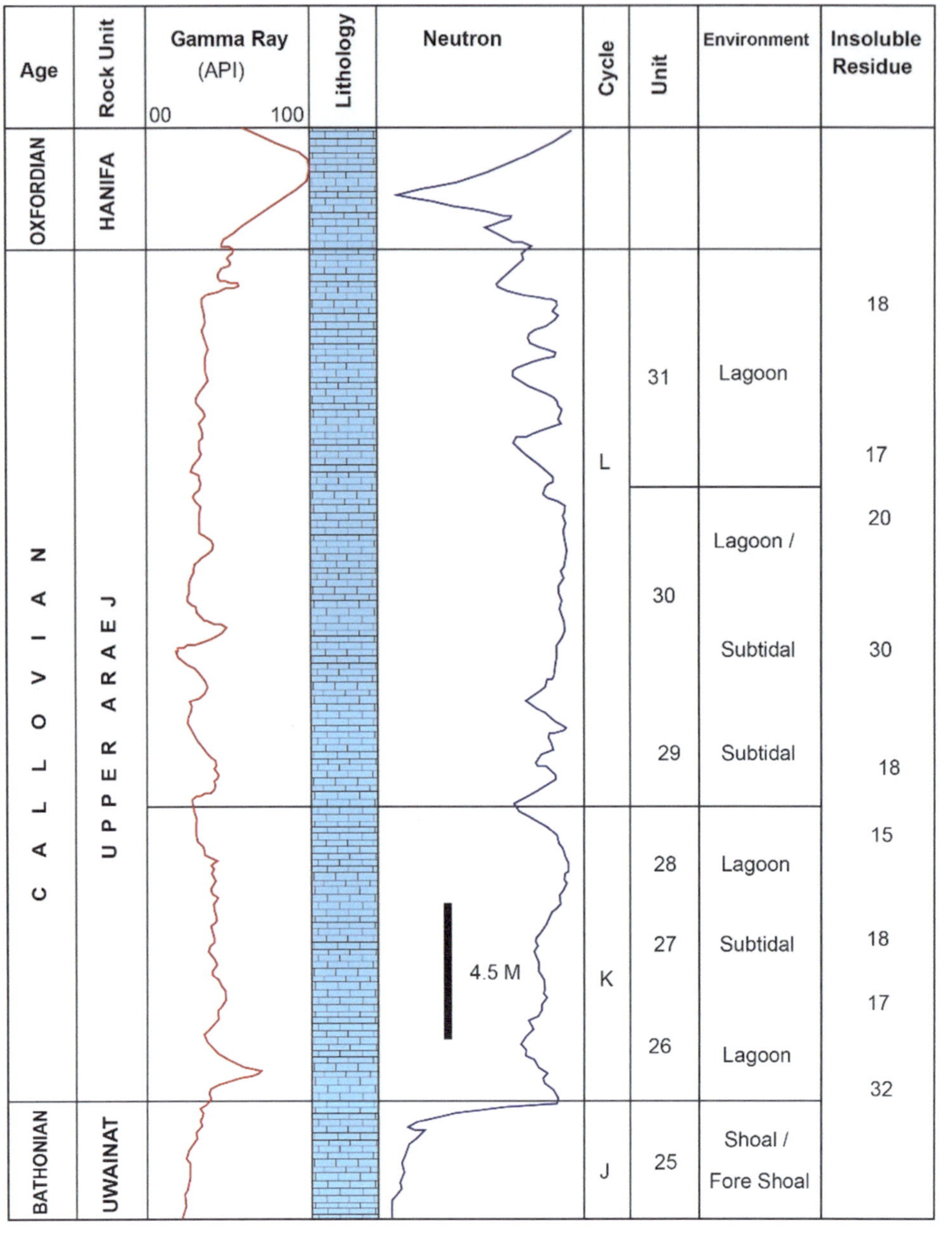

were transported from the nearby dryland. The clay contents are reflected on the Gamma-Ray logs, appearing as narrow peaks at the upper and lower parts of the rock successions (Fig. 7.19).

Al-Saad et al. (1992) divided the rocks of the Uwainat Member into seventeen different facies. Some facies were integrated into several units. From the top, the Uwainat comprises poorly sorted wackestone and packstone with some aggregates of grapestone. Skeletal grains include foraminifera, pelecypods, echinoderms, ostracod fragments, shell debris, rare corals, and sponge spicules. This unit is followed by coarse oncoidal grainstone that gradually changes into peloidal facies. These oncoids are probably remains of green algae. There is some echinoderm debris. Most of the rocks are dolomitized. The third unit is formed of stylolitic, bioclastic packstone. Skeletal grains include abundant foraminifera, pelecypods with rare echinoderm debris, and coral fragments. The fourth unit is formed of peloidal and oncoidal grainstone with remains of dasycladacean algae. Fossils are scarce and include some echinoderm debris with syntaxial rim cement. There are some sparry calcite cement and saddle dolomites. The fifth unit consists of bioturbated, micritized packstone. There are abundant skeletal grains, including dasycladacean algae, blue-green algae, and echinoderm debris. The sixth unit is made up of stylolitic, foraminifera-bearing grainstone. Dolomite fills the inner structures of the oncoid grains. The seventh unit is composed of burrowed wackestone that gradually changes to

Fig. 7.21 Tentative depositional model of the Araej Formation (Alsharhan and Whittle 1995)

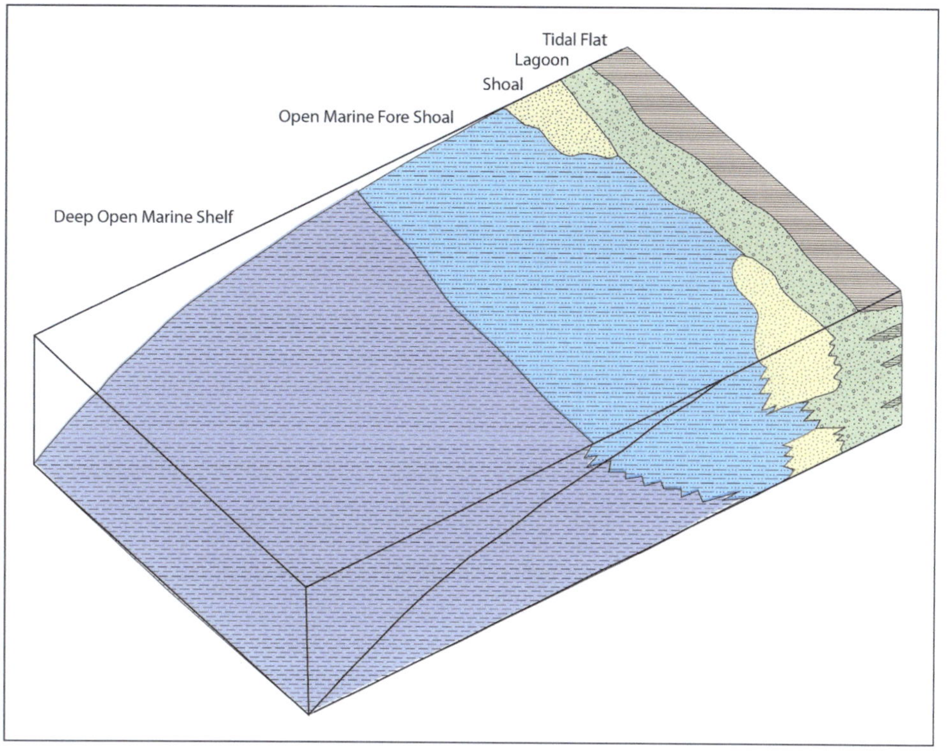

packstone. There are abundant remains of foraminifera and blue-green algae. Most of these grains are replaced by either dolomite or calcite cement.

Sudarsana et al. (2009) described the lithologic characteristics of the Uwainat Member from Bul Hanine and Maydan Mahzam oilfields offshore Qatar. The thickness of the unit in the subsurface is around 55 m, which is uniform all over the area. The Uwainat Member consists of 1.5–5 m thick sedimentary cycles of impermeable wackestone and permeable packstone and grainstone. The major grains are pellets, peloids, benthic foraminifera, echinoderms, and shell fragments. Corals appear in the upper part and oncoids in the lower part. Most of the skeletal grains are extensively micritized. The sediments underwent considerable diagenetic alterations, including compaction, leaching, dolomitization, cementation, and stylolitization.

The Upper Araej Member

This member is the thinnest compared to the other two members and is also characterized by low variability in terms of features and contents. The thickness is between 33 and 38 m. There is a sharp contact between the relatively clean, olive-colored packstone of the underlying Uwainat Member and this member's dark grey, shaly, pyritic packstone as reflected on the Gamma-Ray log (Fig. 7.20).

There is also an abrupt change in lithology between the grey packstone and grainstone of this member and the overlying black mudstone of the Hanifa Formation. This contact is considered unconformable and represents the time boundary between the Middle and Upper Jurassic (Al-Saad et al. 1992).

The Upper Araej member is formed of two minor cycles shoaling upward, where each cycle starts with a peloidal or oolitic grainstone and packstone, changing gradually into wackestone. The thickness of each cycle is around 14 m. Jurassic Al-Saad et al. (1992) assembled the facies of this member into three groups. The first group comprises reddish-to-black wackestone and packstone of fine to medium grains of pelecypod fragments with low diversity. There are abundant pyrite and anhydrite in these sediments. This group is present in the upper and lower parts of this member. The second group is represented by a thin horizon of dark, fossiliferous packstone. The main fauna are echinoderms, algal debris, and benthic foraminifera. The third group consists of oolitic and, pelletal, and skeletal packstone with a local concentration of grainstone. The main biological components are foraminifera and coral fragments. Some of these are replaced by calcite or anhydrite cement.

7.3.2.3 Depositional Environments

The Araej Formation showcases a sequence typical of shallow-shelf carbonate sediments. These deposits form a consistent sheet-like pattern, marked by sedimentary features like bioturbation, burrows, and varying thicknesses. The life forms found in the sediments point to standard marine

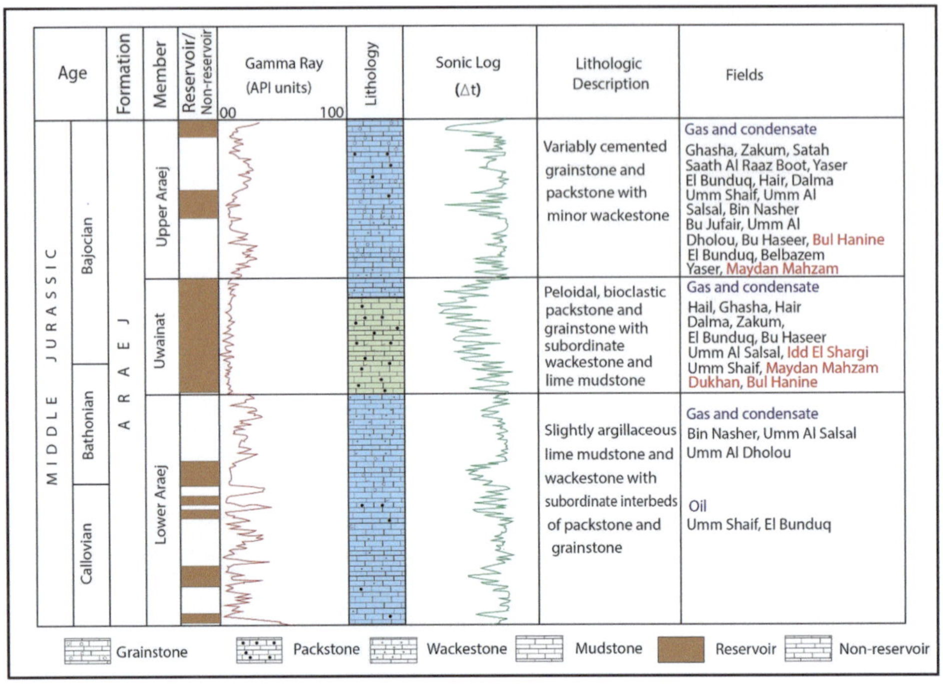

Fig. 7.22 Detailed lithologic description of the Araej Formation in the United Arab Emirates (Alsharhan and Whittle 1995)

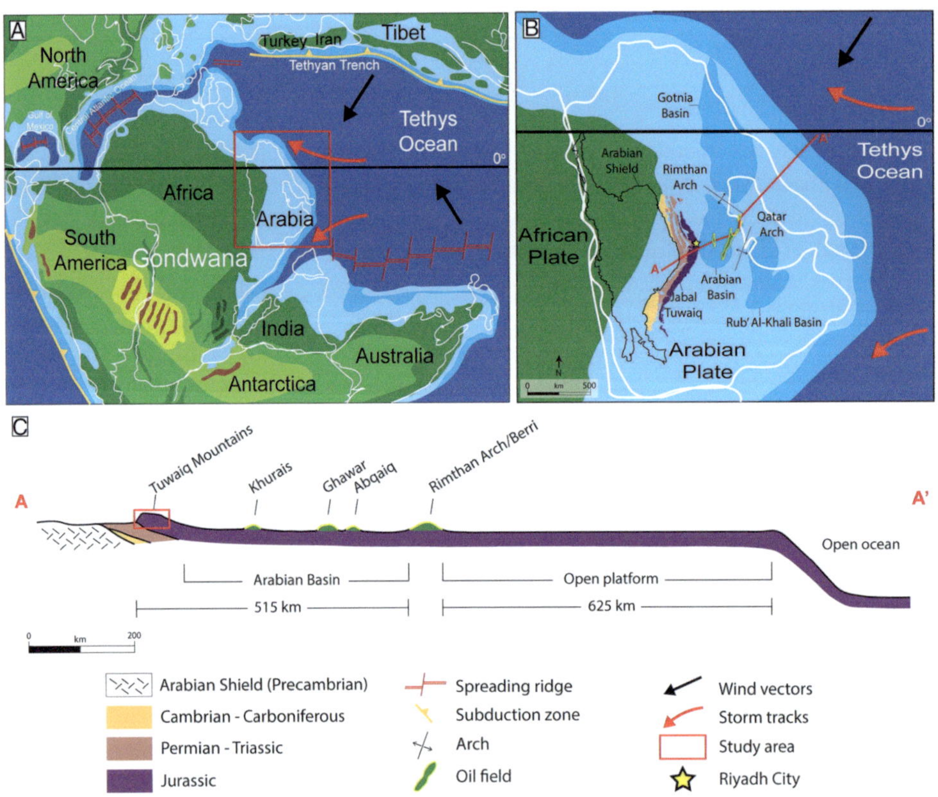

Fig. 7.23 Paleogeographic maps of the Arabian Plate during the Late Jurassic show the establishment of the Arabian Basin in an extension-subsidence setting (Fallatah and Kerans 2018)

salinity levels. Shallow carbonate shelves often feature brachiopods, echinoids, and green algae (dasycladacean). Marine calcareous algae's presence in the Araej Formation indicates water depths within the sunlight-reaching zone. Generally, the different facies of the Araej Formation are largely influenced by sea-level changes.

The restricted nature of both sediments and fauna present, along with a rise in terrestrial material at the Lower Araej Member's top and the presence of anhydrite, point to a depositional setting in a subtidal to lagoon inner neritic environment with energy levels from low to moderate.

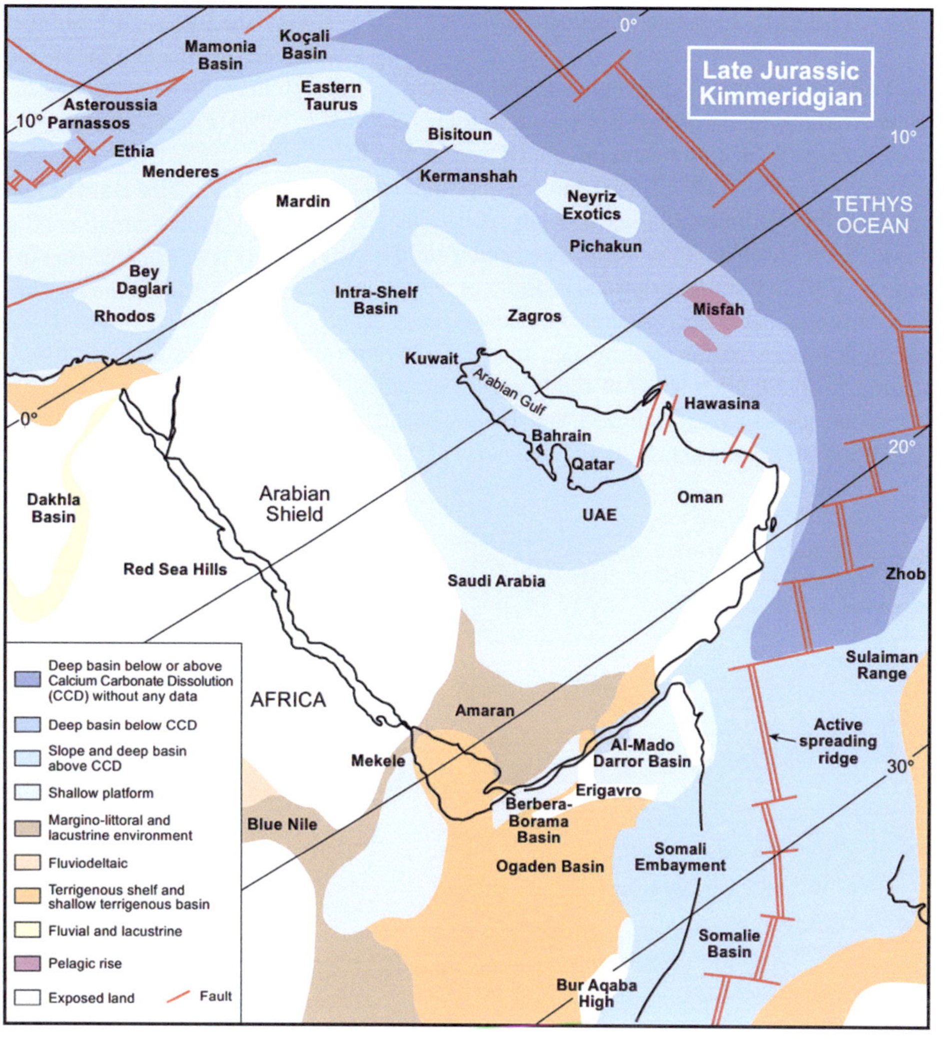

Fig. 7.24 Paleogeographic map of the Arabian Plate during the Late Jurassic (Kimmeridgian) showing the primary depositional environment (Al-Husseini 1997)

The lithological and biological features of the Middle Uwainat Member suggest it originated from brief, transgressive, and regressive sea-level shifts. Each shift alternated bioclastic wackestone/packstone and grainstone on a platform. This pattern supports the concept of recurring sea-level changes. The Uwainat depositional environments had moderate to high energy, evident from the corals, stromatoporoids, and grain-supported sediments. Diagenetic processes like dolomitization, calcite, and anhydrite cementation altered this member's lithological features.

Microfacies features suggest that the upper Araej layer deposited in a shallow subtidal to a somewhat restricted open lagoon setting (inner neritic zone) with moderate to low energy. This environment sometimes experienced an influx of clay and quartz from nearby Saudi Arabian regions. Some porous sections in this bed may be due to leaching in the lagoon sediments during sea-level changes. Indications of pyrite and plentiful anhydrite crystals in the micritic structure, alongside its dark hue, hint at a less oxygen-rich environment. Regarding broad lithological traits and its stratigraphic placement, the upper Araej epitomizes a slight regressive cycle in inner neritic areas.

The Araej Formation reflects oxidizing and reducing conditions in the studied region. The upper and lower Araej layers, with their dark tones, pyrites, limited faunal diversity, and minimal organic content, suggest a low-oxygen environment. Conversely, the Uwainat Member, characterized by lighter layers, abundant faunal diversity, and ferroan calcite in places, seems to have formed in a more oxygen-rich setting Jurassic (Al-Saad et al. 1992) (Fig. 7.21).

Alsharhan and Whittle (1995) (described similar lithologies of the Araej Formation from the United Arab Emirates. The Lower Aerej Member consists of argillaceous mudstone and wackestone with occasional beds of pelletal and peloidal grainstone. The middle Uwainat Member mainly comprises pelletal and peloidal grainstone and packstone with less frequent wackestone and lime mudstone interbeds. The Upper Aerej Member has variably cemented grainstone and packstone with minor wackestone (Fig. 7.22).

7.3.3 The Late Jurassic (Kimmeridgian)

The Late Jurassic basin development was affected by the Indian-African separation and the marine transgression that flooded the Arabian continental margin (Bashri et al. 2022; Ziegler 2001) (Fig. 7.23). This led to the deposition of the organic-richest sediments in north–south basins due to the rifting of the Arabian Plate. The reactivation of the pre-existing faults resulted in extensive subsidence, which, with the extensional tectonics, generated relatively deep south and southwestern basins. During the middle Callovian to Oxfordian, organic-rich strata of the Najmah Formation in Kuwait and the Tuwaiq Mountain and Hanifa Formations were deposited (Aqrawi et al. 2010; Sharland et al. 2001).

The intrashelf basin that deposited these relatively deep-water sediments extends all over the Arabian Gulf region and Iran, with its depocenter in present-day Kuwait (Al-Husseini 1997) (Fig. 7.24).

7.3.3.1 The Hanifa Formation

The Hanifa Formation (Upper Oxfordian-Lower Kimmeridgian) is considered one of the most important hydrocarbon source rocks in the region (Ayres et al. 1982; Murris 1980).

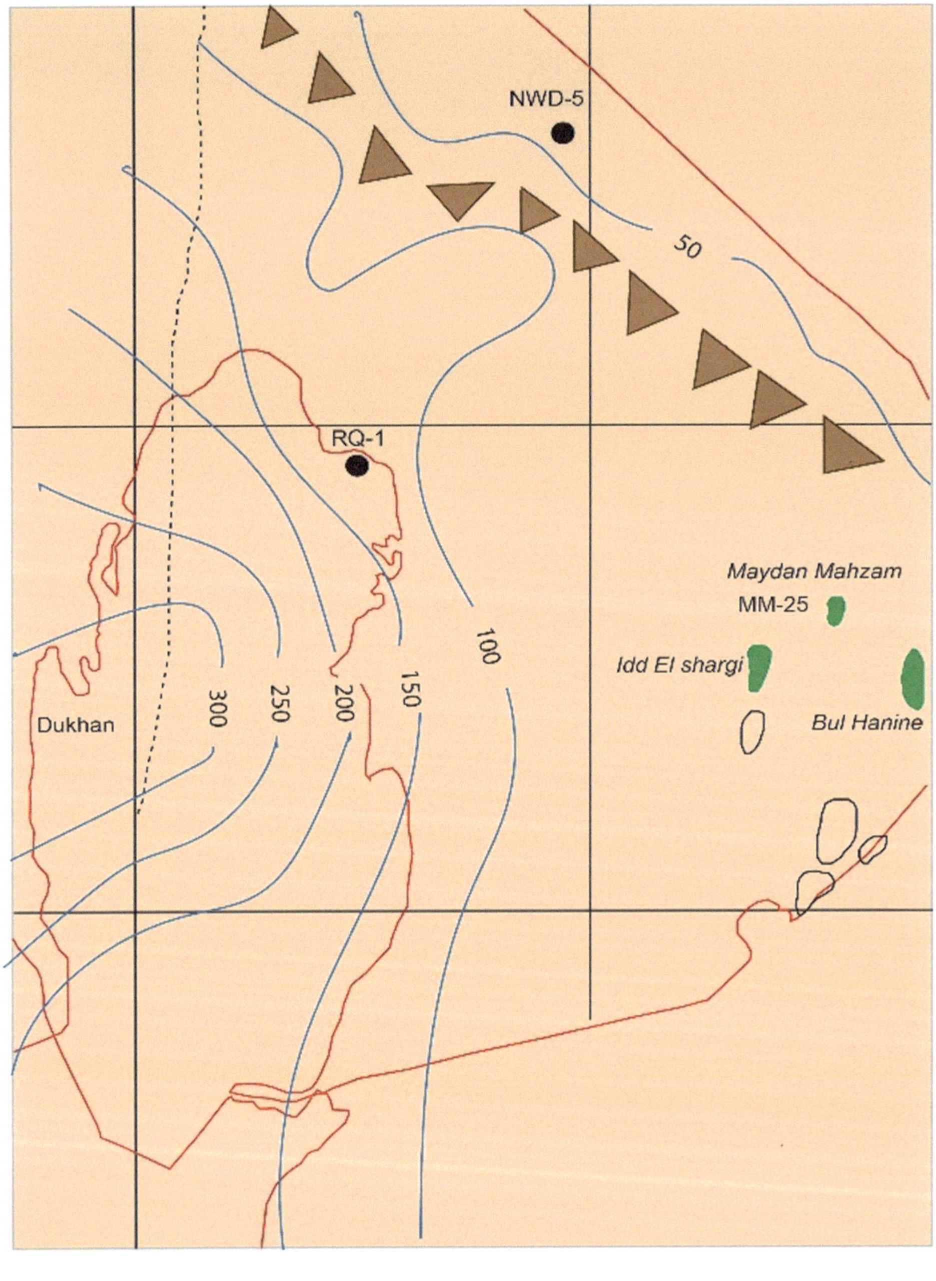

Fig. 7.25 Isopach map of the Hanifa Formation in Qatar. The N–S dashed line delineates Qatar Arch closure. Contour values are in feet (Focke et al. 1986)

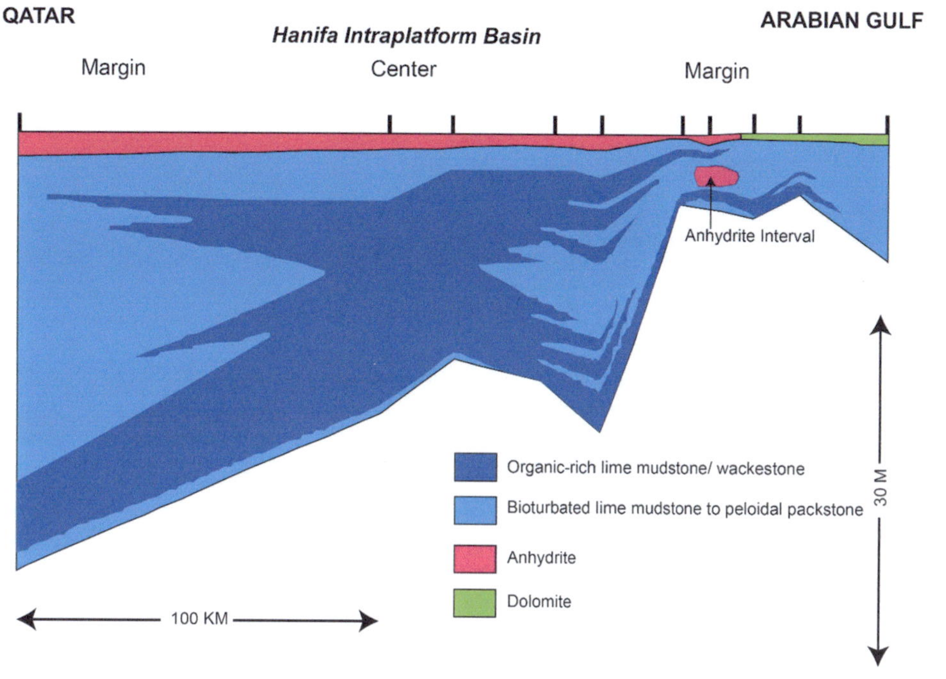

Fig. 7.26 Facies variations of the Hanifa Formation in Qatar between the basin center and its margins (Droste 1990)

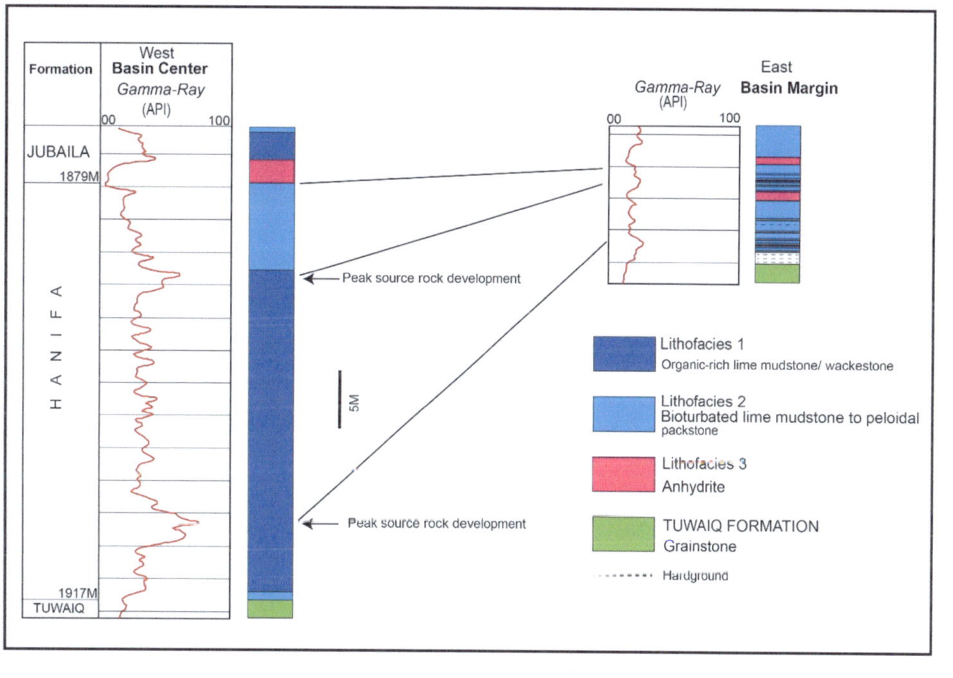

Fig. 7.27 Main lithofacies and regional correlation of the Hanifa sediments in the basin center and margins, Qatar (Droste 1990)

The Hanifa Formation was described in an unpublished report by Steineke as the upper part of the Tuwaiq Mountain Formation in Saudi Arabia. The formal description of Hanifa as an independent formation was published by Bramkamp and Steineke (1952). The type locality of the formation is Wadi Hanifa (Powers et al. 1966). The Hanifa Formation was divided by Vaslet et al. (1983) into the Hawtah and Ulayyah members. The Hawtah member, of middle Oxfordian age, has a thickness of around 57 m and comprises argillaceous material with several grainstone and coral boundstone. The Ulayyah Member is 71 m thick. It is designated as late Oxfordian in age and consists of skeletal grainstone, coral biostromes, bioherms, and stromatoporoid boundstone (Vaslet et al. 1991). In the subsurface sections, the reservoir part of the formation belongs to the upper part of the Ulayyah Member (Hughes 2004).

In Saudi Arabia, the Hanifa Formation is an important economic asset because it represents an important source rocks and a good hydrocarbon reservoir. The formation is exposed in

Fig. 7.28 Contact between the laminated mudstone and wackestone of the Hanifa Formation and the light-colored grainstone of the Araej Formation (Focke et al. 1986)

the central parts of Saudi Arabia along the Tuwaiq Mountain escarpment (Fallatah and Kerans 2018). The Tuwaiq Mountain Formation underlies the formation. The contact between the two formations is a paraconformity surface. The Jubaila Formation overlies it. The contact between the two formations represents a disconformity surface (Fallatah and Kerans 2018; Hughes et al. 2008). The thickness of the Hanifa Formation ranges between 90 and 140 m.

7.3.3.2 The Hanifa Formation in Qatar

The Hanifa Formation is named the Darb Formation in Qatar by Sugden and Standring (1975), but this term is not widely used now, and most workers are using the term Hanifa for the Late Jurassic strata of Qatar also. The Darb Formation was described from the well Dukhan-51 between the drilled depths of 2166–2377 m with a total thickness of around 211 m. The formation consists of four units. From the bottom, it comprises lime mudstone, becoming more argillaceous toward the base. The second unit is also formed of lime mudstone but with multiple thin beds of limestone containing pyrite-stained pellets. The third unit also consists of lime mudstone but with pyritic, pelletal packstone. The top unit is lime mudstone with dolomite streaks and pyrite-stained pelletal packstone. Based on the presence of *Pseudocyclammina jaccardi*, the Darb is considered to be equivalent in part or totally to the Jubaila and

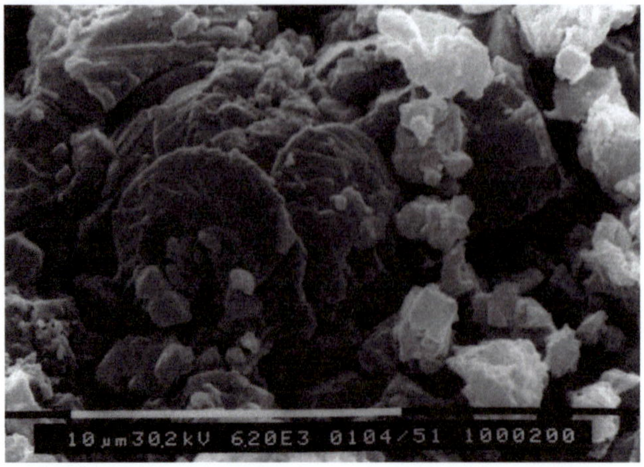

Fig. 7.29 SEM micrograph showing coccolith grains forming parts of the micritic matrix of the lime mudstone lithofacies (Droste 1990)

Hanifa Formations in Saudi Arabia. The formation is considered of late Oxfordian to possibly early Kimmeridgian (Le Nindre et al. 1990). In this work, we will adapt the Saudi nomenclature for this succession to improve regional correlation and eliminate the need for more terms for the same formation.

In the subsurface of Qatar, the formation has a wide thickness variation ranging between 15 and 91 m (Fig. 7.25) and also in facies variations (Fig. 7.26). There is a sharp contact between the deep-water lime mudstone strata of the Hanifa Formation and the porous packstone and grainstone of the Tuwaiq Formation. In the central parts of the basin, the Hanifa Formation is formed of basal bituminous limestone that grades upward into shelf carbonates and ends up with the anhydrite of the Jubaila Formation. The lower part of the formation consists of lime mudstone that changes to bioturbated wackestone and pelletal and peloidal packstone. The upper part of the formation is formed of fossiliferous packstone that is heavily dolomitized with streaks of anhydrite (Ibrahim et al. 2002). The formation generally consists of alternations of laminated, organic-rich, and bioturbated lime mudstone. In some sections, these sediments are associated with subaqueous, probably lowstand evaporites. From log correlation, the top anhydrite beds change gradually into dolomite and dolomitic limestone in the northeastern parts of Qatar (Droste 1990). The Hanifa Formation is overlain by the Jubaila Formation, which comprises two carbonate upward regressive cycles (Droste 1990; Murris 1980).

Lithology

Cores from five wells have been studied by Droste (1990). Three of these wells were located in the basin center, one in the transitional zone, and one at the basin margin. Detailed investigation of the Hanifa Formation in these cores provided a regional understanding of the formation lithologies and facies across the whole basin in Qatar (Fig. 7.27).

The studied sediments exhibit three main lithofacies:

Organic-Rich Lime Mudstone/Wackestone

This lithofacies comprises laminated lime mudstone and wackestone, with some occasional microbial filaments and bioclasts (Fig. 7.28). This facies is interbedded with some mm-thick laminae of coarse-grained sediments. These thin laminations consist of carbonate grains such as peloids, echinoderm debris, benthic foraminifera, lamellibranchs, and other unidentified bioclasts. The laminated nature of these sediments resulted not only from their distinctive lithologies but also from the accumulation of organic matter along the stylolite seams. Sedimentary structures include fining-upward scour-and-fill structures as well as burrowing. There are indications of mechanical and chemical compaction. Other diagenetic processes include pyrite, replacive anhydrite, and some selective dolomitization. Dolomite rhombs seem to follow stylolite seams. Some of the Lamellibranch fragments are replaced by phosphates. The clay content is relatively low and consists of mixed illite/smectite.

The matrix of these sediments is formed of fine micrite with blocky calcite as cement or overgrowth. Some of these micrite grains are formed of coccoliths (Fig. 7.29).

This lithofacies has a considerable amount of Total Organic Carbon (TOC) ranging between 1 and 6% (wt%). In the deeper parts of the basin, this organic matter is concentrated as insoluble residues along the pressure-solution structures (stylolites). On logs, Gamma-Ray logs show high peaks opposite these horizons. The highest TOC values are found in the basal and upper parts of the Hanifa sediments at the basin's center. The organic matter is structureless and formed of Kerogen Types II and I (Droste 1990; Tissot and Welte 2013) (Fig. 7.30).

Bioturbated Lime Mudstone to Peloidal Packstone

This lithofacies consists of lime mudstone changing into peloidal packstone. It is lighter in color with less organic matter content. The skeletal grains are more diverse and include echinoid fragments, foraminifera (*Gubkinella*), dasycladacean algae, sponge spicules, ostracods, lamellibranchs, brachiopods, gastropods, and calcispheres (Fig. 7.31). Extensive bioturbation heavily obliterated the original sedimentary features. Diagenetic processes include selective to extensive dolomitization, replacive anhydrite, chert, and microbial pyrite. There are only a few stylolites and some cemented fractures.

The bioturbated horizons are more porous than lithofacies 1. The main porosity types are bioturbation, inter and intragranular, matrix, and dissolution porosity. The TOC content is relatively low, generally less than one wt%.

The transition between the dark-colored, organic-rich lime mudstone/wackestone lithofacies and the light-colored, bioturbated lime mudstone to peloidal packstone is reflected in the cores recovered from the transition between the two lithofacies (Fig. 7.32).

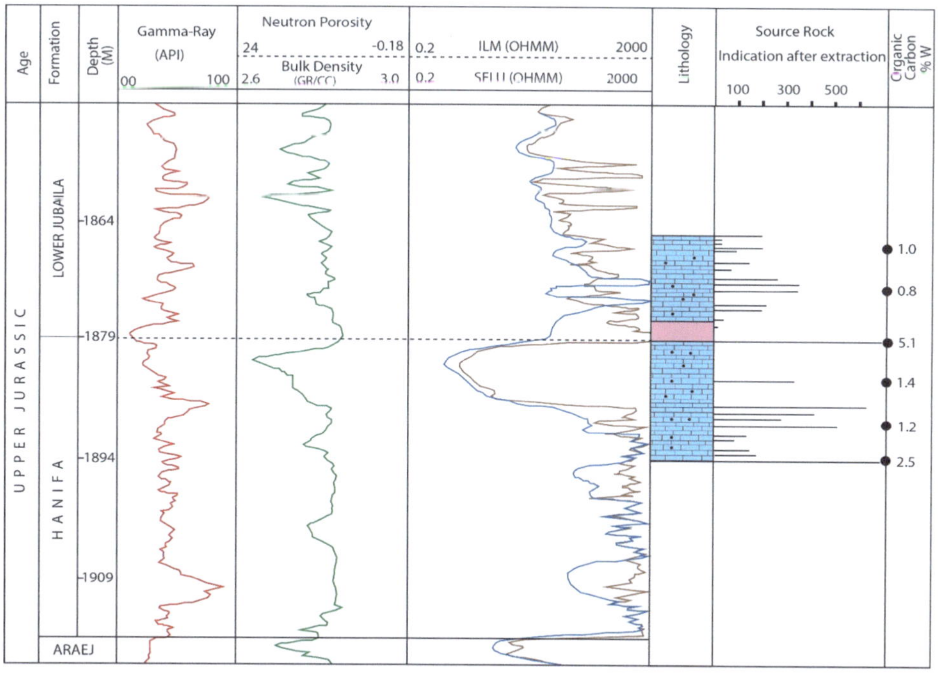

Fig. 7.30 Organic matter contents of the Hanifa Formation and their log signatures (Droste 1990)

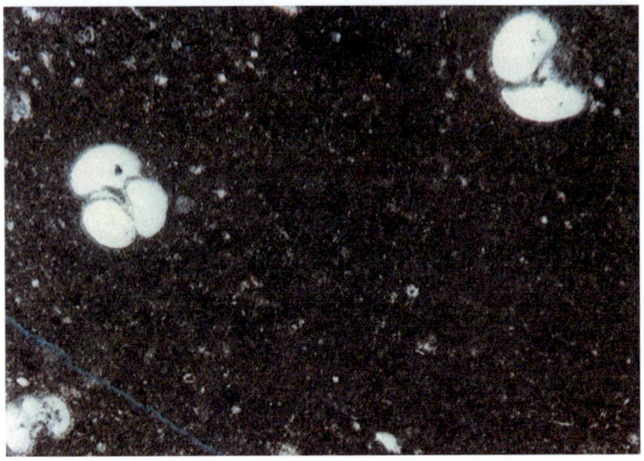

Fig. 7.31 Wackestone with the foraminifera *Gubkinella* sp. With some peloids and small bioclasts (Focke et al. 1986)

Fig. 7.32 Cores showing the difference between the basin margin, light-colored bioturbated lime mudstone, and peloidal wackestone and the basin-center dark, organic-rich lime mudstone of the Hanifa Formation in Qatar (Focke et al. 1986)

The Anhydrite

There are two types of anhydrite textures of anhydrite in the Hanifa Formation. They are either massive or of nodular mosaic texture. According to Droste (1990), the nodules are vertical and occasionally show remains of gypsum crystals. They usually do not show large-scale disturbance of the surrounding sediments. The massive anhydrite intervals are alternating with dolomitic organic-rich lime mudstone. The mudstone is usually burrowed with rare grains of ostracods and sponge spicules. Sometimes, the anhydrite is replaced by length-slow chalcedony or quartz crystal. This anhydrite may represent lowstand evaporites. See the later sections on the depositional environments of these sediments.

Sequence Stratigraphy and Source Rocks' Potential

The Hanifa/Tuwaiq Mountain succession is divided into four sequences by Hohman et al. (2005). These are:

Tuwaiq Mountain Sequence.
Hadriya Sequence.
Lower Hanifa Sequence.
Upper Hanifa Sequence.

They suggested that all four sequences have similar thicknesses. The best development of these strata is in western Saudi Arabia, but they decline in thickness toward Qatar. The Upper Hanifa Sequence is impacted by later erosion and is preserved only in the northeast part of the offshore area of Qatar. Each sequence represents a distinct phase of carbonate shelf and progradation into the Hanifa Basin. Regarding lithology, each sequence is formed of skeletal, oolitic, and peloidal grainstone and packstone in the shallow-shelf area, changing gradually into dark-colored, organic-rich lime mudstone toward the basin center. Each sequence can be divided into a lower Transgressive System Tract (TST) and an upper Highstand System Tract (HST), and the boundary between the two tracts represents the Maximum Flooding Surface (MFS) of that sequence. The Highstand System Tract has the thickest parts of the sediments on the shelf area, thinning out basinward, forming a steep platform margin. Source rocks are deposited in the basinal part of both tracts, although the thicker deposits are also associated with the HST.

In an integrated study on the potentiality of the Hanifa strata as source rocks conducted by Hohman et al. (2005), they found areas of thick and rich source rocks where the source rock sections in the identified sequences are stacked together (Fig. 7.33a). However, rich but thin source rocks may be found where only a single composite source section from an individual sequence is present (Fig. 7.33b).

A thickness map of the Hanifa Supersequence's net source indicates that as much as 44 m of stratified, organic-rich mudstone is present in multiple layered source sections in

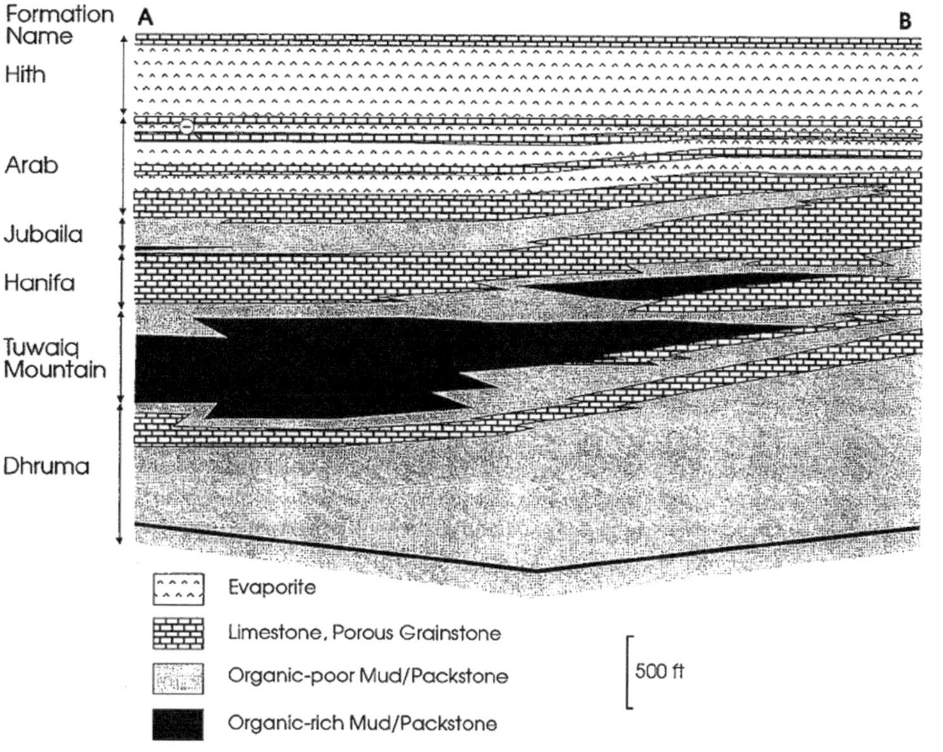

Fig. 7.33 Distribution of the source and reservoir rocks within the Jurassic strata of Saudi Arabia (Carrigan et al. 1995)

the Hanifa Basin. This basin is primarily situated in southeastern Qatar but also extends into offshore Abu Dhabi and southern Qatar (Fig. 7.34). Limited well data also points to the existence of layered source sections in the west, within Saudi Arabia, where the source rock thickness varies between approximately 33 and 44 m. The positioning of geological platforms influences the spatial distribution of these layered source rock sections within individual sequences. These platforms, situated adjacent to the basins, also shifted over time as the Hanifa Basin filled, leading to variations in the location of individual source rock sections.

The Hanifa Supersequence's average Organic Total Organic Carbon (OTOC) distribution reveals that organic enrichment is not limited to areas with multiple layers of source rock (termed thick-rich) but also occurs where a single-layered source section exists (referred to as thin-rich) (Fig. 7.35). Areas near thick source rocks, specifically in the offshore/onshore regions of Abu Dhabi southeast of Qatar, show average OTOC values exceeding 8%. On the other hand, areas in eastern and northern Qatar demonstrate high levels of organic-rich thin rocks, with OTOC values over 8%. This suggests that effective source rocks are not confined to zones with thick source rock layers, and their distribution may be more expansive than traditionally thought (Hohman et al. 2005).

Depositional Environment

The sediments of the Hanifa Formation include three distinctive rock suits that are usually assigned to three formations in the conventional Middle East stratigraphic nomenclature. These are the evaporites, the shelf carbonates, and the basinal facies. Three other factors complicate the picture more. These are (1) the possible presence of lowstand evaporites, (2) the scarcity of planktonic fauna in the basinal sediments, and (3) the presence of laminations of coarse-grained sediments within the dark, organic-rich mudstone.

Droste (1990) suggested that the Hanifa sediments result from two transgressive–regressive cycles. He also suggested that the formation was deposited on a faulted old topographic surface due to the upward movement of the Infracambrian Hormuz Salt. So these sediments were laid down in several subbasins. He believed the entire succession was deposited in relatively shallow marine conditions, mainly below the wave base and up to several tens of meters, with little difference between the basin center and margins. So, in the shelf area of the basin, water salinity oscillated between normal and hypersaline. In the central parts, anoxic to low-oxygen levels prevailed, which may explain the scarcity of their planktonic contents. Coarse materials from the shelf areas were swept downward, probably during storm seasons.

According Droste (1990), the Middle and Upper Jurassic sediments in Qatar represent a single transgressive–regressive "supercycle." Within this context, the Hanifa cycle includes the top of the "transgressive limb" (Fig. 7.36), which

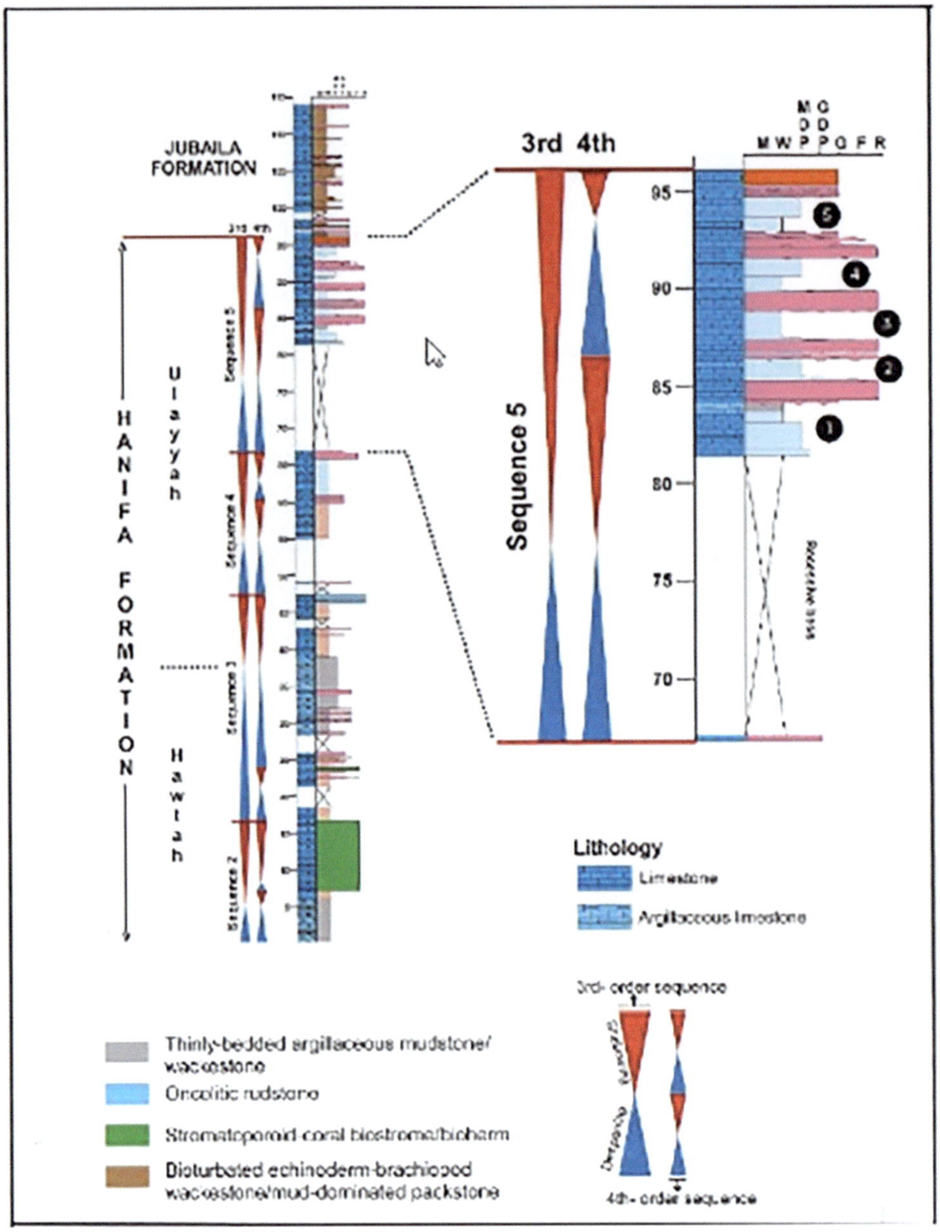

Fig. 7.34 Sequence stratigraphy of the Hanifa Formation in Central Saudi Arabia (Refaat et al. 2024)

explains the development of high-quality, organic-rich sediments. The underlying Tuwaiq and Dhruma strata comprise alternating carbonate sands and lime mudstone. They result from the lower part of the "transgressive limb of the Jurassic supercycle."

Focke et al. (1986) provided a set of environmental environments that explain the different lithologies of the Hanifa Formation. The model suggests a stratified, deep-water basin with planktonic-bearing sediments and the development of an anoxic layer at the bottom. This setting will deposit the organic-rich, dark lime mudstone (lithofacies 1). This normal marine basinal setting is disturbed either by seasonal turnover or longer-term climatic variations that limit the supply of materials from the basin margins. A lowstand conditions prevail with minimum deposition. The anoxic bottom layer will become more saline and depleted of oxygen. These conditions will lead to the deposition of evaporites (lithofacies 3). The third stage represents a well-mixed basin where normal marine conditions cause the precipitation of the bioturbated lime mudstone (lithofacies 2). The proposed model appears in Fig. 7.37.

7.3.4 The Tithonian (Late Jurassic)

The Tithonian stage of the Late Jurassic was characterized by the wide-scale development of evaporites and shallow marine

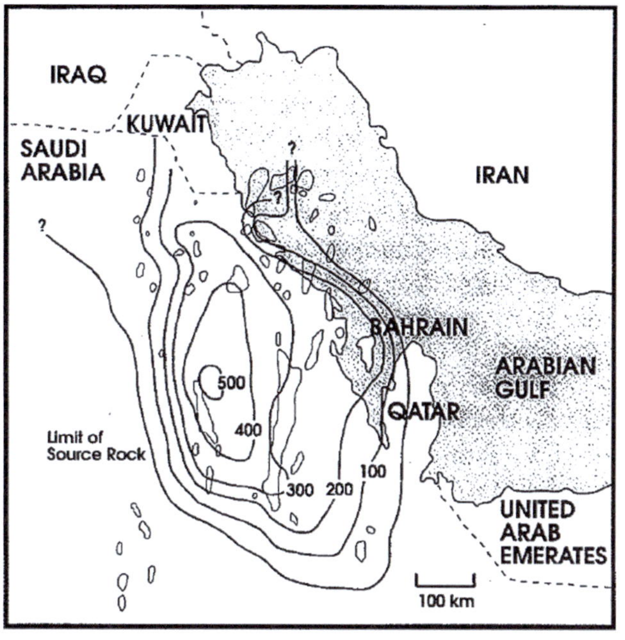

Fig. 7.35 Isopach map of the source rocks with greater than 1% TOC (contours in feet) (Carrigan et al. 1995)

carbonates designated as the Arab, Hith, and Gotnia Formations. Two local salinas were developed: the Gotnia Basin in Kuwait and southern Iraq and the Rub' Al-Khali Basin in Saudi Arabia (Fig. 7.38) (Al-Husseini 1997).

7.3.4.1 The Arab (Qatar) Formation

The Arab Formation is probably the most important rock unit in the world because it contains the highest hydrocarbon reserve ever. The sediments that became known as the Arab Formation were originally assigned as the Riyadh Member, which is the upper member of the Tuwaiq Formation, by Max Steineke in an unpublished report in 1937 in Saudi Arabia. The formal publication of the Arabian stratigraphic nomenclature was achieved by Bramkamp and Steineke (1952), in which the Arab Formation was formally defined. The first type locality of the formation was selected as the general outskirt of the city of Riyadh. Extensive weathering and slumping of the outcrops prevented accurate measurement of the formation thickness and lithologies. Therefore, a subsurface-type section was selected for the Arab Formation. Later, a correlation was achieved between the outcrops and the subsurface sections in the drilled wells in eastern Saudi Arabia. The type section of the formation was described from the well Dammam-7 between the depths 1371.6–1499.1 m. (Fig. 7.39). The lower contact of the formation was placed at the line that separates the packstone-grainstone facies of the Arab Formation from the tight, lime mudstone of the Jubaila Formation. The upper contact was placed at the change from the lime mudstone and wackestone of the Arab Formation to the massive anhydrite of the overlying Hith Formation.

Fig. 7.36 Position of the Hanifa cycle within the Middle and Upper Jurassic "Supercycle" explains the reason for the development of high-quality, organic-rich, Droste (1990)

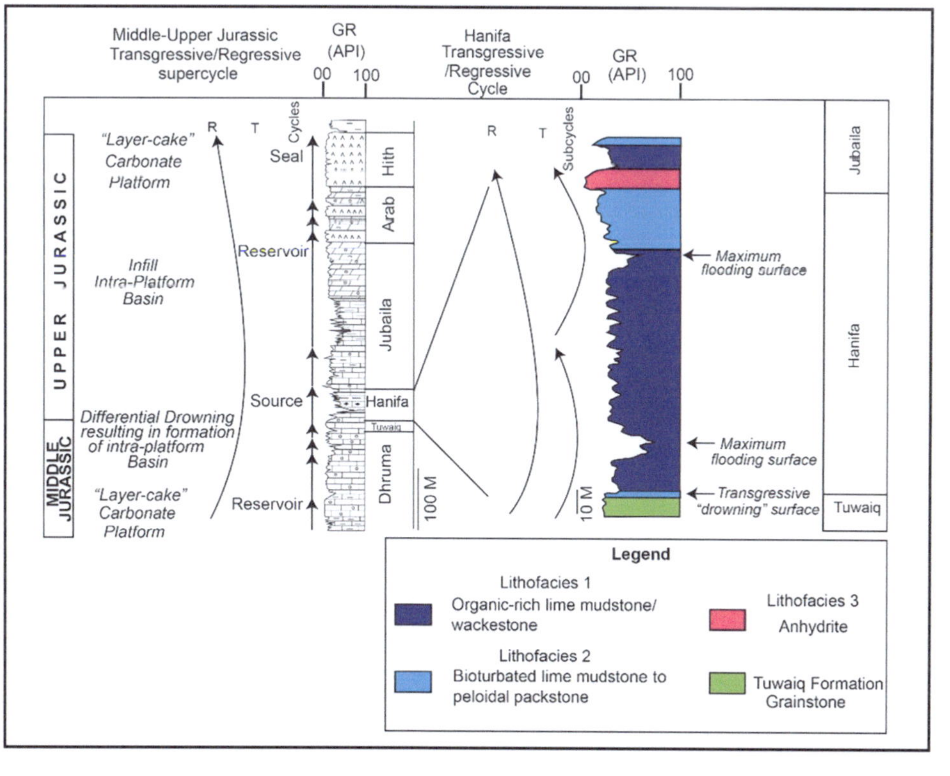

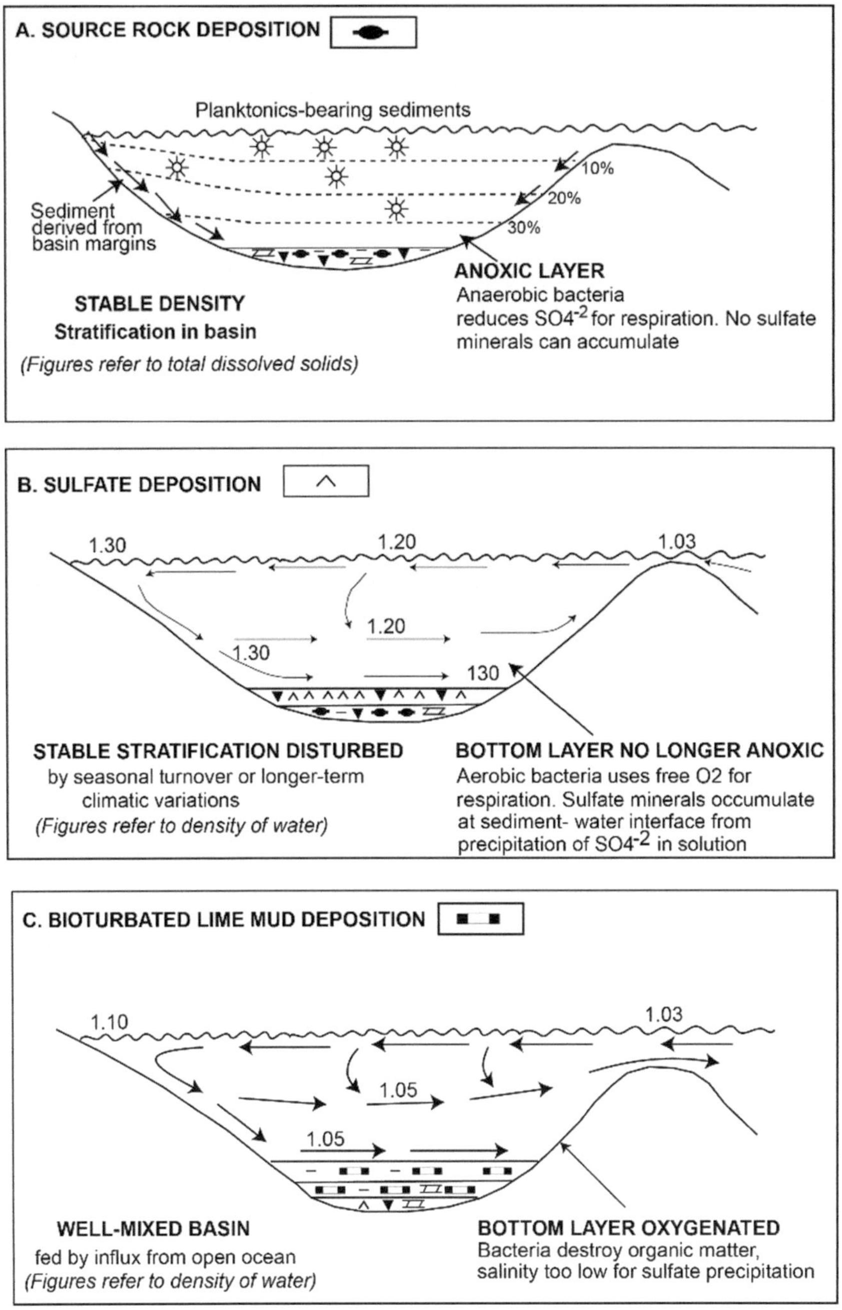

Fig. 7.37 Proposed depositional model to explain the three lithofacies of the Hanifa Formation in Qatar (Focke et al. 1986)

In the subsurface, the alternation of carbonate and evaporite as the dominant lithologies of the formation made it possible to divide the formation into four marine carbonate units separated by thin but laterally continuous anhydrite beds. Consequently, the formation has been divided formally into four members, each consisting of carbonates and evaporites (Powers et al. 1966; Steineke et al. 1958). The well Damam-7 was selected as the type section of the newly divided members in the following manner:

The A Member between 1371.6 and 1388.4 m.
The B Member between 1388.4 and 1399.1 m.
The C Member between 1399.1 and 1440.6 m.
The D Member between 1440.6 and 1499.1 m.

The contact between each member is selected when carbonate changes to evaporite. Hence, each member represents a regressive cycle.

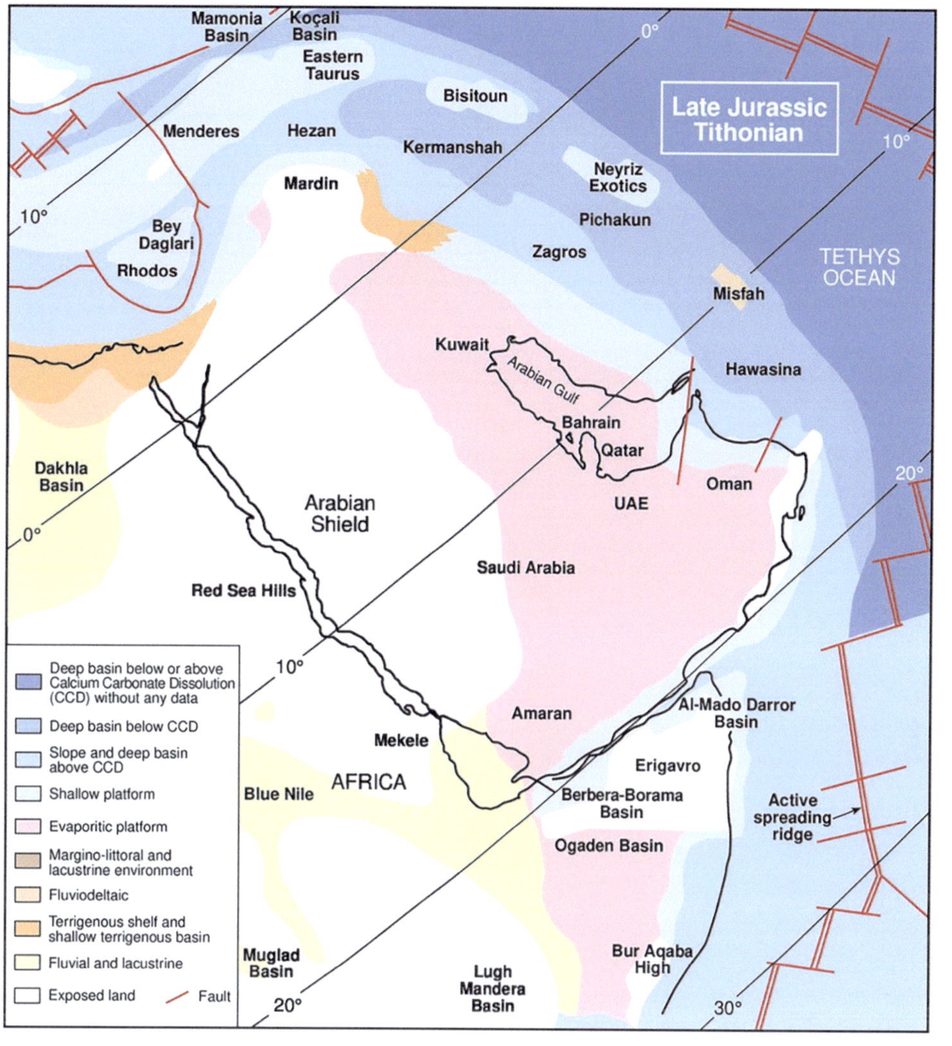

Fig. 7.38 Paleogeographic map and tectonic elements of the Late Jurassic (late Tithonian stage) (Al-Husseini 1997)

Although Focke et al. (1986) separated the anhydrite beds as upper, middle, and lower anhydrite beds (Fig. 7.40), these anhydrites are part of the members because each member terminates with an anhydrite bed.

Regional Correlation

Upper Jurassic sediments across Arabia and the Gulf region had been given different names despite their similar lithologies and stratigraphic status. The Arab "D" of Qatar may be equivalent to the lower part of "unit 3" of the Musandam Group in the northern parts of the UAE (Alsharhan and Kendall 1986); in the central UAE, the Arab "D" may be equivalent to the Kimmeridgian Fateh Formation. Further west (offshore Abu Dhabi), the stratigraphic nomenclature used there is similar to that of the type section in central Saudi Arabia and Qatar (Alsharhan 1989). In Oman, the Arab "D" may be correlated with the upper part of Member F of the Musandam Group. The Arab "D" of Qatar may be equivalent to the lower part of the Gotnia Formation in Kuwait and southern Iraq, which is composed of anhydrite and halite (Sadooni 1997).

While Saudi's Jurassic stratigraphic names, as outlined by Powers et al. (1966), have been accepted in Qatar, and some scholars have suggested local terminologies. In a report referenced by Sugden and Standring (1975) introduced the "Fahahil Formation" to describe the Arab "D" sediments in Qatar. As identified in the Dukhan-66 well, this formation spans 57 m. It features dolomitic limestones at the base, pure limestones at the middle, and beds of dolomitic limestones mixed with rare anhydrite beds above. However, the Qatar General Petroleum Corporation (1981) considered the Fahahil Formation equivalent to the Darb Formation. In the Darb Formation's reference section at the Dukhan-51 well, there is a 211 m section of argillaceous dolomitic limestones. This formation was further divided into two parts, the bottom being the lower Darb, and the top termed the Fourth Limestone Member. Subsequently, Focke et al. (1986) added the "Arab 4 unit" label to the offshore region's stratigraphy, equating it

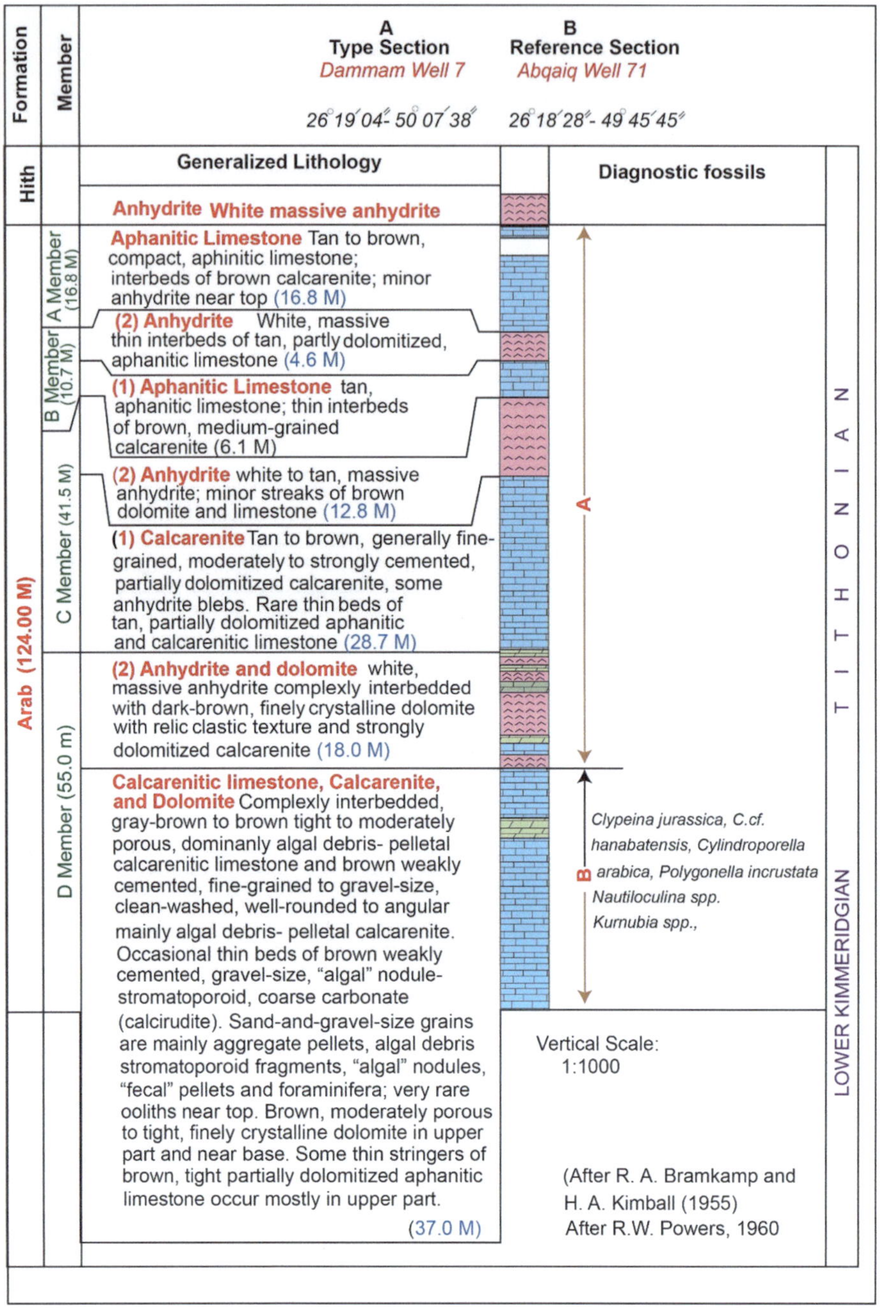

Fig. 7.39 Detailed description of the Arab Formation in its type section at Dammam-7 well (Powers et al. 1966)

to the Fahahil Formation. This work will adapt the nomenclature suggested by Al-Saad and Sadooni (2001), which is equivalent to the work of Powers (1968) (Fig. 7.41).

Sequence Stratigraphy

Based on a literature review of boundary beds in seven regions from five continents, Norris and Hallam (1995) reported the occurrence of a global high sea-level phase during the late Callovian and early Oxfordian. This phase appears to have followed Toarcian rifting and led to the deposition of the Hanifa and Jubaila Formations in Qatar, formations that are correlated with the Najmah Formation of Kuwait and Iraq (Sadooni 1997).

Upper Jurassic sediments can be divided into three major sequences based on large-scale flooding events:

1. *An Oxfordian—early Kimmeridgian* sequence encompassing sediments from the Hanifa and Jubaila Formations. These units consist of laminated, organic-rich, bituminous lime mudstones with some nodular and mosaic anhydrite and were probably deposited in High-stand, stratified, anoxic marine conditions (Sharland et al. 2001).

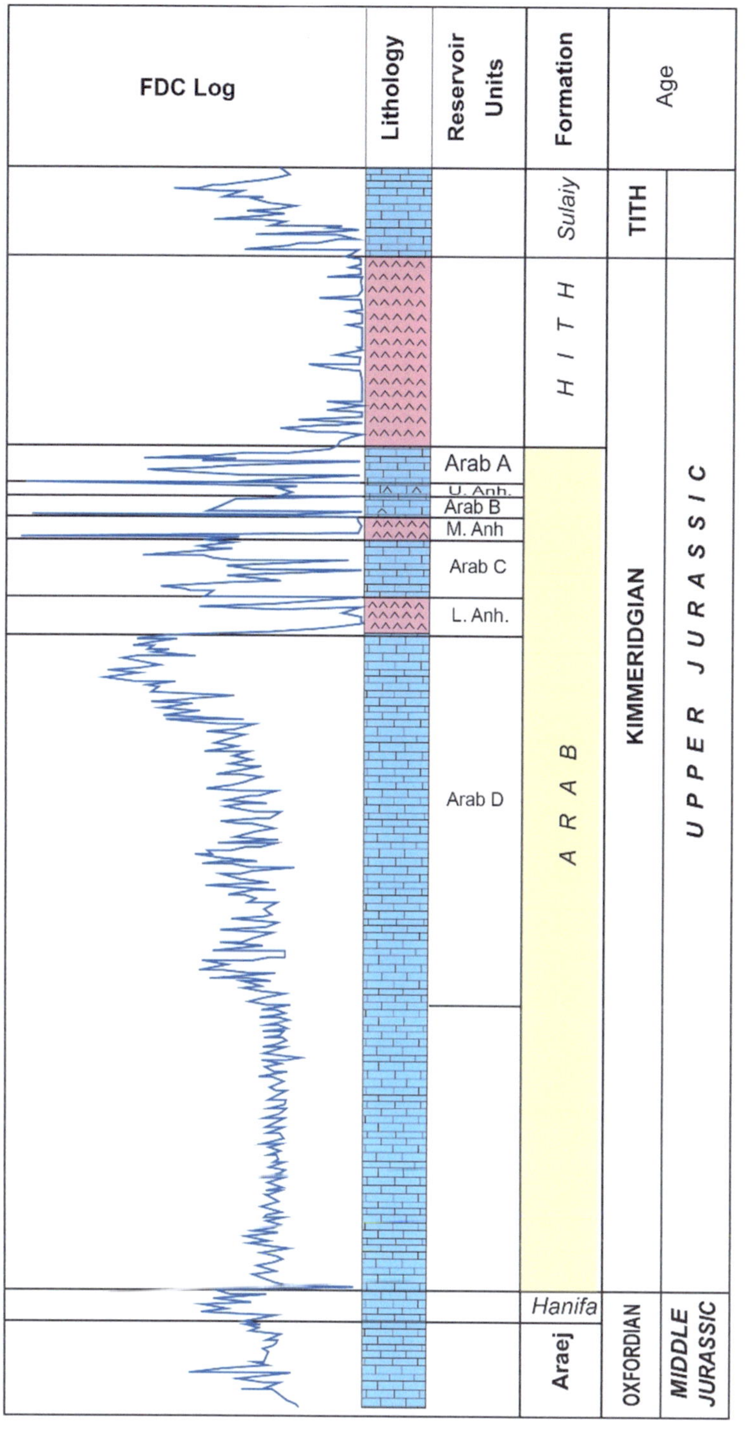

Fig. 7.40 Stratigraphy of the Arab Formation in the subsurface of Qatar (Focke et al. 1986)

2. *A Kimmeridgian-Tithonian* sequence mainly comprises the Arab Formation (D, C, B, and A units). This consists of shelf carbonates alternating with lowstand wedges of gypsum (LSW-G) and laminated basin-center evaporites (probably equivalent to the lower part of the Gotnia Formation in Kuwait and southern Iraq)
3. *A Tithonian sequence*, which consists of the Hith Formation in Qatar, is equivalent to the upper part of the Gotnia Formation in neighboring countries. The Hith Formation consists of Basin-Fill Halite (BFH) and lowstand wedge gypsum (LSW-G).

Al-Husseini (1997) proposed a sequence stratigraphic framework for the Jurassic sediments of the Arabian Gulf.

Fig. 7.41 Comparison between the stratigraphic nomenclature of the Upper Jurassic strata in Saudi Arabia and Qatar (Al-Saad and Sadooni 2011)

Age		Rock Units: SAUDI ARABIA, Powers 1968 FM	Member	QATAR: Sugden & Standring, 1975 FM	Member	OGPC Onshore, 1981 FM	Member	Focke et al. 1986 FM	Member	Al-Saad & Sadooni, 2010 FM	Member
UPPER JURASSIC	Tithonian	Hith		Hith		Hith		Hith		Hith	
UPPER JURASSIC	Tithonian	ARAB	Arab A	QATAR	Limestone 1	QATAR	Limestone 1	QATAR	Arab 1 (A)	ARAB	Arab A
UPPER JURASSIC	Tithonian	ARAB	Arab B	QATAR	Limestone 2	QATAR	Limestone 2	QATAR	Arab 2 (B)	ARAB	Arab B
UPPER JURASSIC	Tithonian	ARAB	Arab C	QATAR	Limestone 3	QATAR	Limestone 3	QATAR	Arab 3 (C)	ARAB	Arab C
UPPER JURASSIC	Tithonian	ARAB	Arab D	Fahahil		Darb	Limestone 4 / Lower Darb	JUBAILA	Arab 4 (D)	ARAB	Arab D
UPPER JURASSIC	Kimmeridgian	Jubaila		Darb		Diyab		JUBAILA	Lower Jubaila	Jubaila	
UPPER JURASSIC	Kimmeridgian	Hanifa		Diyab		Diyab		Hanifa		Hanifa	

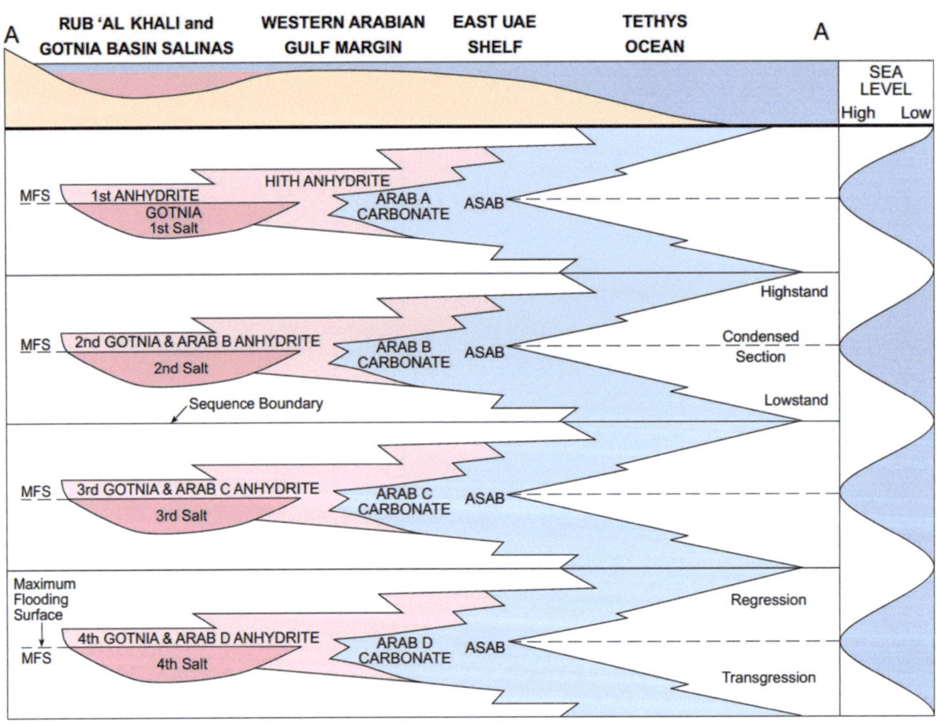

Fig. 7.42 Detailed sequence stratigraphy of the Arab Formation extends from the top of the underlying anhydrite to the top of the major anhydrite bed. The succession consists of (1) Upward shallowing shoals of grainstone (TST); (2) lagoonal mudstone and wackestone representing the Maximum Flooding Surface (MFS); (3) coarse-grained bioclastic grainstone topped by microbial laminations (early HST); (4) anhydrite (later HST). The Gotnia and western Rub' Al-Khali salinas are isolated salt lakes. These successions correspond closely to the eustatic sea level (Al-Husseini 1997)

He suggested that Arab deposition started with a Highstand systems tract followed by a transgressive/shelf-margin wedge and ended with a second Highstand phase. These can be correlated with short-term oscillations in the eustatic curve of (Haq et al. 1988) (Fig. 7.42).

7.3.4.2 The Arab "D" Member

The Arab D is the richest carbonate reservoir in the offshore fields of Qatar. This lower member of the Arab Formation signals drastic changes in the depositional environment with the development of wide-scale evaporites and the predominance of shallow-water carbonates compared to the deeper normal marine conditions that prevailed during the middle Jurassic. These carbonates range from shallow open marine conditions that deposited lime mudstone and wackestone to shallow shoals and buildups that led to the formation of packstone and grainstone and the lagoonal sediments of pelletal limestone. These carbonates are capped with anhydrite, forming a series of regressive cycles (Figs. 7.43 and 7.44). Dolomitization is the main diagenetic process that developed extensively dolomitized rocks where the dolomite

Fig. 7.43 Detailed lithology of the Arab D member from the offshore fields of Qatar (modified from Focke et al. 1986)

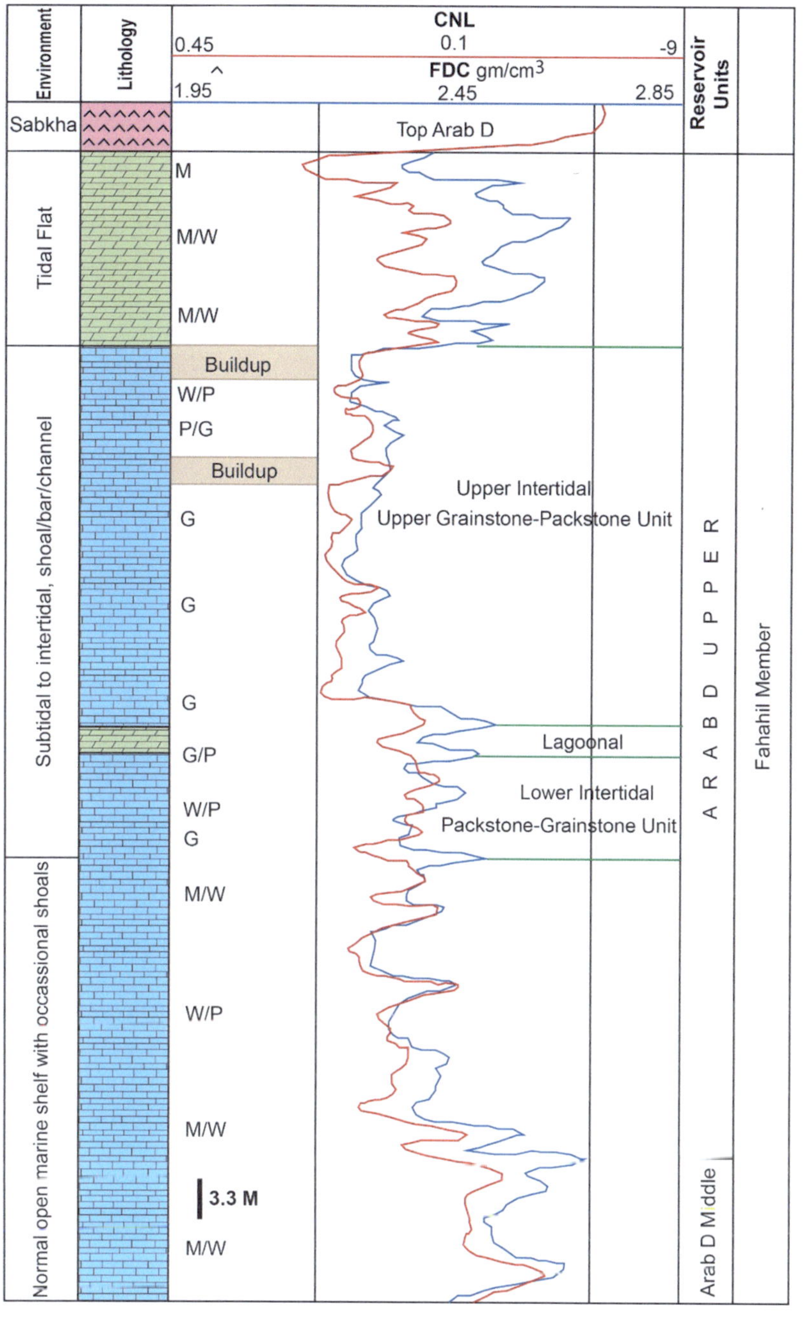

Fig. 7.44 Typical transition of limestone to anhydrite in the cycles of the Arab D from the onshore of Qatar

ranges in size from the aphanitic to well-developed euhedral types.

Rock Types

The Arab D in Qatar is formed of a wide range of rock types that reflect a varied spectrum of depositional environments depositing carbonates and evaporites. Rock types were identified based on their depositional texture, diagenetic signature and porosity, and permeability nature and amount. Accordingly, these rocks were classified into the following rock types:

Mudstone-Wackestone

This rock type is widely distributed and has a lime mudstone or wackestone texture with a considerable amount of argillaceous materials that may be dolomitized at parts (Fig. 7.45a). It contains many microbial threads, bioturbation, and dwarfed echinoderms. These rocks are deposited under shallow lagoons with hypersaline waters that limit their biological components. The presence of microbial mats may indicate upper intertidal conditions. The prevalent diagenetic processes comprise hard grounds, selective and extensive aphanitic and euhedral dolomitization, nodular anhydrite, and anhydrite cementation and dissolution. Dolomite may be microbially mediated similarly to the dolomite found in the intertidal zone of the present sabkha sediments. The relatively high amount of evaporites is associated with low water circulation and increased evaporation. Porosity types include small vugs and both intercrystalline and micrite porosity. Porosity is between 5 and 30% with 1–60 mD permeability. The relation between porosity and permeability was nearly linear, particularly at the lower permeability values (Fig. 7.46). Matrix porosity is the major form of porosity, and intercrystalline porosity is the dominant type in dolomitized sections. Other types were related to the dissolution of the unstable lithologic components by meteoric water. Bioturbation may have also contributed to porosity.

Dolomitic Mudstone and Wackestone

This rock type is restricted to the upper few meters of the succession and consists of euhedral, medium-sized dolomite rhombs replacing what seems to be a peloidal wackestone and featureless mudstone. It was heavily bioturbated and leached. There was also a considerable blocky anhydrite cement filling some pores (Fig. 7.45b). The lithologic nature of the facies may indicate a lagoonal depositional environment. The major porosity types were dissolution, intercrystalline, burrowing, and bird's eye. Porosity values ranged between 6 and 30%, with a permeability of 1–80 mD. There was no clear relation between porosity and permeability values, which may be attributed to the selective dolomitization process (Fig. 7.46).

Microbial Dolomite

This rock type is also found at the upper part of the studied section and may be related to the above rock type. It consists of alternating aphanitic dolomite with dark, organic-rich microbial laminations (Figs. 7.45c and 7.45d), with some hardgrounds and burrows. This rock type resembles the microbial dolomite formed at present-day Sabkhas' upper intertidal zone. The major porosity types are dissolution porosity, which has resulted from the preferential dissolution of microbial materials, and intercrystalline porosity between fine dolomitic rhombs, with bioturbation porosity being a less important form. A porosity average of 16% has been recorded in this rock type, with permeability values of up to 500 mD. There was a considerable correlation between the two parameters (Fig. 7.46) at the relatively low porosity range and higher permeability.

Skeletal Grainstone/Packstone

This is a more common rock type and consists of dasycladacean green algae (e.g., *Clypeina jurassica*), a characteristic fossil of the Arab Formation all over eastern Arabia (Sadooni 1997), with stromatoporoids, sponges, benthonic foraminifera, echinoderms, and gastropods. Some ooliths are entrapped inside the shell fragments (Fig. 7.45e). This mixed-up rock type indicates two-stage development; first, a high-energy system mixed up these skeletal and non-skeletal components, followed by a low-energy system that led to their extensive micritization. As a result, most of the skeletal elements were heavily micritized and relatively well

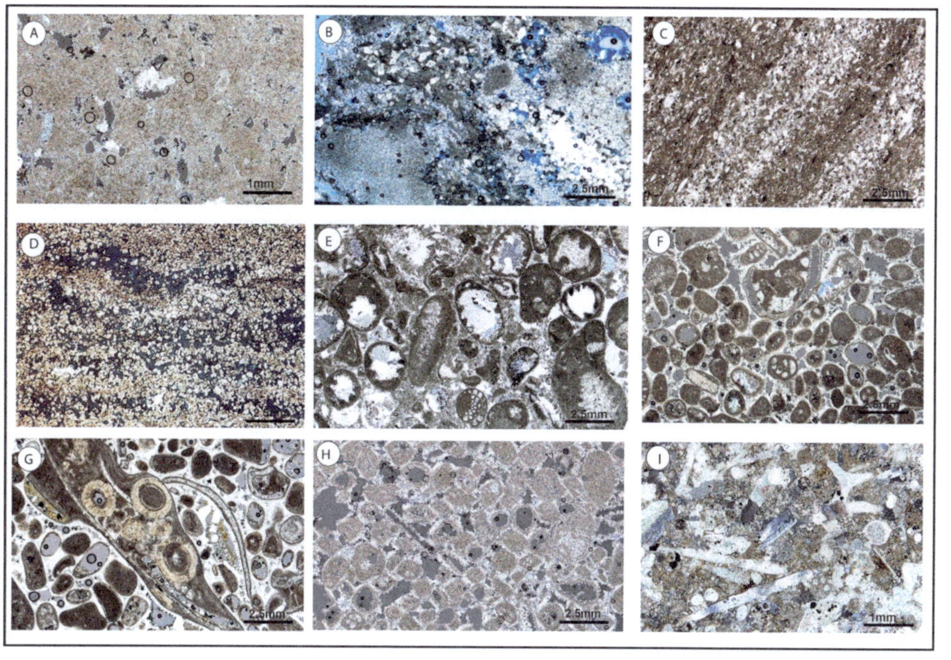

Fig. 7.45 Main rock types of the Arab D Member in the subsurface of Qatar (Sadooni et al. 2023)

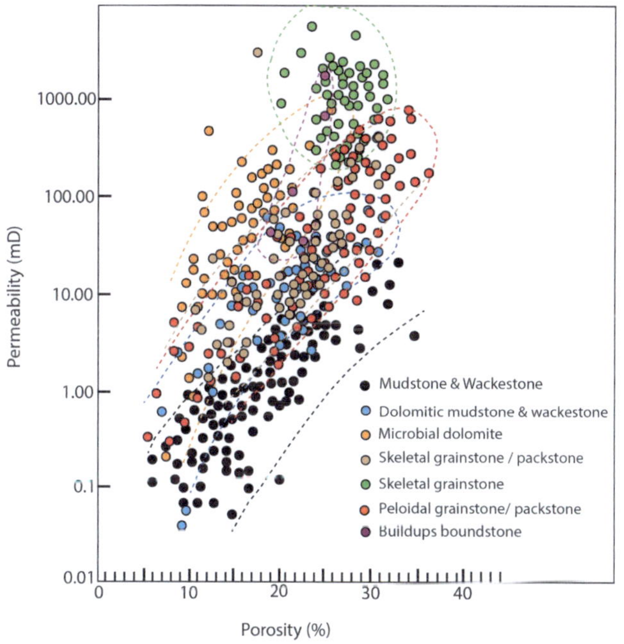

Fig. 7.46 Cross-plot of porosity versus permeability of the Arab D rock types (Sadooni et al. 2023)

sorted. The major diagenetic processes include isopachous rim cement, the dissolution of some grains, and the hosting groundmass. Porosity types include intergranular, dissolution, and vuggy porosity, with porosity values of up to 20% and permeability ranging up to 100 mD. A narrow range of porosity and permeability readings was concentrated in the middle field (Fig. 7.46).

Skeletal Grainstone

This is a significant and most prevalent reservoir unit. It consists of grainstone of mixed skeletal elements, such as benthonic foraminifera, green dasycladacean algae, shell fragments, and echinoderms (Fig. 7.45f, g). Many of these materials were micritized, although the general texture of this rock type suggests deposition under a high-energy system, such as an open marine shelf. Most groundmass is made of granular calcite cement, with a small quantity of micrite to none. Rim and syntaxial rim cement were present, particularly on the edge of some echinoderms or shell fragments. Visual porosity was mainly intergranular, but there were some vugs and moldic porosity. The average porosity was around 27%, with a permeability of more than 1000 mD. The distribution of porosity and permeability did not show any trend. Still, the whole population of the measured samples fell within nearly a circular configuration, with 20–30% porosity and 1000 mD or more permeability (Fig. 7.46).

Peloidal Grainstone/Packstone

This rock type is separated from pure grainstone because of considerable differences in the values and types of its porosity. It consists mainly of peloids with ooliths, small benthonic foraminifera, green algae, and rare echinoderms (Fig. 7.45h). Most of the peloids were either micritized ooliths or skeletal grains. The peloids were well sorted. Some were totally or partially leached with total or partial mosaic calcite cement. This rock type is distinct of high porosity and permeability due to excellent depositional intergranular porosity, enhanced by extensive leaching. The major porosity types are intergranular, vuggy, and micrite porosity. Porosity values of up to 25% were found in this rock type with a permeability of up to 1000 mD. There was a relatively straightforward linear positive relation between porosity and permeability (Fig. 7.46), which may be due to the prevalence of intergranular porosity.

Buildups Boundstone and Packstone

This rock type is of limited distribution and was found in a single well (A) only (Fig. 7.45I). It consists of sponge, stromatoporoids, coral, and algal boundstone with ample sponge spines, probably forming localized mounds in the tidal flat. The major diagenetic processes are the preferential leaching of stromatoporoids and the rim and mosaic cement replacing the sponge spines. The primary porosity was vugs, where the average porosity was around 22%, with permeability up to more than 1000 mD. However, there was no clear relationship between porosity and permeability, probably due to the small number of measurements (Fig. 7.46).

Correlation of the different rock types discussed above across the studied fields (Fig. 7.47) was attempted. There was a considerable consistency of the principal rock types, such as microbial dolomite, peloidal and skeletal grainstone, and packstone. Still, the relative locations of these rock types within the general succession differed from field to field. The other less essential rock types were not consistent due to the complicated depositional environments under which these rocks were formed. The Arab D reservoir was deposited in a shallow carbonate shelf punctuated with many minor depositional settings that do not make regional correlation possible.

Depositional Environments

To explain the carbonate-evaporite cycles of the Arab Formation, Wood and Wolfe (1969) were the first to propose a sabkha model for the depositional setting of the Arab/Darb Formation of the Trucial Coast (UAE). They compared the formation's lithologic succession with the sabkha sediments of the present-day Arabian Gulf. Leeder and Zeidan (1977) also studied what they called the giant Late Jurassic sabkhas of the Arabian Tethys. Lapointe (1991) investigated the sabkha versus salt basin (salina) model as a possible mechanism for the deposition of the Arab Formation in the *Umm Shaif* field, UAE. Many other attempts have been made to reconstruct the depositional regime of the Arab Formation in different parts of Arabia—e.g., Wilson (1985) from the *Qatif* field, Saudi Arabia; Suzuki and Osawa (1987) from Offshore Abu Dhabi; Mitchell et al. (1988) from *Ghawar* field, Saudi Arabia; Bouroullec and Meyer (1995) from Qatar; Alsharhan and Whittle (1995) from southern and SW parts of the Arabian Gulf; and Al Silwa et al. (1996) from offshore Abu Dhabi.

Most of these authors suggested that arid climatic conditions were prevalent during the deposition of the Arab Formation and that the depositional regime was characterized by a cyclic pattern of evaporite and carbonate deposition, which dominated much of the eastern part of the Arabian region (e.g., Alsharhan and Kendall 1986; Mitchell et al. 1988; Murris 1980). Mitchell et al. (1988) noted that the Arab "D" member at *Ghawar* consists of two shoaling-upward cycles that were deposited in a Highstand setting. No explanation was given for the causes of these repetitive cycles. The proposed causes—epeirogenic subsidence, eustatic changes, or variation in carbonate productivity—cannot explain such frequent lithological changes.

According to Wilson (1985) and Alsharhan and Nairn (1997), the Arab "D" Member in Saudi Arabia represents final, upward-shoaling, basin-fill deposition. The Member's sedimentary architecture was controlled by epeirogenic subsidence and eustatic sea-level changes. A similar conclusion was reached by (Hughes 1996), who attributed lithologic variations to either a reduction in subsidence rate, a fall in sea level, increased carbonate productivity, or a combination of these factors.

Problems Associated with the Depositional Model of the Arab Formation

In our view, any model of the depositional history of the Arab Formation should address the following problems:

1. The Arab Formation consists of a repetition of cycles with a frequency of 250 ka. These represent the lowest "order" Milankovitch cycle related to the eccentricity of the Earth's orbit (Fischer et al. 1982; Goldhammer et al. 1990). The Jurassic was, however, a "greenhouse" period when there was no significant sea-level change, and typical cycles were 2–5 m thick (Read and Horbury 1993). However, the Arab Formation cycles are tens of meters thick, and in that sense, they are similar to Vail-type 3rd-order cycles, which are deposited over a time interval of 1–10 Ma. But the whole of the Arab Formation was probably deposited in a considerably shorter time than this (Sharland et al. 2001).
2. The mechanism required to create the accommodation space that fits such abnormal thicknesses should also be considered.
3. The lateral progradation of the Arab system (both Highstand carbonates and gypsum wedges) from the basin margins must be addressed. This kind of progradation cannot be accomplished on a perfectly flat-topped platform.

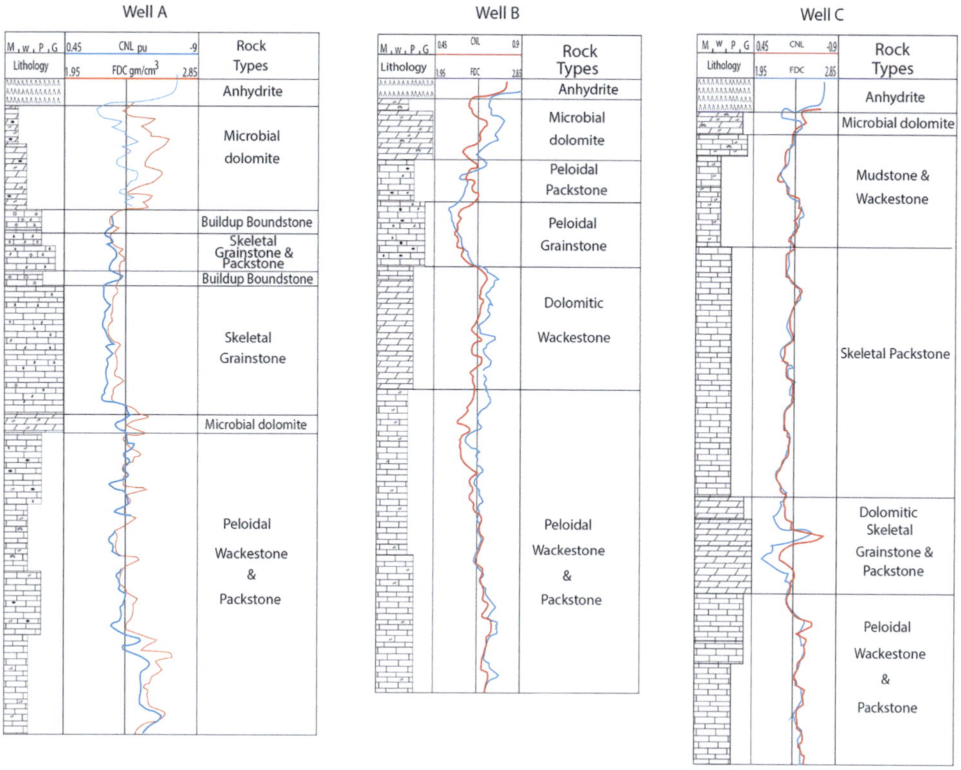

Fig. 7.47 Correlation of the Arab D rock types from three wells in the subsurface of Qatar (Sadooni et al. 2023)

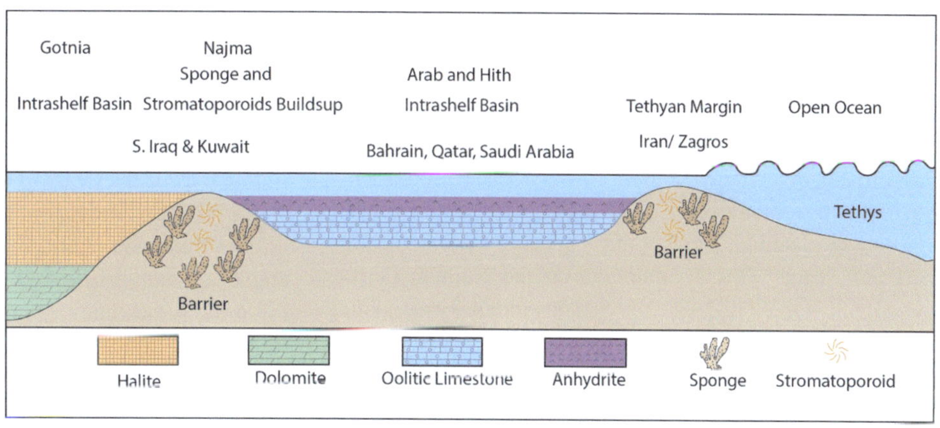

Fig. 7.48 Depositional model of the Arab Formation in the Arabian Gulf region (Sadooni et al. 2023)

The Depositional Model

The Arab Formation was deposited in a series of intrashelf basins (or deeper lagoons) that were separated by reefal barriers both from each other and also from the open Tethys in the Zagros area to the east. These barriers were most probably buildups of sponges, stromatoporoids, and algae. Bouroullec and Meyer (1995) reported the presence of local buildups of stromatoporoids and calcitic spongiomorphs from the Arab "D" of Qatar. Sadooni (1997) described sponge buildups from the Upper Jurassic sediments of northern Iraq. From another part of the Tethys, Schorr and Koch (1985), Wirsing (Wirsing and Koch 1986), and Schweizer (Schweizer 1987) described algal-sponge bioherms from the Swabian Alps in Germany. Schorr and Koch (1985) reported that the bioherms consist of a combination of algae, sponges, ooid-bearing grainstones, and algal stromatolites.

These intrashelf basins were flooded during 4th-order or 3rd/4th-order flooding events, which generally result in 2-5m thick cycles (Read and Horbury 1993), but as noted above, the Arab Formation cycles are much thicker (tens of meters). This difference in thickness could be explained by the possibility that the 4th-order cyclicity "topped up" the intrashelf basin, which was sediment starved, and allowed the cycles to "stretch" in a way that normal 4th-order platform-top cycles (starting with zero accommodation space prior to transgression) cannot. In fact, the Arab Formation cycles essentially "plugged" pre-existing accommodation space and plugged it rapidly.

In such a setting, the Arab "D" Member was probably deposited when the basin was fully connected to the Tethys Ocean and the sea level was well above the barrier height. Therefore, the oolitic and bioclastic limestone units of the Arab "D" represent the transgressive and Highstand systems tracts that were forming on the intrashelf basin margin. The calcispheres-bearing mudstone facies is interpreted as a condensed section deposited in the intrashelf basin center (Bucur 1992; Noel et al. 1991) (Fig. 7.48).

When the sea level fell below the barrier level, the shelf margin would have been exposed and become subjected to dolomitization and dissolution. Evaporites were precipitated around the intrashelf basin margins, forming marginal evaporite wedges. In the intrashelf basin center, carbonates continued to form during the early stages of shallowing but gradually became interlaminated with evaporites, forming "bull's-eye" deposits. Therefore, evaporites were either deposited in sabkhas or in the hypersaline lagoons behind the carbonate shoals or in the basin center. Flooding of the intrashelf basin would have created a new transgressive systems tract, leading to the re-establishment of renewed shelf-margin carbonates. Such a mechanism will cater for the lateral progradation of the Arab Formation sediments. It will also explain the greater thickness of the formation in Qatar (especially in the offshore area to the east) compared to Saudi Arabia. This may suggest that this area was much closer to the area where the main barrier was developed since the initial progradation of shelf sediments away from the barrier will create an isopach thick into pre-existing accommodation space. This is analogous to the Najmah Formation, which is thick because it progrades into the pre-existing accommodation space in the Sargelu Basin in northern Iraq (Sadooni 1997). Subsequent Arab Formation cycles are much thinner but are still thicker than the 2–5 m cycles expected in a "greenhouse" period. The model assumes that there was remnant accommodation space (with some subsidence) following Arab "D" deposition. This can only be achieved if the Arab D and subsequent Arab cycles show an offlapping geometry into the basin, making them late Highstand to the onset of lowstand (i.e., prograding, but with sea-level dropping slightly at every prograde).

The Arab cycle oolites in Saudi Arabia are more probably related to intrabasinal highs but would have developed by essentially the same mechanism.

This depositional model attempts to explain the repetitive nature of carbonate-evaporite beds in the Arab Formation in a more reasonable way than the standard "cycle" theory. The cycle theory assumes that the whole depositional system underwent shallowing up so that evaporites could be deposited, and it requires another marine flooding for carbonates to be deposited. In an area of relatively stable tectonism with no extreme climatic changes, it is difficult to explain the causes of such repetition. By contrast, our proposed model requires only a slight change in the water level to cover or expose the barrier.

7.3.4.3 The Arab C Member

The Arab "C" member was described for the first time by an anonymous author in 1939 and was redefined by Powers et al. (1966) from well Dammam-7 in eastern Saudi Arabia due to the extensive dissolution of the anhydrite beds at the outcrop type section. At its type locality, the member is 41.5 m thick. The lower 29 m is made up of partially dolomitized aphanitic limestones interbedded with porous calcarenites and calcarenitic limestones; the upper 12.5 m consists of massive anhydrites interbedded with dolomites and dolomitic limestones. Based on its stratigraphic position, the Arab "C" member is assigned a Late Kimmeridgian age (Powers 1962).

Upper Jurassic sediments across the Arabian Peninsula and the Gulf region had been given different names in spite of their similar lithologies and stratigraphic status. In Oman, the Arab "C" member of Qatar may be correlated with the upper part of member F of the Musandam Group. It may be equivalent to the middle part of "unit 3" of the Musandam Group in the northern parts of the United Arab Emirates (Alsharhan and Kendall 1986). Further west (offshore Abu Dhabi), the stratigraphic nomenclature used is similar to that of the type section in central Saudi Arabia and Qatar (Alsharhan 1989). The Arab "C" member of Qatar may be equivalent to the middle part of the Gotnia Formation in Kuwait and southern Iraq, which is made up of anhydrite and halite (Sadooni 1997).

The transition from limestones interbedded with thin streaks of anhydrite (Arab "D" member) to the overlying dolomitic limestones of the Arab "C" member seems to be gradual. The Arab "C" is separated from the overlying dolomitic limestones of the Arab "B" reservoir by a horizon of nodular anhydrite that represents the top of the Arab "C" member. In Qatar, the Arab "C" member ranges in thickness from 30 m inshore to about 40 m offshore (Fig. 7.49).

The Arab C is divided into upper and lower parts by a mudstone layer, the "b" Unit. This unit may represent a short hiatus in sedimentation and is marked by the presence of a hardground.

Major Facies

The Arab "C" member consists of a sequence of limestone and dolomite limestone with numerous stylolites. Anhydrite occurs as nodules and in thin streaks as well as forming the reservoir cap rock. Petrographic analysis of the Arab "C" suggests that its microfacies are similar to that of Upper Jurassic sediments in other parts of the Arabian Gulf. The Arab "C" member consists mainly of repeated small cycles

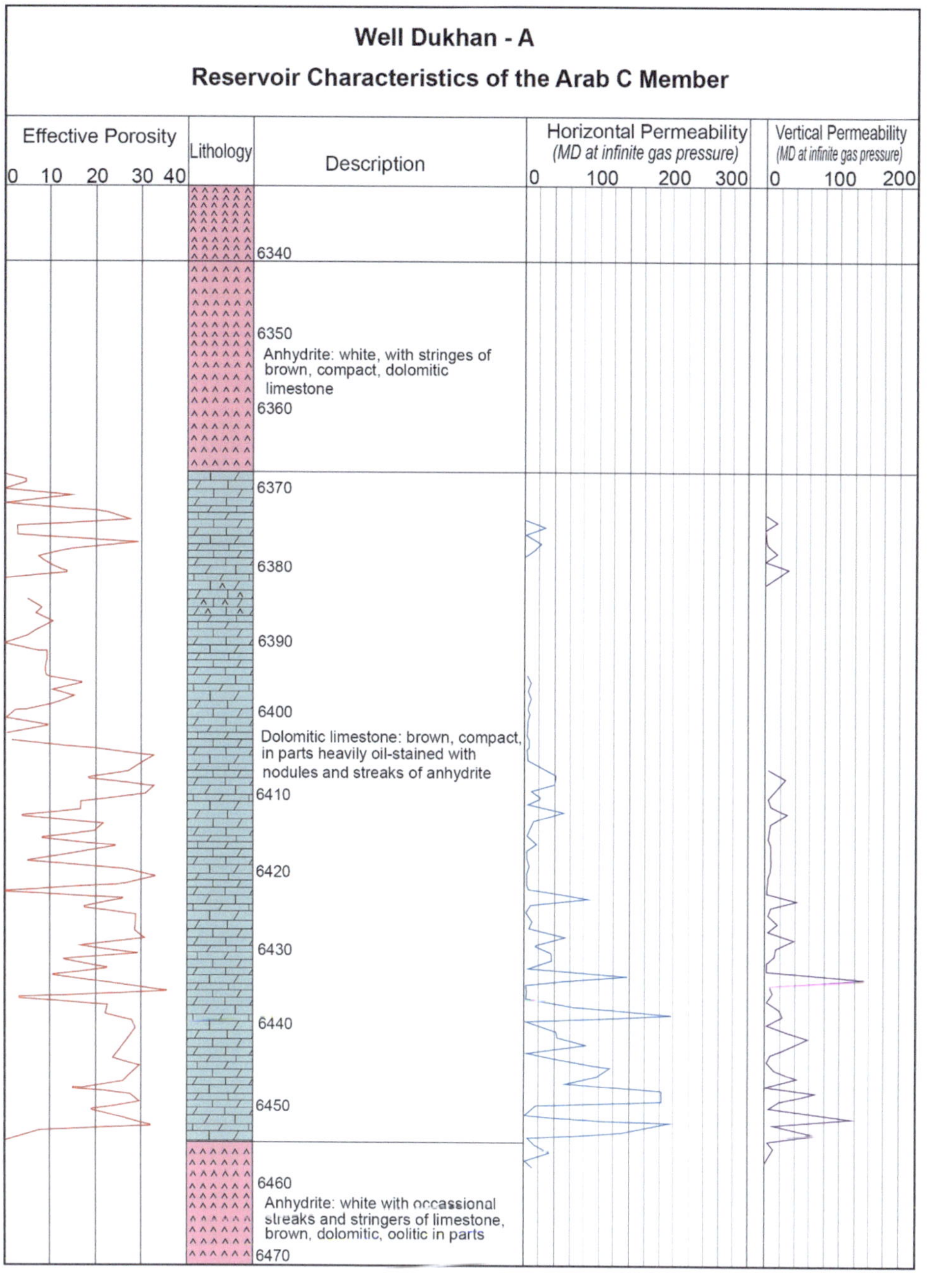

Fig. 7.49 Detailed lithology, porosity, and permeability of the Arab C in Dukhan field, onshore Qatar (Al-Saad and Sadooni 2011)

of micritized grainstone-packstone with some dolomites at the base. This gradually passes upwards to wackestone-packstone with local coral-algal boundstone at the middle part and ends with fossiliferous grainstones-packstone that gradually changes to anhydritic dolomites and anhydrites, which cap the Arab "C" member. The upper boundary of the Arab "C" was picked from a positive gamma-ray deflection. At the base, the gamma-ray change across the boundary is very gradational. The main facies are described in the following section:

Oolitic Grainstone (Fig. 7.50a)

Most of the ooids in Arab C grainstones lack the normally observed regular cortical laminations. Instead, they have a large nucleus or are bound into a grapestone texture, and many are strongly micritized. The ooids appear to float in large crystals of calcite cement. Some nuclei are dissolved, forming partially cemented molds with outer micrite envelopes.

The large nuclei, the thin cortical laminations, the grapestone texture, and the micritization are all interpreted as an indication of water energy, which is lower than that considered for normal ooid shoals. The reduced thickness in circumferential laminations on the ooids is interpreted as an indication of reduced agitation and lower energy than the normal ooid shoals. The grapestone texture usually results from the action

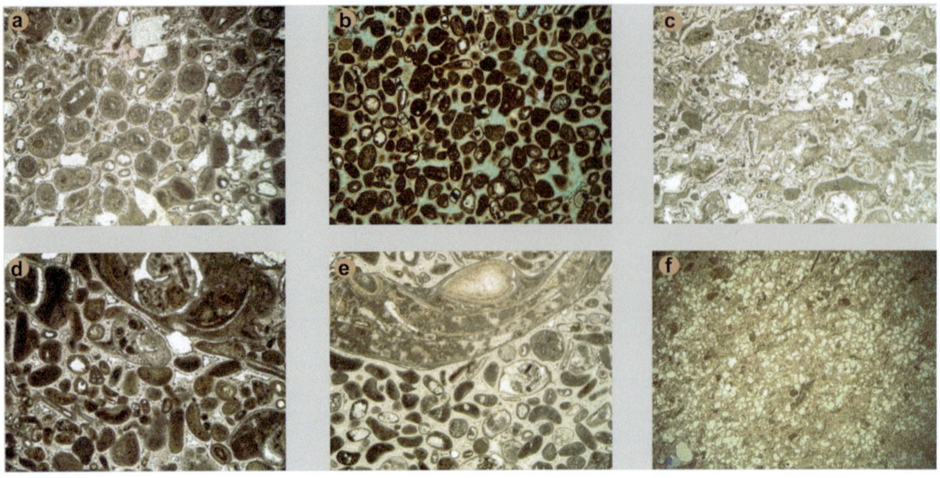

Fig. 7.50 Main microfacies of the Arab C reservoir. **a** Oolitic grainstone comprises different ooliths, peloids, and coated grains; 25X. **b** Peloidal grainstone of micritized grains and clastic debris; 25X. **c** Microbial boundstone formed of threads and debris in a cemented matrix with some micritized peloids; 25X. **d–f** Bioclastic grainstone formed of different types of skeletal grains, including foraminifera and gastropods with peloids and coated grains; 25X (Al-Saad and Sadooni 2011)

of endolithic algae binding the grains together, and the micritization also indicates the presence of strong microbial activity. This facies indicates, therefore, a lower energy, less efficient oolitic shoal or that these ooids were transported to a quieter lagoonal environment after their deposition.

Peloidal Grainstone (Fig. 7.50b)

The peloids in these facies have varied origins, textures, and shapes. Many of them are extensively micritized large benthic foraminifera, shell debris, algae, or ooids. Only ghosts of the original texture can be recognized. Others are just dark mud clasts probably generated by the physical breakdown of lithified rocks. Only a few of them can be identified as fecal pellets. Most of the grains are surrounded by calcite rim cement. Peloid nuclei are completely dissolved or filled with large, granular calcite cement.

The variable nature of the grains indicates that they were generated in different depositional environments, including shallow marine, lagoon, and shelf environments. The extensive micritization of most of the grains suggests that they were all subjected to extensive microbial activity after deposition in sheltered environments such as lagoons or shallow embayment.

Microbial Boundstone (Fig. 7.50c)

This facies is formed mainly of algal debris of indeterminate origin. Peloids, small foraminifera, and shell fragments are among the bound particles. In other parts, they are irregularly shaped, and some are heavily micritized. The boundstone shows signs of dissolution in parts of the filamentous fabric or the mud matrix. Replacive granular calcite cement is also present.

The preservation of these materials and the enclosed materials indicates two aspects: first, these facies must be abundant, forming a considerable portion of the studied rocks, and second, it has been formed in a quiet environment that managed to preserve its original texture. Much of the grapestone texture found in the entire Arab Formation rock suite can probably be attributed to these cyanobacteria.

Bioclastic Packstone and Grainstone (Fig. 7.50d, e).

This facies consists of skeletal packstones and grainstones dominated by large shell fragments of mollusks (mainly gastropods) with large and small benthic foraminifera and micritized grains with micritic envelopes. The shell fragments are either extensively micritized as in Fig. 7.50d or preserve their original features as in Fig. 7.50e. Grains show variable dissolution and pore-filling granular calcite cementation. Some of the grains are totally or partially dissolved, and some are lined with rim cement or replaced by granular calcite cement.

This facies is common in most Arabian Jurassic carbonates, with slight variation depending on the nature of the grains and their sizes. It was deposited in a shallow marine shelf under relatively restricted marine conditions. The presence of gastropods and the extensive micritization of its components suggest the presence of low-energy depositional conditions.

Lime Mudstone (Fig. 7.50f)

Lime mudstone with few grains is a common facies that forms a significant part of the Arab C succession. The mudstone matrix is generally uniform. Sometimes, mudstones are bioturbated or contain clots of darker, possibly organic matter. Some may contain needles or patches of evaporites such as gypsum ghosts or nodules of anhydrite; others may have clastic quartz grains, as in Fig. 7.50f.

This facies is believed to be deposited in shallow intertidal flats where the lime mud was precipitated directly from the shallow seawater due to extensive evaporation or caused by disintegration of the biological components living in the tidal flats, or a combination of the two processes. The presence of evaporite minerals indicates the nearness of the sabkha condition. In contrast, the presence of clastic grains suggests

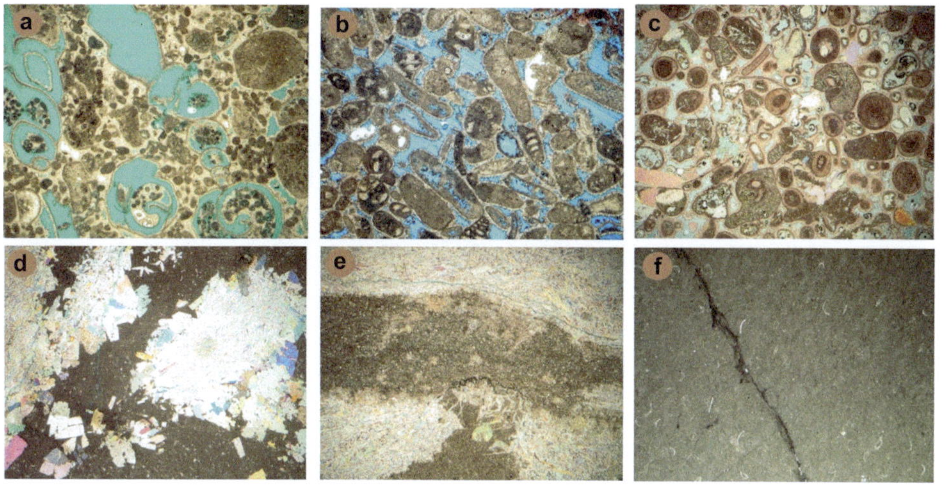

Fig. 7.51 Some of the diagenetic features modify the rocks of the Arab C reservoirs. **a** Extensive dissolution of the shells of gastropods and parts of the matrix; 40X. **b** Rim calcite cement lining the outer boundaries of the skeletal components such as foraminifera and micritized grains 40X. **c** Large, granular calcite cement blocks the porosity in this oolitic-peloidal grainstone and also fills some of the dissolved shell fragments; 40X. **d** Pseudomorph of gypsum filling large patches in a lime mudstone; 40X. **e** Large patches of nodular anhydrite replacing lime mudstone, notice the thin fracture and the small stylolite seam in the middle of the photo; 40X. **f** A microfracture filled with dark material through a lime mudstone; 40X (Al-Saad and Sadooni 2011)

an influx of sand and silt from the nearby dry land either by fluvial flow, aeolian processes, or thunderstorms.

Depositional Environment

During the Early Jurassic, the Arabian Gulf was a vast carbonate-evaporite platform. At that time, evaporitic conditions were limited to the northwestern of this platform. By the end of the Early Jurassic, humid conditions predominated, and evaporite deposits gave way to carbonates. During the Middle Jurassic, carbonate deposits covered most of the Arabian Gulf region (Murris 1980). Mixed clastic-carbonate sediments were deposited toward the north and northwest of the region. By the end of the Middle Jurassic, major flooding of the platform led to the deposition of relatively deep carbonate facies under euxinic conditions. These facies are believed to be the source rocks of much of the oil reservoirs in the Arabian Gulf. In the Late Jurassic, an intrashelf basin extended over most of the platform, and siliciclastics were absent.

In the Arabian Gulf, the sediments of the Upper Jurassic are considered to represent an extensive regressive cycle. Furthermore, the Arab Formation is characterized by short-lived transgression–regression cycles, that can be observed throughout the different members of the Arab Formation (Al-Saad and Sadooni 2001).

The base of the Arab C carbonate in the Arabian Basin (including Qatar) is characterized by the presence of the Maximum Flooding Surface J80. A Kimmeridgian age is assigned to this surface based on the presence of the *Kurnubia jurassica* foraminifera zone in a marine unit within sabkha-dominated sediments (Al Silwa et al. 1996; Bouroullec and Meyer 1995; Sharland et al. 2001). The surface represents the onset of evaporite deposition as a result of the global fall in sea level (Haq et al. 1988; Sharland et al. 2001). The greatest sea-level fall occurred in the upper part of the Arab D and was characterized by a sharp change from normal shallow marine to lagoonal-sabkha conditions (Alsharhan and Nairn 1997). This shift is reflected by more frequent anhydrite beds in the formation.

The Arab C reservoir is an example of the cyclic nature of sedimentation within the Arab Formation. The lower part of Arab C reservoir consists mainly of transgressive grainstone facies while the upper part is characterized by regressive lagoonal sediments.

Diagenetic Processes

The depositional fabric and petrophysical properties of most of the lithologies studied have been extensively modified. Only diagenetic processes associated with the reservoir properties are considered here.

Cementation

Both the grains and matrix of grainstones show signs of variable dissolution associated with meteoric pore waters (Fig. 7.51a). Dissolution was associated with both epitaxial cementation and variable pore filling. There are two major types of cement in grainstones. The first type is rim cement formed of small prismatic calcite crystals lining the grain surface and moldic pores (Fig. 7.51b). The second is an

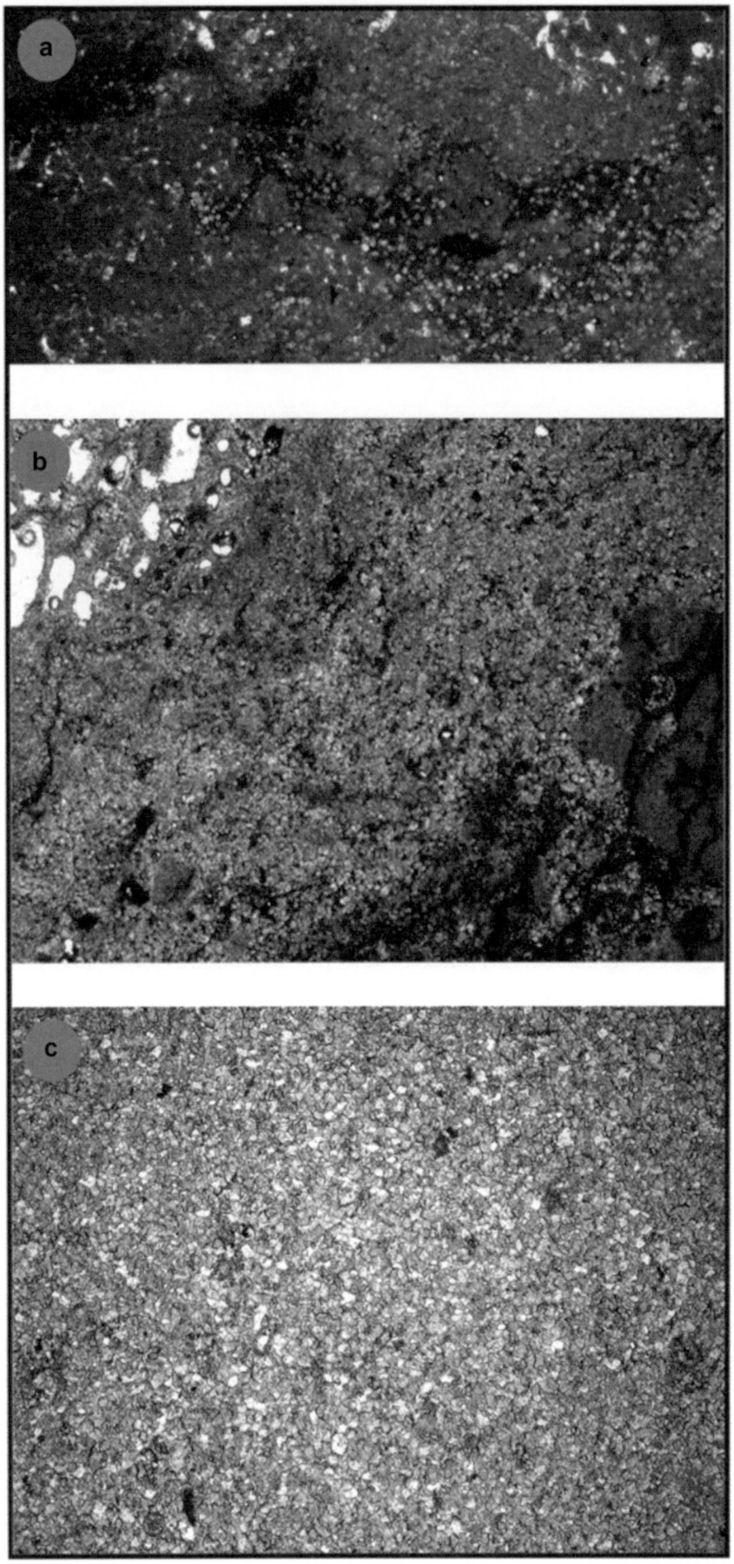

Fig. 7.52 Dolomitization of the Arab C rocks. **a** Aphanitic dolomite following a stylolite seam and associated with oil staining; 40X. **b** Extensive aphanitic dolomitization with only small patches of the original rocks left undolomitized; 40X. **c** Euhedral dolomite rhombs associated with intercrystalline porosity; 40X (Al-Saad and Sadooni 2011)

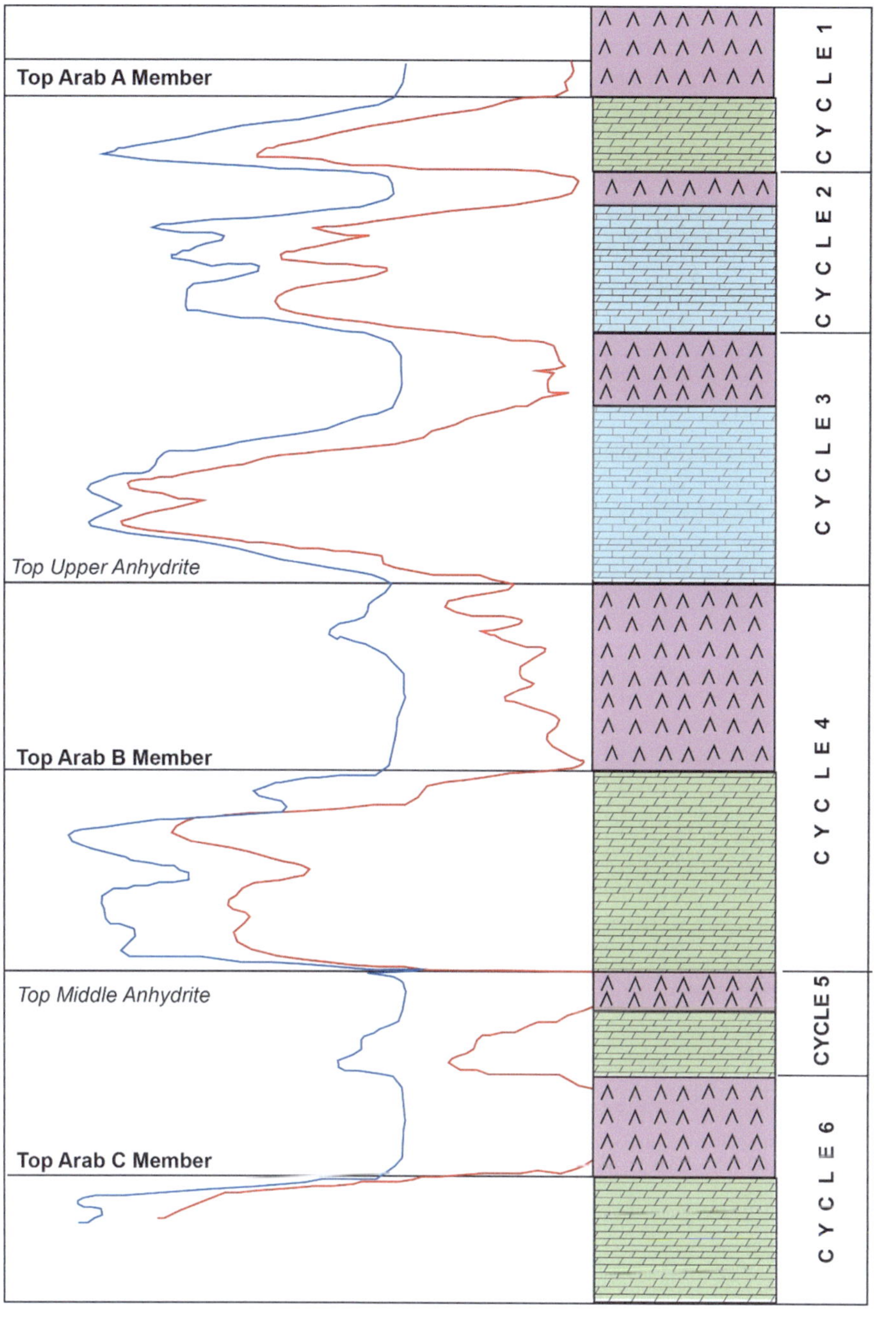

Fig. 7.53 Typical carbonate-evaporite cycles of the Arab A and B members (Focke et al. 1986)

equant granular calcite cement, formed of large crystals filling spaces in the matrix and larger dissolved skeletons of shelly organisms (Fig. 7.51c).

Nodular Anhydrite Replacement

Replacement by nodular anhydrite is a common feature in most of the carbonate sediments of the Arab Formation. The nodules are believed to have formed as small nodules, some of which are enlarged by replacement to their present size. In addition to nodules, there are gypsum needles that are converted to anhydrite by dehydration during shallow burial or upon exposure to atmospheric conditions. Figure 7.51d illustrates patches of gypsum pseudomorphs replacing dark brown micritic limestone. The interior of the patches has the shape of nodular anhydrite, while in Fig. 7.51e, the anhydrite patches replace the depositional fabric.

Fracture

Some lime mudstone facies are characterized by a network of small and microscopic fractures. If present, these fractures probably enhance the permeability of the mudstone-wackestone parts of the reservoir (Fig. 7.51f).

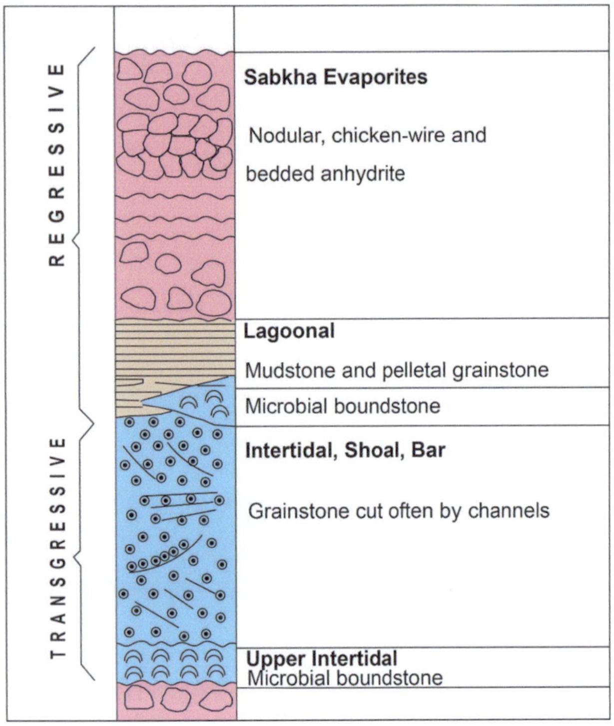

Fig. 7.54 Main sedimentary environments of the Arab A and B members in the offshore subsurface area of Qatar (Focke et al. 1986)

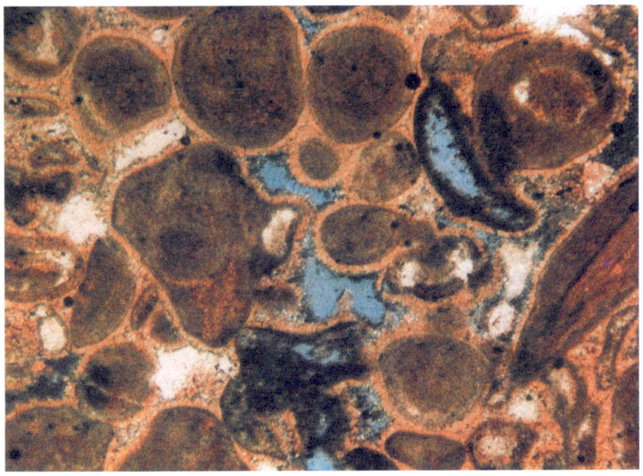

Fig. 7.55 Micrograph of oolitic and peloidal grainstone from the offshore subsurface of Qatar (Focke et al. 1986)

Dolomitization

Dolomite and dolomitic limestone are important lithologies of the Arab C reservoir and contribute to net pay. There is a wide spectrum of grain size and texture in dolomites. The most common is an aphanitic mimetic texture of less than four-micron crystals in size. This fabric sometimes follows microstylolite seams (Fig. 7.52a). Figure 7.52b, c illustrates the other common fabric consisting of a coarse subhedral mosaic. This type of dolomite has intercrystalline and moldic porosity and high permeability. It represents the main reservoir units.

7.3.4.4 The Arab B and A Members

The Arab B and A members represent the upper part of the Arab Formation and its last sedimentary cycles. Their sediments comprise four cycles of carbonate and evaporite (Fig. 7.53). These sediments were formed in a shallow marine setting within subtidal to supratidal environments with some coastal lagoons. Each cycle represents a transgressive–regressive phase. Focke et al. (1986) provided a typical cycle from the subsurface strata of an offshore field in Qatar (Fig. 7.54).

Fig. 7.56 Enterolithic folding of bedded anhydrite from the Arab A member (Focke et al. 1986)

Each cycle starts with a transgression over a supratidal zone of the previous sedimentary cycle and is marked by the deposition of a microbial boundstone in the intertidal zone. This facies has a variable thickness in the subsurface wells due to variations in the old topographic surface and the possible development of buildups, shoals, and bars in the previous cycle. It ranges in thickness between 1.5 and 1.8 m, much less than that in some wells and absent in others.

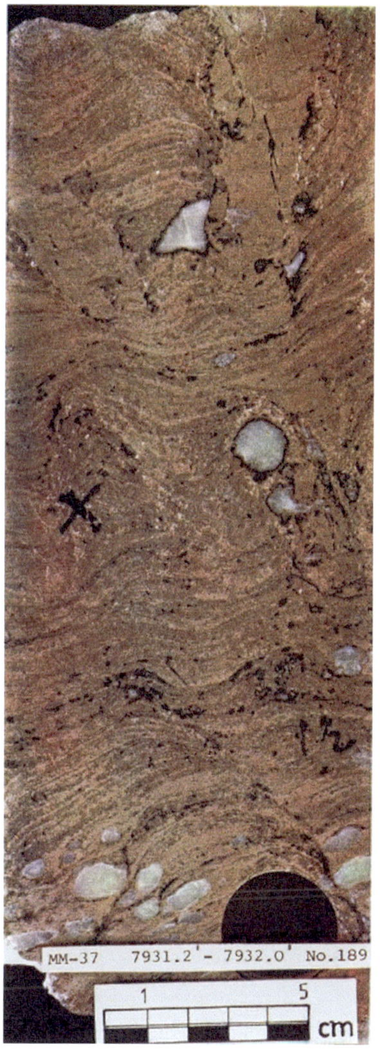

Fig. 7.57 Microbial mudstone with some patches of anhydrite cement (Focke et al. 1986)

Fig. 7.58 Lime mudstone facies with large anhydrite cement patches that may be replacing bioturbation features. The alternation of these facies with anhydrite led to the deposition of the anhydrite cement (Focke et al. 1986)

In some cases, microbial clasts are embedded in the overlying grainstone facies. The microbial boundstone facies is overlain by beds of oolitic grainstone, which were developed in subtidal oolitic shoals. In the second cycle of the Arab A, a lagoonal-channel setting has developed as part of this transgressive cycle. This setting has resulted from restricting water circulation by a complex of shoals and bars. The shoal/bar sediments are formed of well-sorted ooliths with well-developed rim cement (Fig. 7.55).

The regressive phase is characterized by the presence of microbial boundstone followed by the lagoonal lime mudstone and pelletal packstone, and grainstone. They are frequently cemented by anhydrite. Above these lagoonal sediments, other beds of microbial boundstone are formed, indicating a shallowing-up sequence from a subtidal-lagoonal to an intertidal setting. These are difficult to recognize in some cases due to the extensive diagenesis by anhydrite. The final phase of this cycle is the development of widespread deposits of bedded and nodular anhydrite. Nodular and chicken-wire anhydrites are found at the base and the top of the interval. Some of the bedded anhydrite shows enterolithic folding (Fig. 7.56). According to Focke et al. (1986) these structures are due to the growth of anhydrite crystals below the surface of the sabkha. The anhydrite sediments are probably subaqueous and developed in local and isolated pans.

Large parts of the carbonate sediments of both Arab A and B members were subjected to extensive dolomitization, and only thin beds of oolitic grainstone remained intact.

Facies Analysis

As detailed above, the Arab A and B members are formed of carbonates and evaporites. The carbonates are mainly of marine, shallow-water origin and divided into four main facies deposited across a spectrum of depositional environments ranging from intertidal to subtidal.

Fig. 7.59 Schematic vertical lithologic sections showing the main lithologies of the Hith Anhydrite Formation at its type locality at Dahl Hit, Saudi Arabia, as redescribed by Alsharhan and Kendall (1994)

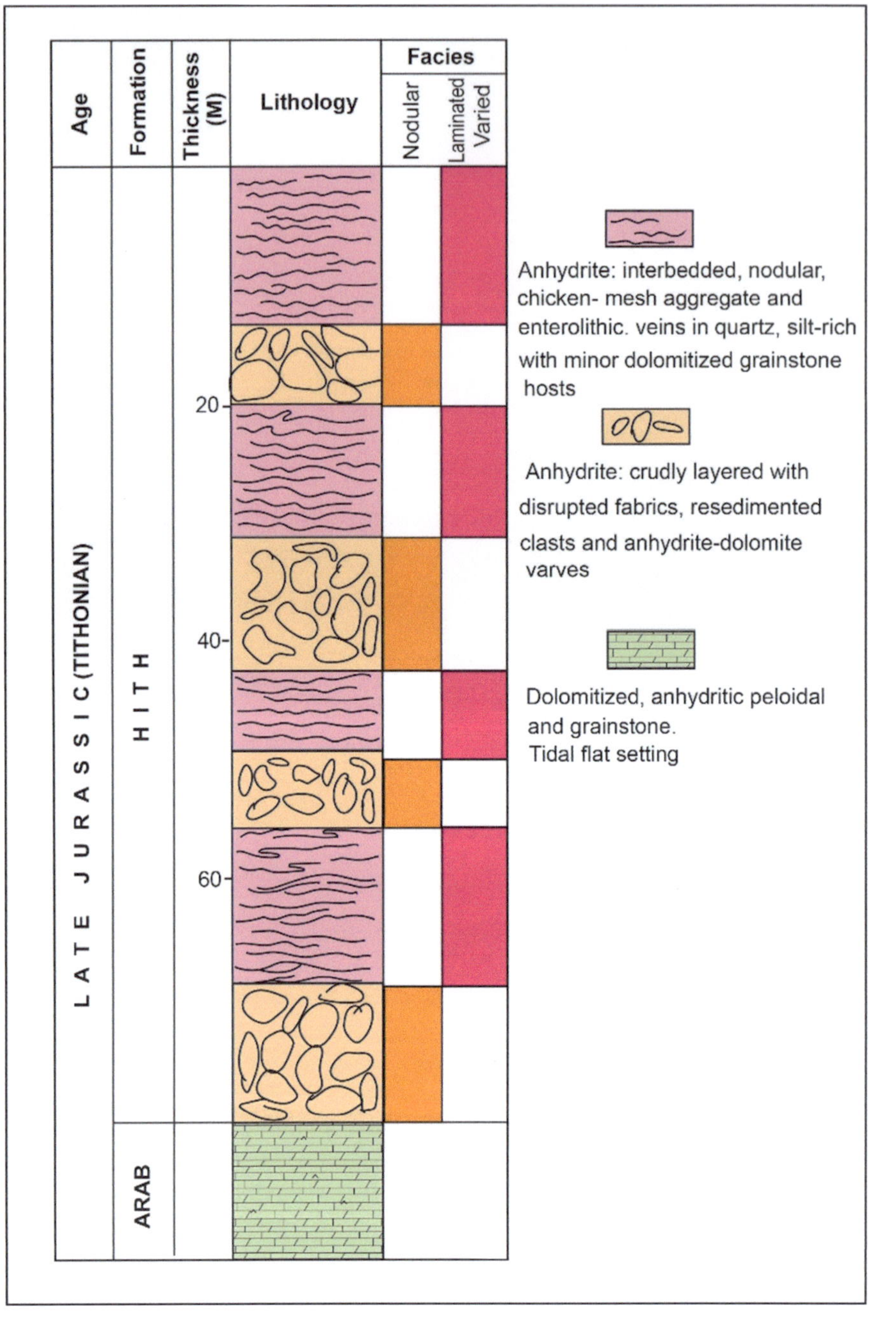

Microbial Boundstone

This facies consists of variable lithologies, and grains bound together by algae or bacteria. These include lime mudstone, wackestone, pelletal packstone, and grainstone. The facies has a laminated texture with distinctive microbial structures similar to those reported from the Recent sabkha sediments of Qatar (See Chap. 2). Most sediments are cemented by anhydrite due to brine water circulation even at later stages, when these sediments are imposed on older evaporite sediments of the older sedimentary cycle (Fig. 7.57). These sediments are deposited in the intertidal zone, where some skeletal and non-skeletal grains are washed by tides to the intertidal zone to be bound together by the bacteria and algae thriving there.

Lime Mudstone

This facies is formed of micrite and very small pellets but underwent extensive dolomitization and dissolution that altered its original textures. It also contains a considerable amount of anhydrite cement. This is probably due to its position below thick anhydrite beds. This facies is formed under lagoonal or tidal flats in a complex of sabkha system. Some

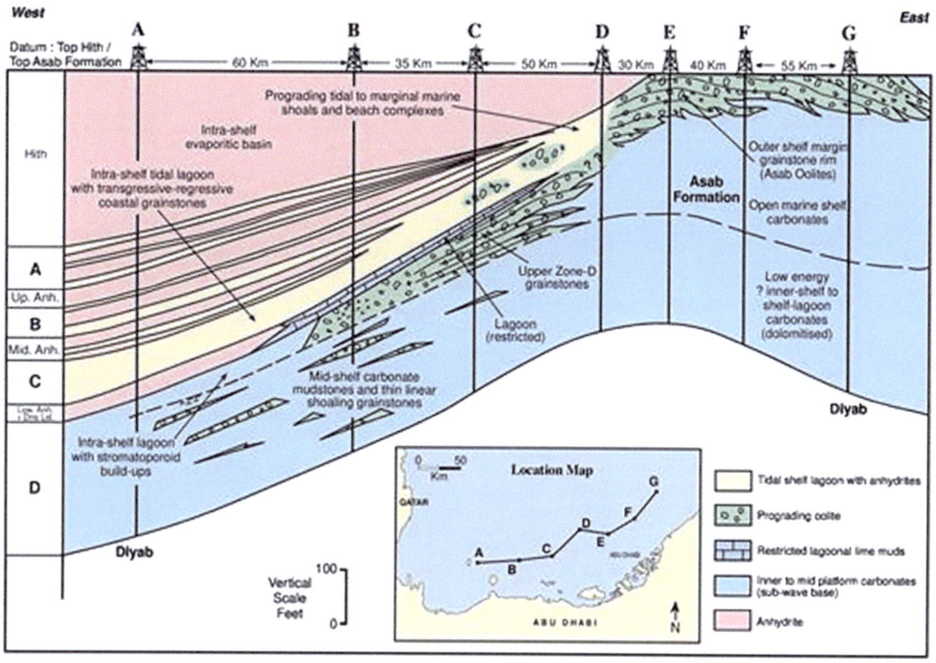

Fig. 7.60 Stratigraphic relationship of the Arab Formation with the Hith Anhydrite Formation along a W-E section of the offshore area of Abu Dhabi illustrating the wedging out of the Hith anhydrite eastward (Al Silwa et al. 1996)

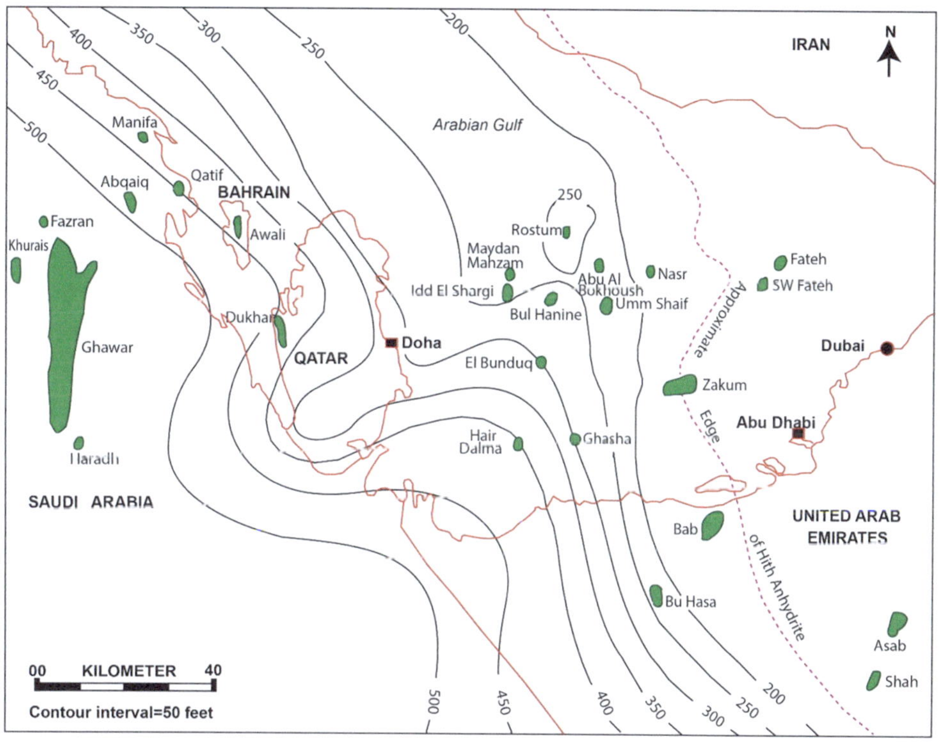

Fig. 7.61 Isopach map of the Hith Anhydrite Formation across the Arabian Gulf region showing the approximate edge of the Hith anhydrite facies (Alsharhan and Kendall 1994)

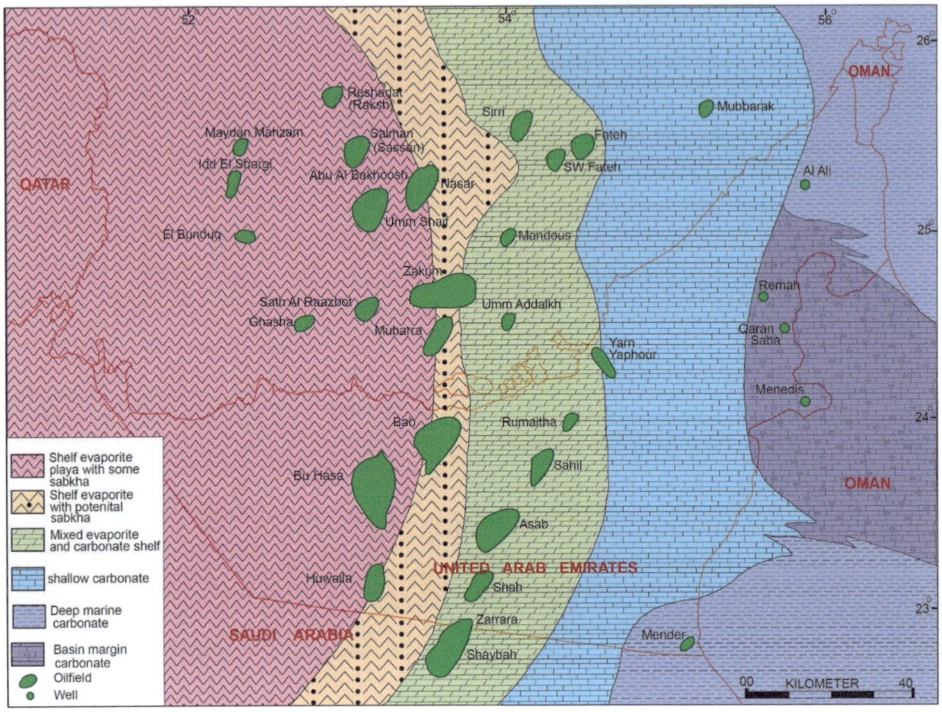

Fig. 7.62 Facies map of the Hith Anhydrite Formation and equivalent sediments across the Arabian Gulf region (Alsharhan and Kendall 1994)

of the anhydrite cement is filling what seems to be bioturbation features, which support the assumption that these sediments were formed in protected environments. These rocks characteristically have low porosity and permeability values (Fig. 7.58).

Oolitic Grainstone

This facies represents the shoals and barriers that punctuated the shallow shelf of the Arab Formation. In some cases, these are buildups of corals, algae, stromatoporoids, and sponges, but in this case, they are formed of oolitic and peloidal shoals. These features differentiate the Arab carbonate shelf into different depositional environments. Since the shelf is relatively shallow, such differentiation may lead to the development of even salina or isolated lakes where evaporites are deposited.

This facies shows considerable variations in components and textures and represents the main reservoir unit in both members. It occurs as thin, sheet-like beds that extend across the whole basin. They are well sorted and have good porosity and permeability values, but some rocks are cemented with anhydrite cement. Others are frequently dolomitized, usually associated with leaching, creating moldic and dissolution porosity. Most of the dissolution porosity is oomoldic in the thick oolitic grainstone units (Fig. 7.55).

7.3.4.5 The Hith Anhydrite Formation

The Hith Anhydrite Formation was named after the Dahl Hit (lat 24° 29′ 18″ N., long 47° 00′ 06″ E). This is a pit 32 km southeast of the city of Riyadh that resulted from the dissolution of evaporite and represents the only exposed section of the formation. The formation was described by Bramkamp and Barger in 1938 and then by Roach in 1952 in unpublished reports. Their description was transcribed by Powers et al. (1966) and appears in Table 7.1.

Accordingly, the Hith Formation in its type section is formed of 90.3 m of bluish-grey anhydrite and anhydrite fragments. The top of the formation was placed at the change from anhydrite to oolitic grainstone, which belongs to the Sulaiy Formation. The base of the formation is placed at the change from anhydrite to limestone and dolomitic limestone of the underlying Arab Formation. The formation is formed mainly of massive and fragmented anhydrite due to dissolution and collapse. Still, it may comprise some interbedded dolomite and limestone in the subsurface, particularly in its upper part (Fig. 7.59).

Also, conformable. It was placed at the appearance of limestone with extensive nodular anhydrite that characterizes the upper unit of the Hith Anhydrite Formation.

In Qatar, a reference subsurface section has been selected in the well Dukhan-25 between drilled depths of 1178–1928 m with a total thickness of 141 m. It consists from top of brown and grey-brown dolomitic limestone with anhydrite nodules (12 m); oolitic, dolomitic limestone (4 m); white, nodular anhydrite with streaks of brown dolomite (35 m); anhydrite, white, massive with rare stringers of dolomite (30 m), white, nodular anhydrite with stringers of brown dolomite with several thin beds of oolitic and pelletal dolomitic limestone (20 m); anhydrite, massive with few

Table 7.1 Type locality detailed description of the Hith Anhydrite Formation in Dahi Hit, Saudi Arabia (Powers et al. 1966)

Unit description	Thickness (m)
Oolite (Upper Jurassic?) of Sulaiy Formation	
Disconformity	
Hith Anhydrite: Limestone breccia, grey-tan, thin to platy, angular block recemented in fine-grained marl matrix; thin beds of grey-tan aphanitic and calcarenitic limestone occur in the lower part. It may represent slumped Sulaiy beds that than residuum of Hith carbonates but interval originally contained at least one beds of anhydrite	12.7
Anhydrite, blue-grey, thin, regularly bedded, common color banding and lamination; some laminae of yellow clay shale in the upper part. Highly irregular sinuous contact at the top with thin beds of anhydrite clearly truncated; knobs and sharp pinnacles of anhydrite extend up to 0.5 m into the overlying breccia	32.0
Anhydrite, blue-grey, massive, finely nodular; few irregular lenses of white gypsum	2.0
Anhydrite, massive; contains alternating bands of blue and white color	9.5
Anhydrite, blue-grey, massive, finely nodular partings	12.0
Anhydrite, blue-grey; some small nodules in a matrix of yellow-weathering, argillaceous dolomite	2.1
Anhydrite, blue-grey, massive, laminated to platy; common sinuous partings and blue and white laminations and banding. Wavy flow structure is prominent in parts of the interval	20.0
Total	90.3

stringers of dolomite (29 m) and anhydrite, while nodular with abundant dolomite stringers changing downward it dolomite with numerous nodules of anhydrite. There are around 2 m of oolitic limestone near the base (10 m) (Sugden and Standring 1975).

The contact with the underlying Arab Formation is conformable and was placed at the top of the clean carbonates of the Member A of the formation. In Qatar, the gradational change to massive anhydrite is considered part of the Hith Anhydrite Formation, although that was not the case with the Saudi type section (Steineke et al. 1958). The contact with the overlying Sulaiy Formation is also conformable. It was placed at the appearance of limestone with extensive nodular anhydrite that characterizes the upper unit of the Hith Anhydrite Formation.

The Hith Formation has a similar distribution to the Arab Formation across the Arabian Gulf region. It was described from the United Arab Emirates by Alsharhan and Kendall (1994) and from the offshore of Abu Dhabi by Azer and Peebles (1995). In this area, the Hith Anhydrite formation wedges out or removed toward the eastern United Arab Emirates and Oman. Equivalent sediments in these areas are oolitic grainstone of the Asab Formation (Fig. 7.60).

Regional variations in both thickness and lithologies of the Hith Anhydrite Formation suggest more complicated depositional environments than being a sabkha system. The two major factors that impacted the whole Upper Jurassic depositional system are the irregular old topographic surface and the shallow depths of the basin. The topographic irregularities led to the formation of isolated salinas and lagoons, while the shallow depth allowed sharp changes in lithologies over a short distance. Figure 7.61 shows an isopach map of the Hith Anhydrite Formation across the Gulf region. The map shows the total thickness of the formation, as well as its anhydrite facies, decline eastward, with the edge of the anhydrite facies passing through the Zakum oilfield in the offshore area of Abu Dhabi. In Qatar, the thickness of the formation ranges between 90 and 150 m, with a thickness of 135 m in the Dukhan oilfield.

These changes in the thickness of the formation also reflect changes in its facies (Fig. 7.62). Evaporites are found mainly in the western part of the basin, where the extreme western areas are characterized by the presence of subaqueous and playa evaporites, followed by supratidal evaporites that give gradually to a mixed carbonate and evaporite facies in the western offshore fields of the United Arab Emirates. In the eastern parts of the UAE and Oman, the evaporites are replaced by shallow and then deep-water carbonates with basin margin shoals of ooliths (Alsharhan and Kendall 1994).

In Kuwait, the Hith Formation consists of two regional lithologies. The lower part comprises clean anhydrite, while the upper part is formed of anhydrite and carbonates. The total thickness ranges from 60 to 200 m, with the carbonate component varying from 15 to 40 m from north to south (Hussein et al. 2015).

As a matter of convenience, most workers in the region put the Jurassic-Cretaceous boundary at the top of the Hith Anhydrite Formation. However, others may argue that the boundary is located slightly higher at the lower part of the Sulaiy Formation. Still, from a sedimentological point of view, it seems more reasonable to consider the upper part of the Hith Formation as an unconformity surface that indicates the end of the Jurassic system across the Arabian Gulf region.

References

Al-Hussaini A, Alnazghah M, Al-Ramadan K, Fallatah M, Polo C (2021) Asymmetrical wave-dominated siliciclastic shorelines with evidence of along-strike variability of sedimentary processes: a revised interpretation for the Toarcian Marrat red beds, central Arabia. Mar Pet Geol 126:104915. https://doi.org/10.1016/j.marpetgeo.2021.104915

Al-Husseini MI (1997) Jurassic sequence stratigraphy of the Western and Southern Arabian Gulf. GeoArabia 2(4):361–382. https://doi.org/10.2113/geoarabia0204361

Al-Mojel A, Razin P, Dera G (2020) High-resolution sedimentology and sequence stratigraphy of the Oxfordian-Kimmeridgian, Hanifa, Jubaila and Arab outcrops along Jabal Tuwaiq, central Saudi Arabia. J Afr Earth Sc 165:103803

Al-Saad H, Sadooni FN (2001) A new depositional model and sequence stratigraphic interpretation for the upper Jurassic Arab "D" reservoir in Qatar. J Pet Geol 24(3):243–264. https://doi.org/10.1111/j.1747-5457.2001.tb00674.x

Al-Saad H, Sadooni F (2011) Stratigraphy, facies analysis and reservoir characterization of the Upper Jurassic Arab "C", Qatar, Arabian Gulf. N Jb Geol Paläont 262(3):359–370

Al-Saad HH, Kendall CG, Hewaidy AG (1992) Lithofacies and depositional setting of the Middle Jurassic Araej Formation in the Dukhan oil field, western Qatar, Arabian Gulf

Al-Saad H, Nasir S, Sadooni F, Alsharhan AS (2002) Stratigraphy and sedimentology of the Hofuf Formation in the State of Qatar in relation to the tectonic evolution of the East Arabian Block, vol 7. Monatshefte. https://doi.org/10.1127/njgpm/2002/2002/426

Al-Silwa MS, Kirkham A, Simmons MD, Twombley BN (1996) New insights into regional correlation and sedimentology, Arab Formation (Upper Jurassic), offshore Abu Dhabi. GeoArabia 1(1):6–27. https://doi.org/10.2113/geoarabia010106

Alsharhan AS (1989) Petroleum geology of the United Arab Emirates. J Petrol Geol 12(3):253–288. https://doi.org/10.1111/j.1747-5457.1989.tb00197.x

Alsharhan AS, Kendall CGSC (1986) Precambrian to Jurassic Rocks of Arabian gulf and adjacent areas: their facies, depositional setting, and hydrocarbon habitat1. AAPG Bull 70(8):977–1002. https://doi.org/10.1306/94886650-1704-11d7-8645000102c1865d

Alsharhan A, Kendall CSC (1994) Depositional setting of the Upper Jurassic Hith anhydrite of the Arabian Gulf: an analog to Holocene evaporites of the United Arab Emirates and Lake MacLeod of Western Australia. AAPG Bull 78(7):1075–1096

Alsharhan AS, Nairn AEM (1997) Sedimentary basins and petroleum geology of the Middle East. Elsevier

Alsharhan A, Whittle G (1995) Carbonate-evaporite sequences of the Late Jurassic, southern and southwestern Arabian Gulf. AAPG Bull 79(11):1608–1630

AlZoukani AM, Babalola LO, Humphrey JD, Kaminski MA, Al-Ramadan KA (2022) Sedimentologic and diagenetic controls on petrophysical properties in the Middle Dhruma Formation (Jurassic), Saudi Arabia. Mar Petrol Geol 146:105935. https://doi.org/10.1016/j.marpetgeo.2022.105935

Aqrawi AA, Goff JC, Horbury AD, Sadooni FN (2010) The petroleum geology of Iraq, vol 424. Scientific Press Beaconsfield

Arkell WJ (1956) Jurassic geology of the world

Ayres M, Bilal M, Jones R, Slentz L, Tartir M, Wilson A (1982) Hydrocarbon habitat in main producing areas, Saudi Arabia. AAPG Bull 66(1):1–9

Azer SR, Peebles RG (1995) Sequence stratigraphy of the Hith/Upper Arab Formations Offshore Abu Dhabi, UAE. In: SPE Middle East oil and gas show and conference

Bashri M, Kaminski MA, Abdullatif O, Humphrey J, Makkawi M, Swennen R, Ali O, Adam A, Salih M, Babiker J (2022) Facies analysis and depositional model for the Oxfordian Hanifa Formation, Central Saudi Arabia. Mar Petrol Geol 146:105940. https://doi.org/10.1016/j.marpetgeo.2022.105940

Beydoun ZR (1988) The Middle East: regional geology and petroleum resources. Scientific Press. https://books.google.com.qa/books?id=oU8ZAQAAIAAJ

Bouroullec J, Meyer A (1995) Sedimentological and diagenetic model of the Arab Formation (Qatar): reservoir implications. Geo 94:236–246

Bramkamp R, Steineke M (1952) Stratigraphical introduction. Jurassic Ammonites from Jebel Tuwaiq, Central Arabia Philos Trans Ser B 236:241–256

Bucur II (1992) Calpionellids and calcispheres from the upper Jurassic-lower Cretaceous deposits in the Resita-Moldova Noua zone, Southern Carpathians, Romania. Cretaceous Res 13(5–6):565–576

Carrigan WJ, Cole GA, Colling EL, Jones PJ (1995) Geochemistry of the Upper Jurassic Tuwaiq mountain and Hanifa Formation Petroleum Source Rocks of Eastern Saudi Arabia. In: Katz BJ (ed) Petroleum source rocks. Springer Berlin Heidelberg, pp 67–87. https://doi.org/10.1007/978-3-642-78911-3_5

Chaube AN, Al-Samahiji J, Al-Husseini M (1995) Jurassic and Cretaceous of Bahrain: geology and petroleum habitat. Middle East Petrol Geosci Geo 94:292–305

Clarke MWH (1988) Stratigraphy and rock unit nomenclature in the oil-producing area of interior Oman. J Pet Geol 11(1):5–60

Cole G, Carrigan W, Colling E, Halpern H, Al-Khadhrawi M, Jones P (1994) The organic geochemistry of the Jurassic petroleum system in Eastern Saudi Arabia

De Keyser TL, Kendall CG (2014) Jurassic and Cretaceous sedimentary fill of intrashelf basins of the Eastern margin of the Arabian Plate. In: 2014 AAPG annual convention and exhibition

de Matos JE, Hulstrand R, Al-Husseini M (1995) Regional characteristics and depositional sequences of the Oxfordian and Kimmeridgian, Abu Dhabi. Middle East Petrol Geosci GEO 94:346–356

Donnadieu Y, Dromart G, Goddéris Y, Pucéat E, Brigaud B, Dera G, Dumas C, Olivier N (2011) A mechanism for brief glacial episodes in the Mesozoic greenhouse. Paleoceanography 26(3)

Dromart G, Garcia J-P, Picard S, Atrops F, Lécuyer C, Sheppard S (2003) Ice age at the Middle-Late Jurassic transition? Earth Planet Sci Lett 213(3–4):205–220

Droste H (1990) Depositional cycles and source rock development in an epeiric intra-platform basin: the Hanifa Formation of the Arabian peninsula. Sedimen Geol 69(3):281–296. https://doi.org/10.1016/0037-0738(90)90054-W

El-Sorogy AS, Galmed MA, Al-Kahtany K, Al-Zahrani A (2017) Microfacies and diagenesis of the middle jurassic Dhruma carbonates, southwest Riyadh, Saudi Arabia. J Afr Earth Sc 130:125–133

Fallatah MI, Kerans C (2018) Stratigraphic evolution of the Late Jurassic Hanifa Formation along the Tuwaiq Escarpment, Saudi Arabia: evidence for a carbonate ramp system. Sed Geol 363:152–180

Farouk S, Al-Kahtany K, El-Sorogy A, El-Motaal EA (2018) High-frequency cycles and sequence stratigraphy of the lower Jurassic Marrat Formation, central Saudi Arabia. Mar Petrol Geol 98:369–383. https://doi.org/10.1016/j.marpetgeo.2018.08.030

Fischer A, Berger W, Crowell J (1982) Climate in earth history. In: Studies in geophysics. Washington, DC, Natl. Academy Press, pp 97–104

Focke J, Munn D, Al Kuwari S, Frikken H, Frei H (1986) Petrographic atlas of rock types, common in the subsurface of Qatar and some recent equivalents. General Petroleum Corporation (Offshore Operations), Qatar, pp 110–189

Goldhammer R, Dunn P, Hardie L (1990) Depositional cycles, composite sea-level changes, cycle stacking patterns, and the hierarchy of stratigraphic forcing: examples from Alpine Triassic platform carbonates. Geol Soc Am Bull 102(5):535–562

Grabowski JGJ, Norton IO (1995) Tectonic controls on the stratigraphic architecture and hydrocarbons systems of the Arabian Plate

Hassan T (1989) The Lower and Middle Jurassic in offshore Abu Dhabi: stratigraphy and hydrocarbon occurrence. SPE Middle East Oil and Gas Show and Conference

Haq BU, Hardenbol J, Vail PR (1988) Mesozoic and cenozoic chronostratigraphy and eustatic cycles. In: In sea level changes: an integrated approach, vol 42. Society of Economic Paleontologists and Minerologists, Special Publication

Hohman J, Guthrie J, Al-Emadi IA, Zahran ME (2005) Sequence stratigraphy of the Jurassic Hanifa Supersequence: a stratigraphic framework for understanding the distribution of marine restricted source rocks in the Hanifa intrashelf basin. In: International petroleum technology conference

Hughes GWaG (2004) Middle to Upper Jurassic Saudi Arabian carbonate petroleum reservoirs: biostratigraphy, micropalaeontology and palaeoenvironments. GeoArabia 9(3):79–114. https://doi.org/10.2113/geoarabia090379

Hughes GW (1996) A new bioevent stratigraphy of Late Jurassic Arab-D carbonates of Saudi Arabia. GeoArabia 1(3):417–434

Hughes GWaG, Varol O, Al-Khalid M (2008) Late Oxfordian micropalaeontology, nannopalaeontology and palaeoenvironments of Saudi Arabia. GeoArabia 13(2):15–46. https://doi.org/10.2113/geoarabia130215

Hussain F, Husain R, Singh P, Rabie A, Prokoso A, Rahman M, Al-Fares A (2015) Hith and gotnia formations: emerging frontier tight oil exploration targets in kuwait. In: International Conference and Exhibition, Melbourne 339–339 https://doi.org/10.1190/ice2015-2204676

Ibrahim MIA, Al-Saad H, Kholeif SE (2002) Chronostratigraphy, palynofacies, source-rock potential, and organic thermal maturity of Jurassic rocks from Qatar. GeoArabia 7(4):675–696. https://doi.org/10.2113/geoarabia0704675

Klemme HD (1993) World petroleum systems with Jurassic source rocks. Oil Gas J (United States) 91:45, Medium: X; Size: Pages: 96–99 2009–2012–2016. https://www.osti.gov/biblio/5816598

Klemme HD (1994) Petroleum systems of the world involving upper Jurassic source rocks: Chapter 3: Part I. Introduction

Lapointe P (1991) Sabkha vs. salt basin model for the Arab Formation understanding in the Umm Shaif field, UAE. In: SPE Middle East oil and gas show and conference

Le Nindre Y-M, Manivit J, Manivit H, Vaslet D (1990) Stratigraphie sequentielle du Jurassique et du Cretace en Arabie Saoudite. Bull Soc Geol Fr 6(6):1025–1034

Leeder M, Zeidan R (1977) Giant Late Jurassic sabkhas of Arabian Tethys. Nature 268(5615):42–44

Lindsay RF, Cantrell DL, Hughes GW, Keith TH, Mueller III HW, Russell SD (2006) Ghawar Arab-D reservoir: widespread porosity in shoaling-upward carbonate cycles, Saudi Arabia

Manivit J, Le Nindre Y-M, Vaslet D (1990) Le jurassique d'Arabie centrale, vol 194. Editions du BRGM

McGuire M, Koepnick R, Markello J, Stockton M, Waite L, Kompanik G, Al-Shammery M, Al-Amoudi M (1993) Importance of sequence stratigraphic concepts in development of reservoir architecture in Upper Jurassic grainstones, Hadriya and Hanifa reservoirs, Saudi Arabia. In: SPE Middle East oil and gas show and conference

Mitchell J, Lehman P, Cantrell D, Al-Jallal I, Al-Thagafy M (1988) Lithofacies, diagenesis, and depositional sequence; Arab-D member, Ghawar field, Saudi Arabia

Murris RJ (1980) Middle East: stratigraphic evolution and oil habitat1. AAPG Bull 64(5):597–618. https://doi.org/10.1306/2f918a8b-16ce-11d7-8645000102c1865d

Noel D, Busson G, Cornee A (1991) The middle-upper Jurassic, pelagic fine-grained limestones are mainly made of oligospecific calcareous nanofossils (cuff-link shaped Coccoliths). Comptes Rendus De l Academie des Sciences Serie II 313(12):1455–1462

Norris MS, Hallam A (1995) Facies variations across the Middle-Upper Jurassic boundary in Western Europe and the relationship to sea-level changes. Palaeogeogr Palaeoclimatol Palaeoecol 116(3–4):189–245

Philby HSJB (1922) The heart of Arabia: a record of travel and exploration, vol. 2. Constable and Co., Ltd.

Powers RW (1968) Asie: Arabie Saoudite. Lexique Stratigraphique Internationale, Paris

Powers RW, Ramirez LF, Redmond CD, Elberg Jr EL (1966) Geology of the Arabian peninsula: sedimentary geology of Saudi Arabia [Report] (560D). Professional Paper, Issue. U. S. G. Survey. https://pubs.usgs.gov/publication/pp560D

Qatar General Petroleum Corporation (EGPC) Staff (1981) Geology of Qatar Schlumberger well evaluation conference. UAE Qatar, Paris 23–32

Read J, Horbury AD (1993) Eustatic and tectonic controls on porosity evolution beneath sequence-bounding unconformities and parasequence disconformities on carbonate platforms: Chapter 11: diagenesis, sequence stratigraphy, and changes in relative sea level

Refaat A, Eltom HA, El-Husseiny A, Humphrey JD (2024) Modeling porosity and permeability evolution of burrow fillings with packstone fabric in the Upper Jurassic Hanifa Formation, central Saudi Arabia: a digenetic backstripping study. Mar Petrol Geol 170:107075. https://doi.org/10.1016/j.marpetgeo.2024.107075

Sadooni FN (1997) Stratigraphy and petroleum prospects of Upper Jurassic carbonates in Iraq. Petrol Geosci 3(3):233–243. https://doi.org/10.1144/petgeo.3.3.233

Sadooni FN, Al-Kuwari HA-S, Sakhaee-Pour A, Mattar WS (2023) Microporosity evolution and destruction in the Jurassic Arab D reservoir, Qatar. Carbonates Evaporites 38(2):45. https://doi.org/10.1007/s13146-023-00869-7

Saint-Marc P (1978) Arabian peninsula. In: The Phanerozoic geology of the world II: The Mesozoic, Amsterdam, Elsevier, pp 435–462

Schorr M, Koch R (1985) Facies zonation of an Upper Jurassic algal-sponge bioherm (Swabian Alb, Germany). Facies 13:227–269

Schweizer V (1987) The Upper Jurassic sponge-algal-facies of the western Swabian Alb (Southwest Germany). Facies 17:197–202

Sharland PR, Casey DM, Davies RB, Simmons MD, Sutcliffe OE (2001) Arabian plate sequence stratigraphy, vol 9. Gulf Petrolink

Steineke M, Bramkamp RA, Sander NJ (1958) Stratigraphic relations of Arabian Jurassic oil. In: American association of petroleum geologists symposium, Tulsa, U.S.A.

Sudarsana A, Abdelouahab M, Chanpong R, Fryer V, Hall J, Vizcarra E (2009) A mid-jurassic carbonate reservoir case study, offshore Qatar; how to capture high permeable streaks in a 3D reservoir model. https://doi.org/10.3997/2214-4609-pdb.151.iptc13581

Sugden W, Standring AJ (1975) Qatar Peninsula. In: Lexique Stratigraphique International. Aise, Fasc. lOa, III, 120

Suzuki M, Osawa M (1987) Sedimentary facies in the Arab formation offshore Abu Dhabi, UA E. Sekiyu Gijutsu Kyokaishi (Japan) 52(1)

Tissot BP, Welte DH (2013) Petroleum formation and occurrence. Springer Science and Business Media

Torsvik TH, Cocks LRM (2017) Earth history and palaeogeography. Cambridge University Press

van Bellen RC, Dunnington HV, Wetzel R, Morton D (1959) Lexique stratigraphique international: Asie. Iraq, vol 3C

van Buchem FSP, Pittet B, Hillgärtner H, Grötsch J, Al Mansouri AI, Billing IM, Droste HHJ, Oterdoom WH, van Steenwinkel M (2002) High-resolution sequence stratigraphic architecture of Barremian/Aptian carbonate systems in Northern Oman and the United Arab Emirates (Kharaib and Shu'aiba Formations). GeoArabia 7:461–500. https://doi.org/10.2113/geoarabia0703461

Van Buchem F, Svendsen N, Hoch E, Pedersen-Tatalovic R, Habib K (2014) Depositional history and petroleum habitat of Qatar

Vaslet D, Al-Muallem MS, Maddah SS, Brosse J-M, Fourniguet J, Breton J-P, Nindre Y-ML (1991) Geologic map of the Ar Riyadh Quadrangle, Sheet 24I [Geoscience Map GM-121, with explanatory notes]. Kingdom of Saudi Arabia, Ministry of Petroleum and Mineral Resources

Vaslet D, Delfour J, Manivit J, Le Nindre Y, Brosse J, Fourniguet J (1983) Geologic map of the Wadi ar Rayn quadrangle, sheet 23H, Kingdom of Saudi Arabia. Saudi Arabian Deputy Ministry for Mineral Resources, Jeddah, Geosciences Map, GM-63A

Wilson A (1985) Carbonate petroleum reservoirs, casebooks in Earth science. Springer-Verlag, New York

Wilson AO (2020) Chapter 1: Introduction to the Jurassic Arabian Intrashelf Basin. Geol Soc London, Memoirs 53(1):1–19. https://doi.org/10.1144/M53.1

Wirsing G, Koch R (1986) Shallow marine algal-sponge bioherms (Upper Jurassic, Middle Kimmeridgian: Swabian Alb). Facies 14:285–307

Wood GV, Wolfe MJ (1969) Sabkha cycles in the Arab/Darb formation off the Trucial coast of Arabia. Sedimentology 12(3–4):165–191. https://doi.org/10.1111/j.1365-3091.1969.tb00873.x

Yousif S, Nouman G (1997) Jurassic geology of Kuwait. GeoArabia 2(1):91–110

Ziegler MA (2001) Late Permian to Holocene paleofacies evolution of the Arabian Plate and its hydrocarbon occurrences

8 The Cretaceous

Abstract

The Cretaceous period represents considerable geological and climatic changes throughout the Arabian Plate, profoundly affecting depositional ecosystems. This chapter examines the Cretaceous formations with significant hydrocarbon reserves in Iraq, Kuwait, Bahrain, Qatar, and the UAE. During this age, the tropical placement of the Arabian Plate inhibited evaporite formations, promoting wetter conditions that resulted in substantial clastic and carbonate deposits. The Neo-Tethys rifting culminated, extending the Cretaceous Basin boundaries beyond the Jurassic coastlines. Carbonate sedimentation prevailed, characterized by intricate ramp and intrashelf basin formations mixed with clastic influxes. Numerous reef-building creatures, especially rudists, became dominant, forming carbonate platforms that diversified vertically and horizontally due to tectonic movements. Tectonic action, such as southwest tilting and salt tectonics along the Qatar Arch, affected regional stratigraphy. This chapter analyzes important formations within this period, specifically the Sulaiy and Yamama formations of the Early Cretaceous, formed of continuous shallow-water carbonates with local intrashelf basins. The sedimentological evolution of formations such as Ratawi, Kharaib, Shu'aiba, Nahr Umr, Mauddud, Mishrif, and Simsima is elaborated, with each formation enhancing the region's hydrocarbon potential. Focus is directed toward facies analysis and depositional settings, demonstrating transitions from carbonate to clastic-dominated systems and the intricate interactions among tectonics, sea-level fluctuations, and climatic factors that influenced the Cretaceous strata of the Arabian Plate.

8.1 Introduction

The Cretaceous time of Arabia was a time of profound changes in climate and depositional settings, leading to the deposition of thick and varied sediments across the Arabian Plate. These strata host large hydrocarbon reserves in some parts of Arabia, such as Iraq, Kuwait, Bahrain, and the United Arab Emirates (Fig. 8.1).

The Cretaceous time witnessed the positioning of the Arabian Plate in the Tropics, thus reducing the amount and extent of the evaporites that characterized the Triassic and Jurassic strata of Arabia. The rifting of the Neo-Tethys reached a new peak, where the limits of the Cretaceous Basin extended beyond most of the Jurassic shorelines. The relatively wet climate led to the deposition of thick clastic sediments in some parts of the plate that were derived from western Arabia, while other areas were characterized by varied carbonates with apparent vertical and horizontal variations due to tectonic differentiation (Sadooni and Aqrawi 2000). Several types of reef-builders, particularly rudists, became predominant during that time. Others include sponges, stromatoporoids, and algae. The dominant sedimentation regime is complex carbonate ramps with platforms and intrashelf basins occasionally receiving clastic influxes (Van Buchem et al. 2014). The advent of the Cretaceous in Arabia was also associated with the regional tilt and subsidence of the Qatar Arch in a south-southwest direction, and increased salt-associated tectonics resulted from the accelerated upward movement of the Hormuz Salt (Alsharhan and Nairn 1994).

The Jurassic-Cretaceous boundary in Qatar is abrupt and marked by the change from the evaporites of the Hith Anhydrite Formation to the argillaceous sub-basinal facies of the

F. N. Sadooni and H. A. Al-Kuwari, *Geological Evolution of Qatar and the Arabian Peninsula*, Regional Geology Reviews,
https://doi.org/10.1007/978-3-031-94394-2_8

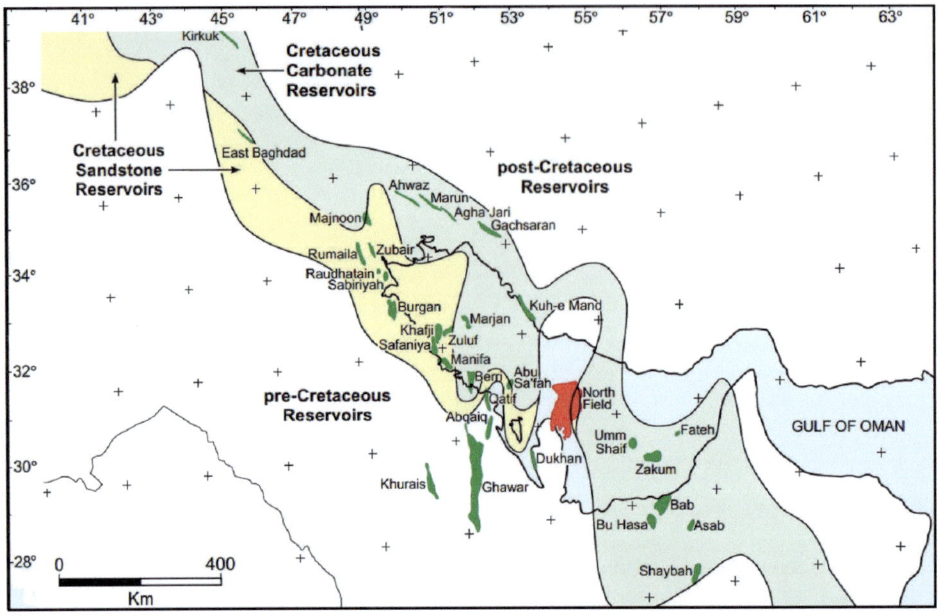

Fig. 8.1 The main Cretaceous oilfields in the Arabian Plate (Christian 1997)

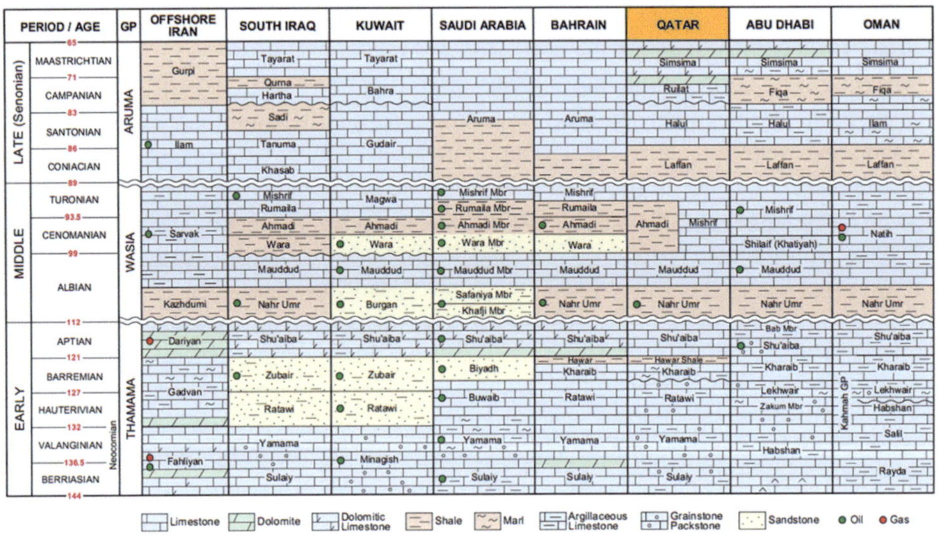

Fig. 8.2 Regional Cretaceous lithostratigraphy of the Arabian Gulf (Christian 1997)

Sulaiy Formation after the early Late Jurassic flooding that covered most parts of the Arabian Plate. That shift resulted from climatic, eustatic sea-level changes, regional and local tectonic changes, and the possible dissolution of the Upper Jurassic evaporitic topographic surface that led to the development of local basins.

The Arabian Cretaceous basin is divided into Lower, Middle, and Upper Cretaceous time slices, and their strata were assigned to three stratigraphic groups. These groups are Thamama, Wasia and Aruma (Fig. 8.2).

8.2 The Early Cretaceous (Berriasian-Valanginian)

The deposition of continuous shallow water characterized this interval of carbonates over most parts of the Arabian Plate (the Yamama, Minagish, and Habshan formations), while the previous areas that were covered by the Gotnia Basin and other similar residual local depressions witnessed the deposition of the sub-basinal, argillaceous limestone of the Sulaiy and Makhul formations. These sediments are wedging eastward

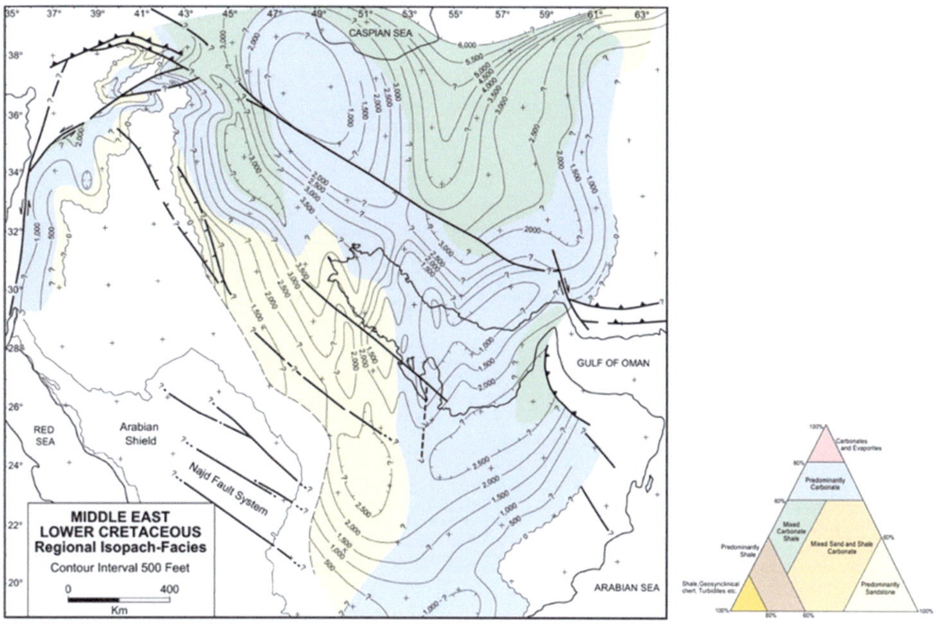

Fig. 8.3 General isopach map of the Lower Cretaceous strata over the Arabian Plate. Different lithologies are shown in the accompanying triangle. The contour interval is 500 ft. (Christian 1997)

and absent in Oman (Ziegler 2001). A general isopach map of the Lower Cretaceous sediments over the Arabian Plate is shown in Fig. 8.3.

The map shows the clastic sediments that were derived from the Arabian Shield in yellow (these are assigned to the Zubair Formation, which is a major hydrocarbon reservoir in Iraq and Kuwait), the carbonate sediments appear in blue, and the mixed carbonate and shale strata appear in green. These are assigned to the Thamama Group, which has various formation names across political boundaries in some cases. The average thickness of the Lower Cretaceous sediments in Qatar ranges between 300 and 450 m.

The Berriasian-Valanginian interval was characterized by the deposition of open-shelf carbonates over most parts of the Plate, with some shale and argillaceous materials in local intrashelf basins. In some parts of Oman, deep-water marine clastics were deposited, and some clastics in the Levant and western Iraq. According to Ziegler (2001), many parts of the Arabian Plate were affected by the late Valangian unconformity (Fig. 8.4).

In Qatar, this interval is characterized by the deposition of the argillaceous limestone of the Sulaiy Formation and the relatively cleaner limestone of the Yamama Formation. This section is poorly documented in Qatar because it has no commercial hydrocarbon reserves.

8.2.1 The Sulaiy Formation

The Sulaiy Formation was originally defined by H. L. Burchfiel and F. W. Hoover in an unpublished report in 1935 from Wadi, "As Sulayy," which is, according to Powers et al. (1966), a channel filled with gravels at the foot of the Hit escapement, Central Saudi Arabia. In 1938, Bramkamp and Barger defined the Riyadh Formation to include (in our present terms) from the top down the Yamama detrital limestone, Sulaiy limestone, Hith Anhydrite, and the Lower Riyadh members. When certain Cretaceous fossils were reported from the Sulaiy and Yamama members, they were removed from the Riyadh Formation and defined as a new formation, the Thamama, to include, in addition to these two members, a third unnamed member. The Thamama was upgraded into a group by Steineke and Bramkamp in 1938 (Bramkamp & Steineke, 1952), and its members were named and raised to the status of formations. These were named the Buwaib, Yamama, and Sulaiy from the top. Later, the upper part of the Thamama Group was redefined, and another formation, the Biyadh Sandstone, was included.

In its type locality, the top of the formation is placed at the change from the chalky lime mudstone of the Sulaiy Formation to the packstone and wackestone facies of the Yamama Formation. The lower contact is placed at the change from the limestone breccia of the Hith Anhydrite Formation to the calcarenite of the Sulaiy. The formation is outcropping an area that extends for around 350 km. The relationship between the Sulaiy and Yamama formations remained controversial till recent years (Sadooni 1993). It is agreed that the Sulaiy characteristically has predominant argillaceous limestone lithologies, while the Yamama Formation has more grain-supported textures. Furthermore, the top of the Sulaiy Formation marks an important faunal division. The benthonic foraminifera *Pseudocyclammina sulaiyana* Redmond is found to be confined to the Sulaiy Formation. In the subsurface sections, the top of the Sulaiy Formation

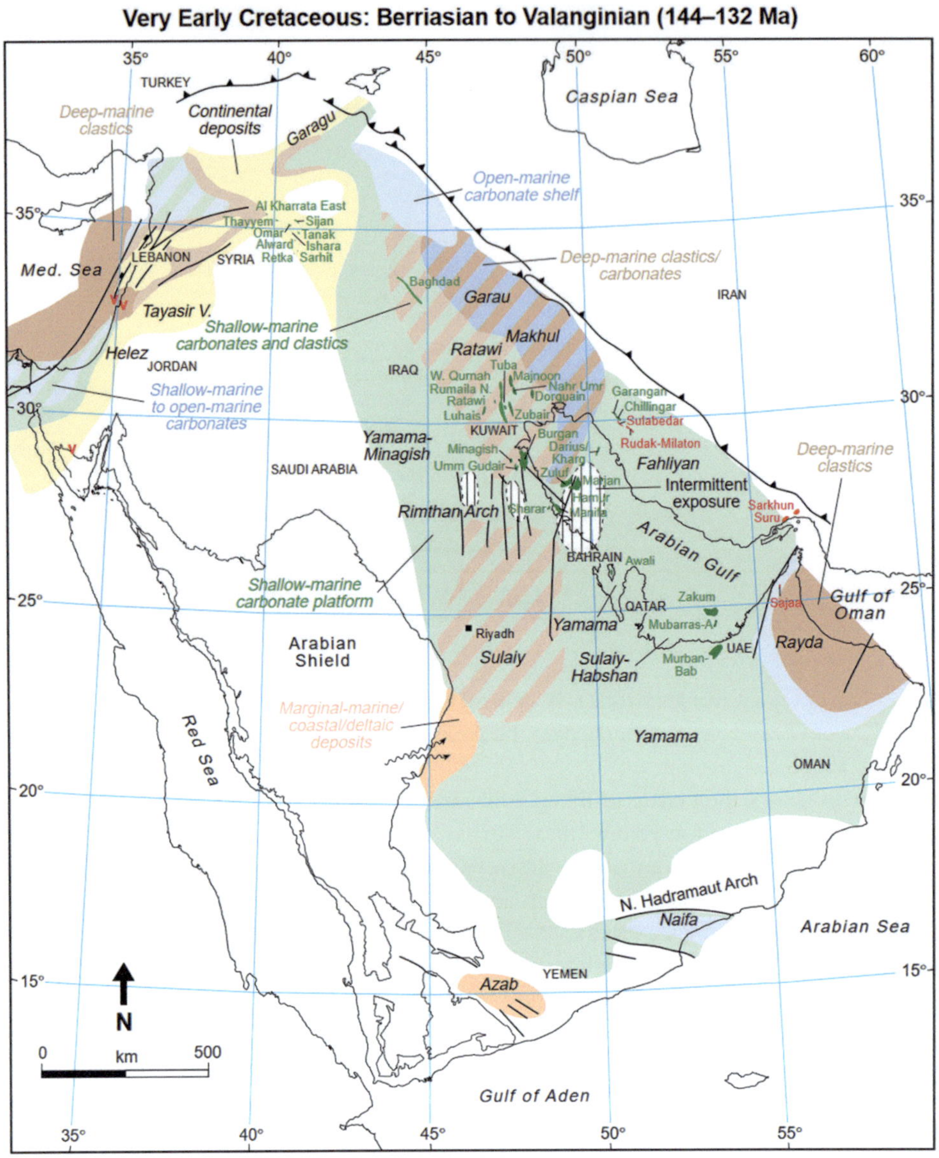

Fig. 8.4 General facies map of the Berriasian-Valanginian interval of the Arabian Plate. The green areas are carbonates, and the dark brown and yellow areas are deep and shallow clastics (Ziegler 2001)

coincides with a major lithologic change characterized by a unique log signature.

In the type locality, the Sulaiy comprises alternations of lime mudstone to pelletal, oolitic, and lithoclastic wackestone and packstone, sometimes partially dolomitic and recrystallized or tightly cemented. There are some horizons of molluscan coquina and scattered quartz grains at the base of the formation. According to the present population of large benthonic foraminifera, the age of the Sulaiy Formation is considered to be Tithonian to early Valanginian (Fig. 8.5).

The presence of a certain gastropod Pterocera and coccoliths, including the species *Conusphaera mexicana minor,* may indicate a Tithonian age to the lower part of the formation (Hughes and Naji 2008; Vaslet et al. 1991; Wolpert et al. 2015). A probable Berriasian age is assigned to the upper part of the formation in some areas of Saudi Arabia based on the presence of some rare nannoflora, including *Watznaueria barnesae* and *Haqius circumradius* (Vaslet et al. 1991). These age assignments may suggest that the Jurassic-Cretaceous boundary is located within the lower part of the Sulaiy Formation and not at the top of the HithAnhydrite Formation.

The type section of the Sulaiy Formation in the Dahal Hit has been re-examined in detail by Wolpert et al. (2015) (Fig. 8.6). They reported that the section above the Hith Anhydrite is around 116.2 m thick. The Sulaiy Sequence Boundary (SB) was located at the base of the well-bedded Sulaiy Formation. The Lower Sulaiy Sequence is a transgressive–regressive sequence, with the transgressive part characterized by the presence of bioclastic packstone, which was considered a storm deposit with oolitic grainstone. There is a conspicuous biostromes around 8 m below the Maximum Flooding

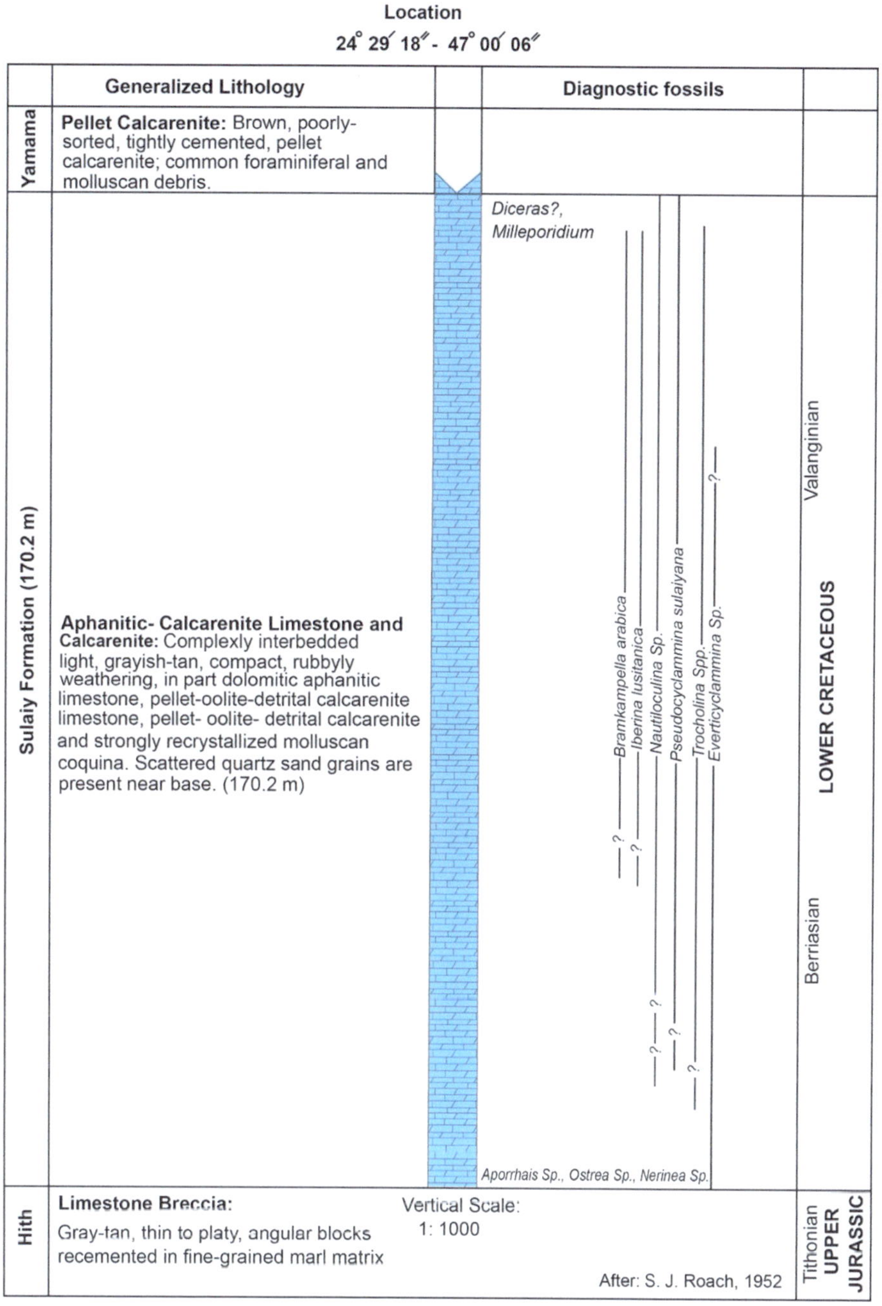

Fig. 8.5 Lithologic section and fossil content of the Sulaiy Formation in its type locality in Saudi Arabia (Powers et al. 1966)

Surface (MFS), representing the MFSK10 of the Lower Berriasian. The Maximum Flooding Surface (MFS) is selected in a section dominated by lime mudstone and waclestone. The regressive leg of the sequence is formed of tempestites (storm deposits) with a larger amount of coarse bioclastic packstone toward the top.

8.2.1.1 Facies Analysis

Facies analysis of the Sulaiy Formation from its type locality in central Saudi Arabia and other outcrops was conducted by Wolpert et al. (2015). They recognized nine distinctive facies. The following summary is based on their work.

Laminated and bioturbated lime mudstone: This microfacies comprises heavily bioturbated lime mudstone. The bioturbated areas are filled with wackestone and packstone with fine skeletal debris. This facies is probably formed in a distal offshore setting or subbasin.

Graded wackestone: The main skeletal components of this facies are molluscan shells and shell debris with some bioclasts and peloids. Some burrows are filled with coarser materials, probably derived from the overlying storm deposit. This facies represents a distal storm deposit with an eroded base and fining-up succession.

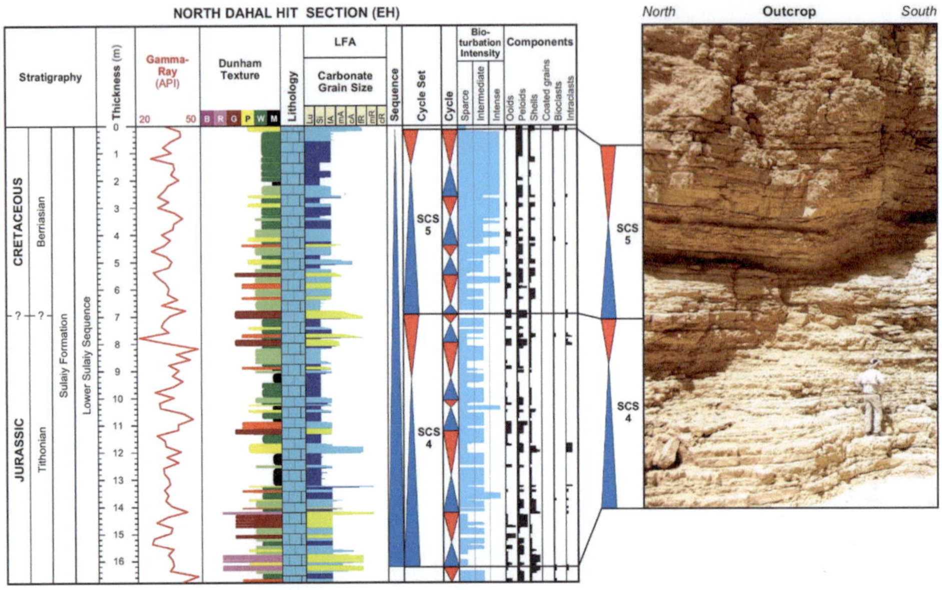

Fig. 8.6 A detailed description of the Sulaiy Formation in a location just north of its type locality (Wolpert et al. 2015)

Fig. 8.7 The buildup of the Sulaiy Formation as a well-recognized marker in the field (Wolpert et al. 2015)

Bioturbated wackestone—Packstone: This facies is formed of nodular, coarser skeletal grains such as gastropods, bivalves, brachiopods, and terebratula. Some burrows are filled with wackestone and packstone, and other bioturbation features disturb the original textures, although some traces of primary layering can be noticed. These sediments may represent storm deposits heavily reworked by burrowing and bioturbation.

Buildup Boundstone: This facies is a well-recognized marker within the Sulaiy Formation and can be easily spotted in the field by its big molds and may extend for several kilometers. It represents a complex buildup of corals, bivalves, gastropods, echinoderm, shell fragments, and bioclasts (Fig. 8.7).

Bioturbated packstone-grainstone: This facies is formed of macro and micro-shell fragments with large gastropods and some echinoderm debris. It can be recognized in the field by the nodular texture due to the extensive burrowing and bioturbation. It represents storm deposits in a foreshoal setting. It was probably bioturbated extensively in an early diagenetic phase.

Graded packstone-grainstone: This facies is recognized by its graded bedding. Its base is formed of packstone and grainstone of shell fragments, gastropods, bioclasts, and lithoclasts changing upward into a finer grainstone with shell debris and peloids. The contact between the two units is erosive. Bioturbation is observed in the upper parts of the facies with gutter casts that characterize storm deposits in deep sedimentary settings. These sediments are storm deposits in a foreshore setting.

Bioclastic floatstone-rudstone: This facies is characterized by large skeletal grains (up to 8 cm), mainly bivalves and gastropods with lithoclasts and bioclasts. There is considerable bioturbation that disturbs the original bedding. The components' large size and sorting suggest storm deposits that formed under high-energy conditions.

Well-sorted packstone-grainstone: The main component of this facies is well-sorted shell fragments with low-angle cross-bedding. It also includes some bivalves, lithoclasts, and peloids that are occasionally bioturbated. The overall well-sorting and cross-bedding indicate deposition under high-energy conditions. It may represent deposition in a shoal margin setting. These components were derived from the nearby shoal by waves and submarine erosion.

Well-sorted, cross-bedded oolitic and peloidal grainstone: This a typical oolitic shoal facies formed of well-sorted oolites and peloids with high-angle cross-bedding. Other skeletal components are limited, including some brachiopods and bivalves that can survive in the strongly agitated conditions of the oolite shoal. Sediments are partially dolomitized also.

8.2.1.2 The Sulaiy Formation in Qatar

The reference section of the Sulaiy Formation in Qatar was selected at the well Dukhan-27 between the drilled depth of 1615–1753 m with a total thickness of 138 m. The Hith Anhydrite Formation underlies the formation, and the contact between the two formations is conformable and was placed at the first appearance of nodular anhydrite. The upper contact with the Yamama Formation is conformable also and was put at the change from the lime mudstone of the Sulaiy to the porous pelletal limestone of the Yamama Formation (Sugden and Standring 1975).

The formation is poorly studied in Qatar, and generally, it consists of dense grey mudstone at the bottom, changing into slightly porous at the top. The middle and upper parts are slightly dolomitic. A thin bed of pelletal or oolitic limestone has been reported from the lower part of the formation (Fig. 8.8).

Depositional Environments

The above-mentioned facies described by Wolpert et al. (2015) from the type locality of the formation in central Saudi Arabia include facies that are usually assigned to the Yamama Formation, mainly the oolitic shoal and the well-sorted "detrital" limestone. However, this is a detailed description of the type locality, and the Sulaiy Formaion may have more varied facies that are encountered in the subsurface of eastern Arabia, where the dominant facies are argillaceous and coccolith-bearing limestones.

Figure 8.9 represents the Sulaiy Formation shelf as described by Wolpert et al. (2015). Accordingly, the facies of the Sulaiy Formation occupy a carbonate ramp and span a wide spectrum of different depositional environments ranging from supratidal to offshoal with an oolitic shoal with back and foreshoal facies. Deposition occurred under the influence of storm and wave energy, which may explain the presence of the large floatstone and rudstone rock assemblages.

8.2.2 The Yamama Formation

The Yamama Formation type section is around 20 km south of Al Yamamah town in Saudi Arabia. The nomenclature of this formation was controversial and changed from an un-named part of the Tuwaiq Formation, then defined as a combined formation with the Sulaiy Formation due to their interrelated lithologies and then as an independent formation within the Thamama Group. The type section was described from outcrop patches in the area.

The first formal definition of the formation was published by Steineke et al. (1958) from a section between the lime mudstone of the Sulaiy Formation below and the pre-Buwaib unconformity above. The thickness of the formation in its type section is around 45.5 m. The lower contact is placed at change from the chalky mudstone and calcarenite of the Sulaiy Formation to the golden-brown calcarenite of the Yamama above. The upper contact is unconformable and marked the change from the golden-brown pelletal calcarenite below to the mollusks-bearing calcirudite of the Buwaib above (Fig. 8.10). The general difference between the Sulaiy and the Yamama formation is that the Yamama has more grain-supported limestone compared to argillaceous mudstone of the Sulaiy. All over the Gulf region, the facies of the two formations are mixed up or misidentified, but this is the basic difference.

The Sulaiy and Yamama formations are equivalent to the Minagish Formation in Kuwait and the Habshan Formation in Abu Dhabi.

8.2.2.1 The Yamama Formation in Qatar

The Yamama Formation is poorly studied in Qatar. The reference section of the formation is selected in the well Dukhan-26 between the drilled depths of 1553–1674 m with a total thickness of 120 m. The Formation consists at the base of around 25 m of light grey, chalky limestone with variable amounts of pellets, followed by 66 m of pelletal mudstone and wackestone with occasional oolitic horizons, including some relatively coarse to fine lithoclasts made of carbonate materials. The upper part comprises around 29 m of mudstone, which is pelletal at the lower part with some local porosity development. The overlying formation is the Ratawi. The contact seems conformable, but regional data may suggest the presence of a hiatus between the two formations. The boundary is placed at the change from porous pelletal and oolitic packstone and wackestone of the Yamama to the argillaceous limestone at the base of the Ratawi Formation. The underlying formation is the Sulaiy. The contact is conformable and placed between the porous pelletal wackestone and packstone of the Yamama Formation and the lime mudstone of the Sulaiy Formation (Sugden and Standring 1975).

Facies Analysis

The facies of the Yamama Formation have not been studied in Qatar because these strata have not been tested for their hydrocarbon potential. However, judging from the few available data, they seem similar to the facies of the formation in Kuwait (Minagish Formation) and the Yamama Formation in southern Iraq. The following synthesis attempts to integrate the facies of the formation in these two countries to provide an approximate conception of its facies in the subsurface of Qatar.

In southern Iraq, the Yamama Formation is divided into three reservoir units (YRA-C) separated by two barrier units

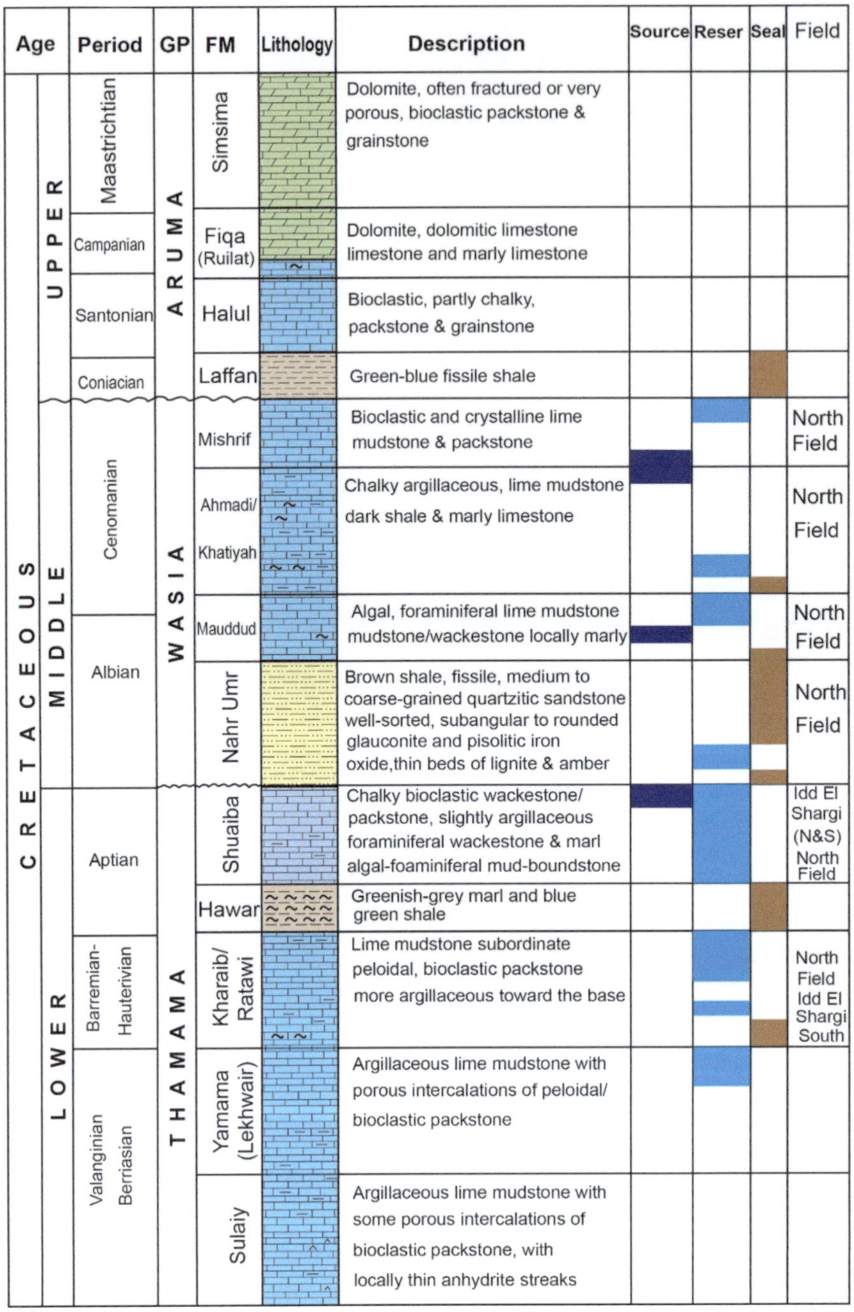

Fig. 8.8 The stratigraphic column of the Cretaceous strata in Qatar (Alsharhan and Nairn 1994)

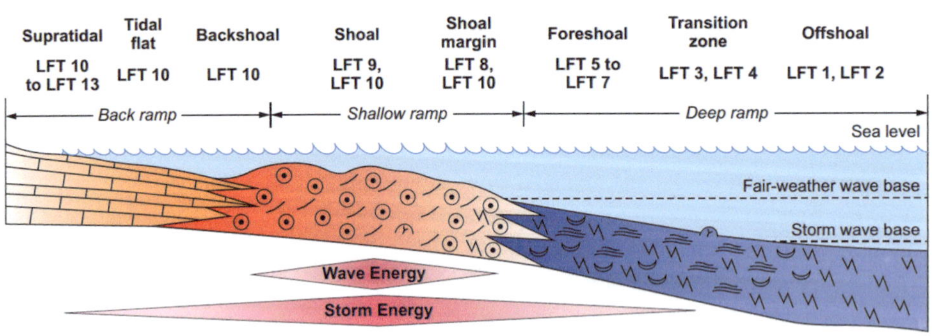

Fig. 8.9 Depositional environment of the Sulaiy Formation in its type locality in central Saudi Arabia (Wolpert et al. 2015)

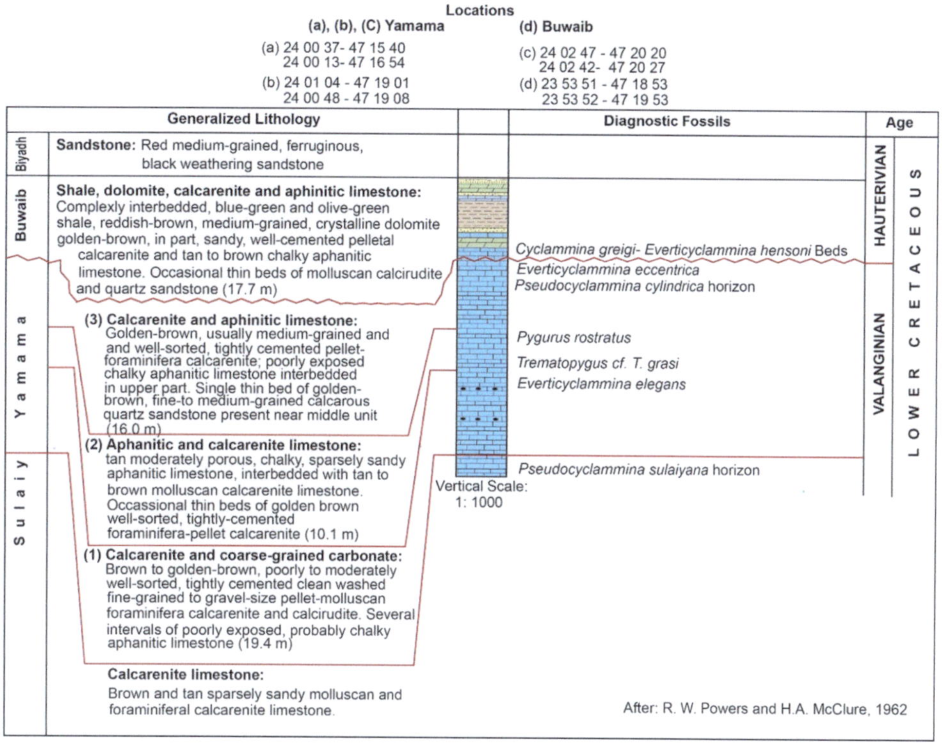

Fig. 8.10 Description of the Yamama Formation in its type section in Saudi Arabia (Powers et al. 1966)

formed of dense mudstone (YB-1 and 2). Six microfacies have been described from the oil wells in southern Iraq (Sadooni 1993). Most of these are similar in nature to those described in Kuwait.

Algal-Bearing Wackestone and Packstone

Two variants of this facies have been reported depending on the type of the present algae. The dasycladacean wackestone and packstone subfacies consists of green algae. These were extensively dissolved and replaced by saprry calcite cement. Usually, this subfacies also contains some shell fragments and micritized benthic foraminifera. The matrix is slightly argillaceous. The second subfacies is the red algae wackestone-packstone, which contains mainly fragments of the *Permocalculucs*, which is the most predominant red algae genus in the Yamama Formation characterized by their perforated skeleton.

The green algae usually flourish in partially protected subtidal environments, preferably in depths ranging between 3–5 m under tropical to subtropical conditions. Generally, lagoons and protected reef flats are their favored environments. The red algae prefer normal marine conditions in tropical and subtropical areas. They are found in lower tidal areas a few meters below sea level, but their water depth does not exceed 90 m. Their favored environments are hard sea floors such as reefs and open marine environments.

Large Benthic Foraminifera-Bearing Wackestone and Grainstone

The dominant grains in this facies are large benthic foraminifera that are either micritized, encrusted, or pyritized with shell fragments and green algae. Some of these foraminifera are *Pseudocyclammina* spp., *Everticyclammina* spp., *Brakmpella* spp., *Orbitolina* spp., and many others. These foraminifera prefer shallow, normal marine shelves and forereef environments.

Oolitic Packstone-Grainstone

These oolitic sediments are found as longitudinal bodies parallel to the shorelines or on the crest of the growing structures in the basin. Some ooliths were transported either to the deeper parts of the basin, where they were abraided and deformed, or to the back lagoons, where they were micritized and pyritized according to the prevalent wind directions during deposition. This facies was deposited in shoals or barriers in the basin margins or the crests of the growing structures when the water depth and energy were favorable for their formation.

Peloidal Packstone-Grainstone

Peloids in the Yamama sediments are of different sizes, shapes, and sorting. Some were ooliths moved to the more sheltered parts of the basin and got micritized; others were fecal pellets formed and preserved in back lagoons. The considerable thickness and wide distribution of these sediments suggest that they were accumulated in depressions or protected areas subjected to microbial activities that led to their micritization and boring.

Lithoclastic Wackestone-Packstone

The early reports about the Yamama Formation always mentioned that the Yamama Formation contains "detrital or fragmental limestone" (Al-Siddiki 1978; Powers et al. 1966; van Bellen et al. 1959). This facies consists of varied lithoclasts; some are genuine intraclasts formed of typical Yamama materials, and others are probably diagenetic, resulting from micritization or pressure solution. Intraclasts are probably the result of wave action or gravitational flow of unconsolidated sediments on slopes and flanks of growing structures.

Stromatoporoid-Sponge-Coral Boundstone

This facies represents local buildups formed of boundstone of a combination of stromatoporoid, sponges, corals, and even rudist, as mentioned by Nath et al. (2014) from Kuwait. Most of these organisms are builders and can generate wave-resistant structures. They are capable of baffling mud, forming mud mounds or buildups. So, these are probably isolated buildups scattered in the Yamama basin.

Depostional Environment

The Yamam Formation has a relatively wide distribution over most of the Arabian Gulf region. It was deposited in a shallow-water carbonate ramp, with the major deposition occurring in the inner ramp in the form of oolite shoals. Storms and waves may transport these ooliths to the back of the shoals, where they were accumulated and micritized, forming the peloidal facies. The ramp was punctuated with buildups wherever conditions allowed. There are small mounds of sponges, corals, coralline algae, and even rudists, as reported by Nath et al. (2014) from Kuwait. Slowly growing structures as a result of the Hormuz halokinesis create mounds that may influence deposition and energy. In these high areas, some buildups may flourish. Sadooni (1993) reported striking facies changes on the same structure, where grain-supported facies were described from the crest of the anticline, while flankish wells show more argillaceous, mud-supported carbonates. The outer ramp sediments are formed of lime mudstone with silt-size quartz and shale. These sediments are extensively bioturbated.

8.2.3 The Ratawi Formation

The Ratawi Formation was described for the first time in an unpublished report by Sami Nasr (1950) and published by Owen and Nasr (1958) from the Basra Petroleum Company well Ratawi-1 between the drilled depths 3293 and 3643.4 m (thickness 350 m) and was named after this well (Fig. 8.11). In its type locality, the Ratawi Formation consists of greenish-black shale, massive at the upper part and interbedded at the lower part. It also has some layers and beds of pyritic, peloidal, and detrital limestone. These beds contain fossils such as *Pseudocyclammina lituus, Ostrea rectangularis*, and *Terebratula* cf. *squamosa*.

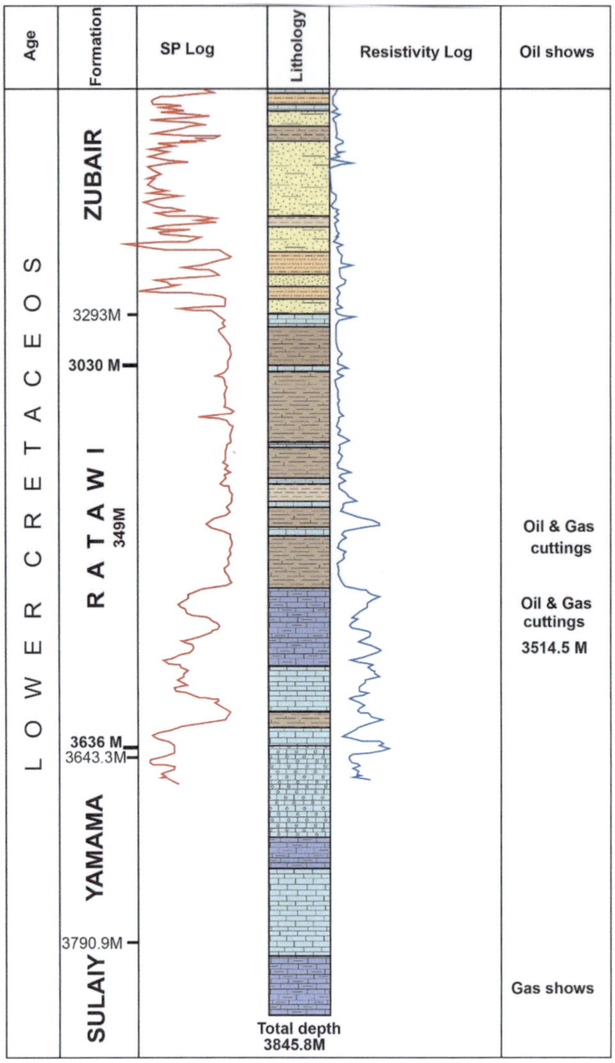

Fig. 8.11 Lithology of the Ratawi Formation in its type locality at the well Ratawi-1 in southern Iraq

The formation was also described from other countries, as in Kuwait, in K.O.C. Wells Burgan No. 113, Magwa No. 1, and Umm Gudair No. 1 (Owen and Nasr 1958), and wells in Qatar (Sugden and Standring 1975).

The reference section of the Ratawi Formation in Qatar is in the well Kharaib-1, between the drilled depths of 1208 and 1352 m, with a total thickness of 144 m. The Formation consists, from the bottom, of 18 m of grey, compact, argillaceous pelletal limestone, grey argillaceous limestone, pelletal in part, detrital (22 m), buff, porous limestone (3.3 m), blue-grey marl (3 m), buff porous limestone (4 m), blue-grey mark (7.5 m), pelletal limestone (7 m), grey marl with interbeds of argillaceous, detrital, pelletal limestone (36.5 m), grey marl interbedded with grey, argillaceous pelletal limestone (16 m) and grey limestone, argillaceous, detrital, and pelletal in part (26 m) (Sugden and Standring 1975).

The contact with the overlying Kharaib Formation is conformable, but some regional evidence suggests a possible break between the Ratawi and Kharaib over Qatar. The contact was placed at the change from the porous, clean limestone of the Kharaib and the argillaceous and marly limestone of the Ratawi. The contact with the underlying Yamama Formation also appears to be conformable, but there is evidence of an existing break between the two formations. The line was placed at the change from the grey, argillaceous, pelletal limestone of the Ratawi to the lime mudstone of the Yamama Formation. Generally, the Ratawi is characterized by higher argillaceous contents than the Yamama.

Sugden and Standring (1975) suggested that the Ratawi Formation in Qatar is similar to the Ratawi in its type locality. This formation primarily comprises argillaceous sediments interbedded with carbonates that underlie limestone containing the large foraminifera *Orbitolina discoidea* and overlie clean carbonates of the Yamama Formation.

They also reported that an accurate correlation of this formation with the Saudi stratigraphic system is impossible. They suggested that the Ratawi of Qatar is probably equivalent to the Buwaib Formation of the nearby subsurface sections. *Cyclammina greigi*, typical of the lower Buwaib Formation in nearby areas, occurs in unit 3 of the Qatar reference section.

The argillaceous limestone in the lowest part of the Ratawi Formation in Qatar may be equivalent to the upper part of the "Mid Thamama Limestone" of Saudi Arabia. This unit was originally named the Rakan Formation in Qatar.

The fossil recorded as *Pseudocyclammina kelleri* from the Ratawi Formation of Qatar is probably not the true *P. kelleri* of Iraq. In appearance, it has some affinity with *Cyclammina* and could be similar to *Everticyclammina hensoni* Redmond from the Buwaib Formation of Saudi Arabia.

Current usage in Qatar's subsurface geology assigns Ratawi to the combined Hawar, Kharaib, and Ratawi Formations, as described here. Sugden and Standring (1975) elaborated that the unit between the base Shu'aiba and top Yamama Formations is the lateral equivalent of the Zubair-Ratawi clastic sequence of areas to the north.

8.2.3.1 Facies Analysis

The Ratawi Formation is an important hydrocarbon reservoir in the Kuwaiti-Saudi Arabia Divided Neutral Zone in fields such as the offshore fields of Khafji, Hout, Dorra fields, and the onshore Wafra Field. The Ratawi Formation is divided into lower oolitic grainstone and upper bioclastic packstone-grainstone units. This synthesis of the Ratawi facies is based on the work of Albehbehani (1988), Al-Dwaish et al. (2016), and Bushang (2002).

Algal wackestone: This facies has a wide distribution and a total thickness of 12–21 m. Codicean algae are the major skeletal grain of this facies, and dasycladacean algae are also common. Other fossils include mollusks, echinoderms with fewer corals, and benthonic foraminifera. The considerable presence of mollusks and echinoderms with the algae may indicate deposition in a middle ramp environment. In areas where dasycladacean algae become dominant, and the facies is more argillaceous, this suggests more sheltered, possibly lagoonal, conditions.

Skeletal wackestone-packstone: This is a repeated microfacies throughout the formation ranging in thickness between 9–30 m. It is sometimes intercalated with two relatively thick shale horizons. The major skeletal grains are benthic foraminifera (both large and small) and include but are not limited to orbitolinids, *Pseudocyclammina*, milliolids, and rotallids with ostracods, calpionellids, echinoderms, and mollusks. Dasycladacean and codiacian algae are also common, with rare calcispheres, corals, and bryozoans. Non-skeletal grains are dominantly peloids. This facies is also deposited in a middle ramp environment under normal shallow marine conditions with slightly deeper conditions when calcispheres and calpionellids become abundant and more sheltered if small benthic foraminifera and dasycladacean algae become the major components.

Intercalations of shale and limestone: This facies is difficult to sample due to a considerable proportion of shale, which is difficult to correlate with depth. In some areas, it forms continuous shale; in others, it represents alternations of shale units around 30 cm thick with shaly limestone or calcareous shale. This zone is around 20 m. The facies is deposited nearer to the shoreline where the shallow carbonate shelf receives clastic materials from the nearby dry lands.

Algal boundstone: The main components of this facies are calcareous green, codicean phylloid algae with corals. Others include mollusks, echinoderms, and micritized foraminifera. The non-skeletal grains are peloids and lithoclasts. The thickness of this facies may reach 15 m. It represents algal biostromes developed on possibly growing structures or basin margins.

Bioclastic-peloidal packstone-grainstone: This facies represents the main bulk of the Ratawi rocks in the subsurface. It reaches a thickness of around 120 m in the Wafra Field of Kuwait. The main skeletal components are mollusks, echinoderms, micritized foraminifera such as *Pseudocyclammina*, milliolids, and rotalids, and micritized codiacean algae such as *Lithocodium* and *Polygonellea*. Peloids are the most dominant non-skeletal grains. This facies represents basin margin deposits under high-energy, normal marine conditions.

Oolitic-bioclastic peloidal grainstone: This facies comprises well-sorted and well-developed fine to coarse oolitic grainstone. Most ooliths have many cortical laminations, but others have fewer. Spherical ooliths and coated bioclasts are also present. Most ooliths are selectively dolomitized and contain organic matter and pyrite. Skeletal grains are less abundant and include crinoids, echinoids, mollusks, calcareous algae, and foraminifera. Micritization is common in this facies, and many grains are converted to peloids. This is an oolitic shoal or barrier that forms in agitated shallow water.

Planktonic-skeletal wackestone: This facies consists of thin-layered wackestone that becomes argillaceous in some horizons. The argillaceous wackestone is massive lime mud containing a few fossils, such as calpionellids, ostracods, sponges, and calcareous algae. The normal wackestone comprises bioturbated mudstone and wackestone with coarse fragments of ostracods, mollusks, crinoids, and calpionellids in a burrowed matrix. This facies represents deeper marine conditions, probably sub-basinal.

8.2.3.2 Depositional Environments

The Ratawi Formation represents a carbonate ramp. This ramp deposition is impacted by many factors. These include the following:

1. The upward movement of the Infracambrian Hormuz salt.
2. The northeast wind direction from the open Tethys Ocean.
3. The greenhouse weather of the Lower Cretaceous.
4. The Gotnia intracratonic basin in the north (Kuwait and Southern Iraq).

This ramp is divided into an inner ramp with restricted coastal lagoons developed at the back of the sand (oolites or buildups) barriers. The facies described above suggest that the grain-supported facies (packstone-grainstone) were deposited above a fairweather wave base in a medium to high-energy setting. The mud-supported facies (wackestone and lime mudstone) were deposited in a low-energy setting below the storm wave base or above the fairweather wave base.

This depositional setting formed of a sand barrier and patchy skeletal buildups that differentiate a carbonate ramp into distinctive facies belts is a characteristic feature of the Cretaceous stratigraphy of the eastern Arabian Plate.

The Ratawi formation is associated with the deposition of a clastic front resulting from the incursion of clastic materials to the carbonate basin. These clastic sediments range from well-developed, thick sandstone of the Zubair Formation in Kuwait and Iraq, the Biyadh Formation in Saudi Arabia, or shale and marl as in the Hawar Formation in Bahrain and Qatar.

In studying the Ratawi Formation in the oilfields of the Neutral Zone between Kuwait and Saudi Arabia, Bushang (2002) suggested that the shale part of the Ratawi Formation represents the prodelta of the Zubair Formation, while the limestone part represents the deep-water limestone in front of the Zubair Formation, but the nature and distribution of the Ratawi limestone facies suggest different environmental settings as detailed above.

8.2.4 The Kharaib Formation

The Kharaib Formation represents the Barremian supersequence and was described for the first time by Walter Sugen in an unpublished report in 1953 from the well Kharaib-1 between the drilled depths 1125 and 1208 with a total thickness of 83 m (Sugden and Standring 1975). From the base, the formation consists of 18 m of light grey porous limestone followed by 41 m of grey to cream to grey limestone, occasionally detrital with large pellets alternating with argillaceous limestone (Fig. 8.12). The top unit of the formation consists of light grey limestone, partially recrystallized and porous. The upper contact is conformable with the overlying Hawar Formation and was placed at the change from the limestone of the upper Kharaib to the shale and marl of the Hawar Formation. The lower contact is conformable, although the regional data may suggest some missing strata between the two formations. The contact was placed at the change from the limestone of the basal Kharaib and the marl and shale of the Ratawi Formation (Fig. 8.12). The formation represents an important oil reservoir in Abu Dhabi (Strohmenger et al. 2004). Sugden and Standring (1975) suggested that the upper and lower members of the Kharaib Formation in Qatar are equivalent to the first and second *Orbitolina* limestones of the Biyadh Formation in Saudi Arabia.

According to Focke et al. (1986), the deposition of the Kharaib Formation in Qatar started with the C Member of the formation, which consists of around 15 m of transgressive chalky limestone (Fig. 8.13) that overlies the erosional surface with the Ratawi (Lekhwair or Fahia formations) (Fig. 8.14). The C member is overlain by the Middle Marly Member, which represents a seal bed. The marl changes gradually upward into the Kharaib-B member, which is sealed by the Hawar shale and marl. This is the most important member of the Kharaib because it is an oil-bearing reservoir in the Idd El Shargi field offshore Qatar.

8.2.4.1 Facies Types

Focke et al. (1986) identified two major rock types in the Kharaib B and C members. These are the Peloidal

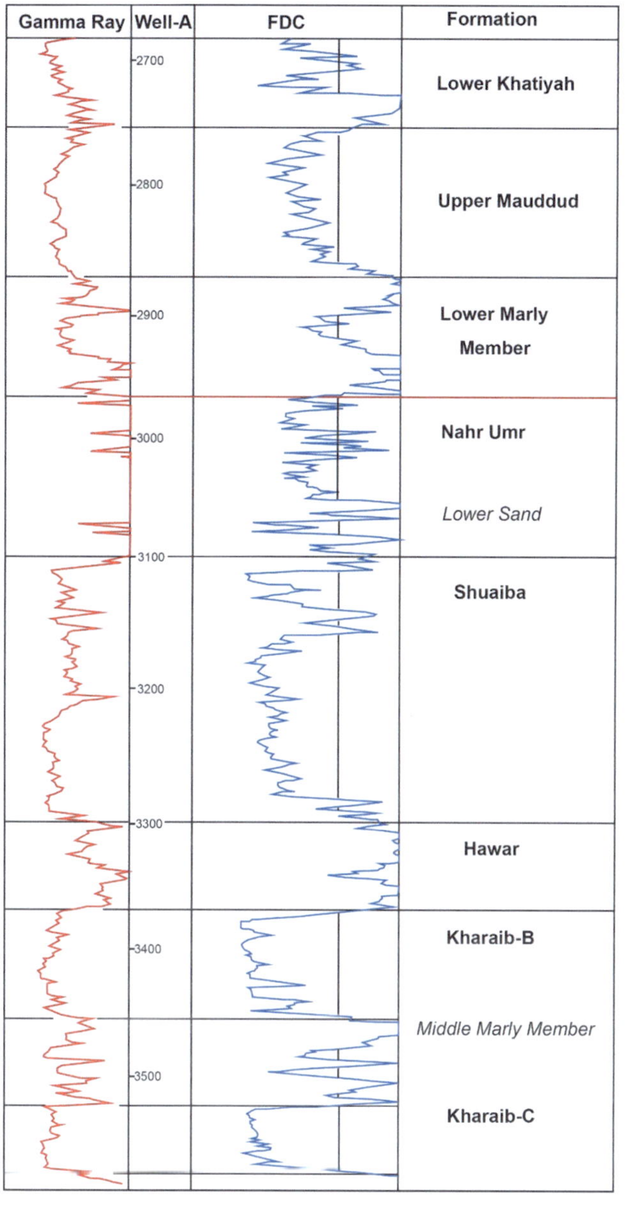

Fig. 8.12 Stratigraphy of the Cretaceous strata in the subsurface of Qatar (Focke et al. 1986)

Fig. 8.13 Chalky limestone that marks the start of deposition of the C Member of the Kharaib Formation shows a considerable amount of bioturbation (Focke et al. 1986)

Bacinella debris grainstone and packstone in the upper part and the peloidal *Orbitolina* wackestone in the lower part. They suggested that these sediments were the product of shallow marine open-shelf (the lower facies) to an inner shelf (the upper facies). The grainstone shows indications of bottom current and higher energy levels. The relatively large *Bacinella* debris may indicate local erosion that led to the accumulation of these algae. They proposed that the B-member sediments represent a shoaling-upward succession.

In his study on the Kharaib Formation in Abu Dhabi, Strohmenger et al. (2004) reported that the formation comprises two reservoir units (Lower and Upper Kharaib) separated by three barrier units called the sense zones (Lower, middle, and Upper Dense Zones). The reservoir units are formed of coated-grain grainstone and rudist floatstone and rudstone. They divided the formation into five facies sets based on their fauna, texture, structure, and lithology.

Peritidal Platform Restricted Lagoon with Orbitolinid and Skeletal Components (Lower, Middle, and Upper Dense Zone)

These rocks are formed of *Orbitolina*-bearing limestone (Upper Dense Zone) with small mollusks and calcareous algae with some siliciclastic influx. They are intensively bioturbated and burrowed. These sediments were deposited in low to moderate energy-restricted lagoons with less than 10 m water depths.

Shallow Subtidal Open Platform and Upper Ramp with Rudist, Miliolids, and Peloids (Lower and Upper Reservoir Unit)

This suite consists of rudist rudstone and floatstone with miliolids, peloids, coated grains packstone, and grainstone. It is characterized by planar lamination and low-angle cross-bedding. There are no siliciclastic materials except some burrow fills. The water depth is between 0–10 m. The sediments were deposited in a moderate to high-energy depositional system. A rudist buildup subjected to strong winnowing leads to the formation of rudist debris. The mixture of miliolids with coated grains was probably deposited in a sheltered coastal lagoon behind the rudist buildup.

Fig. 8.14 The contact between the Kharaib C member and the underlying Ratawi Formation is erosional, characterized by lithification, stylolitization, and boring (Focke et al. 1986)

Shallow Subtidal Open Platform and Upper Ramp with Coated Grains and Algae (Upper and Lower Reservoir Units)

This is a grain-supported facies with mixed biota, including foraminifera, echinoderms, algae, and mollusks with coated grains such as ooids, oncoids, and peloids. The finer grains form mud-rich packstone to grainstone, and the coarser components form floatstone to rudstone. There are rare cross-bedding and no clastic materials. The water depth ranges between 1–10 m.

Shallow Subtidal Open Platform and Upper Ramp with Lithocodium and Bacinella (Lower and Upper Reservoir Units)

This facies is of moderate energy with wackestone to packstone texture for the fine grains and floatstone to rudstone in the coarser fractions. A diverse fauna includes foraminifera, bivalves, gastropods, rudists, and echinoderms. Microbial activities heavily encrust the sediments, and there are no siliciclastic materials. The water depth is between 3–20 m.

Restricted Lagoon and Subtidal Open Platform with Orbitolinid and Diverse Biota, Upper to Lower Ramp (Lower and Upper Reservoir Units)

These sediments are formed of skeletal, peloidal wackestone to rare packstone with a wide range of fauna such as orbitolinids, bivalves, gastropods, rudists, and echinoderms. The bioturbated areas and the burrow-filling materials are dolomitized. They were deposited in low to moderate energy settings with water depth ranges between 3–40 m.

8.2.5 The Hawar Formation

The Hawar Formation may represent the distal part of the clastic front that deposited the Zubair Formation in Kuwait and Southern Iraq or the Biyadh Formation in Saudi Arabia (Fig. 8.2). The formation was described in an unpublished report by Sugen in 1953 and officially defined by Sugden and Standring (1975). The formation's type locality is in the well Kharaib-1, between the drilled depths of 1109 and 1125 m, with a total thickness of 16 m. The main lithologies of the formation are shale and blue-grey marl. The underlying formation is the Kharaib, with a conformable contact placed at the change from limestone below to the shale and marl above, and the overlying formation is the Shu'aiba Formation, with a conformable contact between the limestone of the Shu'aiba and the shale of the Hawar.

The Hawar Formation extends to the offshore areas of Qatar. It changes lithology into foraminiferal, pelletal wackestone, and packstone with argillaceous lime mudstone with some glauconite. It extends to Abu Dhabi and Oman. In Bahrain, the formation is made up of shale and represents the higher member of the Biyadh Formation in eastern parts of Saudi Arabia.

Sugden and Standring (1975) justified the assignment of a formation rank to this relatively thin clastic unit because of its value as a distinctive rock unit that extends for a wide area separating the Shu'aiba and Kharaib formations (Fig. 8.8).

8.2.6 The Shu'aiba Formation

The Shu'aiba Formation was described for the first time in an unpublished report by P.M.V. Rabanit in 1951 and studied by Owen and Nasr (1958) (Fig. 8.15), who documented the formation as forming of dolomitic and crystallized limestone with some rudist fragments and benthonic foraminifera

Fig. 8.15 Iraq Petroleum Company geologists in the field at Palestine, from left to right: Sami Nasr, F. E. Wellings, Norval Baker, and Mike Morton, December 1945 (Micheal Quentin Morton archives with permission)

such as *O. discoidea* and *Choffetella decipiens*. van Bellen et al. (1959) described the formation from its type locality at the well Zubair-3 between 2991 and 3070 m drilled depths with a total thickness of around 80 m but with completely different lithologies. The formation consists, from the base, of pseudo-oolitic limestone with scattered sand grains, changing upward into argillaceous limestone, then chalky limestone with around 10 m of crystallized limestone containing the planktonic foraminifera globigerinids and glauconite. This is followed by alternations of limestone and shale. The underlying formation is the Zubair Formation, and the contact is conformable and gradational. It was placed at the top of the first shale bed below the *Orbitolina*-bearing limestone. The overlying formation is the Nahr Umr Formation, and the contact is conformable also. It was placed at the base of the first shale above the limestone (Fig. 8.16). Both Hawar and Shu'aiba sediments represent the Aptian supersequence. The Upper Shu'aiba and the lower part of the Nahr Umr Formation are assigned to the Upper Aptian-Lower Albian supersequence (Raven et al. 2010).

Later drilling in other parts of Iraq and the southern parts of the Arabian Plate and the discovery of oil in the Shu'aiba Formation in the United Arab Emirates (Alsharhan 1987) have indicated that the lithologies of the Shu'aiba Formation in its type locality were not conclusive of the main lithologies, and the formation section drilled in southern Iraq probably represents a deeper part of the basin or mixed lithologies of other formations.

The reference section of the Shu'aiba Formation in Qatar is in the well Dukhan-11, between the drilled depths of 1006 and 1135 m, with a total thickness of 132 m.

8.2.6.1 Regional Distribution

The Shu'aiba Formation (Aptian) is an important hydrocarbon reservoir in most Gulf countries and has been studied extensively from many subsurface sections (e.g., (Al-Ghamdi and Pope 2020; Alsharhan 1987; Alsharhan and Nairn 1986, 1997; Borgomano et al. 2002; Hughes 2000; Immenhauser et al. 2004; van Buchem et al. 2002, 2010).

The Aptian basin of Eastern Arabia and Iran is formed of a shallow marine carbonate shelf that extended from central Iraq to Oman with some deeper intrashelf basins such as the Bab Basin (Fig. 8.17). The formation has a complex lithology generated from an interplay of tectonic (such as growth faulting), eustatic, and local depositional topography. The most important unit of the formation is the rudist buildups, which represent the major hydrocarbon-producing unit.

The lower part of the Shu'aiba Formation represented a continuation of the usual shallow-water carbonate deposition. However, early workers (e.g., Focke et al. 1986) had noticed that the upper part of the formation and its contact with the overlying Nahr Umr Formation is characterized by erosion, bioturbation, stylolitization, and oxidation (Fig. 8.18). These changes were attributed to the upward movement of the Infracambrian Hormuz salt (Focke et al. 1986) or climatic changes (Wilson 1975).

The isopach map of the Shu'aiba formation across the Arabian Gulf appears in Fig. 8.19. It shows that the formation appears in the middle and southern parts of Qatar with a total thickness of around 300 m. Hohman et al. (2005) suggested that the thickness differentiation of the formation is not depositional (i.e., the development of local basins such as the Bab Basin) but rather due to uplift and erosion that removed

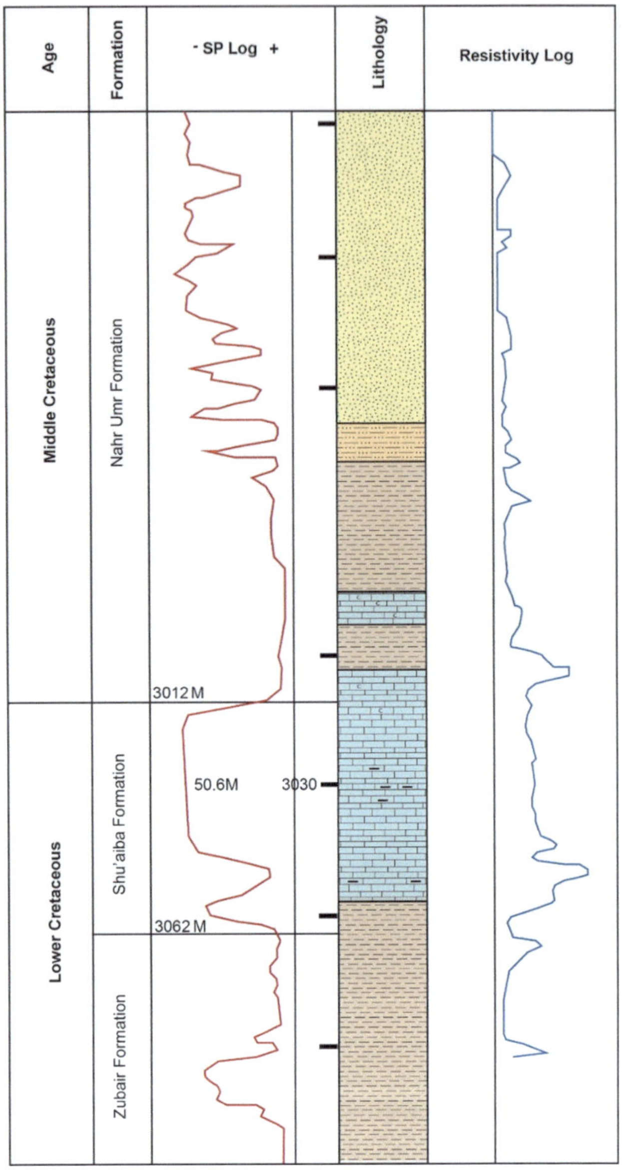

Fig. 8.16 The lithology of the Shu'aiba Formation in the well Zubair-31, southern Iraq

parts of the Shu'aiba sediments (Fig. 8.19). This assumption seems practical since it may represent the start of the large-scale uplift and erosion associated with the Lower-Middle Cretaceous break or the so-called Late Aptian glacio-eustatic lowstand that caused the demise of the Shu'aiba carbonate shelf and the deposition of the Nahr Umr clastics (Al-Husseini 2010; Maurer et al. 2013; Moosavizadeh et al. 2015; Raven et al. 2010; van Buchem et al. 2010, 2014).

More evidence of the erosion and truncation of the upper Shu'aiba Formation was reported by Raven et al. (2010) from offshore Qatar. Using horizontal well results and seismic mapping in the offshore areas of Qatar, they found a conspicuous east–west meandering, terraced, and incised valley at the top of the Shu'aiba Formation. The main valley is around 8 km wide and 30 m deep. They also reported indications of subaerial exposure features, such as clay and sand-filled

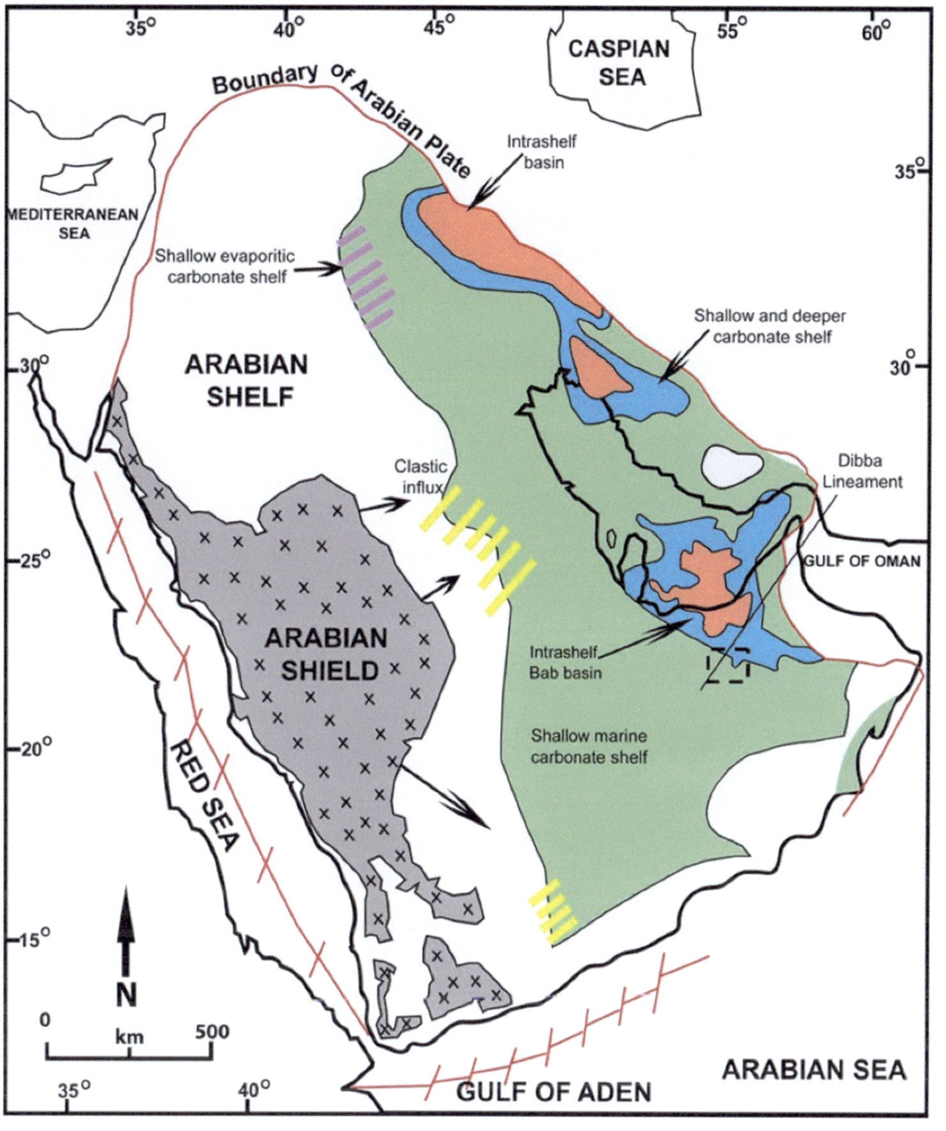

Fig. 8.17 Paleogeographic map of the eastern Arabian Plate during the Aptian (Al-Ghamdi and Pope 2020)

karst fractures at the top 25 m of the formation. The channel-fill sediments are estuarine sandstone with scattered plant remains, coal, and amber at the lower part, changing upward to sand interval with glauconite and ferruginous ooids that cover most parts of the Shu'aiba Formation. They suggested that these sediments represent a transgressive, back-fill system tract of a valley system. The oolitic ironstone and glauconite-bearing sandstone are condensed facies. They concluded that these observations may indicate one or all of the following:

1. The upper part of the Shu'aiba Formation underwent subaerial exposure for a considerable time.
2. At least a 30 m sea-level drop during the early Late Aptian.
3. An equivalent sea-level rise may happen at the latest Aptian.
4. A possible sea-level stillstand may have occurred during the Early Albian and led to the deposition of the condensed section.

These deep changes to the depositional regime of the upper Shu'aiba Formation were attributed to a Late Aptian glacio-eustatic lowstand in most parts of the Arabian Plate and other parts of the world (Maurer et al. 2013). Worldwide evidence from published literature, fauna, and chemostratigraphy cooling trends suggests a glacio-eustatic cause for these changes (Fig. 8.20). This is in contrast to the old assumption that the Cretaceous had a greenhouse climate with no ice (e.g., Norris et al. 2002). Maurer et al. (2013) argued that many data pieces suggest a short cooling period and even ice cap formation. They suggested that the ice cap

Fig. 8.18 The contact between the Shu'aiba (below) and the Nahr Umr Formation in the subsurface of Qatar (Focke et al. 1986)

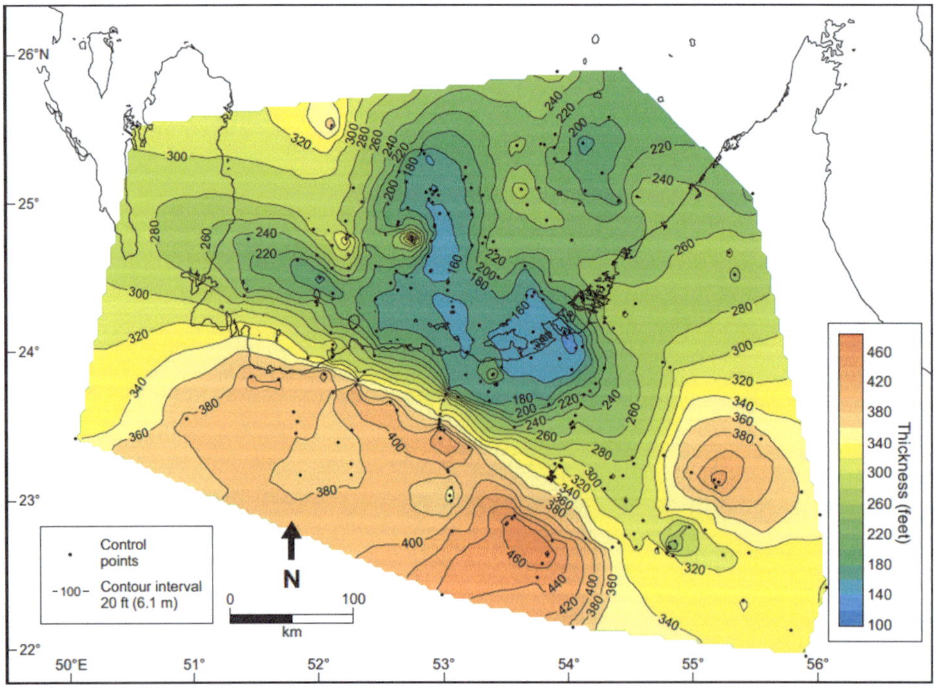

Fig. 8.19 The isopach map of Shu'aiba Formation across the Gulf region (Pierson et al. 2010)

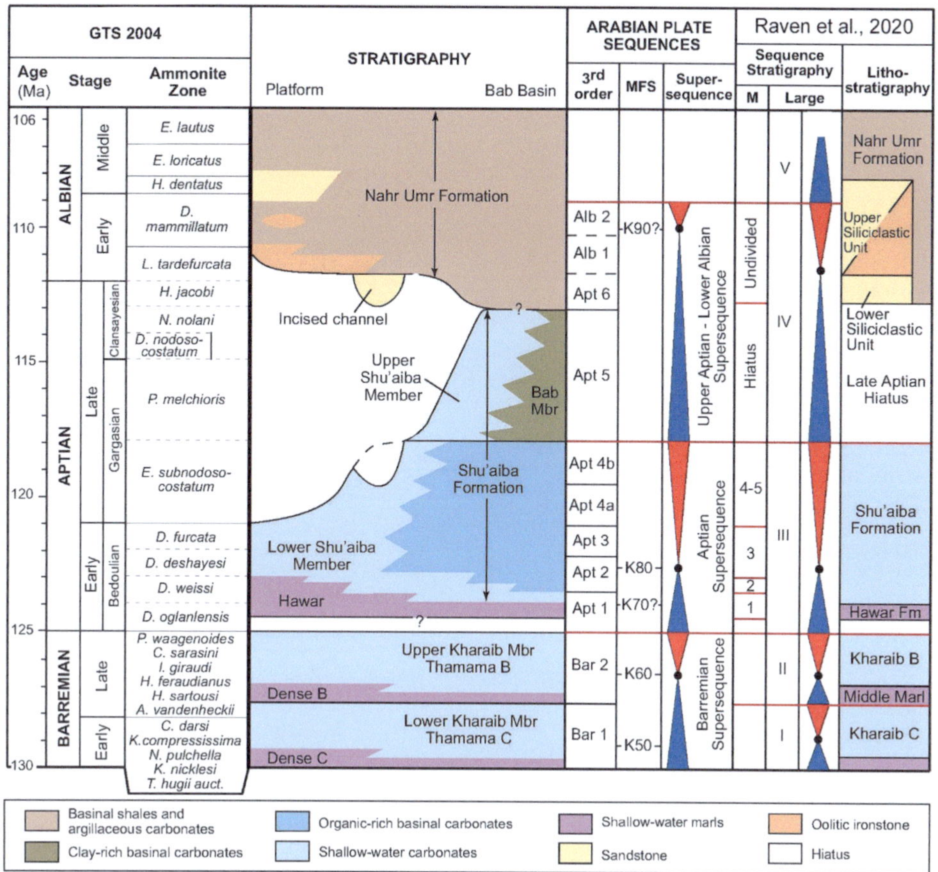

Fig. 8.20 Litho and sequence stratigraphy of the Arabian Plate during the Barremian-Albian showing the interruption of the carbonate system deposition during the Aptian-Albian (Raven et al. 2010)

was fast-growing and lasted for around five million years, reaching its peak during the latest Aptian *H. Jacobi* ammonite zone (Fig. 8.20).

8.2.6.2 Lithology of the Shu'aiba Formation in Qatar

The Shu'aiba Formation has been described from several wells in onshore and offshore areas of Qatar. The formation ranges in thickness between 60 and 106 m (Fig. 8.21).

The general lithology of the Shu'aiba Formation in the subsurface of Qatar is dominated by fine-grained carbonates (mainly chalky lime mudstone) as described from its type locality in southern Iraq (van Bellen et al. 1959). The Shu'aiba section has been divided into four zones (A-D) by Focke et al. (1986). From the bottom, Zone D is the most regular unit in the Gulf region in terms of lithology and thickness. This zone represents a deepening upward algal boundstone that can be easily delineated on well logs. Zone C is an argillaceous and compact limestone with stylolites and other pressure solution features. Zone B is a burrowed, featureless, chalky mudstone, and wackestone. The top A zone shows a wide range of lithologies and variable diagenetic features due to the subaerial exposure of the upper part of the formation (Fig. 8.22a, b).

8.2.6.3 Facies Analysis

The facies of the Shu'aiba Formation in the offshore wells of Qatar have been studied by van Buchem et al. (2009). They identified around eighteen facies. They grouped these into five distinctive facies associations deposited in a carbonate shelf with possible rudist basin-margin buildups. This synthesis is based on their work and Focke et al. (1986) study on

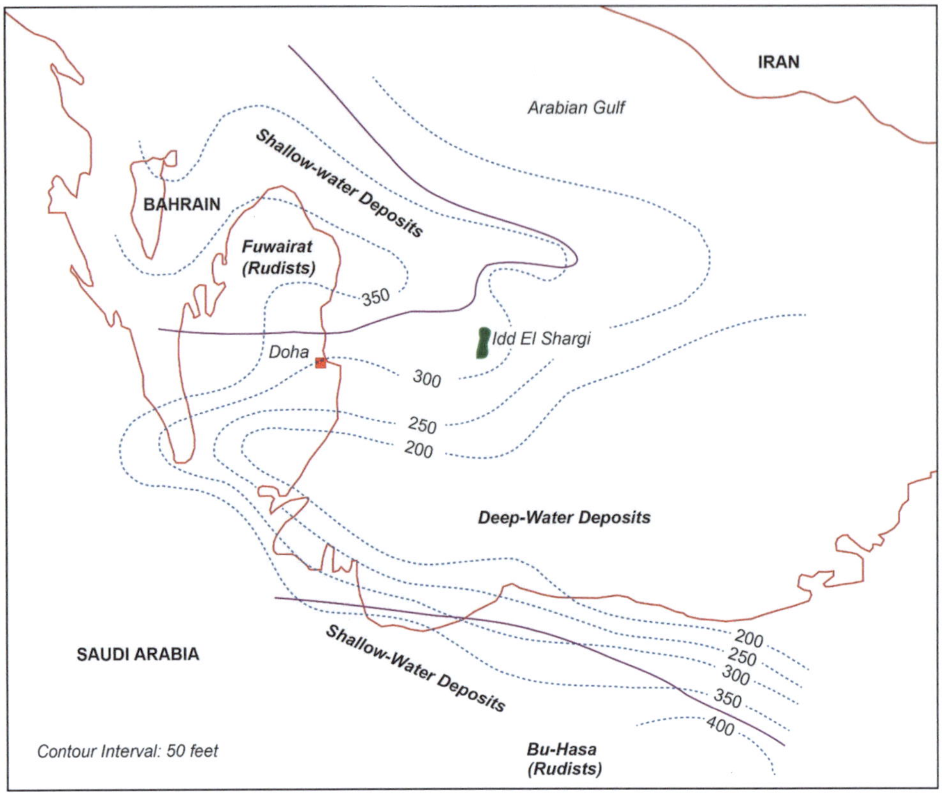

Fig. 8.21 Tentative isopach map of the Shu'aiba Formation in the subsurface sections of Qatar (Focke et al. 1986)

the Shu'aiba in the subsurface of Qatar. The major depositional environment and their characteristic facies include the following:

Inner platform: This is a relatively quiet environment with rich microbial contents that bound sediments. The major facies include *Lithocodium* and *Bacinella*-bearing and sponge-coral boundstone, oncoidal/Lithocodium floatstone, and algae and *Orbitolina*-bearing wackestone to packstone. These facies were deposited on the platform top, representing small buildups punctuating the Shu'aiba carbonate shelf. Focke et al. (1986) described this facies from the offshore area of Qatar as consisting of "reef-like" fabric formed of *Bacinella* and rudist boundstone. There are irregular patches with cemented organic framework. They reported that this facies is heterogeneous due to the combination of depositional textures and the diagenetic processes. It is characterized, however, by the presence of biomoldic porosity resulting from the leaching of the *Bacinella* algae (Fig. 8.23).

Tidal flat: This is a normal marine setting with slightly protected lagoons. The main skeletal components are the benthic foraminifera *Orbitolina* with gastropods. This foraminifera is a characteristic form of the Cretaceous and is adapted to a wide range of environments, from lagoon to foreslope environments (Fig. 8.24).

Buildups: These relatively wave-resisting skeletal structures separated the tidal shelf and lagoons from the foreslope areas. The major components are rudists, echinoderms, *Orbitolina*, and miliolids. Rudists buildups in the Shu'aiba Formation have been described for the first time from the Bu Hasa field in the United Arab Emirates by Twombley and Scott (1976) and by many other researchers, including Alsharham (1985) and Hamdan and Alsharhan (1991). In that field, these facies consist of coarse grainstone and boundstone of grave-size rudist debris; some were in growth patterns with corals and stromatoporoids. Focke et al. (1986) described a well-developed rudist buildup from the Fuwairat area in northern Qatar. The facies consists of large, poorly sorted

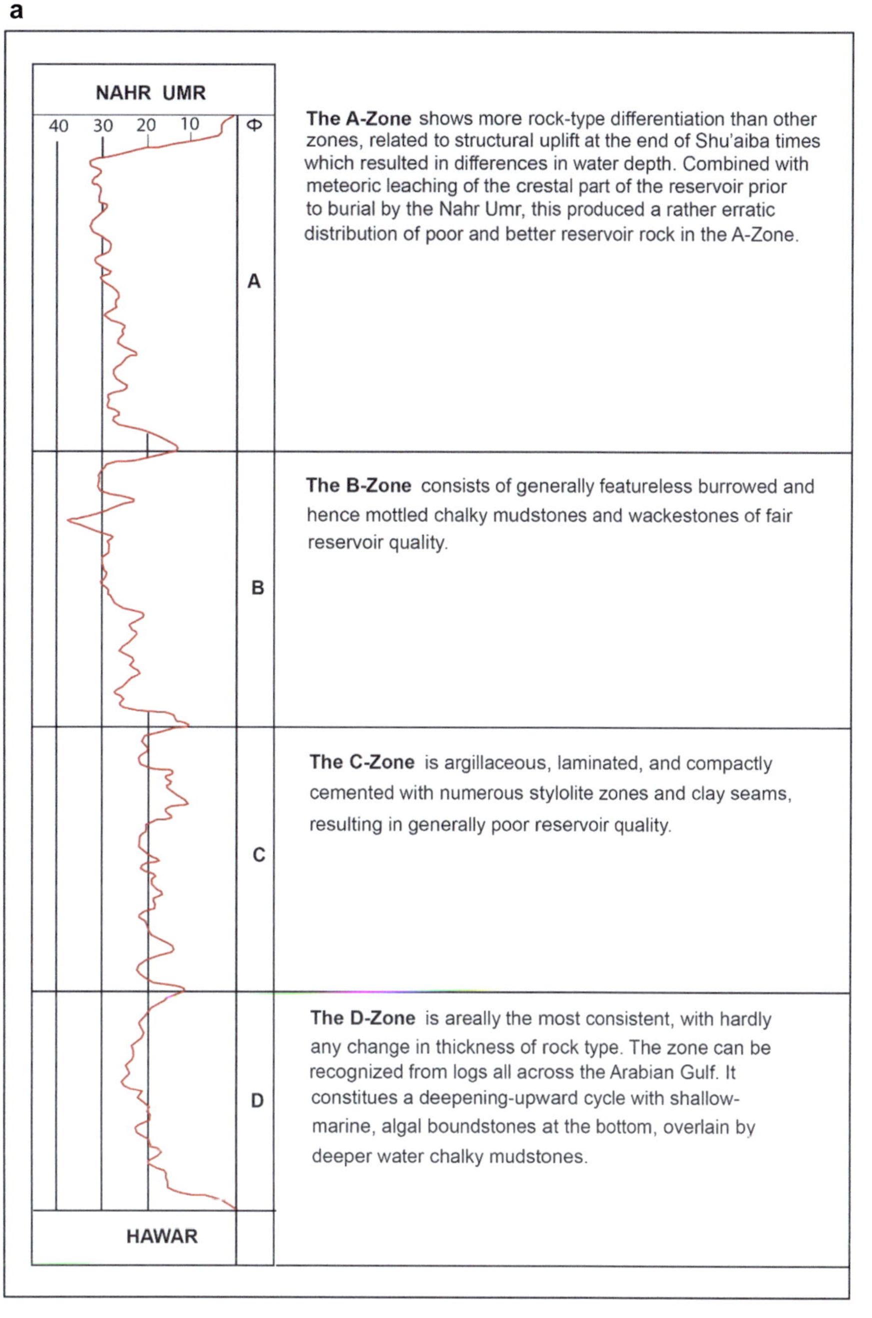

Fig. 8.22 **a** Zonation of the Shu'aiba Formation in the subsurface sections of Qatar, **b** lithologies of the different zones (Focke et al. 1986)

Fig. 8.22 (continued)

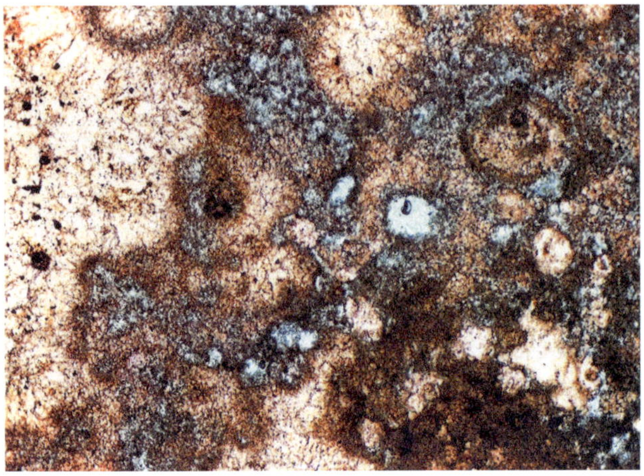

Fig. 8.23 Boundstone of the algae *Bacinella* with moldic porosity (Focke et al. 1986)

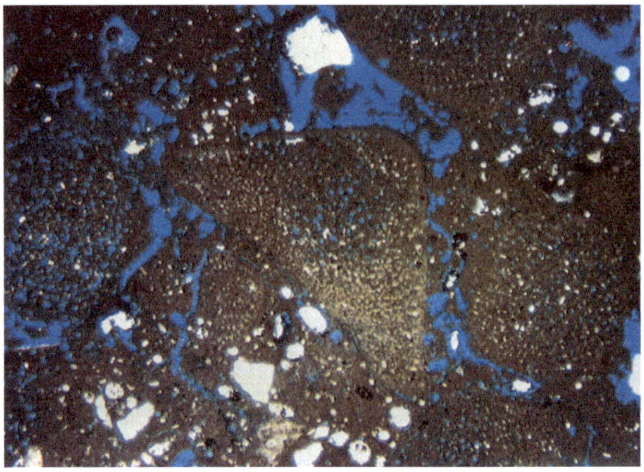

Fig. 8.24 *Orbitolina*-bearing packstone. The *Orbitolina* is a large benthic foraminifera that characterized the Cretaceous shallow-water carbonates in different parts of the world (Focke et al. 1986)

rudist fragments with large cavities filled with sediments (Fig. 8.25).

Foreslope: This is the deeper-water part of the basin, formed of mudstone and wackestone with mixed pelagic and benthic fauna.

The Shu'aiba Formation depositional environments in the United Arab Emirates have been studied in detail due to their importance as a hydrocarbon reservoir by Hamdan and Alsharhan (1991). They reported that the Shu'aiba sediments were deposited in five major depositional zones. These include the shelf lagoon, back-barrier slope, rudist barrier (or buildup), barrier foreslope, and open marine (Fig. 8.26). They suggested that the skeletal buildups started with an algal platform that was colonized later by rudists, forming the rudist buildups, which represent the reservoir rocks of the formation. The major rudist "reef" builders are the *Caprinidae*. Some of these skeletal materials were destructed and transported to the back barrier slope and the foreslope zones. The Nahr Umr Formation clastics that invaded the upper Shu'aiba carbonate shelf may have led to the demise of the rudist buildups (Sadooni 2018).

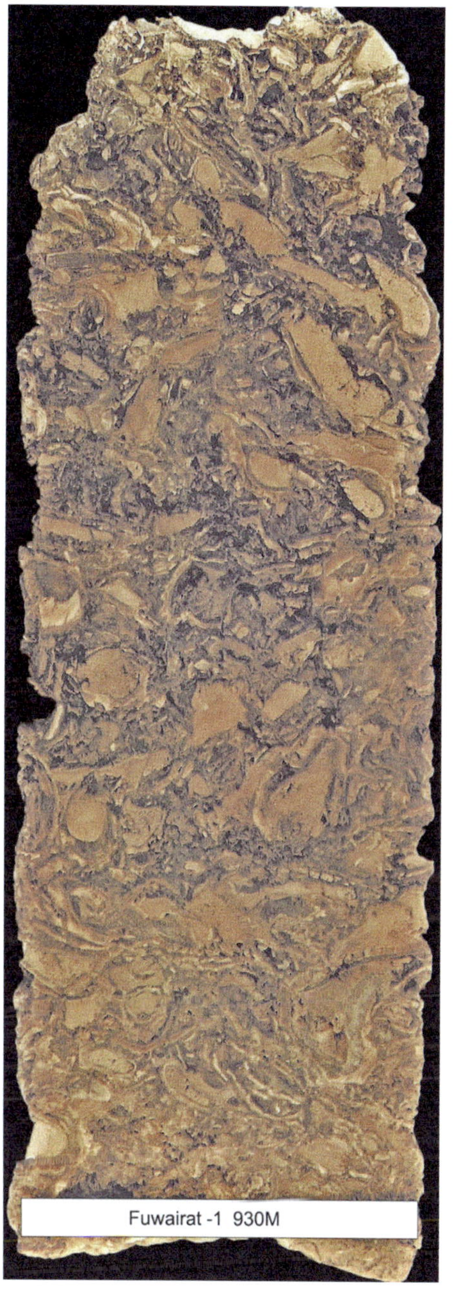

Fig. 8.25 A core piece from the Shu'aiba Formation in the well Fuwairat-1 showing rudist boundstone facies (Focke et al. 1986)

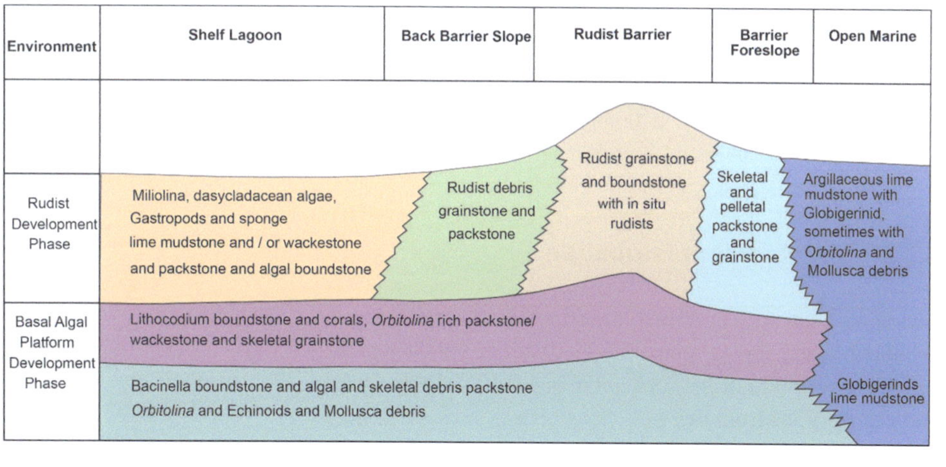

Fig. 8.26 The different depositional zones and major facies of the Shu'aiba Formation in the Arabian Plate (Hamdan and Alsharhan 1991)

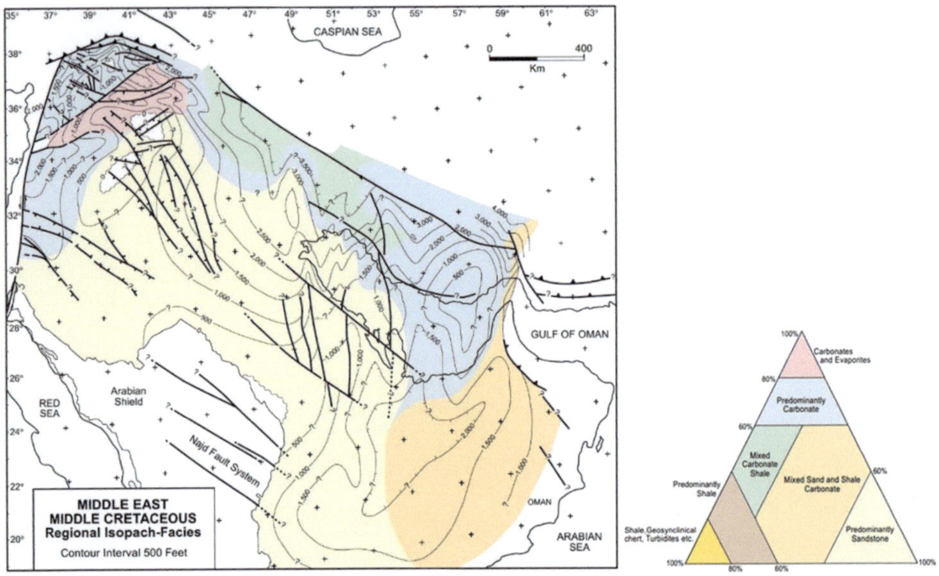

Fig. 8.27 Isopach and facies map of the Middle Cretaceous sediments across the Arabian Plate. Different lithologies are shown in the accompanying triangle. The contour interval is 500 ft. (Christian 1997)

8.3 The Middle Cretaceous

As detailed above, the Middle Cretaceous rocks of the Arabian Plate represent a major shift in the depositional system, with clastics becoming the predominant sediment type. They host a large hydrocarbon reserve in most Gulf countries and Iraq. Clastic derived from the Arabian Shield forming country-transcending major deltas across the western parts of the Arabian Plate in Kuwait, southern Iraq, and eastern parts of Saudi Arabia (Fig. 8.27). In Qatar, the thickest part of this clastic front is represented by the Nahr Umr Formation (sandstone and shale) and, to a lesser extent, the Ahmadi Formation (mainly shale and argillaceous limestone). There are, in fact, two clastic fronts during the Middle Cretaceous with different intensities. The lower Albian front was widespread and extended throughout the Arabian Plate and Iran. It consists of shale and sandstone that given different formation names across the region. It is called Kazhdumi Formation in Iran, Burgan in Kuwait, and Nahr Umr in Iraq, Qatar, Abu Dhabi, and Oman. The second front is of late Albian-Cenomanian and has less extent and was replaced by carbonates in Iran, Abu Dhabi, and Oman. The clastic sediments of this cycle are assigned to the Wara and Ahmadi formations. The thickness of the Middle Cretaceous sediments in Qatar is around 450 m.

8.3.1 The Nahr Umr Formation

The Nahr Umr Formation was described for the first time in an unpublished report by D. Glynn Jones in 1948, but the official description of the formation was published by Owen and Nasr (1958) from the well Nahr Umr 2 between 2632.7 and 2824.5 m. The Formation consists of black shales alternating with fine to medium sandstones. It also contains lignite, amber, and pyrite. The sand to shale ratio is around 40/60. In other areas in southern Iraq, there is a limestone member in the upper third part of the formation. This member is called the Dair Limestone and contains the benthic foraminifera *Orbitolina* spp., algae, and mollusks (van Bellen et al. 1959). The main fossils of the Nahr Umr Formation are *O. discoidea*, *Cythereis* sp., and *Haplophragmoides* sp. The Underlying formation is the Shu'aiba, and they suggested that the contact is conformable, although regional evidence suggests a big cutoff, as discussed above. The overlying formation is the Mauddud Formation, and the contact is conformable and gradational. It was placed at the base of the limestone of the Mauddud Formation and the black shale of the Nahr Umr Formation (Al Naqib 1967; Ibrahim 1981).

The formation thickens eastward reaching a total thickness of around 270 m along the Iraqi-Iranian border. The formation also has been described from other wells in southern Iraq and it was suggested that the formation may pinch out against the eastern flank of the Hail-Rutbah Arch (van Bellen et al. 1959). The section of the formation in the well Ratawi-1 has been included in the IPC Black Book (no date) as more representative of the formation. The thickness of the formation in that well is around 225 m and consists, from the base, of a thin shale bed, sitting on the Shu'aiba limestone, followed by a thick sandstone unit. The middle part of the formation is formed of sandstone and shale units of nearly the same thickness. The upper part is characterized by alternation of limestone and shale beds. The top of the formation is characterized by a relatively thick sandstone that changes upward to the limestone of the Mauddud Formation (Fig. 8.28).

In southern Iraq, the formation was deposited in a wide range of depositional environments including alluvial to lower

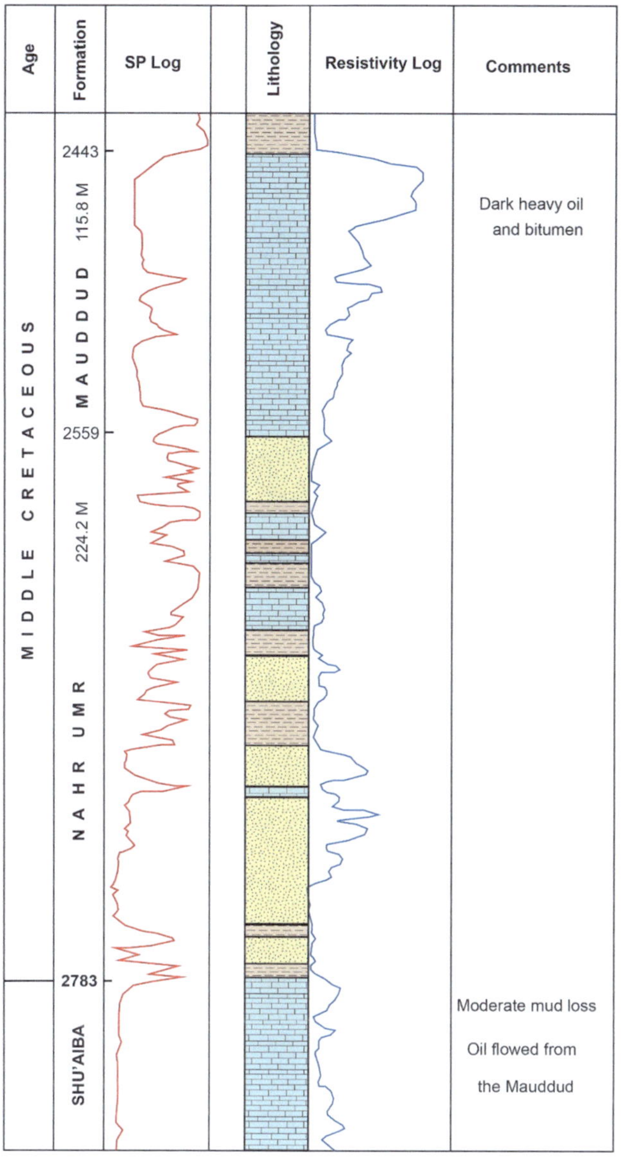

Fig. 8.28 The Nahr Umr Formation section in the well Ratawi-1, southern Iraq

coastal plain and deltaic deposits with shallow marine and aeolian influences (Ibrahim 1983). He suggested that the Nahr Umr Formation in the western parts of southern Iraq (the well Sawafi-1), the Formation is composed of fine-grained sandstone with frosted surfaces, rounded to subrounded with some fossil-free shale and siltstone. He suggested a continental environment for these sediments. In the central parts of southern Iraq (e.g., the well Rachi-1), the Nahr Umr Formation consists of medium to coarse-grained sandstone with angular to subangular, poorly sorted argillaceous and carbonaceous sandstone with black carbonaceous shale that was probably formed under unidirectional subaqueous transportation with abundant vegetation at a flood-plain or delta plain environments. In the central northwestern area of southern Iraq, the formation comprises lower to medium to fine-grained sandstone with subangular to rounded grains that change upward to *Orbitolina*-bearing limestone. These sediments were deposited under shallow marine water with clastic influx from the nearby land.

In Kuwait, the Nahr Umr Formation is called the Burgan Formation, named after the famous oilfield in Kuwait. The Burgan Formation also consists of well-bedded, well-sorted medium—to coarse-grained siliciclastic strata deposited in a near delta front on a sinking shelf. These sediments graded upward to an alternation of fine-grained sandstone and shale. The shale may be estuarine and contains a significant amount of plant remains, lignite, amber, and glauconite with no fossils. The same has been reported from southern Iraq (Abd el-aal et al. 2023; Alsharhan and Nairn 2003). The formations attain a thickness of around 380 m in the Burgan field to around 275 m in the northern fields.

Two sandstone units have been identified by Owen and Nasr (1958) and designated as the "Third" and "Fourth" Sand units. The Third Sand is equivalent to the Safaniya Member in Saudi Arabia, while the Fourth Sand is equivalent to the Khafji Member (Fig. 8.2).

8.3.2 The Nahr Umr Formation in Qatar

During the deposition of the Nahr Umr Formation (middle Albian), Qatar was located in a transitional zone between a continental depositional system (alluvial and lower coastal plain) to the west and a shallow marine, shale-dominated shelf to the east (Fig. 8.29). There are thin, shallow marine sheet sand beds at the base of the Nahr Umr Formation in the offshore area of Qatar. In this area, the basal part of the formation is heterogeneous and comprises a mixture of shale, siltstone, sandstone, and limestone. Near the contact with the underlying Shu'aiba Formation, there is a friable sand unit which is 3–5 m thick and called the Lower Sand. This is the most important unit in terms of hydrocarbon potential in the Nahr Umr Formation. Most of the work on the Formation is focused on the sandstone lithologies only due to their economic importance.

Focke et al. (1986) divided the sandstones of the Formation into three types:

1. Relatively clean and well-sorted sandstone friable to slightly cemented.
2. Relatively clean, well-sorted sandstone, friable to moderately to well-cemented.
3. Argillaceous quartz sandstone.

All the sandstone of the formation is very fine to fine-grained. The sand grains are generally angular indicating short transportation (distribution by tidal and storm-generated

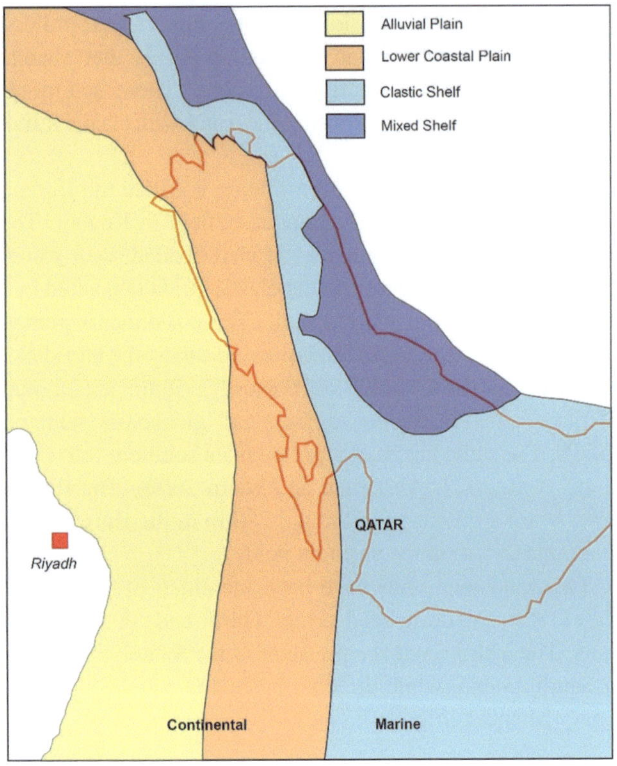

Fig. 8.29 Depositional environments of the Nahr Umr Formation in Qatar and the Arabian Gulf region (Murris 1980)

Fig. 8.30 Ironstone ooliths below the Lower Sand unit. These were probably normal carbonate ooliths that were replaced later by chamosite (Focke et al. 1986)

waves). The original bedding planes are frequently disturbed by bioturbation and burrowing. The presence of ferruginous minerals (Fig. 8.30), glauconite, chamosite/ goethite ooliths, siderite, ankerite, pyrite, and hematite may suggest transportation from the nearby dry land to the west. The different facies of the Nahr Umr Formation appear in Fig. 8.31.

The lower part of the Nahr Umr Formation in the Qatari offshore area was studied by Van Buchem et al. (2009). They divided that succession into two sequences. The basal Sequence 1 represents a large incised channel system up to eight kilometers in width, cutting through the upper part of the underlying Shu'aiba Formation to a depth of around 40 m (Fig. 8.20). Later, a rise in sea level led to filling that incised valley with transgressive, clean, tidally influenced estuarine back-fill sediments. The Highstand part of this sequence comprises oolitic ironstone alternating with sandy and silty claystone with siderite and glauconite. The overlying Sequence 2 consists of a block-wide sheet of sand with a significant amount of glauconite. Its deposition was associated with uplift and erosion as reflected in the facies nature and distribution. Iron-oolitic facies is the main lithology in the northern area, while the southern area was characterized by glauconitic sandstone alternating with carbonates. As the transgression proceeded, the sediments became marl and limestone.

8.3.3 The Mauddud Formation

The Mauddud Formation was described for the first time by the IPC geologist Francis Henson in an unpublished report in 1940 (cited by Sugden and Standring 1975) from the Dukhan-1 well in Qatar (Fig. 8.12). The formation is named after the locality of Ain Mauddud near Dukhan. Older names of the formation before its formal definition include the *Orbitolina concava* limestone by Barber (1948) or the *Second Pay* in Bahrain. The original description was updated by Sugden in an unpublished report in 1958. According to this report, which was transcribed by Bellen et al. (1959, p. 178), the formation comprises around 55 m of "limestone, light gray, earthy, mostly of fairly high porosity except for the bottom few feet, which are rather marly. Much of the limestone appears to be silty, reflecting the presence of fine calcareous detritus." The formal description of the formation in Qatar was published by Sugden and Standring (1975) as consisting of foraminifera-bearing limestone with lime mudstone at the bottom grading upward to wackestone, pelletal and skeletal packstone, and wackestone at the top. In some horizons, the formation consists predominantly of tests of *Orbitolina* and *Trocholina*. In the northern offshore areas of Qatar, the Mauddud Formation is formed of rudist grainstone and packstone with peloidal, *Orbitolina*-bearing limestone (Fig. 8.32) (Focke et al. 1986).

In its type locality in Qatar, the lower contact with the Nahr Umr Formation is conformable and was placed at the change from the clastic rocks of the Nahr Umr to the limestone of the Mauddud. In the subsurface of the offshore area, the lower part of the Mauddud Formation is called the Lower Marly Member which is a transition zone between the Nahr Umr clastics and the limestones proper of the so-called "Upper Mauddud" (Fig. 8.11) (Focke et al. 1986). The overlying formation is the Ahmadi Formation, and the contact is also conformable

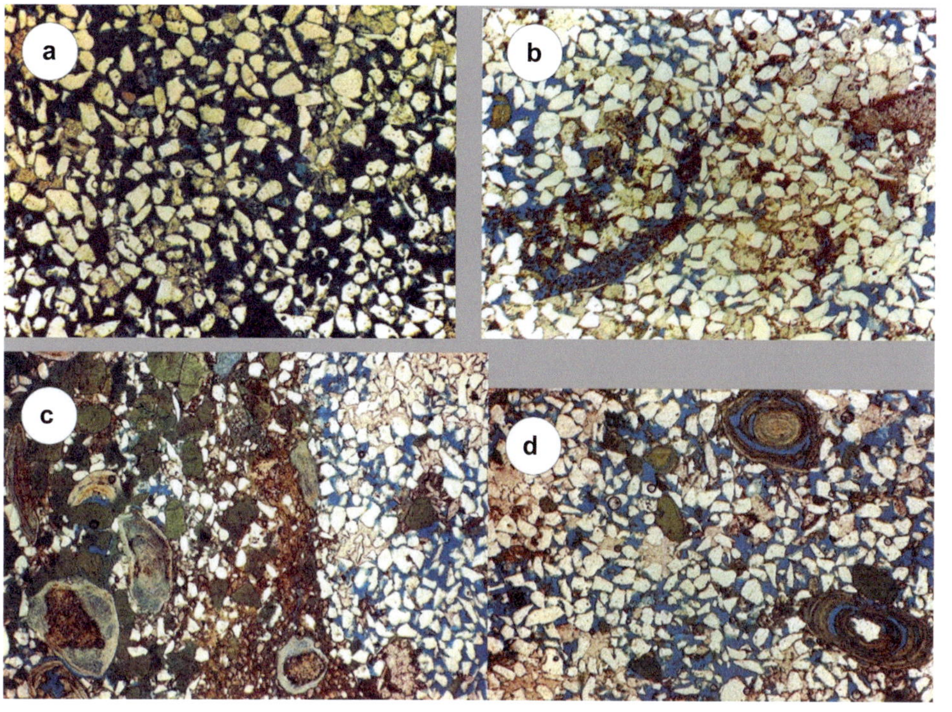

Fig. 8.31 Representative sand facies of the Nahr Umr Formation in the subsurface of Qatar **a** Slightly cemented, well-sorted, angular sandstone. **b** Moderately cemented sandstone with some lithoclasts and dissolution features. **c**, **d** Sandstone, cemented with calcite and siderite with glauconite (green crystals) and some leached chamosite ooliths. The blue area is porosity (Focke et al. 1986)

Fig. 8.32 Rudist grainstone (in cores to the left and thin sections to the right) from the subsurface offshore area of Qatar (Focke et al. 1986)

and marked by a sharp change from the *Orbitolina*-bearing limestone of the Mauddud Formation to the clastics of the Ahmadi Formation. In some offshore areas, the overlying formation is the lower Khatiyah Formation, where the contact is unconformable, and the soft marl of the Khatiyah Formation is filling the fractures within the limestone of the Mauddud Formation (Fig. 8.33).

In Kuwait, the reference section of the Mauddud Formation is in the Burgan oil field. It comprises a single limestone unit ranges in thickness from less than 2 m in the Umm Gudair field to more than 90 m in the Raudhatain field in northern Kuwait (Abd el-aal et al. 2023; Adasani 1967; Al Shamlan 1975). The Mauddud Formation is composed of dolomite and dolomitic limestone with *Orbitolina* and *Trocholina* foraminifera and dasycladacean algae. It is overlain by the Wara Formation and underlain by the Burgan Formation. Strohmenger et al. (2004) divided the formation into seven high-frequency depositional sequences. The lower transgressive sequence changes

Fig. 8.33 The unconformity between the hard limestone of the Mauddud Formation and the soft marl of the Khatiyah Formation filling the fractures between the limestone pieces (Focke et al. 1986)

from limestone in northern Kuwait to siliciclastic deposits of the Burgan Formation in the south. The upper Highstand sequence is carbonate-dominated and commonly thins southward because of depositional thinning and post-depositional erosion.

Powers et al. (1966) identified the Mauddud as a member in the Safaniya Well 17 in Saudi Arabia, with a thickness of 5.2 m. The unit, however, increases in thickness as it extends toward the southeast and attains a maximum thickness of approximately 60 m in the central Rub' Al-Khali. The Mauddud Formation in that region primarily contains marly dolomite containing *Orbitolina.* However, toward the western edge, it transitions into coarse dolomite or pyritic, glauconitic lime mudstone with occasional shale and silty sandstone layers. The prevailing lithology in the southeastern region of Qatar and the United Arab Emirates is a fossil-rich wackestone/packstone composed mainly of carbonate. This lithology becomes more argillaceous toward its lower section and has a thickness of approximately 61 m. In Oman, the Mauddud Formation, equivalent to Members E, F, and G of the Natih Formation, is found further east. It has a thickness of approximately 200 m and comprises skeletal and peloidal wackestones and lime mudstones. The upper part of the formation consists of packstone, with some areas containing rudist rubble (Alsharhan and Nairn 1988).

Henson (1948) suggested that the Mauddud Formation in Qatar is of the Cenomanian age, based on the presence of *O. concava* and *Praealveolina cretacea.* Bellen et al. (1959) proposed that *P. cretacea* is a result of contamination from caving of the overlying beds, and they suggested that the formation is of Albian age. Ibrahim et al. (2000) proposed that the formation in the Dukhan field, Qatar has an age ranging from late Albian to early Cenomanian. This is based on the simultaneous presence of the pollens *Florentinia radiculata, Florentinia resex,* and *Florentinia* berran. There are numerous arguments supporting the idea of this diachronous age of formation. These are some of the proposed arguments:

(a) The carbonate sediments of the Mauddud Formation were deposited on the clastic front of the Nahr Umr Formation as it retreated. This deposition occurred in areas far from the source of the clastics, along the present-day Iraq-Iran border. Over time, the deposition gradually moved toward the southern and southwestern parts of the Arabian platform as the clastic source was cut off. There are many examples of this, such as the occurrence of limestone containing *Orbitolina* that is mixed with clastic facies in the Halfaiya field near the border of Iraq and Iran and the occurrence of *Orbitolina* in the exposed rock formations of the Nahr Umr Formation in Oman.

(b) In Saudi Arabia, the upper boundary of the Mauddud Formation is considered diachronous due to the presence of clastic components within the formation. The stratigraphy demonstrates that an increase in thickness occurs by adding layers at the uppermost part. *Hemicyclammina whitei* and *O. concava* are found at the uppermost part of the Mauddud Formation in the Dammam field. When the carbonate component alternates with the sandier beds, *H. whitei* replaces *O. concava* in the sandier interval. Both forms have similar ranges and can be used interchangeably (Powers 1968). Thus, the occurrence of *H. whitei* in sedimentary layers above the Mauddud limestone units in wells located south of Safaniya suggests a partial similarity with the limestone units found in the upper section of the Mauddud Formation at the northern boundary of that field. In other fields, such as Khursaniyah, Qatif, and Dammam, similar variations have been reported in the position of the carbonate beds that indicate the top of the Mauddud Formation. Powers (1968) determined that while a faunal zone linked to the uppermost part of the Mauddud Formation can be identified across the fields situated in the coastal regions of Saudi Arabia, the specific lithologic top exhibits significant variability.

(c) The Mauddud Formation is terminated by regional conformity, which resulted in the erosion of different formation thicknesses based on the shape of the ancient paleo-topographic surface.

8.3.3.1 Facies Analysis

The Mauddud Formation has more or less consistent lithologies all over the eastern parts of the Arabian Plate (Sadooni and Alsharhan 2003). This is a characteristic feature of most of the Cretaceous carbonate sediments due to the prevalence of a shallow-water carbonate ramp all over the area. The following are the most common facies of the Mauddud Formation:

Rudist Skeletal Grainstone and Packstone

This facies is similar to the facies reported from the Shu'aiba formation (see above). It consists of large rudist fragments, some in a growth position. It has been reported from different parts of the basin. Mohammed and Al-Sayyab (1993) documented the presence of a rudist biostrome measuring up to 40 m in the Mauddud Formation. This biostrome was observed in the Abu Ghirab-5 well, on the Iraq-Iran border in southern Iraq. The facies consists of bioclastic rudist limestone, occasionally interspersed with rudist boundstone. Focke et al. (1986) reported the presence of limestone containing abundant large fragments of caprinid-caprotinid rudists in the upper section of the Mauddud Formation in the northern regions of Qatar (Fig. 8.32). Radiolitid rudists are also found in the southern regions of the Qatari offshore fields. Additionally, rudist debris has been documented in the Mauddud member of the Natih Formation in Oman, as described by Harris and Frost (1984) and Alsharhan and Nairn (1997). This facies may represent basin-margin rudist buildups that were fragmented and scattered by wave and wind action.

Orbitolina-Bearing Limestone

Some parts of the Mauddud Formation in the Arabian Gulf basin are formed of limestones containing *Orbitolina* fossils. This facies has been documented early in the region by Henson (1948), in Qatar by Focke et al. (1986), in Kuwait by Al Shamlan (1975), in several oil wells in southern Iraq by Mohammed and Al-Sayyab (1993), in northern Iraq by Sadooni (1978, as cited by Sadooni and Aqrawi 2000) and Al Shdidi et al. (1995), and in other areas of Arabia by Alsharhan and Nairn (1988, 1997). Multiple species were documented, such as *O. concava* (Lamarck), *O. concava var. qatarica* Henson, *O. concava var. iraqi* Henson, *O. concava var. sefini* Henson, *O. discoidea* Gras, and *Orbitolina aff. curdica* Henson (1948).

Orbitolina is found in various environments, with varying abundance levels and accompanying components. The *Orbitolina* rudstone predominantly consists of *Orbitolina* tests, which comprise 80–90% of the entire rock, with minimal or no matrix present (Fig. 8.34). It was reported from Kuwait by Al Shamlan (1975) and by Sadooni and Alsharhan (2003) from the Kirkuk field in northern Iraq. The *Orbitolina*-rudist limestone is the second most prevalent association. This facies consists of a mixture of *Orbitolina*, large shell fragments of rudists, other mollusks, and echinoderms.

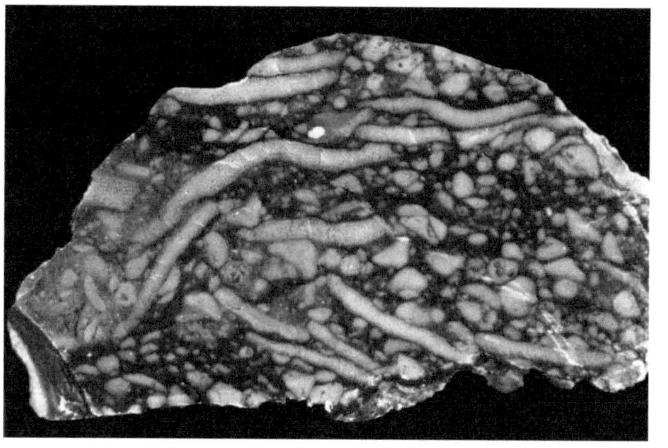

Fig. 8.34 *Orbitolina* rudstone formed mainly of large benthic foraminifera (Sadooni and Alsharhan 2003)

Focke et al. (1986) documented this facies in the North field of Qatar, and it was observed in the Jambur field of northeast Iraq by Sadooni and Alsharhan (2003). The third facies consists of limestone containing *Orbitolina*, accompanied by dasycladacean algae, pellets, and small benthonic foraminifera, including miliolids. The *Orbitolina* present in this facies have undergone extensive micritization and have been darkened by organic pyrite. This facies has been documented in the Jawan Formation, which is the lagoonal counterpart of the Mauddud Formation in the oil fields of northern Iraq. And also, from Kuwait by Al Shamlan (1975).

The *Orbitolina* thrived in tropical to subtropical waters near shallow coastlines with temperatures ranging from 15° to 35 °C (Douglass 1960). The water depths associated with *Orbitolina* exhibit variability and are most accurately deduced through inference from the accompanying organisms. The presence of *Orbitolina* alongside miliolids and green algae implies a shallow-water environment, while its association with rudists may suggest a deeper and more open marine water setting.

Peloidal and Pelletal Grainstone

This facies consists of pellets, peloids, green algae, micritized miliolids, and *Orbitolina* within a slightly argillaceous micritic matrix. Sadooni and Alsharhan (2003) described the facies from the western areas of the Kirkuk field. In this region, the Mauddud Formation transitions into the Jawan Formation, which represents a lagoonal environment. Focke et al. (1986) reported two comparable facies found in the Mauddud Formation in the northern fields of Qatar: the peloidal *Orbitolina*/echinoid packstone and the pelletoidal grainstone (Fig. 8.35). It is possible that many of the peloids in this facies are benthic foraminifera that have undergone micritization. The facies experienced partial leaching and cementation by blocky calcite cement. The present peloids, regardless of whether they are fecal pellets or micritized small benthic foraminifera,

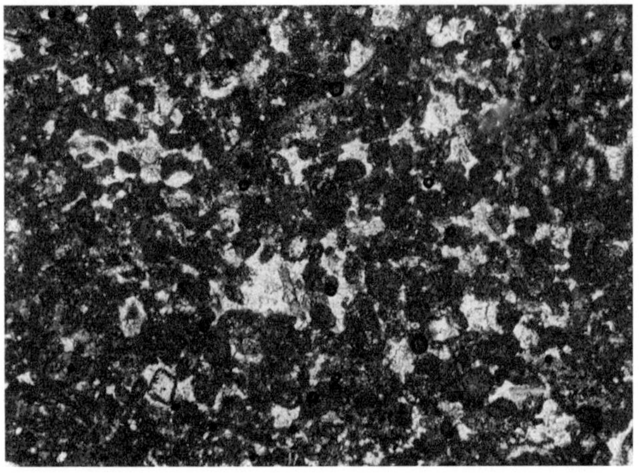

Fig. 8.35 Peloidal and pelletal grainstone from the subsurface of the offshore area of Qatar. Some of the peloids may be micritized benthic foraminifera. Blue areas are porosity and light areas are blocky calcite cement (Focke et al. 1986)

most likely originated in a protected region and could indicate the deposition in a confined, shallow marine shelf or a lagoon.

8.3.3.2 Depositional Environment

In the majority of the basin, the Mauddud Formation is situated on top of the Nahr Umr Formation. The contact between the two formations may be attributed to a stratigraphic discontinuity that formed during the flooding of the clastic-dominated shelf, leading to the accumulation of shallow-water carbonates. The cessation of the transport of clastic sediments from the western source area may be attributed to either a shift in climate (Murris 1980) or a glacio-eustatic lowstand (Raven et al. 2010). The lower section of the Mauddud Formation can be described as a transgressive system that reached its peak and became stable as a High-stand systems tract, characterized by limestone containing *Orbitolina* fossils. The widespread distribution of these facies, both vertically and horizontally, across the Arabian Basin signifies the enduring stability of the basin over a significant period. Rudist banks were formed on the basin margin in areas with higher energy levels, gradually transitioning into deep-water carbonates to the east. Local lagoons were also created where geographical or hydrodynamic conditions permitted. These were distinguished by the presence of algal-pelletal facies. Clastic sediments remained the primary constituent in certain regions.

The presence of the Ahmadi Formation (or the Khatyia) above indicates that clastic sediments once again prevailed on the shelf. Dunnington (1967) reported the existence of a regional unconformity at the uppermost part of the Mauddud Formation in Kuwait and southern Iraq. Sugden and Standring (1975) stated that there is no tangible evidence of such unconformity in Qatar. Nevertheless, the upper section of the Mauddud Formation in the northern fields of Qatar is fractured and pierced by the soft marls of the lower Khatiyah (Fig. 8.32), indicating the existence of a depositional discontinuity. The Mauddud Formation in certain areas of central Iraq, such as the Awasil-5 well, and in northwestern Iraq, is directly covered by the Cenomanian Mishrif Formation or its equivalents. Consequently, the Ahmadi/Khatiyah formations in these regions were either eroded or never deposited.

The extensive dispersion of the Mauddud rocks signifies the conclusion of the clastic phase that predominantly prevailed in most of the basin. The late Albian transgression caused a swift cessation of the clastic materials. The occurrence of clastic materials in the western desert of Iraq and Saudi Arabia, as observed by Powers et al. (1966), indicates the presence of a nearby but limited source of clastic sediments within the predominantly carbonate regime.

8.3.4 The Khatiyah Formation

Although the Qatari geologic lexicon Sugden and Standring (1975) considered the Khatiyah Formation as an obsolete name to describe the Cenomanian shale, marl, and limestone of the Ahmadi Formation, the term is still lingering in the published literature (Van Buchem et al. 2014). The original term was introduced to encompass a succession of predominantly argillaceous strata encountered in the Dukhan-1 well. This succession was divided later into the Laffan, Mishrif and Ahmadi formations. The Laffan shale was the first unit to be defined separately after the discovery of significant unconformity at its bottom. The Khatiyah section is similar to that of the Ahmadi Formation in southern Iraq and therefore the Sugden and Standring (1975) recommended stop using the term Khatiyah in favor of the Ahmadi Formation.

Focke et al. (1987) described what called the Lower Khatiyah from the subsurface of the offshore area of Qatar. In these sections, the unit comprises compact to chalky mudstone, wackestone and packstone. The main fauna are large benthic foraminifera such as *Praealveolina* and codiacean algae (Fig. 8.36). They suggested a shallow marine, low-energy depositional system for the deposition of its sediments. The unit is overlain by the Laffan Formation and the contact is erosional. The lower contact with the underlying Mauddud Formation is sharp.

8.3.5 The Mishrif Formation

The Mishrif Formation was described for the first time by P. M. V. Rabanit in 1952 and officially defined by Owen and Nasr (1958). The type locality is in the well Zubair-3, southern Iraq, between the drilled depth of 2183–2339.3 m with a total thickness of 156.4 m. From top to

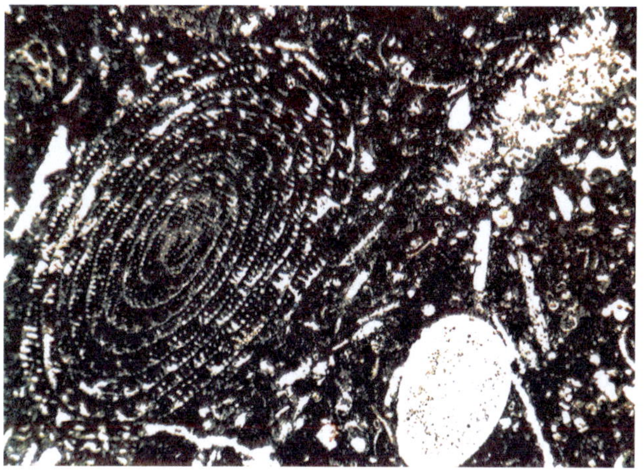

Fig. 8.36 The main facies of the Lower Khatiyah unit consisting of large benthic foraminifera and some algae (Focke et al. 1986)

base, the formation consists of freshwater limestone with iron oxide and the freshwater algae *Charophytae*. This section is followed by algal limestone with gastropods and shell fragments with fractures and stylolites. The following lithology comprises detrital, porous limestone with shell fragments and foraminifera with rudist buildups changing downward into marly limestone. The Rumaila is the underlying formation with a conformable contact placed at the change from shallow-water carbonates of the Mishrif formation to the planktonic foraminifera-bearing marly limestone of the Rumaila Formation. The Khasib is the overlying formation, and the contact may involve some hiatus, as proposed by van Bellen et al. (1959).

Aqrawi (1998) provided a more detailed description of the formation from subsurface sections in southern Iraq. The Mishrif in these sections is formed of shallow-water carbonates formed of large foraminifera such as Alveolinids and *Nezzazata* sp. With green algae. The carbonate shelf is punctuated with rudist buildups (barriers) formed of large fragments of rudist materials. These shallow-water carbonates alternate with marly and chalky limestone with planktonic foraminifera. These sediments are assigned to the Rumaila Formation in the Iraqi stratigraphic nomenclature system, representing the Mishrif Formation's basinal equivalent. The upper part of the formation comprises argillaceous limestone with iron staining and the freshwater algae *Charophytes*. These sediments indicate subaerial exposure and the development of freshwater lakes that ended the marine deposition of the Mishrif Formation (Sadooni 2018) (Fig. 8.37).

8.3.5.1 The Mishrif Formation in Qatar

The Mishrif Formation in Qatar has been described by Sugden and Standring (1975) from the well Dukhan-28 between the drilled depth of 646 and 727 m with a total thickness of 81 m. It consists of soft and friable wackestone and packstone containing shell fragments and foraminifera with a thin shale bed at the base. The underlying formation is the Ahmadi, with a conformable contact at the top of a well-developed shale unit. The overlying formation is the Laffan with an unconformable contact. The Mishrif Formation was originally included within the Khatiyah Formation until a thick section of the formation, similar to the formation in southern Iraq, was encountered in Dukhan wells. Later, the formation was found in the offshore areas, too. Hewaidy and Al-Hitmi (1999) described the formation from the Dukhan field as forming of grey limestone intercalated with shale bands. They suggested that the change in lithology from limestone to shale and the moderate biodiversity of the formation may indicate deposition in "an oscillating shallower inner shelf environment."

In the subsurface of the offshore area, the Mishrif Formation was studied by Deville de Periere et al. (2017), focusing on the upper part of the formation due to its better reservoir characteristics. They suggested that the Mishrif represents a wide carbonate platform that progrades eastward to the Shilaif intrashelf basin (Fig. 8.38). The studied section is around 100 m thick with an oil column of 45–50 m. It consists of mud-supported and bioturbated limestone deposited on a platform top. The main skeletal components include benthic foraminifera, such as prealeveolinids and chrysalidinids, miliolids and rotalids, echinoderm fragments, and green dasycladacean algae.

Facies Analysis

The Mishrif Formation represents a complete carbonate shelf with local rudist buildups. It exhibits a wide range of facies. These are the most persistent and characteristic facies (Fig. 8.39).

Lime Mudstone and Wackestone (Fig. 8.39a, d).

This facies consists of dark micritic limestone rich in organic materials. The wackestone variant contains dwarf rotaliids and miliolids, small green algae debris, and some microbial materials. This facies represents the lagoonal restricted shelf of the Mishrif Formation. In western southern Iraq, the Mishrif carbonate changed gradually to the evaporites and pelletal limestone of the Kifl Formation (Sadooni and Aqrawi 2000).

Large Benthic Foraminifera Wackestone and Packstone (Fig. 8.39b, e).

This is a common facies in the Mishrif Formation across the Arabian Plate. It consists of large benthic foraminifera such as *Praelveolina*, *Orbitolinella*, and *Cycledomia*. In some cases, these fauna are mixed with green dasycladacean algae or rudist fragments. This facies was deposited in a normal shallow marine shelf. The existence of rudist debris may suggest a foreslope environment where these debris were fragmented from a local rudist buildup. Green algae may indicate

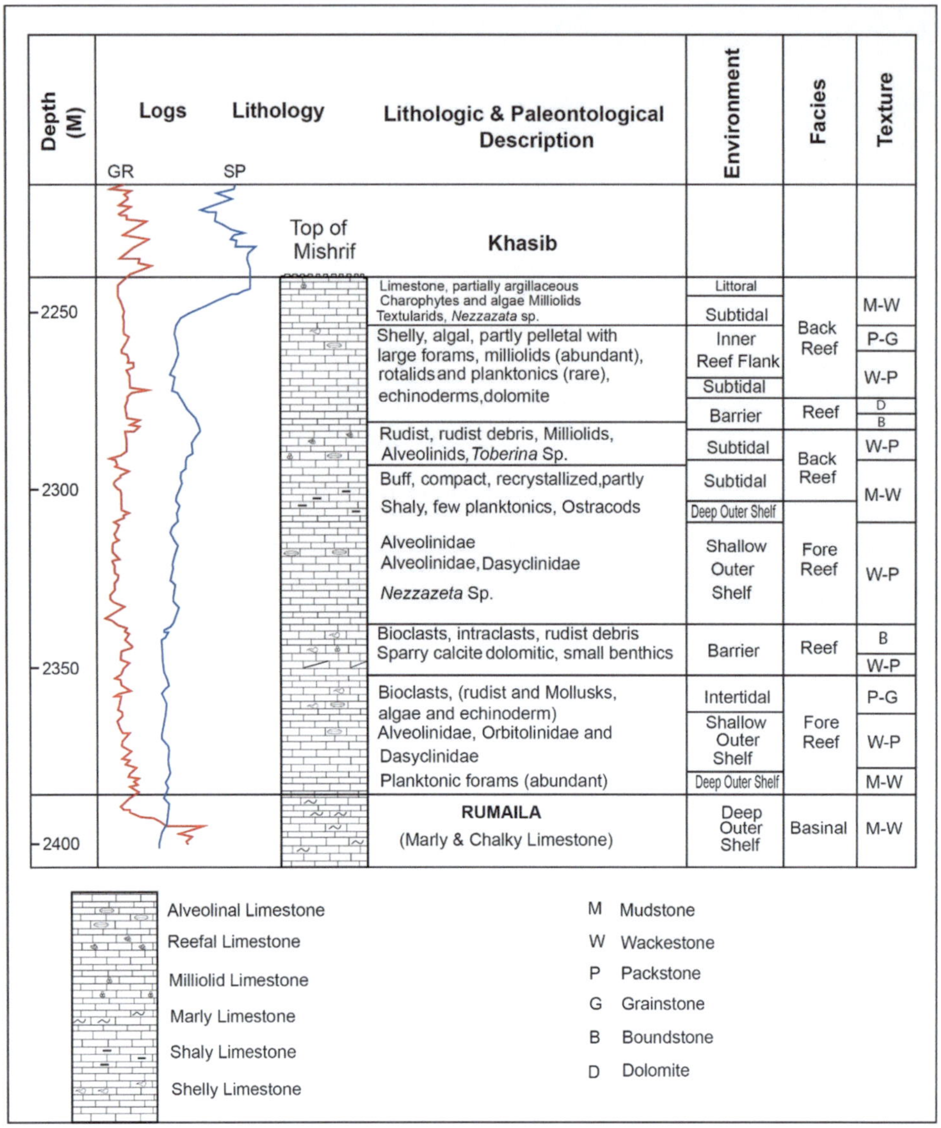

Fig. 8.37 Stratigraphy, facies, and depositional environments of the Mishrif Formation in the subsurface of southern Iraq (Aqrawi et al. 1998)

a more restricted shelf or a coastal lagoon. The Cretaceous large benthic foraminifera showed a higher resilience to environmental changes and may exist in conditions that range from lagoonal to foreslope environments.

Dasycladacean Algae Wackestone and Packstone (Fig. 8.39f)

This is an interesting facies because of its high content of a relatively well-preserved dasycladacean green algae debris. The dark color of the matrix may suggest high organic content. This facies has been described from other Cretaceous formations in the region (Sadooni and Aqrawi 2000). It is a characteristic of the lagoonal environment.

Echinoderm Wackestone-Packstone (Fig. 8.39g)

This facies comprises echinoderm debris, mainly with some shell fragments. There is a syntaxial rim cement on most of the grains. This facies was deposited in a foreslope environment.

Rudist Packstone, Floatstone, and Rudstone (Fig. 8.39c, h)

There are two variants of this facies. The rudist fragments packstone comprises small rudist debris (less than 2 mm) with shell fragments, foraminifera, and echinoderms. The second is formed mainly of large rudist fragments. The first variant suggests deposition in the foreslope area in front

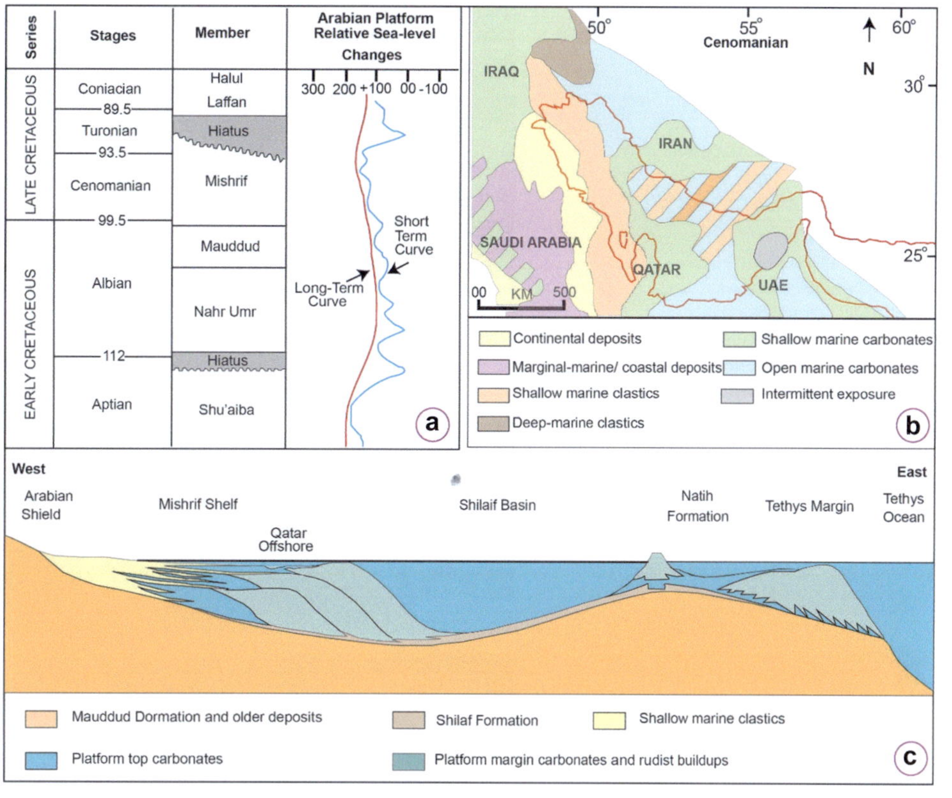

Fig. 8.38 The Mishrif Formation in the offshore area of Qatar **a** the stratigraphic succession, **b** facies map of the Cenomanian, **c** stratigraphic section through the Cenomanian Basin (Deville de Periere et al. 2017)

of the rudist buildups. The second subfacies represent rudist buildups, which punctuated the carbonate shelf of the Mishrif Formation.

Figure 8.38c shows the whole spectrum of the different environments in which the sediments of the Mishrif Formation were deposited.

8.4 The Upper Cretaceous

The Upper Cretaceous transgressive system reversed the widespread clastic system that characterized the Middle Cretaceous. Clastic sediments are now restricted to narrow basin margins in the western parts of the basin. Open marine shelf and deep-water shale and carbonates become the dominant sediment types in most parts of the basin. In some areas, like northern Iraq, basement faults reactivated, causing basin differentiation (Aqrawi et al. 2010). This helped the development of basin margin buildups and obstructed the widespread carbonate sheet-like deposition that characterized the Lower Cretaceous strata. Deep-water turbidites subducted for the first time beneath Iran. During the Coniacian–Maastrichtian, early phases of the Zagros Orogeny with its associated folding and thrusting started in some parts of the region (Fig. 8.40).

The Upper Cretaceous sediments in Qatar are poorly studied. They are categorically assigned to the Aruma Group and subdivided into four formations. These are, from the base, Laffan, Halul, Ruilat, and Simsima (Fig. 8.41). These strata were deposited during two sedimentary cycles. The Coniacian-Santonian cycle witnessed the deposition of the Halul-Laffan formation during relatively deeper water conditions, followed by the Maastrichtian cycle in which the Ruilat and Simsima formations were deposited in shallow water conditions (Hewaidy and Al-Hitmi 1999).

8.4.1 The Laffan Shale Formation

The first mention of the Laffan Formation is in an unpublished report by Sugden in 1953 but officially announced by Sugden and Standring (1975). The original description was from the well Dukhan-25 between 610 and 641 m. Later, a new type section was selected at the well Dukhan-55 between the drilled depths of 584 and 613 m with a total thickness of 29 m. The formation is named after the area of Ras Laffan on the northeastern coast of Qatar. The formation is formed totally of olive-green shale. The contact with the underlying Mishrif Formation is unconformable, placed at the change from the

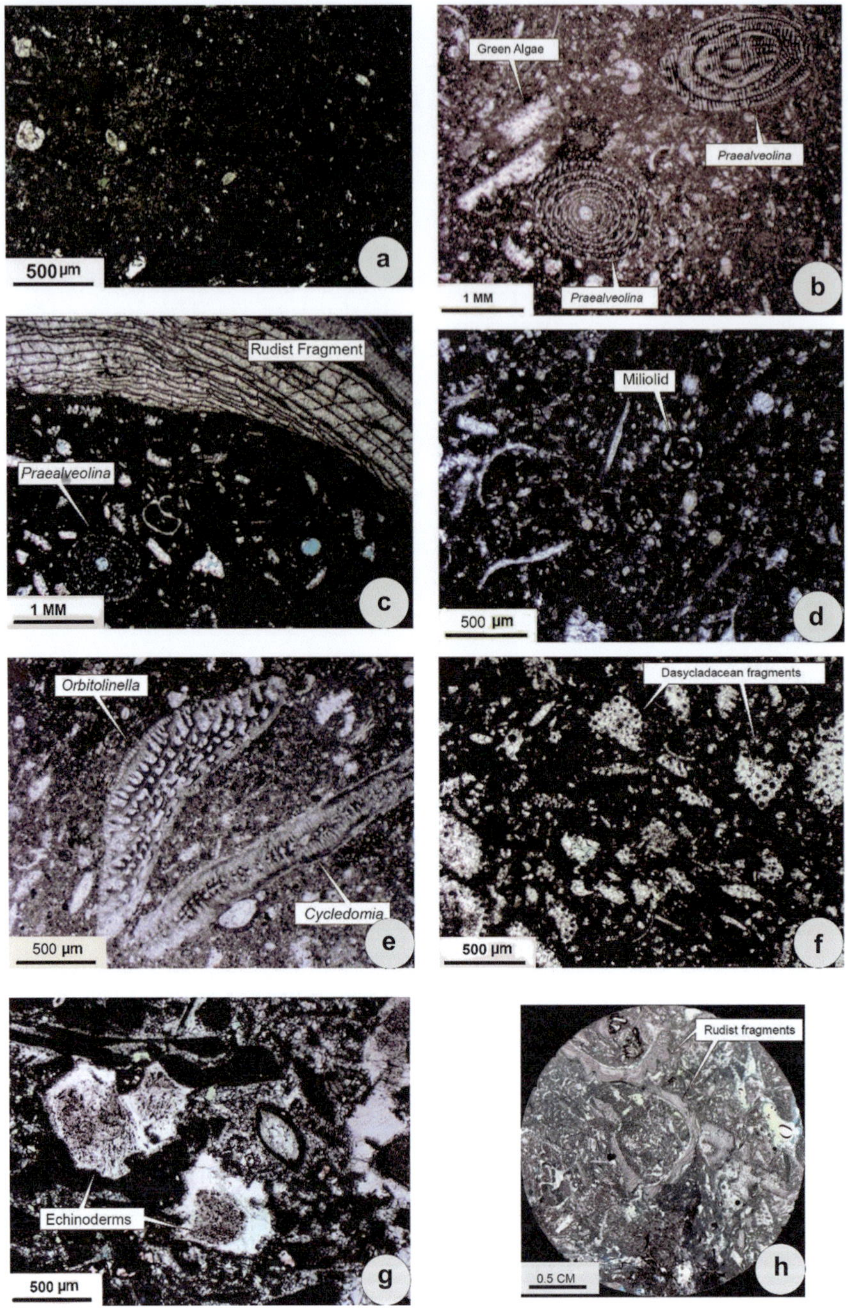

Fig. 8.39 The main microfacies of the Mishrif Formation in the offshore area of Qatar. **a** lime mudstone to wackestone with dwarf rotaliids, **b** large foraminifera and algae packstone, **c** rudist debris and large foraminifera packstone, **d** miliolids wackestone-packstone, **e** large foraminifera packstone, **f** dasycladacean algae wackestone-packstone, **g** echinoderm-bearing packstone, **h** rudist buildup (Deville de Periere et al. 2017)

fossiliferous limestone of the Mishrif to the shale of the Laffan Formation. The Overlying formation is Da'asah, a Member of the Halul Formation. The contact is conformable and placed at the change from shale to argillaceous lime mudstone (Sugden and Standring 1975). The formation has been encountered in the offshore areas of Qatar, Iran, Bahrain, and Abu Dhabi. Hewaidy and Al-Hitmi (1999) described the formation from Dukhan Field as around 20 m of olive-green shale. The Laffan Formation was studied by Ghajari et al. (2013) from the Sirri oil field offshore in Iran. They reported that the formation is 17m thick and consists of 30–70% clay minerals (Fig. 8.42).

Kaolinite is the main mineral, and its fibers are closely cemented by calcite and Quartz (Fig. 8.43).

8.4.2 The Halul Formation

The first mention of the Halul Formation was in an unpublished report by M. W. Hughes-Clarke in 1963. Sugden and Standring (1975) officially published the formation from the well Id-al-Shargi-1 between the drilled depths of 1009 and 1081 m with a total thickness of 72 m. The name is derived

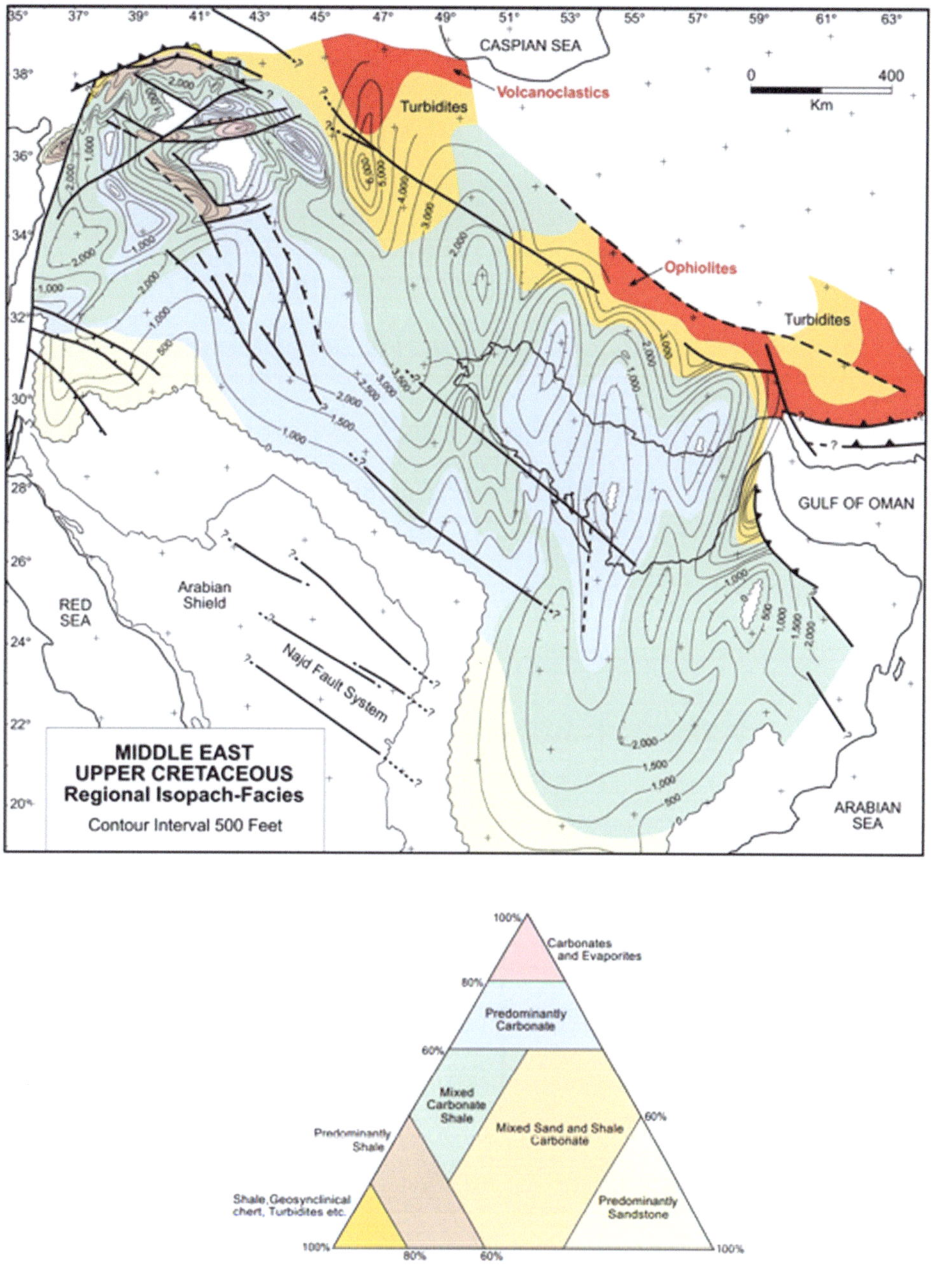

Fig. 8.40 Isopach and facies map of the Upper Cretaceous sediments across the Arabian Plate. Different lithologies are shown in the accompanying triangle. The contour interval is 500 ft. (Christian 1997)

from the Island of Halul, the eastern coast of Qatar (Sadooni et al. 2004).

The formation consists of four distinctive lithologies. These are from its base:

(a) Light grey, soft chalk, marly with chalky limestone with some shell fragments (15 m).
(b) Grey-green marl with blue-grey shale and limestone, which is chalky and recrystallized in part (8.6 m).
(c) Chalky, bioclastic wackestone. Skeletal grains include spicules, shell fragments, and small and large foraminifera. There are some pyrite and dolomite (21 m).
(d) Chalky, bioclastic packstone with large and small foraminifera, ostracods, and shell fragments (27 m).

Athersuch (1987) studied the fossil contents of the Halul Formation from the offshore area of Abu Dhabi and reported that the formation has been assigned to the Coniacian based on the planktonic foraminifera *Whiteinella archaeocretacea*, *Marginotruncana Renzi* and *Dicarinella imbricata*. He also suggested that the lower part of the formation, which he assumed to be developed in "a marine reefal lime mudstone, contains the Ostracoda *Brachycythere* spp.

Fig. 8.41 Stratigraphy of the upper cretaceous sediments across the Arabian Gulf region (El-Nakhal and El-Naggar 1994)

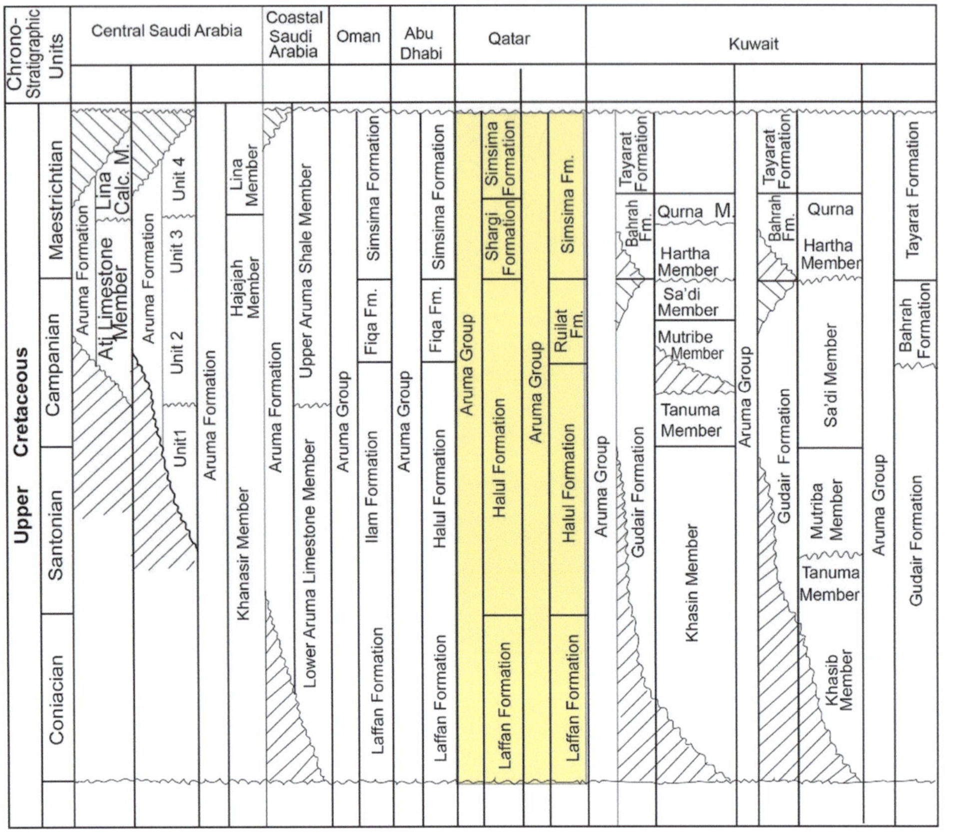

Fig. 8.42 Logs of the Laffan Formation in Sirri oil field, offshore Iran (Ghajari et al. 2013)

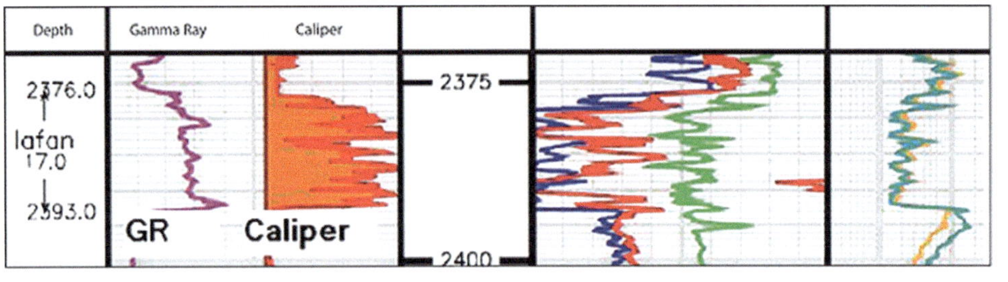

8.4.3 The Ruilat Formation

This formation was described in an unpublished report by Walter Sugden in 1953 and was officially defined by Sugden and Standring (1975). The formation was first described from the well Dukhan-25, but its type section was selected at the well Dukhan-55 between the drilled depths of 501 and 570 m with a total thickness of 69 m. The formation comprises two major lithologies: the lower succession consists of around 3 m of mudstone and wackestone with rounded phosphatic nodules and glauconite grains, and the upper part is formed of the same lithologies with calcareous silt and elongated spicules. It also has some chert nodules in the upper part. The skeletal components are only small foraminifera obliterated by extensive recrystallization. The total thickness of the upper part is around 66 m. The underlying formation is Halul, with an unconformable contact placed at the change from the lime mudstone to the glauconitic limestone at the base of the Ruilat Formation. The overlying formation is the Simsima, with a conformable contact placed at the change from the fossiliferous packstone of the Simsima Formation to the specular mudstone of the Ruilat Formation.

Hewaidy and Al-Hitmi (1994) studied the fossil content of the Ruilat Formation from Dukhan Field. She divided the formation into two units. Unit B consists of fine-grained, silicified limestone with some black chert nodules. This unit is free of fossils. Unit A comprises limestone with intercalations of silicified limestone in the upper part. The lower part is characterized by marly limestone, marl, and glauconite grains. This unit also has no fossils.

Fig. 8.43 SEM image of Laffan Formation in Sirri Oilfield, Iran, showing hexagonal crystals of the clay mineral kaolinite cemented by calcite and quartz (Ghajari et al. 2013)

8.4.4 The Simsima Formation

There is a common mix-up regarding the use of the term Simsima in the prevalent Qatari geology practice, particularly among engineering geologists and civil engineers. So, some workers (e.g., (Karagkounis et al. 2016)) used the term Simsima for the Umm Bab Member, the upper part of the Dammam Formation. This nomenclature violates the stratigraphic code and should be discarded.

The Simsima Formation is the last formation deposited during the Cretaceous period on the eastern Arabian Plate. The Formation is equivalent to the Tayarat Formation in Kuwait and southern Iraq (Owen and Nasr 1958). In Qatar, the formation was described for the first time in an unpublished report by Walter Sugden in 1956 and officially announced by Sugden and Standring (1975). First, the formation was described from the well Dukhan-28, but its type section has been selected in the well Dukhan-55 between the drilled depths of 353 and 503 m with a total thickness of 150 m. Alsharhan and Nairn (1997) stated that the Simsima Formation is divided into two members. These are the lower Jana'an and the upper Salwa. Each was divided into three and two parts, respectively.

8.4.4.1 Salwa Member (21–33 m)

The Salwa Member comprises the following two main lithologies:

(a) Argillaceous limestone changing upward into bioclastic packstone that shows some microporosity.
(b) Calcareous shale, green to grey, containing pyrite crystals, with small foraminifera and ostracods.

8.4.4.2 Jana'an Member (95–180 m)

This member consists of three distinctive lithologies. These are:

(a) Argillaceous limestone gradually changes upward into dolomitized wackestone alternating with bioclastic packstone.
(b) Bioclastic wackestone, selectively dolomitized.
(c) Slightly dolomitized, microporous limestone alternating with bioclastic packstone.

Hewaidy and Al-Hitmi (1999) studied the fossil content of the Simsima Formation in the Dukhan field and reported that the Simsima Formation is formed of around 150 m of light grey limestone, which is chalky at the base (the Janaan (sic) Member) and argillaceous at the top (the Salwa Member). They suggested that the Formation has no planktonic foraminifera, but each member has its fossil assemblages.

They identified five benthic calcareous species in the Jana'an Member: Ornatanomalina, Fissurina, and Orbitoides, and one arenaceous species, Loftusia morgan. They also identified ten benthic species in the Salwa Member. The calcareous fauna includes two species of *Discorbis* and one of *Woodella*, *Fissoelphidium*, *Omphalocyclus*, *Elphidiella*, and others. They concluded that increased biodiversity and fauna density are associated with a change from the chalky limestone at the base to the more argillaceous limestone at the top.

Based on the nature and diversity of the described fauna, they suggested that the sediments of the Jana'an Member were deposited in a shallow-water carbonate platform with warm and clear water, while the Salwa Member sediments were laid down in a deeper, open platform environment.

8.4.4.3 Facies Analysis

The Simsima Formation is poorly studied in Qatar, but due to its hydrocarbon potential in Abu Dhabi, it has been investigated in more detail. In the United Arab Emirates, the Simsima Formation was described from outcrops by Glennie et al. (1974). Still, the site became inaccessible, and a new reference section was selected in Jabal Al Faiyah, on the Northern Oman Mountains, by Nolan et al. (1990). Alsharhan et al. (2000) divided the formation succession in eastern UAE into a Lower Member consisting of 50 m of chalky limestone with chert nodules and bands. This member is rich in skeletal components such as planktonic and benthic foraminifera, algae, rudists, corals, and echinoids. The Upper Member is 28 thick and consists of marl containing planktonic and benthic foraminifera topped by siltstone. In other sections, it comprises nodular, bioturbated, and *Orbitolina*-bearing limestone. El-Gawad et al. (2010) studied the Simsima Formation in five exposures in the UAE. These are Jabal Qarn El Bar, Jabal Buhays (Al Faiyah Range Mountains (Fig. 8.44), Jabal El Rawadah, Jabal Malaqet and Jabal Mundassah (Fig. 8.45).

They classified the sediments of this formation into two main facies: the Planktonic Foraminiferal Biomicrite and Rudistid Algal Foraminiferal Biomicrite (Fig. 8.46). They

Fig. 8.44 Outcrops of the Simsima Formation, Jabal Qarn El Barr, near Al Dhaid city, Sharjah Emirate (in the middle of the image Dr. Osman Abdelghany) (Courtesy Hamdi Kandil)

Fig. 8.45 Macrofossils of the Simsima Formation in its outcrops at the United Arab Emirates. **a** Rudistid bivalves, Jabal Buhays. **b** The gastropod *Acteonella* sp. And algae **c** Algal limestone (white patches). **d** The contact between the Qahlah Formation (top)/Simsima Formation. Rudistid bivalves at the lower part (Courtesy of Hamdi Kandil)

proposed that the first facies was deposited in open marine and represents a source rock, while the second facies was formed in shallow marine conditions and represents the reservoir rocks.

8.4.4.4 The Depositional Environments

According to Hozayen et al. (2008), the Simsima carbonate shelf passes into a deep-water marly limestone and shale of the Fiqa Formation and restricted lagoons and marshes to the west. The Loftusia-bearing, bioturbated wackestone and packstone of the lower Simsima Formation form several shoaling-upwards cycles. These cycles are characterized by low diversity, suggesting restricted depositional conditions. These Loftusia shoals were accumulated in restricted to open ramp and lagoonal conditions during the early transgression. The presence of more diverse fauna and dasycladacean algae may suggest that some of these shoals were formed in open marine conditions. The middle part of the formation is

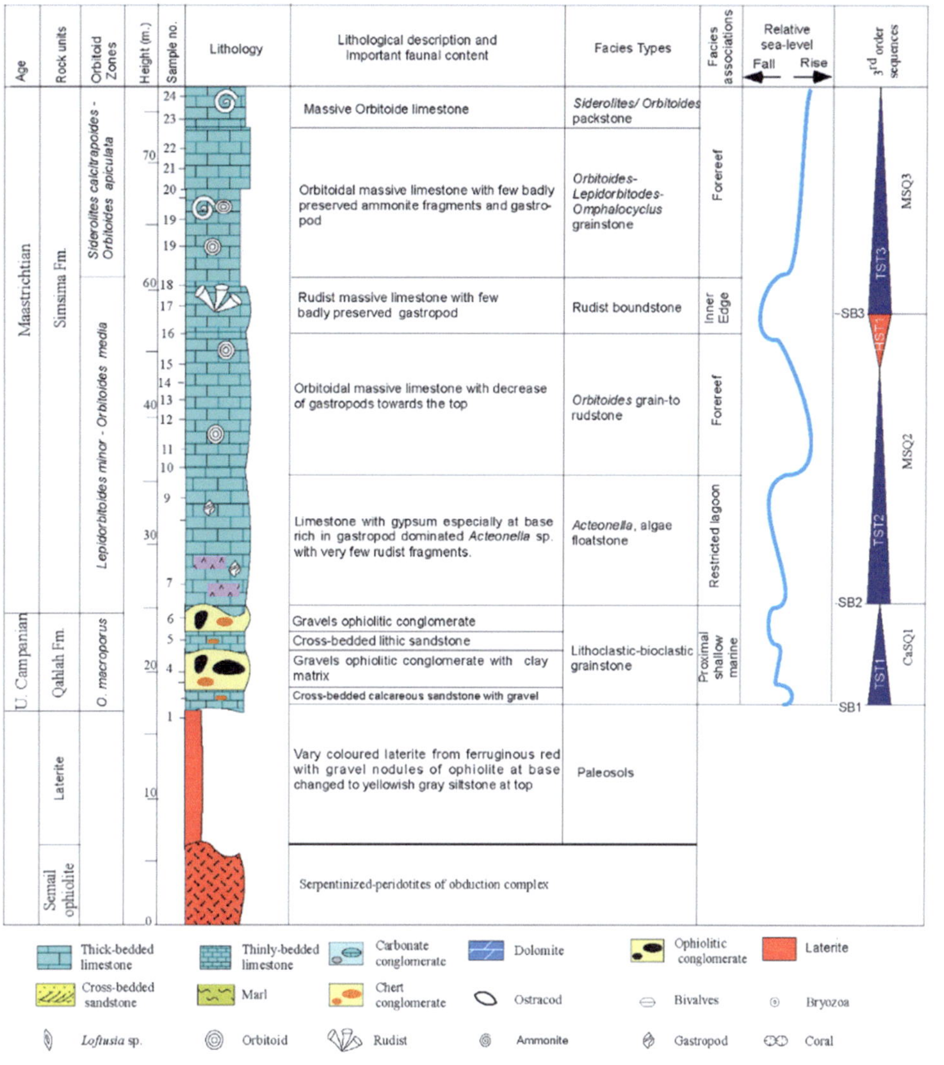

Fig. 8.46 Lithostratigraphic section and facies association of the Simsima Formation in Abu Dhabi (Mohammed et al. 2021)

formed of more open marine conditions than the lower part. This unit has more diverse fauna, including bivalves, echinoids, rudists, algae, large benthic foraminifera, and dasycladacean algae. These shoaling-upward cycles represent peloidal, bioclastic, and rudist shoals deposited in shoreface environments. The bioturbated and laminated peloidal strata, probably represent shoals or shoreface settings. The rudist buildups and the bioclastic shoals may provide lee-ward protected areas where dasycladacean algae meadows flourished.

Mohammed et al. (2021) studied the wedge-top to foredeep architecture of the Simsima formation in the Faiyah Mountain (Fig. 8.47). They suggested that Oman's ophiolite obduction resulted in the development of an irregular topographic surface, which led to the generation of rapid lateral facies variations amounting to a total of 19 different facies types. They concluded that the Simsima sediments are formed of distal turbidites with carbonate conglomerate and marl. These strata were deposited in a foredeep environment. The facies variation of the formation is associated with high global sea level.

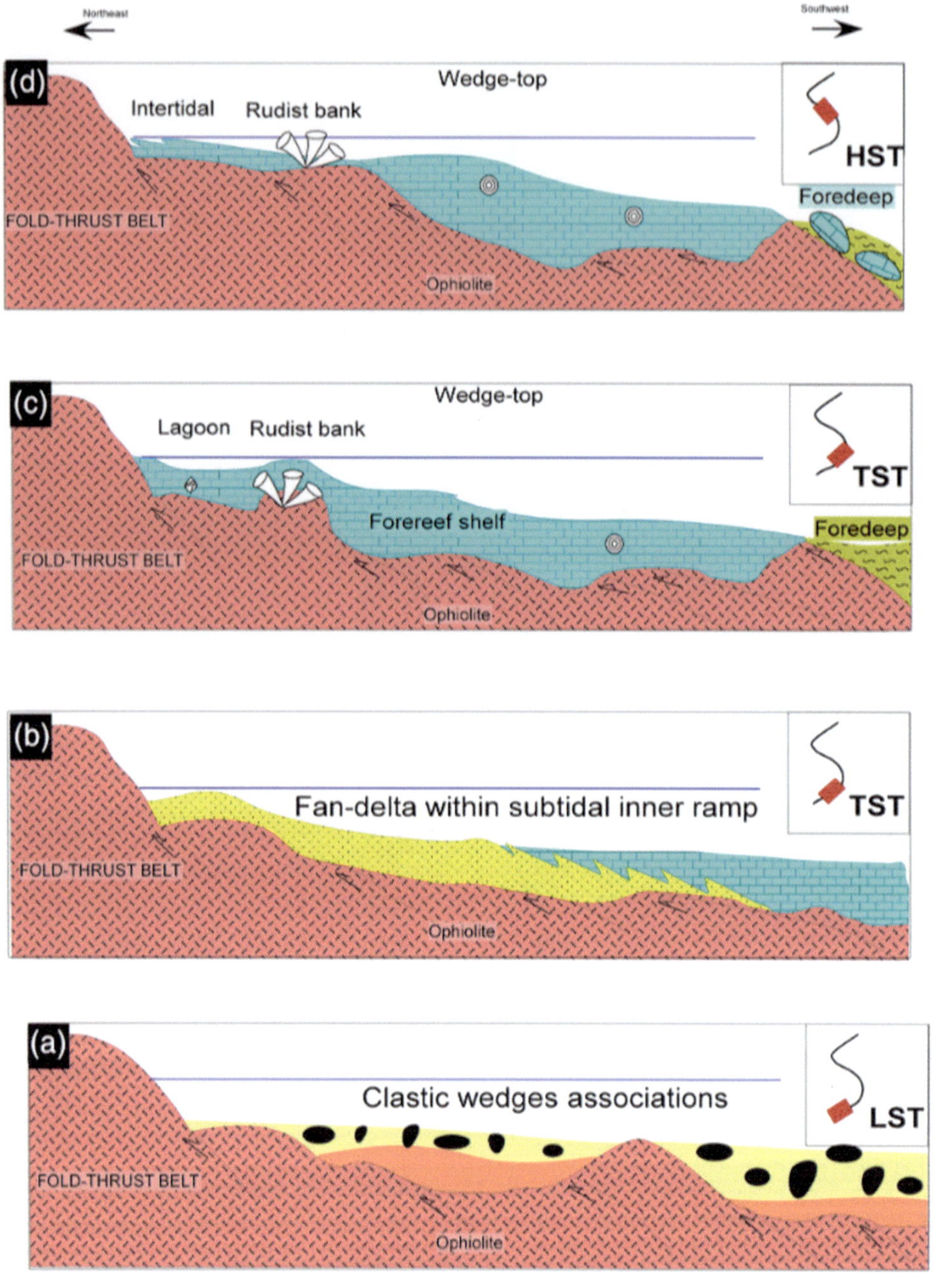

Fig. 8.47 Depositional environment of the Simsima Formation in Abu Dhabi (Mohammed et al. 2021)

References

Abd el-aal Ae-aK, Al-Awadhi JM, Al-Dousari A (2023) The geology of Kuwait. Springer International Publishing.https://doi.org/10.1007/978-3-031-16727-0

Adasani M (1967) The North Kuwait oil fields. Ministry of Finance and Oil. https://books.google.com.qa/books?id=B5JMHQAACAAJ

Al-Dwaish M, Al-Azmi FM, Al-Houli M, Alrubaiyea J, Al-Harbi M, Al-Najim A, Sikdar K, Al-Jamaan A, Lahmar HA, Bora K, Chakraborty P, Roy DB, Goswami R (2016) Integrated multiwell approach for heterogeneity mapping and production optimization: a case study from Ratawi formation of South Fuwaris field, partitioned zone between Kuwait and Saudi Arabia. Abu Dhabi International Petroleum Exhibition & Conference

Al-Ghamdi N, Pope M (2020) High-resolution stratigraphic architectures, facies anatomies of the lower cretaceous Biyadh and Shu'aiba formations, and their implications on platform evolution and global correlation. In: Khomsi S, Roure FM, Al Garni M, Amin A (eds) Arabian plate and surroundings: geology, sedimentary basins and georesources. Springer International Publishing, pp 3–39. https://doi.org/10.1007/978-3-030-21874-4_1

Al-Husseini MIaRKM (2010) Tuning late Barremian—Aptian Arabian plate and global sequences with orbital periods. In: van Buchem MIA-HFSP, Maurer F, Droste HJ (eds) Barremian—Aptian stratigraphy and hydrocarbon habitat of the eastern Arabian Plate, vol 1. GeoArabia Special Publication 4, pp 199–228

Al Naqib KM (1967) Geology of the Arabian Peninsula; southwestern Iraq [Report](560G). (Professional Paper, Issue). https://pubs.usgs.gov/publication/pp560G

Al-Siddiki A (1978) Subsurface geology of southeastern Iraq. 10th Arab Petroleum Congress, Tripoli, Libya

Al Shamlan AA (1975) Microfacies analysis of the Mauddud formation in Kuwait: University of Kuwait]. Kuwait

Al Shdidi S, Thomas G, Delfaud J (1995) Sedimentology, diagenesis, and oil habitat of lower Cretaceous Qamchuqa Group, Northern Iraq1. AAPG Bull 79(5):763–778. https://doi.org/10.1306/8d2b1b9c-171e-11d7-8645000102c1865d

Albehbehani ASK (1988) Carbonate lithofacies, diagenesis, porosity relationships and crude oil geochemistry, Ratawi Formation, Kuwait-Saudi Arabia divided neutral zone. Texas Tech Univ., Lubbock, TX (USA), United States

Alsharham AS (1985) Depositional environment, reservoir units evolution, and hydrocarbon habitat of Shuaiba formation, lower Cretaceous, Abu Dhabi, United Arab Emirates. Am Assoc Pet Geol Bull (United States) 69:6, Medium: X; Size, pp 899–912, 2009–2012–2016. https://www.osti.gov/biblio/5172674

Alsharhan A (1987) Geology and reservoir characteristics of carbonate buildup in giant Bu Hasa oil field, Abu Dhabi, United Arab Emirates. AAPG Bull 71(10):1304–1318

Alsharhan A, Nairn A (1986) A review of the Cretaceous formations in the Arabian Peninsula and Gulf: Part I. Lower Cretaceous (Thamama Group) stratigraphy and paleogeography. J Pet Geol 9(4):365–391

Alsharhan A, Nairn A (1994) Geology and hydrocarbon habitat in the Arabian Basin: the Mesozoic of the State of Qatar. Geol Mijnbouw 72(3):265–294

Alsharhan AS, Nairn AEM (1988) A review of the Ctretaceous formations in the Arabian Peninsula and Gulf: PARTII. Mid-cretaceous (WASIA GROUP) stratigraphy and paleogeography. J Pet Geol 11(1):89–112. https://doi.org/10.1111/j.1747-5457.1988.tb00803.x

Alsharhan AS, Nairn AEM (1997) Sedimentary Basins and petroleum geology of the Middle East. Elsevier

Alsharhan AS, Nairn AEM (2003) Chapter 9—the latest part of the Zuni and TEJAS cycles in the MIDDLE EAST: THE CENOZOIC. In: Alsharhan AS, Nairn AEM (eds) Sedimentary basins and petroleum geology of the Middle East. Elsevier Science B.V, pp 393–465. https://doi.org/10.1016/B978-044482465-3/50010-3

Alsharhan AS, Ziko A, Shebl HT, Whittle GL (2000) Microfacies and microfabrics of Maastrichtian carbonates, Northwestern Oman Mountains, U.A.E. In: Alsharhan AS, Scott RW (eds) Middle East models of Jurassic/Cretaceous carbonate systems, vol 69. SEPM Society for Sedimentary Geology, pp 0. https://doi.org/10.2110/pec.00.69.0129

Aqrawi AA (1998) Paleozoic stratigraphy and petroleum systems of the western and southwestern deserts of Iraq. GeoArabia 3(2):229–248

Aqrawi AAM, Thehni GA, Sherwani GH, Kareem BMA (1998) Mid-cretaceous Rudist-bearing carbonates of the Mishrif formation: an important reservoir sequence in the Mesopotamian basin, IRAQ. J Pet Geol 21(1):57–82. https://doi.org/10.1111/j.1747-5457.1998.tb00646.x

Aqrawi AA, Goff JC, Horbury AD, Sadooni FN (2010) The petroleum geology of Iraq, vol 424. Scientific press Beaconsfield

Athersuch J (1987) Ostracod faunas from the Halul, Laffan and Nahr Umr formations of offshore Abu Dhabi, UAE. J Micropalaeontol 6(1):1

Barber CT (1948) Review of Middle East oil, vol 10a. https://books.google.com.qa/books?id=tZWswgEACAAJ

Borgomano J, Masse J-P, Al Maskiry S (2002) The lower Aptian Shuaiba carbonate outcrops in Jebel Akhdar, northern Oman: impact on static modeling for Shuaiba petroleum reservoirs. AAPG Bull 86(9):1513–1529

Bramkamp R, Steineke M (1952) Stratigraphical introduction. Jurassic ammonites from Jebel Tuwaiq, central Arabia philosophical transactions Series B, 236:241–256

Bushang IO (2002) Lower cretaceous Ratawi formation Saudi Arabia-Kuwait partitioned neutral zone: deposition, diagenesis and petroleum geology in the framework of sequence stratigraphy. Durham University

Christian L (1997) Cretaceous subsurface geology of the Middle East region. GeoArabia 2(3):239–256

Deville de Periere M, Durlet C, Vennin E, Caline B, Boichard R, Meyer A (2017) Influence of a major exposure surface on the development of microporous micritic limestones—example of the Upper Mishrif formation (Cenomanian) of the Middle East. Sediment Geol 353:96–113. https://doi.org/10.1016/j.sedgeo.2017.03.005

Douglass RC (1960) The foraminiferal genus Orbitolina in North America: geological survey professional paper 333. United States Government Printing Office

Dunnington H (1967) Stratigraphic distribution of oil fields in the Iraq–Iran–Arabian basin: Journal of the Institute of Petroleum, v 53

El-Gawad E, Abdelghany O, Lotfy M, Saima M, Hashem W (2010) Basin analysis of the late cretaceous sediments in United Arab Emirates

El-Nakhal HA, El-Naggar ZR (1994) Review of the biostratigraphy of the Aruma Group (Upper Cretaceous) in the Arabian Peninsula and surrounding regions. Cretac Res 15(4):401–416. https://doi.org/10.1006/cres.1994.1023

Focke J, Munn D, Al Kuwari S, Frikken H, Frei H (1986) Petrographic atlas of rock types, common in the subsurface of Qatar and some recent equivalents. General Petroleum Corporation (Offshore Operations), Qatar, pp 110–189

Ghajari A, Kamali M, Mortazavi SA (2013) A comprehensive study of Laffan Shale Formation in Sirri oil fields, offshore Iran: implications for borehole stability. J Petrol Sci Eng 107:50–56. https://doi.org/10.1016/j.petrol.2013.03.030

Glennie KW, Boeuf MG, Hughes-Clarke MHW, Moody-Stuart M, Pilaar WF, Reinhardt BM (1974) Geology of the Oman Mountains. Verhandelingen Koninklijke Nederlands Geologisch Mijnbouw Kundig Genootschap 31:423

Hamdan ARA, Alsharhan AS (1991) Palaeoenvironments and palaeoecology of the rudists in the Shuaiba formation (aptian) United Arab Emirates. J Afr Earth Sci (and the Middle East) 12(4):569–581. https://doi.org/10.1016/0899-5362(91)90018-T

Harris PM, Frost SH (1984) Middle cretaceous carbonate reservoirs, Fahud Field and Northwestern Oman1. AAPG Bull 68(5):649–658. https://doi.org/10.1306/ad461359-16f7-11d7-8645000102c1865d

Henson FRSM (1948) Larger imperforate foraminifera of South-Western Asia: families Lituolidae, Orbitolinidae and Meandropsinidae. British Museum (Natural History)

Hewaidy A, Al-Hitmi H (1994) Cretaceous-Early Eocene biostratigraphy of the Dukhan oil field, west Qatar, Arabian Gulf. Earth Science Series, 8:1–24

Hewaidy A, Al-Hitmi H (1999) Paleoecology of the cretaceous—early tertiary sequence in the Dukhan Oil Field, west Qatar. Arabian Gulf Inter Conf Geol Arab World (GAW 4)

Hohman J, Al-Emadi IA, Zahran ME (2005) Sequence stratigraphic analysis of the Shuaiba formation: implications for exploration potential in Qatar. International petroleum technology conference

Hozayen MA, Strohmenger CJ, Mueller K, Al-Marzouqi SH, Shuaib M (2008) Seismic and high-resolution sequence stratigraphy of the upper Cretaceous Simsima formation, onshore Abu Dhabi Oil Field, United Arab Emirates. Abu Dhabi International petroleum exhibition and conference

Hughes GW (2000) Bioecostratigraphy of the Shu'aiba Formation, Shaybah field, Saudi Arabia. Geoarabia 5(4):545–578

Hughes G, Naji N (2008) Sedimentological and micropalaeontological evidence to elucidate post-evaporitic carbonate palaeoenvironments of the Saudi Arabian latest Jurassic. Volumina Jurassica 6(1):61–73

Ibrahim M (1981) Lithostratigraphy and subsurface geology of the Albian rocks of south Iraq. J Pet Geol 4(2):147–162

Ibrahim MW (1983) Petroleum geology of Southern Iraq1. AAPG Bull 67(1):97–130. https://doi.org/10.1306/03b5accd-16d1-11d7-8645000102c1865d

Immenhauser A, Hillgärtner H, Sattler U, Bertotti G, Schoepfer P, Homewood P, Vahrenkamp V, Steuber T, Masse J-P, Droste H (2004) Barremian-lower Aptian Qishn formation, Haushi-Huqf area, Oman: a new outcrop analogue for the Kharaib/Shu'aiba reservoirs. GeoArabia 9(1):153–194

Karagkounis N, Latapie B, Sayers K, Mulinti SR (2016) Geology and geotechnical evaluation of Doha rock formations. Geotech Res 3(3):119–136

Maurer F, van Buchem FSP, Eberli GP, Pierson BJ, Raven MJ, Larsen P-H, Al-Husseini MI, Vincent B (2013) Late Aptian long-lived glacio-eustatic lowstand recorded on the Arabian Plate. Terra Nova 25(2):87–94. https://doi.org/10.1111/ter.12009

Mohammed MU, Al-Sayyab AS (1993) Paleoenvironment of the Mauddud limestone formation at south and southeast Iraq. Iraqi Geol J 26(2):159–173

Mohammed IQ, Farouk S, Lawa F, Alsuwaidi M, Morad S (2021) Upper Cretaceous wedge-top to foredeep architecture in the United Arab Emirates: insights from the Faiyah anticline. Geol J 56(5):2602–2624. https://doi.org/10.1002/gj.4057

Moosavizadeh SMA, Mahboubi A, Moussavi-Harami R, Kavoosi MA, Schlagintweit F (2015) Sequence stratigraphy and platform to basin margin facies transition of the lower Cretaceous Dariyan formation (Northeastern Arabian Plate, Zagros fold-thrust belt, Iran). Bull Geosci 90(1):145–172

Murris RJ (1980) Middle East: stratigraphic evolution and oil Habitat1. AAPG Bull 64(5):597–618. https://doi.org/10.1306/2f918a8b-16ce-11d7-8645000102c1865d

Nath PK, Singh SK, Ye L, Al-Ajmi AS, Bhukta SK, Al-Otaibi AH (2014) Reservoir characterization and strati-structural play of Minagish formation, SE Kuwait. International petroleum technology conference

Nolan SC, Skelton PW, Clissold BP, Smewing JD (1990) Maastrichtian to early tertiary stratigraphy and palaeogeography of the Central and Northern Oman Mountains. Geol Soc London, Special Publ 49(1):495–519. https://doi.org/10.1144/GSL.SP.1992.049.01.31

Norris RD, Bice KL, Magno EA, Wilson PA (2002) Jiggling the tropical thermostat in the cretaceous hothouse. Geology 30(4):299–302

Owen RMS, Nasr SN (1958) Stratigraphy of the Kuwait-Basra area. In: WLG (ed) Habitat of oil. The American Association of Petroleum Geologists, pp 1252–1278

Pierson BJ, Eberli GP, Al-Mehsin K, Al-Menhali S, Warrlich G, Droste HJ, Maurer F, Whitworth J, Drysdale D (2010) Seismic stratigraphy and depositional history of the Upper Shu'aiba (Late Aptian) in the UAE and Oman. GeoArabia Special Publication, Gulf PetroLink, Bahrain 4(2010):411–444

Powers RW (1968) Asie: Arabie Saoudite. Lexique Stratigraphique Internationale, Paris

Powers RW, Ramirez LF, Redmond CD, Elberg Jr EL (1966) Geology of the Arabian Peninsula: sedimentary geology of Saudi Arabia [Report](560D). (Professional Paper, Issue. U. S. G. Survey). https://pubs.usgs.gov/publication/pp560D

Raven MJ, van Buchem FSP, Larsen P-H, Surlyk F, Steinhardt H, Cross D, Klem N, Emang M, van Buchem FSP, Al-Husseini MI, Maurer F, Droste HJ (2010) Late Aptian incised valleys and siliciclastic infill at the top of the Shu'aiba formation (Block 5, offshore Qatar). In: Barremian—Aptian stratigraphy and hydrocarbon habitat of the Eastern Arabian plate, vol 2. Gulf PetroLink, pp 0

Sadooni FN (1993) Stratigraphic sequence, microfacies, and petroleum prospects of the Yamama formation, Lower Cretaceous, Southern Iraq. AAPG Bull 77(11):1971–1988

Sadooni FN (2018) Impact of the demise mechanisms of the Cretaceous rudist buildups in the Arabian Plate on their reservoir characteristics. Carbonates Evaporites 33(3):465–476. https://doi.org/10.1007/s13146-017-0355-4

Sadooni FN, Aqrawi AAM (2000) Cretaceous sequence stratigraphy and petroleum potential of the Mesopotamian Basin, Iraq. In: Alsharhan AS, Scott RW (eds) Middle East models of Jurassic/cretaceous carbonate systems, vol 69. SEPM Society for Sedimentary Geology, pp 0. https://doi.org/10.2110/pec.00.69.0315

Sadooni FN, Alsharhan AS (2003) Stratigraphy, microfacies, and petroleum potential of the Mauddud formation (Albian–Cenomanian) in the Arabian Gulf basin. AAPG Bull 87(10):1653–1680. https://doi.org/10.1306/04220301111

Sadooni FN, Al-Saad H, Nasir SJ (2004) Halul and sharao islands, offshore Qatar: remnants of the great infracambrian Hormuz Salt Basin. Carbonates Evaporites 19(1):17–27. https://doi.org/10.1007/BF03175193

Steineke M, Bramkamp RA, Sander NJ (1958) Stratigraphic relations of Arabian Jurassic oil. American Association of Petroleum Geologists Symposium, Tulsa, U.S.A.

Strohmenger CJ, Weber LJ, Ghani A, Rebelle M, Al-Mehsin K, Al-Jeelani O, Al-Mansoori A, Suwaina O (2004) High-resolution sequence stratigraphy of the Kharaib Formation (Lower Cretaceous, UAE). Abu Dhabi international petroleum exhibition and conference

Sugden W, Standring AJ (1975) Qatar Peninsula. In: Lexique Stratigraphique International. Aise, Fasc. lOa,, III, 120

Twombley B, Scott J (1976) Application of geological studies in the development of the Bu Hasa field, Abu Dhabi. 9th Arab Petrol Congr (United Arab Emirates) 133

van Bellen RC, Dunnington HV, Wetzel R, Morton D (1959) Lexique stratigraphique international: Asie. Iraq, vol 3C

van Buchem FSP, Pittet B, Hillgärtner H, Grötsch J, Al Mansouri AI, Billing IM, Droste HHJ, Oterdoom WH, van Steenwinkel M (2002) High-resolution sequence stratigraphic architecture of Barremian/Aptian carbonate systems in Northern Oman and the United Arab Emirates (Kharaib and Shu'aiba Formations). GeoArabia 7:461–500. https://doi.org/10.2113/geoarabia0703461

Van Buchem F, Maurer F, Hoch E, Habib KH (2009) Shuaiba formation, depositional facies distribution revealed from extended reach horizontal wells in Al Shaheen Field, Offshore, Qatar. International petroleum technology conference

van Buchem FS, Al-Husseini MI, Maurer F, Droste HJ, Yose LA (2010) Sequence-stratigraphic synthesis of the Barremian–Aptian of the eastern Arabian Plate and implications for the petroleum habitat

Van Buchem F, Svendsen N, Hoch E, Pedersen-Tatalovic R, Habib K (2014) Depositional history and petroleum habitat of Qatar

Vaslet D, Al-Muallem MS, Maddah SS, Brosse J-M, Fourniguet J, Breton J-P, Nindre Y-ML (1991) Geologic map of the Ar Riyadh Quadrangle, Sheet 24I [Geoscience Map GM-121, with explanatory notes.]. Kingdom of Saudi Arabia, Ministry of Petroleum and Mineral Resources

Wolpert P, Bartenbach M, Suess P, Rausch R, Aigner T, Le Nindre Y-M (2015) Facies analysis and sequence stratigraphy of the uppermost Jurassic—lower Cretaceous Sulaiy formation in outcrops of central Saudi Arabia. GeoArabia 20(4):67–122. https://doi.org/10.2113/geoarabia200467

Ziegler MA (2001) Late Permian to Holocene paleofacies evolution of the Arabian Plate and its hydrocarbon occurrences

The Cenozoic

9

Abstract

This Chapter of this work examines the Cenozoic evolution of the Arabian Plate, focusing on its stratigraphy and depositional history in Qatar, Saudi Arabia, Oman, and Iraq. The Cenozoic era commenced with notable tectonic and structural settings resulting from the collision of the Arabian Plate with Eurasia, culminating in developing the Zagros Mountains and related foreland basins. The stratigraphic succession is categorized into two main systems: the Paleogene and Neogene, each characterized by unique tectonic activity and sedimentary processes, where the Paleogene period, marked by marine transgressions, witnessed the deposition of significant carbonate sediments such as the Umm er Radhuma, Rus, and Dammam Formations. These strata comprise a blend of shallow marine, evaporitic, and carbonate-dominated lithologies, representing vital groundwater aquifers in the Arabian Gulf region. The Neogene period marked the Arabian Plate's progressive separation from the Tethys Sea, leading to limited depositional environments characterized by continental effects. The Miocene period witnessed the deposition of the Dam and Hofuf Formations, shaped by the emergence of the Red Sea and the commencement of rifting in the Gulf of Aden. The Dam Formation consists of mixed marine and evaporitic facies, whereas the Hofuf Formation is mostly characterized by terrestrial clastic deposits originating from the Arabian Shield. These episodes reflect the intricate interactions of tectonics, sea-level fluctuations, and sedimentation patterns that influenced the Arabian Plate during the Cenozoic. This chapter thoroughly examines lithological variations, depositional settings, and paleoenvironmental reconstructions to elucidate the geological history of the Arabian Plate during the Cenozoic era.

9.1 Introduction

During the Late Cretaceous, with the continued subduction of the Arabian Plate under Eurasia, the Neotethys narrowed to the extent that two plates collided, leading to significant compression and structural deformation. Compression between the two plates generated the Zagros Orogeny. The Arabian plate convergence and subduction beneath Iran led to a slight northwest tilt and created a sequence of anticlines and thrusts in the Zagros Mountains. The major Hercynian structures continued to grow, leading to the final formation of the major hydrocarbon traps in the region (Beydoun 1991). During the peak of this phase, ophiolites were obducted along the margins of the region, mainly in Oman. Later deformation led to the uplifting of the Zagros Mountains in present-day Iran. This was accompanied by the development of the Zagros Foreland Basin on the margin of the old carbonate platform during the Late Cretaceous and the Cenozoic.

The advent of the Arabian Cenozoic was associated with a progressive restriction of the Arabian platform. Alsharhan and Nairn (1995) divided this event into two phases, the Paleogene and Neogene, separated by a Neogene tectonic activity and illustrated by the sediment distribution. The depositional history in the Gulf region, during the Cenozoic, began with a regression at the Maastrichtian/Danian boundary, leaving most of the Arabian Peninsula exposed except basinal areas in the northern United Arab Emirates (the Ras al Khaimah Basin) and the Zagros Basin in Iran. This hiatus is frequently overlooked because it was brief, and an early Paleogene major transgression re-established marine conditions over eastern Arabia and the entire northern part of the shelf extending over the Zagros trough (Alsharhan and Nairn 1995). These conditions persisted into the early to mid-Eocene, with only western Saudi Arabia, parts of high areas in Iraq and Iran,

F. N. Sadooni and H. A. Al-Kuwari, *Geological Evolution of Qatar and the Arabian Peninsula*, Regional Geology Reviews,
https://doi.org/10.1007/978-3-031-94394-2_9

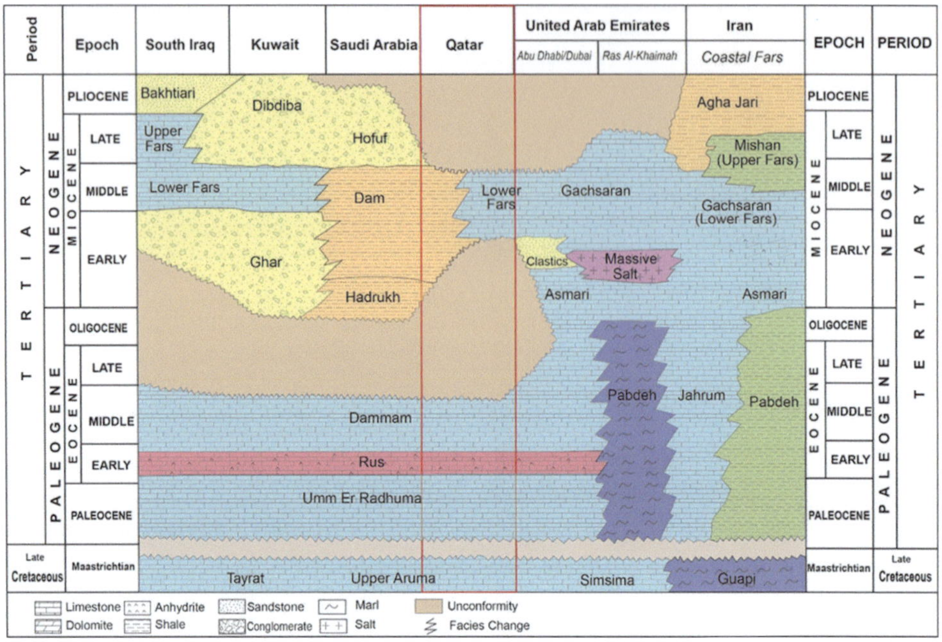

Fig. 9.1 The Tertiary stratigraphic system in the Arabian Plate and Iran (Alsharhan and Nairn 1995)

and the Semail and Hawasina nappes in the Oman Mountains remaining emergent.

The troughs, although not exposed, vanished as distinct depositional features during the second, late Eocene/Oligocene regression, and the upper two-thirds of the Paleogene sedimentary succession in the troughs merges with the sediments on the bordering shelves. The stratigraphic chart (Fig. 9.1) shows the regions affected by the regression and its duration. Once marine conditions were established, paleogeography did not vary significantly during the interval of relative tectonic calm between the two regressions. The broad pattern featured a stable land mass in the west surrounded by an extensive carbonate shelf that deepened toward two eastern troughs adjacent to tectonically active areas in Iran and Oman (Alsharhan and Nairn 1995; Murris 1980).

The Neogene sediments of eastern Arabia are only found in scattered areas, as recent dunes or sabkha deposits cover the majority. Outcrop distribution reflects the continued shrinkage of the deposition area. The main orogenic folding during the late Tertiary displaced the axis of sedimentation southwestwards toward the Zagros trough, and as water depth decreased, the Neogene sequence, consisting of basinal shales, marls, and argillaceous limestones, thins. However, thick flysch deposits near the northeastern margin of the Zagros trough in Lurestan reflect the rapid erosion of an uplifted Late Cretaceous radiolarite sequence that was originally deposited in a deep trough (Alsharhan and Nairn 1995; Beydoun 1988, 1991; Buday 1980; Dunnington 1958). The flysch is followed by coarsening red molasse beds (James and Wynd 1965).

As a result, a distinction can be made between the Paleogene and Neogene depositional settings in the region. While the Paleogene conditions are essentially a continuation of those that existed for most of the Mesozoic, the Neogene marks the beginning of a new tectonic environment today. A foredeep is imposed over the old Mesozoic shelf margin, gradually filling with sediments transported along its axis.

The Paleogene strata of Qatar comprise three formations. These are the Paleocene-Lower Eocene (Ypresian) Umm er Radhuma Formation, the Lower Eocene (Ypresian) Rus Formation, and the Middle Eocene (Lutetian/Bartonian) Dammam Formation. The Neogene strata include the Miocene Dam Formation and the Miocene-Pliocene Hofuf Formation (Fig. 9.2). Except for the Umm er Radhuma Formation, these formations are exposed at different parts of Qatar. The Paleogene carbonate strata are important water aquifers in most parts of the Arabian Gulf region. Ziegler (2001) divided the Cenozoic strata into five time—slices. This practice will be followed in this treatment.

9.2 Late Paleocene to Early Eocene: Selandian-Ypresian (60.9–49 Ma)

Deformation, resulting from the collision of Arabia with Eurasia, is reflected by a regional unconformity at the base of the Cenozoic strata, namely the lowest beds of the early Tertiary Umm Er Radhuma Formation (Johnson 2008). According to Ziegler (2001), this period witnessed the deposition of the Umm Er Radhuma and Rus Formation for most

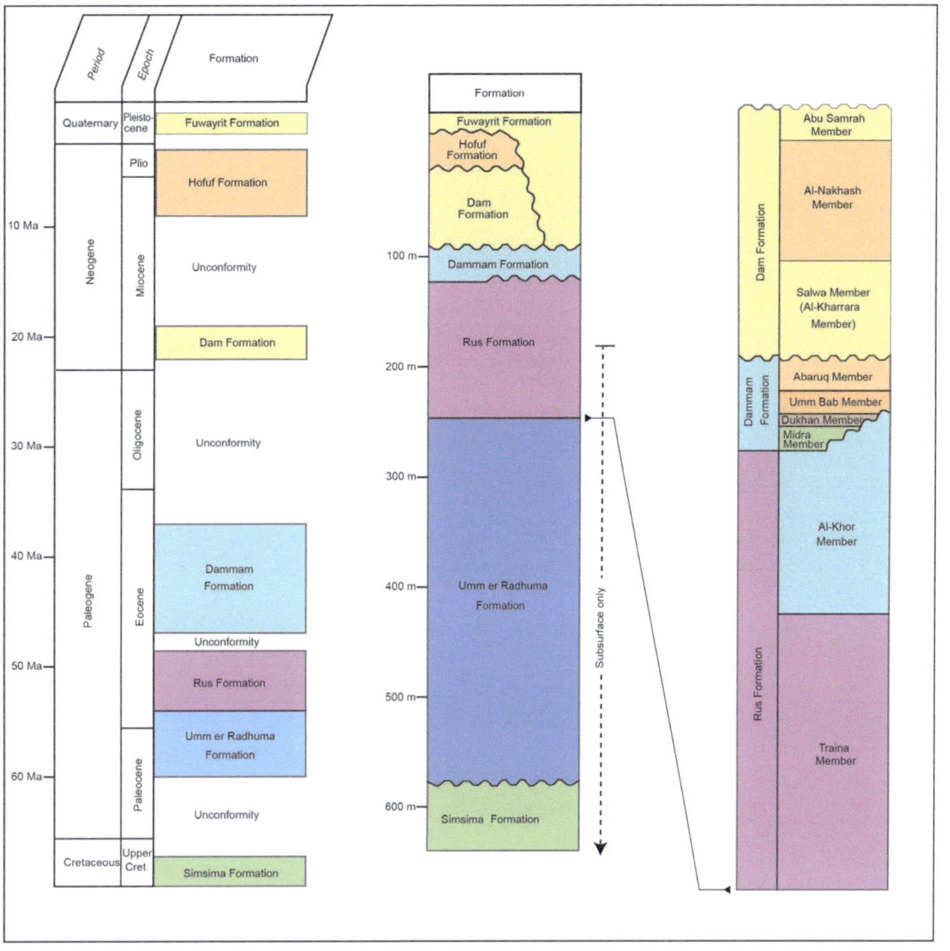

Fig. 9.2 The main formations and unconformity surfaces of the Tertiary strata in Qatar (LeBlanc 2021)

parts of eastern Arabia and the Pabdeh Formation in Iran. These carbonate sediments were deposited within a "remnant" foredeep setting associated with rapid erosion, lowering of sea level, and subsidence of the emergent ophiolites along the eastern margin (Fig. 9.3).

During the Paleocene, shallow marine conditions were re-established over most parts of eastern Arabia (green color in Fig. 9.3) and continued to the early to mid-Eocene. The eustatic sea level was consistently high during that time, accompanied by frequent regression episodes (Haq et al. 1988). The Umm er Radhuma Formation, located above the Maximum Flooding Surface (MFS) Pg10, signifies a significant environmental shift characterized by the absence of the usual rudist-bearing marine-platform carbonates. The end of the Paleocene witnessed a new depositional phase characterized by sea transgression covering the Arabian craton with shallow marine carbonates, which extended far beyond its boundaries. The maximum flooding level was reached during the Early Eocene period. The deposited sediments were light-colored foraminifera-bearing, aphanitic to calcarenitic limestones and dolomites.

9.2.1 The Umm er Radhuma Formation

The Umm er Radhuma Formation was described for the first time from eastern Saudi Arabia by Steineke et al. (1958) (Fig. 9.4). The original type locality is located near Umm Radmah wells (lat. 28° 41 N′, long 44° 41′ E). The full succession could not be described from this location; a new reference section was selected by Steineke et al. (1958) from Wadi Batin, at a distance of around 108 km to the southwest of the original site. The whole section was pieced together from a dry riverbed (Hasson 1985). The base of the formation is placed at the contact of a fossiliferous limestone of the Upper Cretaceous Aruma Formation. The upper contact is selected at the chalky limestone and evaporites of the Rus Formation.

Sander (1952) described the lithology of the Umm er Radhuma reference section as 217 m of grey to cream limestone, partly silicified in the upper 140 m, followed by chalky limestones and about 15 m of hard dolomitized limestones. There are no well-preserved fossils in the outcrop sections; however, drilled wells contain intervals of well-preserved

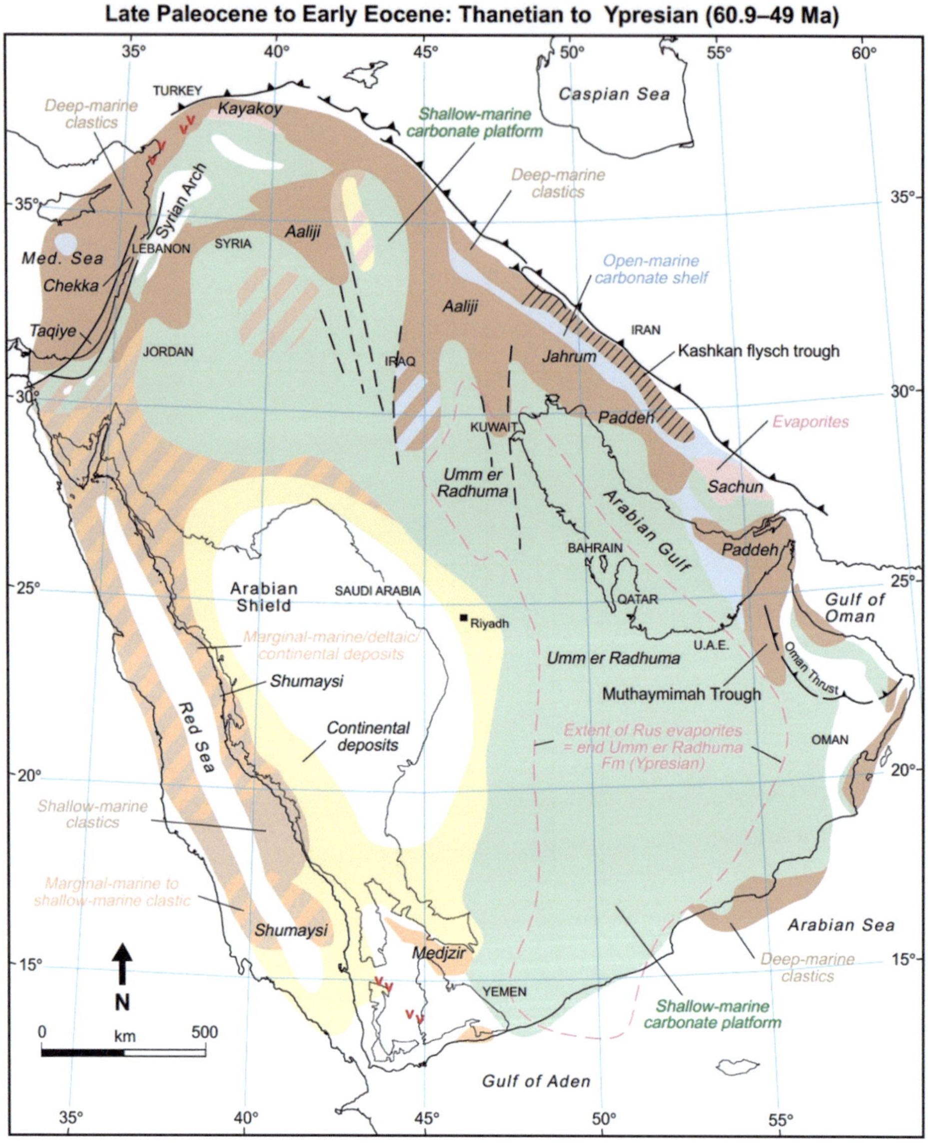

Fig. 9.3 Paleofacies map of the Late Paleocene-Early Eocene. The map shows that at that time, Qatar was covered by shallow water carbonates deposited within a "remnant foredeep" setting (Ziegler 2001)

microfossils that can be correlated with the outcrop section. Hasson (1985) provided a detailed description of the formation using cutting samples from four wells in the Rub' al Khali area in Saudi Arabia. Figure 9.5 shows the general lithologies of the Umm er Radhuma Formation. The main lithology is limestone, ranging in texture from lime mudstone to packstone and grainstone with some dolomitic and silicified limestone.

Dolomite is more common in the upper parts of wells A and B, while dolomitic limestone is present through most parts of well C, but dolomite becomes rare in the eastern wells of D and E. The wells A–D are found in the deeper parts of the basin, while well E is located at the edge. Under the anhydrite and dolomitic limestone of the Rus Formation, wells A, B, C, and D comprise hundreds of meters of packstone and grainstone, forming a shelf complex. These lithologies overlie a chert-bearing, pelagic limestone and a relatively thin succession of argillaceous limestone. A thin shale bed is just above the contact with the underlying Cretaceous Aruma Formation.

9.2.1.1 The Umm er Radhuma Formation in Qatar

The Umm er Radhuma Formation was described by Sugden and Standring (1975) from a reference section in the well Dukhan 22, between the drilled depths of 61–389 m with a total thickness of 328 m. The formation consists, from bottom to top, of around 12 m of pyritic, blue-grey marl, 53 m of bluish-grey, argillaceous limestone with beds of blue marl, 131 m of light-grey to grey-brown, dolomitic limestone with bands of argillaceous limestone toward the bottom, 79 m of brown, and dark grey saccharoidal dolomite and

Fig. 9.4 From left to right: Max Steineke (1898–1952) (Wikipedia 2024), Richard Allan Bramkamp (1910–1958) (Conchology 2024), and Nestor John Sander (1914–2012) (Geol. 2024). *Source* public domain

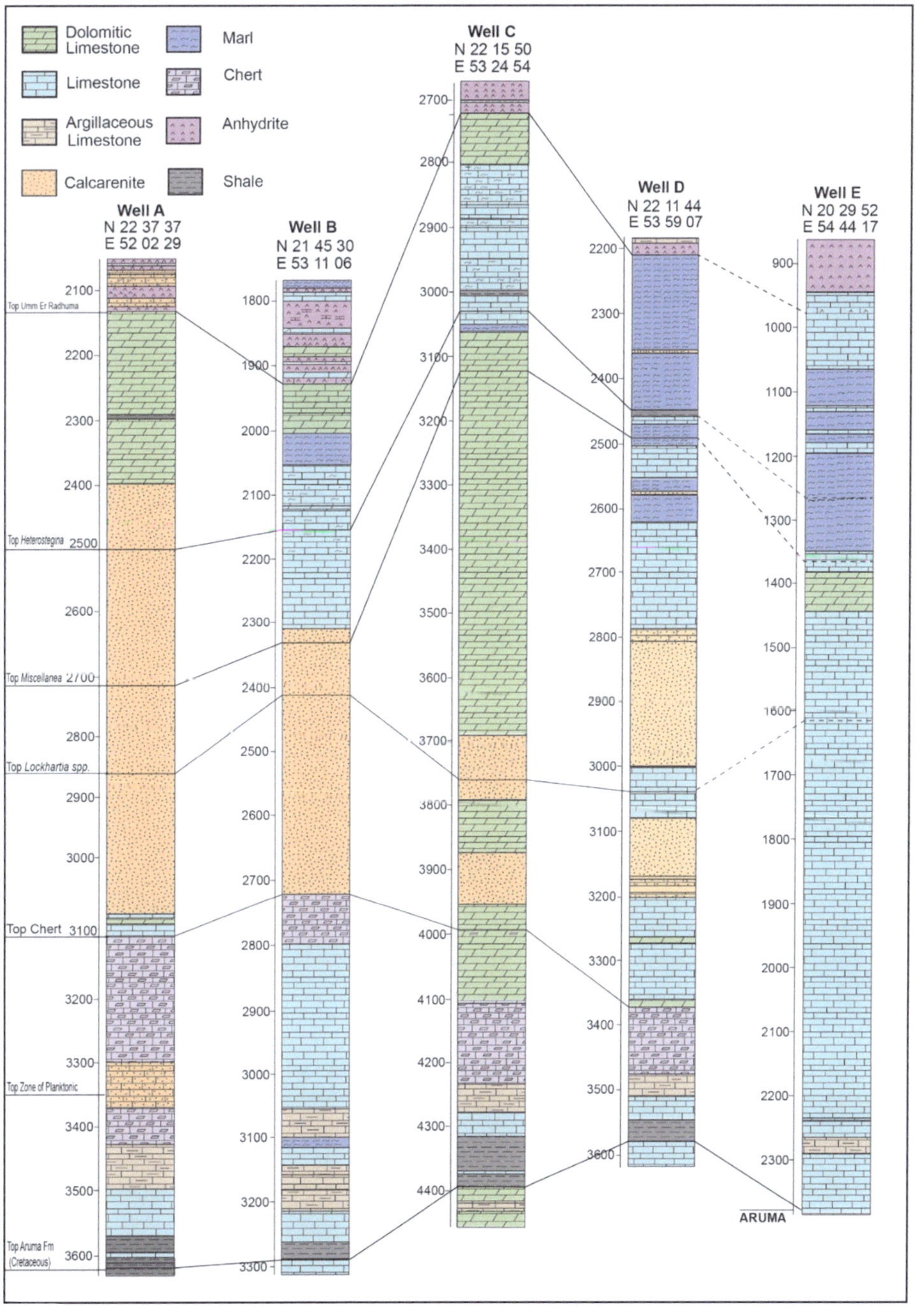

Fig. 9.5 General lithologies of the Umm er Radhuma formation from five wells in the Rub' al Khali area, Saudi Arabia (Hasson 1985)

50 m of light-brown or light-grey, porous, dolomitic limestone with small patches of quartz and chalcedony. The underlying formation is the Simsima with a local conformable contact but disconformable regionally. The contact is placed at the change from the grey marl or shale of the Umm er Radhuma Formation to the argillaceous limestone and marl of the Simsima Formation. The contact with the overlying Rus Formation is conformable and placed at the change from the dolomitic limestone of the Umm er Radhuma Formation to the white chalk at the base of the Rus Formation. The age of the formation is Paleocene to Lower Eocene; both Sander (1952) and Smout (1954) argued that the whole Paleocene is presented in the Umm er Radhuma Formation. Still, there is no evidence of Danian sediments in the Simsima or the Umm er Radhuma formations. Rivers et al. (2019) also reported that the formation age is Paleocene-Lower Eocene (Ypresian).

Boukhary et al. (2011) studied the foraminifera and ostracods of the Umm er Radhuma Formation in the Dukhan field. They reported that the thickness of the formation in the Dukhan area ranges between 64.5 and 127.5 m and comprises alternations of chert-bearing limestone, marly limestone, dolomitic limestone with some evaporite intercalations, marl, and pyritic shale. According to these lithological variations, they suggested dividing the formation into three members. These, from bottom to top, are:

Member A (Shammar Shale): This member consists mainly of marl and shaly marl. However, some wells consist of 3–5 m of limestone with chert nodules or around 12 m of pyritic shale.

Member B: This member comprises varied lithologies, including shale, marl, and limestone. Its thickness ranges between 3.6 and 42.6 m.

Member C: The third member comprises around 42 m of limestone with chert nodules changing into dolomitic limestone (60 m thick) with increasing evaporite contents toward the top.

Hewaidy and Al-Hitmi (1999) studied the Early Eocene rocks (including the lower part of the Umm er Radhuma Formation) in the Dukhan field. They subdivided the studied sections into three informal members: UR I, UR II, and UR III. The basal member UR I consists of black lignite shale barren of fossils. It is called the Shammar Shale and is present all over the Arabian Gulf region. It is around 12 m in thickness. A considerable amount of lignite in the shale suggests the presence of a sea-margin sea swamp environment. This may represent the first face of the transgression during the Eocene. The second UR II is formed of grey fossiliferous limestone. Most of the fossils are planktonic foraminifera. The presence of such fauna may indicate open marine conditions. The type of these planktonic foraminifera suggests shallow open sea. The third UR III member comprises limestone, dolomite, and chalk. The fauna in this part is mainly benthic, calcareous, small, and large foraminifera, which may point to an open platform setting with normal marine, warm, and clean water rich in nutrients. Abu-Zeid (1991) noticed that the Umm er Radhuma Formation is mainly carbonate in the eastern and central parts of Qatar, becoming more argillaceous, gypsiferous, and dolomitic in the west.

Hewaidy (1994) studied the biostratigraphy of the formation from a well located in south-eastern Qatar around 5 km east of the Traina Farm. In this well, the Umm er Radhuma Formation is overlain by the Rus Formation. There is a sharp contact between the white gypsum with chalk streaks of the Rus and the porous dolomitic limestone of the Umm er Radhuma. The lower boundary with the underlying Simsima Formation is gradational and was placed at the change of the black pyritic shale of the Umm er Radhuma and the argillaceous limestone of the Simsima Formation. The detailed lithology of the Formation in the studied well is summarized in Table 9.1. These lithologies are also shown in Fig. 9.6.

Table 9.1 Detailed lithological description of the Umm er Radhuma formation from a well in south-east Qatar (Hewaidy 1994)

Member	Depth (in m)	Thickness	Lithology
5	155–196	41	White, friable, chalky, and anhydritic limestone with foraminifera in certain horizons. Grey friable, very porous, very dolomitic limestone with high carbonate content (90%) and a moderate content of highly dolomitized foraminifera
4	196–275	79	Dark grey, compact limestone with 85% carbonate content and a rich foraminiferal fauna
3	275–400	125	Light to dark grey, compact, and massive limestone, with massive white gypsum. The carbonate content is about 85%, and the foraminifera is rich
2	400–466	66	Grey, compact, argillaceous limestone with 60% carbonate content and a rich foraminiferal fauna
1	466–478	12	Black, pyritic shale to calcareous shale with about 30% carbonate content and very rare foraminifera, much of which is excavated from the upper horizons

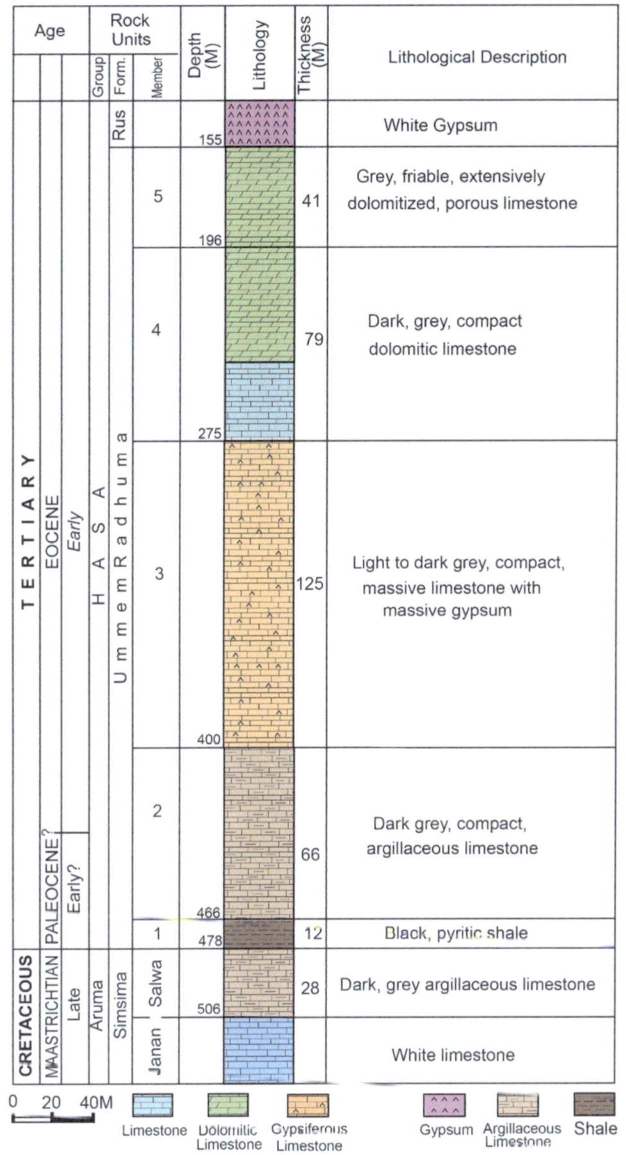

Fig. 9.6 Detailed lithologies of the Umm er Radhuma formation from a well in the south-east of Qatar (Hewaidy 1994)

Lithologic Characterization

In their study of the near-surface Tertiary water aquifers from three wells in Qatar, Rivers et al. (2019) investigated the lithology and depositional environment of the upper unit (Member C of Boukhary et al. 2011). In Borehole 1, this unit is formed of light to dark-brown fining-upward cycles of dolomitic mudstone to packstone with a mottled appearance due to the extensive bioturbation of these sediments. These cycles are floored by lag deposits and capped by dolomite crusts. Palygorskite is the main clay mineral. There are also replacive and, to a lesser extent, displacive gypsum nodules that become more abundant toward the cycle tops. Although grains are deformed by extensive dolomitization, it is possible to identify some dasycladacean green algae and large benthic foraminifera such as *Nummulites* and *Lockhartia sp.* (Fig. 9.7).

The Formation has more varied lithologies in Borehole 2. The C Member is encountered between the 120–65 m drilled depths. The main rock types include:

Depth Interval 120–105.4 m comprises brown, bioturbated, extensively dolomitized packstone. Most fauna are obliterated by dolomitization, and only some large benthic (*Nummulites sp.*) foraminifera with rare dasycladacean green algae, echinoderm debris, and mollusks. A thin layer of breccia terminates this succession.

Depth interval 105.4–73.2 m consists of friable sediments and consolidated rocks. The friable sediments are formed of poorly sorted, sand-sized dolomite crystals with cobble-sized dolomite clasts. The solid rocks are formed of dolomitic packstone that is burrowed and intercalated with bioturbated and burrowed wackestone and mudstone. The Dolomite destroyed most of the present fauna, and only ghosts of benthic foraminifera and green algae can be recognized. There is a 3 cm thick bed of chert at a depth of 88.3 m.

Depth interval 73.2–62.4 m is similar to the above interval. Both friable sediments and rocks are present. The sediments are formed of fine sand-sized dolomite. The lithified part is dolomitic wackestone and mudstone, with some breccia probably resulting from in-situ reworking. The main grains are green algae. The main difference between this section and the above is the lighter color of sediments and rocks.

Figure 9.8 shows the detailed lithologic characterization of Member C of the Umm er Radhuma Formation in borehole 2.

In Borehole 3, the C member of the Formation is encountered between the drilled depths of 123–50 m and formed of two main lithologies:

Depth interval 122.2–99.7 m is formed of light to dark-grey, meter-scale beds of bioturbated dolomitic packstone capped by cm-thick dolomitic packstone and wackestone. Some mollusk molds were found in some parts of the cores. In thin sections, it is possible to identify some skeletal grains, such as small benthic and planktonic foraminifera and dasycladacean algae. A 20-cm thick brecciated dolomitic layer caps this section.

Depth interval 99.7–60.2 m comprises friable sediments and lithified rocks. The main lithology is dolomite, which is lithified below 90.6 m. It shows bioturbation in the precursor limestone, interpreted as packstone before dolomitization.

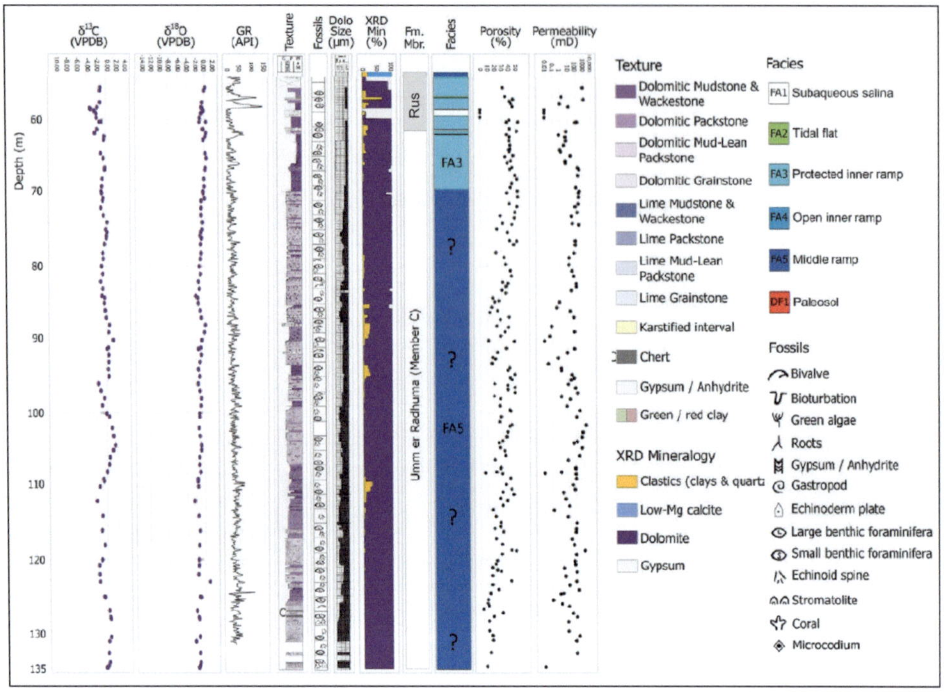

Fig. 9.7 Lithological characteristics of the Member C of the Umm er Radhuma formation from Borehole 1 in Qatar Rivers et al. (2019)

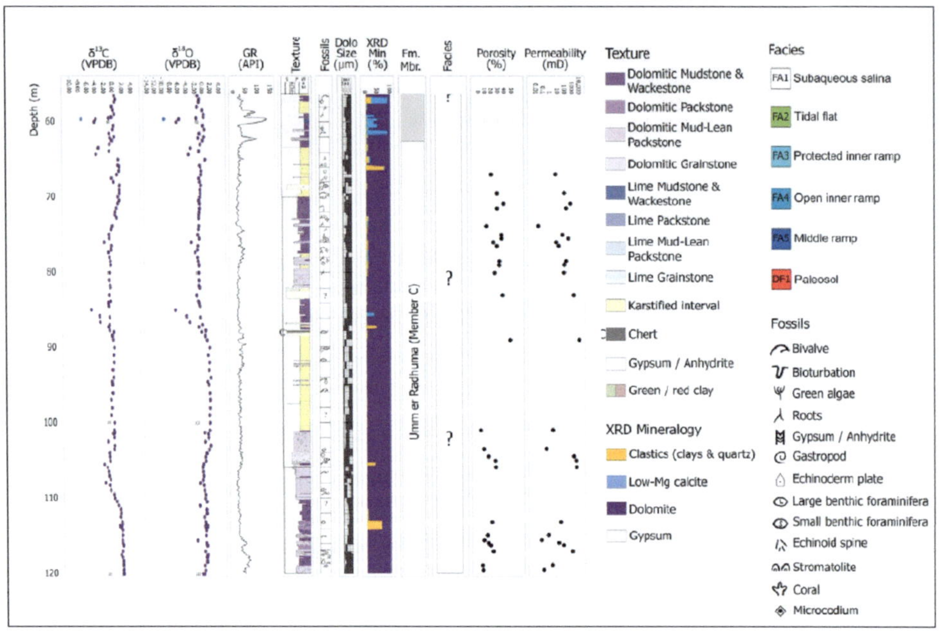

Fig. 9.8 Lithological characteristics of the Member C of the Umm er Radhuma formation from Borehole 2 in Qatar (Rivers et al. 2019)

Most recovered materials are poorly cemented, and the section between 90.6 and 74.2 m is formed of fine sand and silt composed of dolomite crystals. Lithified rocks above this section consist mainly of bioturbated dolomitic packstone and wackestone. A conspicuous 10 cm band of silicified rock is found at 71.35 m. Rare ghosts of dasycladacean green algae and some molds of large benthic foraminifera and bivalves were identified in thin sections. The top 10 m of this section contains a cm-scale bed of calcite crystals and is capped by a 2 cm thick palygorskite clay deposit.

Figure 9.9 shows the detailed lithologic characterization of Member C of the Umm er Radhuma Formation in Borehole 3.

Depositional History

The Umm er Radhuma Formation has similar lithologies to the overlying Rus and Dammam formations. All formations are shallow water carbonates containing large benthic foraminifera, dasycladacean green algae, and mollusks, with some less common sections of shale, evaporites, and clay

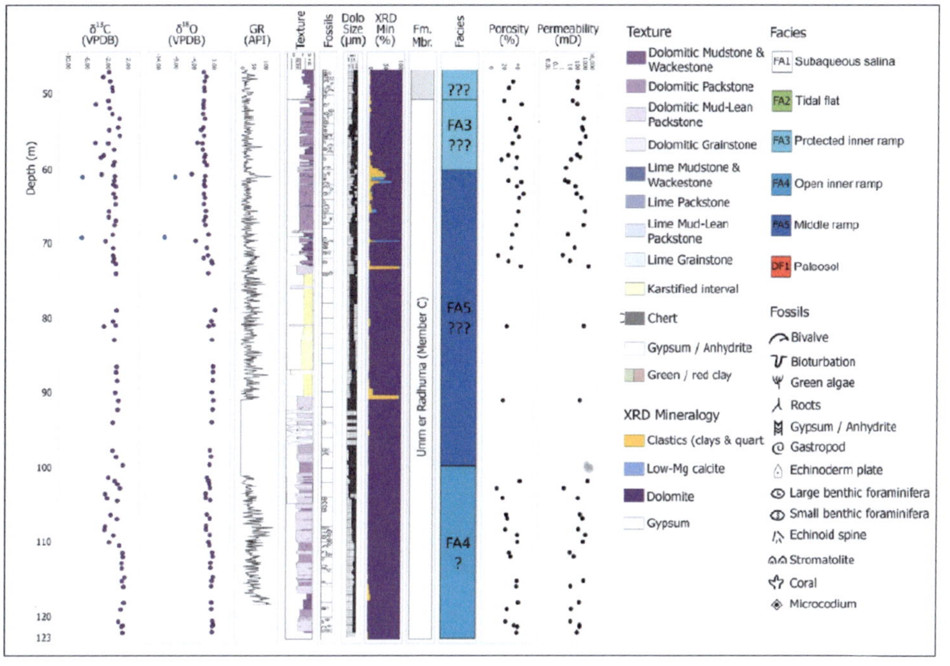

Fig. 9.9 Lithological characteristics of Member C of the Umm er Radhuma formation from Borehole 3 in Qatar (Rivers et al. 2019)

minerals. This similarity caused workers to deal with these formations as a single succession (e.g., Rivers et al. 2019). So, according to Hewaidy (1994), the Umm er Radhuma Formation sediments were deposited in a warm, shallow-water carbonate shelf. However, the basal Shammar Shale Member (Member A) was deposited in a starved basin with a slow sedimentation rate under probably low-stand conditions. The succession that contains planktonic foraminifera was deposited under relatively deep water or basinal and sub-basinal conditions. This section is overlain by shallow-water carbonates with thin evaporite streaks. These sediments were formed under shallow evaporitic or closed lagoonal conditions. This evaporitic succession is overlain by shallow shelf carbonates rich in large, benthic foraminifera, typical of a warm climate. The upper part of the Umm er Radhuma Formation comprises a porous dolomitic limestone with scarce fauna. Upward shallowing and increased evaporation led to the deposition of the overlying Rus Formation.

Across the Arabian Gulf region, sulfate-rich carbonates are found in the northern parts of the Gulf. In contrast, the middle and southern parts are characterized by shallow shelf carbonates with thin horizons of shale or planktonic foraminifera-bearing limestone (Fig. 9.10).

9.2.2 The Rus Formation

The Rus Formation was described for the first time by Steineke et al. (1958) from the southeast flank of the Dammam Dome below Umm Er Rus Hill (or the Umm Er Ru'us according to Alsharhan and Nairn 1997). The total thickness of the Formation in its type locality is 56 m. There is a considerable variation in the lithological components of the formation, even at its type locality. The formation is divided into three units, with variations in lithology occurring mostly in the middle unit. These units are:

(a) White, friable, chalky, and porous carbonates with one or more beds of grain-supported texture at the top (3.6 m).
(b) Light-colored marl with occasional patches of gypsum and thin, hard limestone beds containing Quartz geodes at different levels. In other locations, this unit shows a wide spectrum of lithologic variations, including (a) white, compact, anhydrite interbedded with green shales and lesser amounts of dolomitic limestone, (b) grey marl with coarse calcite interbedded with shale and limestone (31.8 m).
(c) Grey to buff-colored, compact, crystallized, partially dolomitized limestone with minor amounts of soft limestone with dissolution porosity. Rare Quartz geodes are found at the lower part and are typical of the uppermost part (21.0 m).

The lower contact of the Formation is placed at the change from the dolomite containing the fossil *Lockhartia tipperi* of the underlying Umm er Radhuma Formation to the leached dolomitic limestone with small mollusks of the basal Rus Formation. The upper contact is selected at the change from the light-colored, grain-supported carbonates of the Rus to the overlying thin-bedded limestone and shale of the Dammam

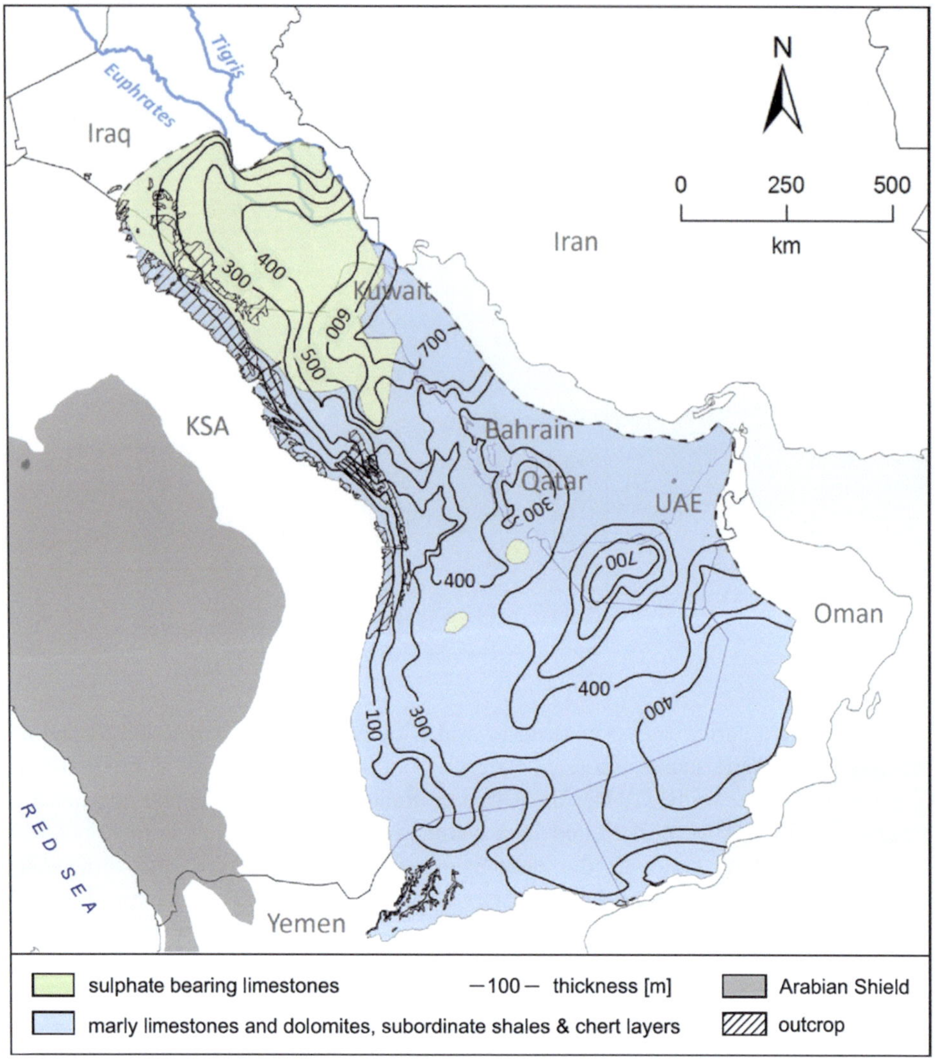

Fig. 9.10 Regional tentative isopach and facies map of the Umm er Radhuma formation across the Arabian Gulf area (Dirks et al. 2018)

Formation. The authors did not find any important fossils in the type locality, but the formation is underlain and overlain by rocks of the lower Eocene.

Tleel (1973) used the same tripartite subdivision to describe the Formation from eastern Saudi Arabia. The lower unit consists of 20–23 m of dolomitic limestone that ranges from chalky to coarse-grained and limestone with gypsum streaks at the top. The middle unit is 10 m thick and comprises calcite geodes and mud-balls in massive greyish limestone. The upper unit is 21 m thick and consists of quartz geodes and chert nodules (Fig. 9.11).

The Rus Formation outcrops in Jabal Hafit, near Al Ain city in the United Arab Emirates, have been studied by Whittle and Alsharhan (1994) and Alsharhan. They reported that the lower part of the formation comprises dolomitic limestone with quartz geodes at the top. The middle part is formed of marl and limestone with some irregular patches of gypsum, and thin limestone beds contain some quartz geodes, anhydrite, and green shale beds. The upper part is mostly microporous limestone with thin sandy beds (Fig. 9.12).

The outcrops of the Rus Formation in Bahrain have been studied by Amer et al. (2023) from the Horse Racing Clube site. The strata are exposed along the northern edge of the Awali anticline and represent the upper part of the Rus Formation. They briefly studied the lower and middle units of the Formation in the Jabal al Dukhan, but their investigation was incomplete due to limited access to this site. They reported that the upper part of the formation in the Horse Racing Club consists of five facies association, and each association is formed of several facies. This synthesis is based on their work.

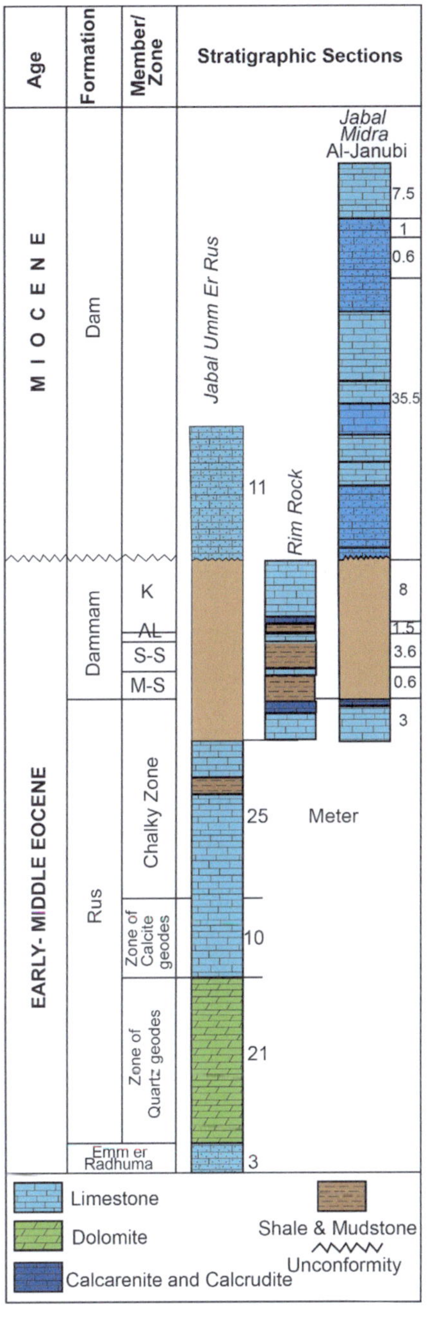

Fig. 9.11 Lithostratigraphy of the Early Eocene-Miocene strata at the Dammam Dome, Eastern Saudi Arabia (Alsharhan and Nairn 1997)

9.2.2.1 Facies Association

Facies Association 1 is formed of four facies:

Chalky dolostone facies: This facies consists of dolomudstone to wackestone with scattered calcite geode. It also has some micrite clots with a poloidal texture. There are no fossils, but bioturbation in the form of *Thalassinoides* burrows is common.

Lithoclastic chalky dolostone: white to grey dolomudstone to floatstone with large, angular micrite clasts floating in a micritic matrix. The clasts contain foraminifera molds.

Peloidal dolostone facies: well-bedded, off-white to yellow packstone containing micrite clots and peloids, with occasional oscillation ripple marks. Foraminifera molds are present with burrows at the lower part of the unit.

Foraminiferal dolostone facies: This facies consists of dolostone ranging in texture from mudstone to floatstone with calcite geodes, shell fragments, and gastropod molds. Chert nodules are common in the upper part. The main skeletal grains are foraminifera, including *Alveolina* and *nummulites*. This facies have well-preserved fossil mangrove trunks and roots (rhizoliths).

Facies Association 2 consists of two main facies:

Claystone facies: buff to green-yellowish, soft to elastic clays, and marls with oxidized red laminations. It also has some breccia and dolomitic beds. Some layers exhibit high concentrations of halite in the desiccation cracks with quartz pseudomorphs after gypsum.

Evaporite facies: This facies is of limited occurrence and characterized by elongated crystals of quartz pseudomorphs after gypsum. It also contains rare halite, and desert rose quartz pseudomorphs after gypsum.

Facies Association 3 comprises two major facies also:

Argillaceous dolostone facies: white to ivory Packstone with micrite clots, foraminifera molds, and subangular lithoclasts. The sediments are bioturbated and dominated by *Ophiomorpha* burrows. The main mineral is dolomite with illite.

Foraminiferal dolostone facies: This facies is described above.

Facies Association 4 comprises four facies:

Dolomitic marl facies: Yellowish green to reddish mudstone becoming harder upward due to the association with fossil mangrove rhizoliths. The main lithologies are marl and argillaceous dolostone. It sometimes has fossil-rich beds such as bivalves, shell fragments, and *Nummulites*. The main minerals are dolomite and illite.

Bioturbated chalky dolostone facies: well-bedded to massive chalky dolomite with a peloidal wackestone

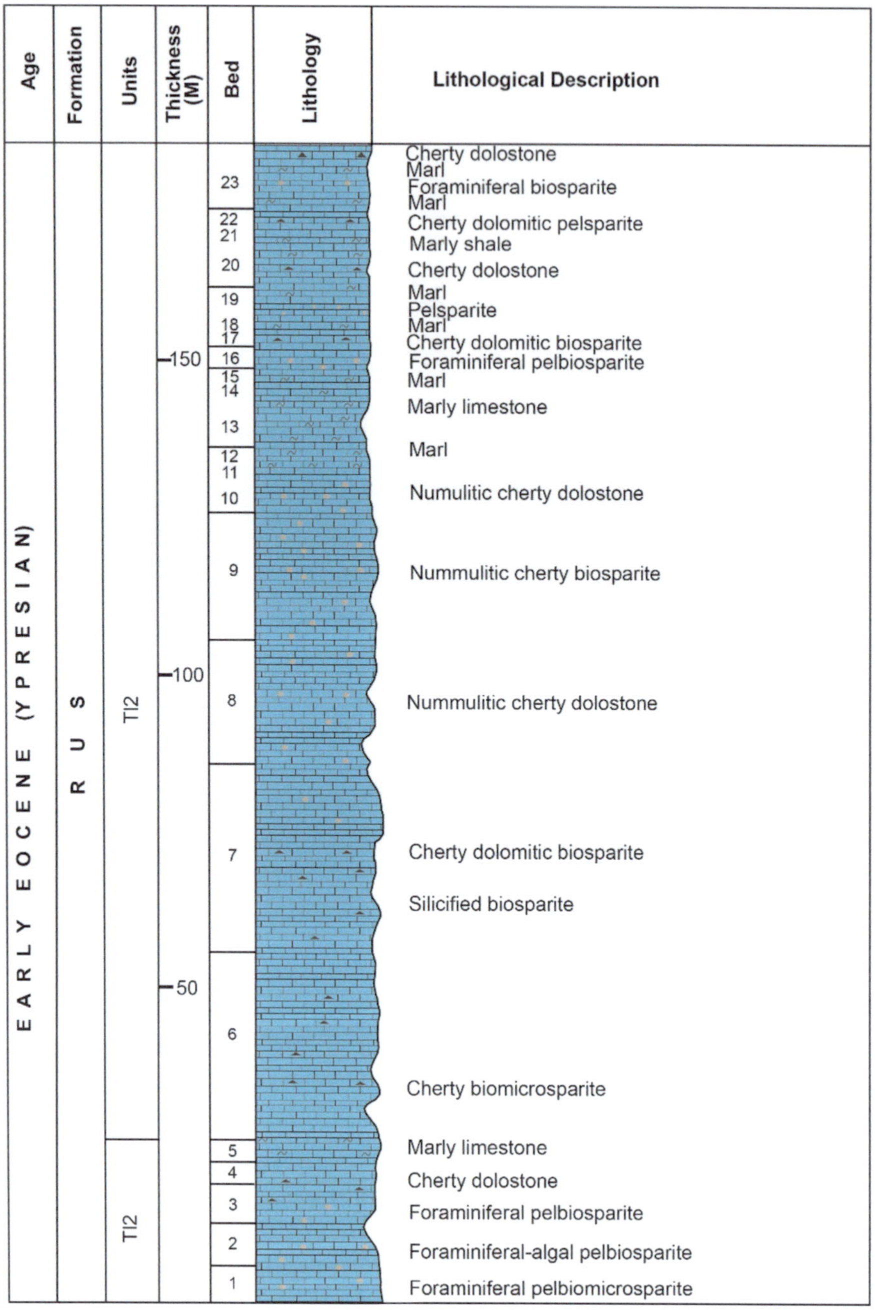

Fig. 9.12 Lithostratigraphy of the Rus formation in the United Arab Emirates (Whittle and Alsharhan 1994)

and packstone and foraminiferal molds. Bioturbation is common at the base associated with *Thalassinoides* burrows. The base of this facies is erosional and characterized by lag deposits.

Cross-bedded dolostone facies: Ivory to cream dolomite of a grainstone texture with trough cross-bedding. The main grains are foraminifera, echinoderm spines, shell fragments, ooliths, and pellets. Some of these grains are extensively leached. There are some quartz grains also. *Ophiomorpha* burrows are common and usually filled with overlying sediments obliterating the bedding. The main mineral is dolomite.

Microbial dolostone facies: Bright white to grey dolomite with wavy microbial laminations, wackestone to floatstone with fenestral porosity, peloids, micrite clots, abundant foraminifera (mainly *Miliolids* and *Alveolina*), ostracods, and shell fragments. The base is characterized by brecciated peloidal lithoclasts, whereas the upper part has stromatolites.

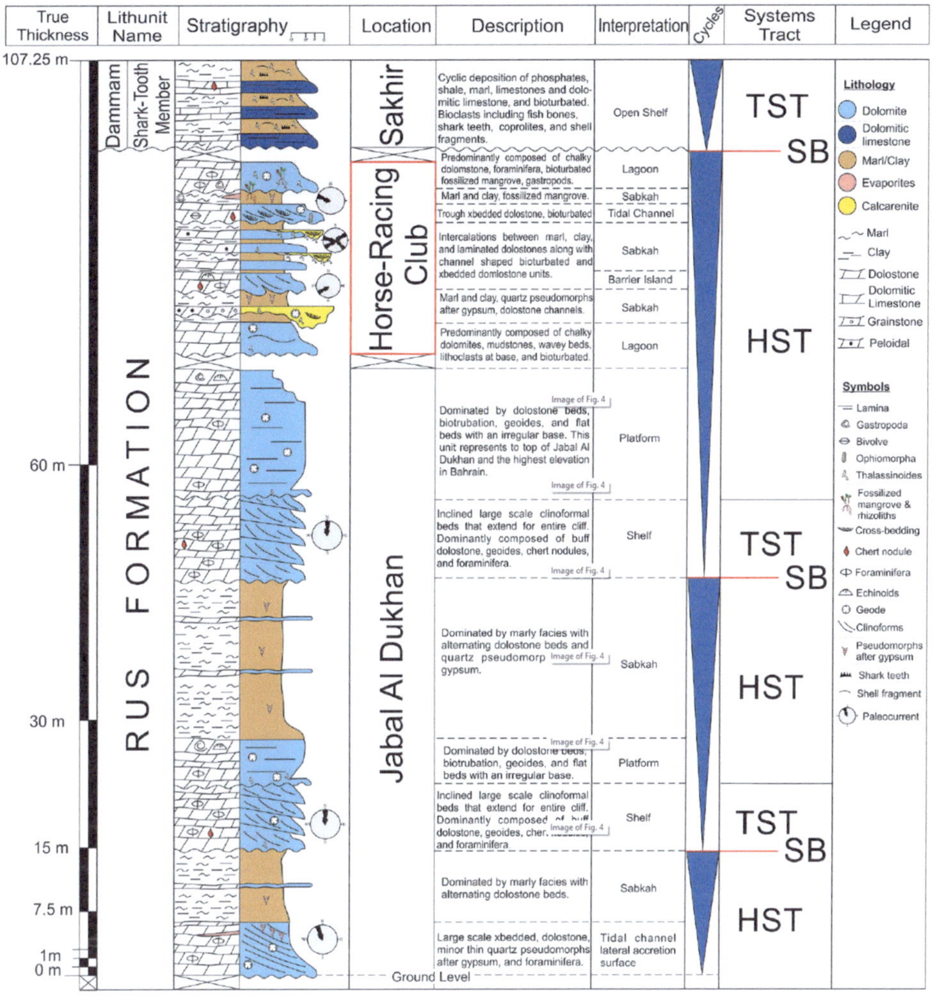

Fig. 9.13 Facies association and depositional environments of the Rus formation from two outcrops (Horse-Racing Club and Jabal Al Dukhan) in Bahrain (Amer et al. 2023)

Facies Association 5 is formed of two main facies:

Bedded dolostone facies: Buff to white, well-bedded dolomite with a wackestone to packstone texture. There are also lithoclasts and quartz grains, dark brown to black hematite spots, peloids, and foraminiferal molds. Dolomite is the main mineral forming in this facies.

Clinoformal dolostone facies: Grey to yellow mudstone to wackestone, bivalve molds, subangular to subrounded detrital sand grains, with water-escape structures. These sediments are well-bedded and contain large-scale cross-bedding that looks like clinoform structures. Dolomite and quartz are the main minerals of this facies.

These different facies associations are shown in Fig. 9.13.

9.2.2.2 The Rus Formation in Qatar

The reference section of the Rus Formation, as described by Sugden and Standring (1975), is located in the exposed cliffs of Jebel Dukhan and west of this area to the coast. The total thickness is around 36 m. The formation comprises dolomitic chalk with relatively harder limestone beds containing pellets or oolitic debris. There are some chert nodules and aggregates of quartz and chalcedony. The lower part of the formation is not exposed, and the overlying formation is the Dammam. The contact between the two formations is conformable and placed at the basal shale member of the Dammam Formation and the underlying chalky limestone of the Rus Formation. The Rus formation contains some bedded or massive gypsum in the subsurface sections. A soft, pale-colored limestone bed also occurs between the gypsum of the Rus and the base of the Dammam Formation. The gypsum member of the Rus increases in thickness southeastward from Qatar, becoming anhydrite in the deeper sections and reaching a thickness of around 200 m in Abu Dhabi. The absence of evaporites in the Rus outcrops is due to weathering and erosion, as reflected by many karst and sinkholes in the formation.

LeBlanc (2021) revised the Cenozoic stratigraphy of the surface formations of Qatar, including the Rus Formation, based on extensive fieldwork and sampling. He provided more

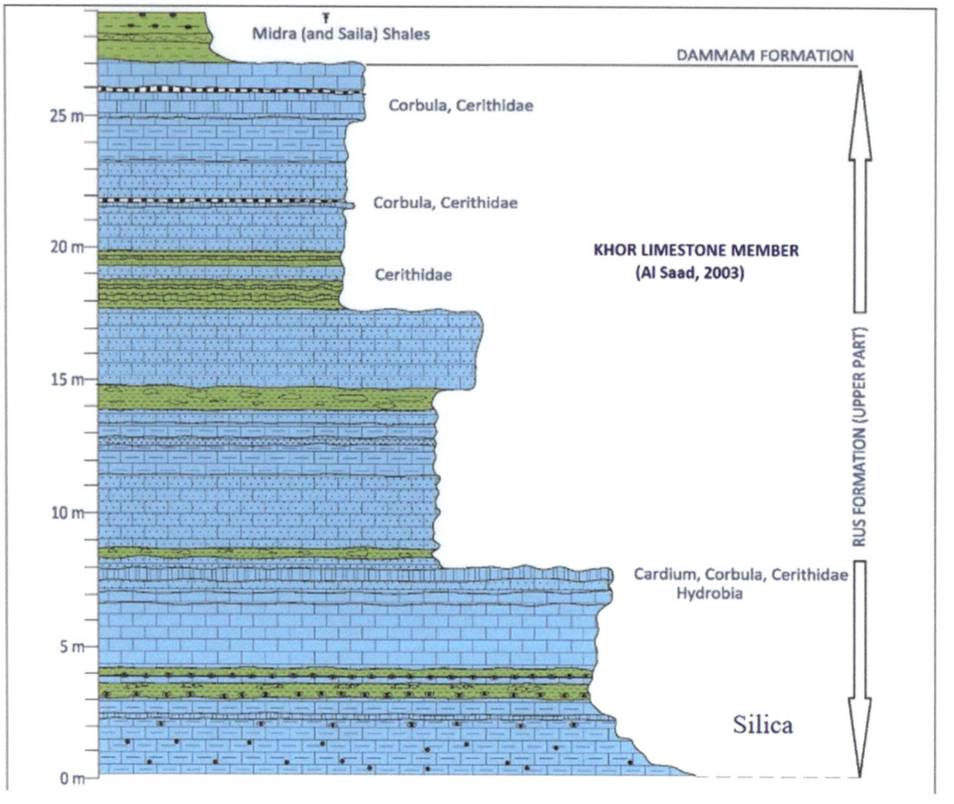

Fig. 9.14 The reference section of the Rus formation in Qatar at the Dukhan Hills (LeBlanc 2021)

details on the lithology of the Rus formation in the reference section at Dukhan. It is formed of soft, whitish, dolomitized limestone with small argillaceous intercalations associated with brown dolomitic marl. Sometimes, these lithologies are interbedded with harder, greyish dolomitic limestone. These are the only sediments containing fossils in the formation (Fig. 9.14). There are siliceous materials in the forms of chert and quartz crystal geodes at the basal part. He attributed much of this silica content to the possible presence of tubular conduits of freshwater springs that used to contain high sulfide contents. He suggested that these springs were active during the Miocene, similar to the present-day water–gas-oil seepages in Azerbaijan.

The Rus and the overlying Dammam formations outcrops cover around 80% of Qatar's total area, of which the Rus represents 10% of the total Eocene sediments. The distribution of these sediments is controlled to a large extent by Qatar Arch with small longitudinal folds such as the Dukhan anticline. According to Sugden (1962), these features may represent an east-north extension of the Central Arabian Arch. Al-Saad (2003) studied the Rus Formation from the surface outcrops and two wells. The surface exposures are around Dukhan field, Umm Salal Ali (central), Al-Khor, and Al-Thakhira (northeast). The wells are from the Traina Farm (south) and Ras Laffan (northeast). He suggested that the Early-Middle Eocene strata of Qatar were the product of two major "second-order cycles." The strata of the Umm er Radhuma Formation were laid down under a Highstand System Tract (HST), which led to the deposition of the Rus Formation. The upper part of the Rus Formation was deposited in a Transgressive System Tract (TST), followed by the deposition of the Midra Shale, the basal member of the Dammam Formation, a condensed section that was formed during transgression (Fig. 9.15).

Cavelier (1970) studied the formation in different parts of the country. He also named the part of the formation the Khor Limestone Bed. This suggestion was not adopted in any of the later studies. Abu-Zeid and Boukhary (1984) subdivided the exposed section of the formation into two informal members: Member (A) comprises chalky limestone with marly claystone, and Member (B) is made up of limestone intercalated with gypsum and clay. Abu-Zeid (1991) studied the formation in the northern parts of the country and divided the studied strata into three informal members. Member (A) is formed of gypsiferous dolomitic limestone; Member (B) consists of dolomitic limestone, and Member (C) comprises dolomitic chalky limestone.

The main fossils reported from the Rus strata include foraminifera, mollusks, echinoid debris, ostracods, and shark teeth. These are detailed in Table 9.2.

In general, two facies provinces characterize the strata of the Rus Formation. These are the southern evaporitic dolomitic limestone province and the northern carbonate province, formed mainly of carbonate rocks (Fig. 9.16). The boundary between the two provinces is a topographic hinge

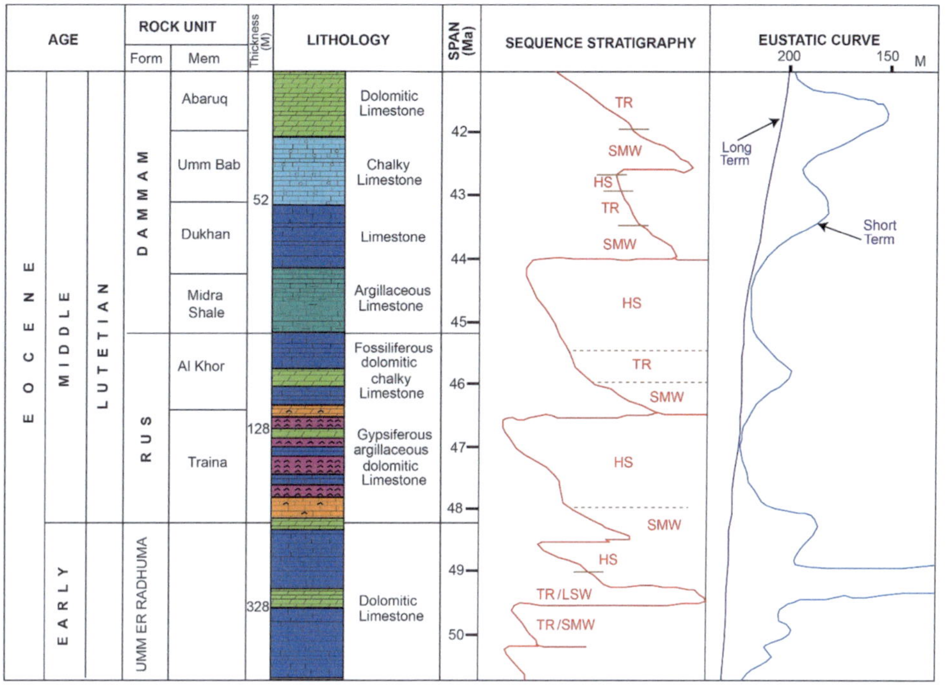

Fig. 9.15 Sequence stratigraphy of the Early and Middle Eocene strata of Qatar (Al-Saad 2003)

Table 9.2 The main fossil assemblages of the Rus Formation (Al-Saad 2003)

Foraminifera	Mollusks	Echinoid debris	Ostracods	Shark teeth
Lituonella douvillei	*Cardium sp.*	Crustaces	Present	Present
Nummulites aff. atacicus	*Carbula sp.*			
Nummulites globulus	*Cerithidae*			
Ammobaculites sp.	*Terebellum sp.*			
Veneuilina sp.	*Velates Schmiedeli*			
Gaudryina sp.				
Eggerella sp.				
Texturlaria				

line that can be easily picked from satellite images. It takes the shape of a V, with its apex at the Qatar Arch.

The Southern Province

The Rus Formation in this province comprises greyish, green, off-white, massive, gypsiferous, argillaceous, dolomitic limestone alternating with clay and marl. The whole succession reflects dominant cyclic sedimentations. There are at least eight short-lived cycles, each starting with marly limestone and/or dolomitic limestone capped by gypsum. The evaporites dominate the lower part of the formation. The argillaceous limestone gradually changes into marly limestone, and the major carbonate rocks are formed of dolomitic limestone intercalated with a few intervals of limestone. Later, the dissolution of evaporites led to the formation of karst and collapse features. These are called "Dhool single Dahl" in the local slang. They are common features in many parts of Qatar (see Chap. 2).

The Northern Province

The lower part of the Rus Formation in this province consists of dolomitic limestone with rare occurrences of marl and gypsum. The upper part is formed of chalky limestone with thin streaks of gypsum. The carbonate sediments are formed of wackestone to packstone in the lower part, changing into lime mudstone in the upper part. The rare occurrence of evaporites in the northern provinces is attributed to non-deposition due to gentle structural folds, such as the Qatar and Simsima Arches. Another possible cause is the probable extensive dissolution of evaporites in this area (Al-Saad 2003).

In general, these two different depositional provinces, with their characteristic suits of sediments, have led to the differentiation of the Rus facies into gypsiferous and argillaceous facies termed the Sulfate Facies, and the Carbonate Facies formed mainly of limestone, dolomitic limestone, and dolomite. The boundary between the two facies is shown in

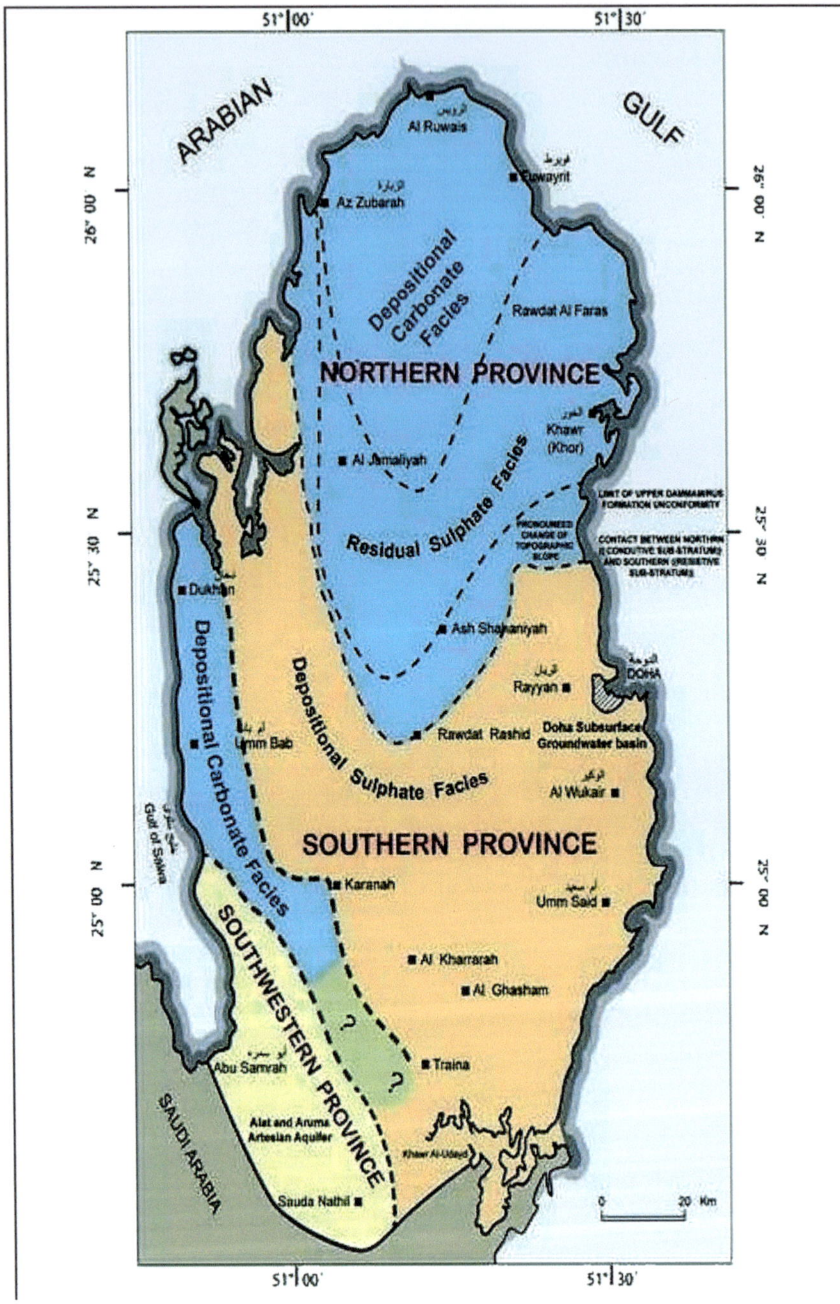

Fig. 9.16 The two depositional provinces of the Rus formation in Qatar (Al-Saad 2003; LeBlanc 2021)

Fig. 9.16. The regional distribution of these two facies across Qatar is shown in Fig. 9.17. This figure shows two cross-sections in the strata of the Rus Formation from different wells and outcrops. The figure shows the prevalence of Sulfate Facies in the southern part of the country compared to the northern part, where these facies are restricted to the extreme western parts only. The absence of evaporites at the center of the country is due either to later dissolution and removal of the evaporites or no deposition due to the impact of the structural arches, such as Qatar Arch, on the depositional settings.

Vertically, the formation is divided into two members: the Traina and the Al-Khor (Fig. 9.18) (Al-Saad 2003).

The Traina Member

This member was described from the Traina well Dw-4, 50 km south-east Doha city, between the 30 and 108 m drilled depths with a total thickness of 78 m. This is the lower part of the formation, which is made of white to light-grey limestone with thin beds of gypsiferous, marly, argillaceous, and dolomitic limestone. The lower boundary is placed at the change from the dolomitic limestone of the Umm er Radhuma Formation to the gypsiferous limestone of the Traina Member. The upper contact is selected at the change from the grey gypsiferous dolomitic limestone of the Traina Member to the white chalky

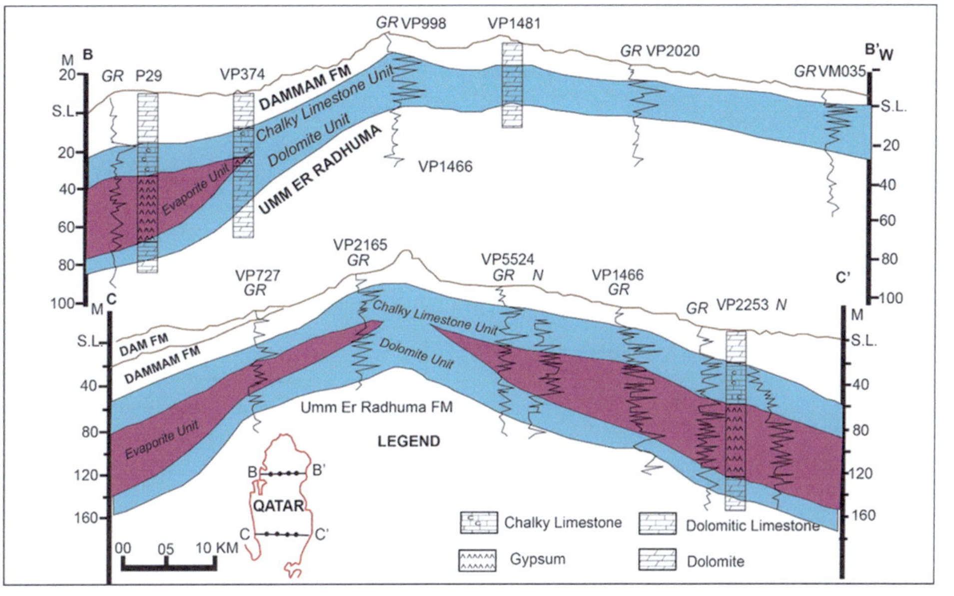

Fig. 9.17 East–west sections through north and south Qatar show the regional distribution of the Sulfate and the Carbonate facies across the country (Al-Hajari and Kendall 1992; LeBlanc 2021)

limestone of the Al-Khor Member. These sediments have no important fauna.

Gypsum is the main lithologic component of this member. However, it shows significant variations across the country. The amount of gypsum decreases northward, changing into dolomitic cherty limestone. This member represents a typical sabkha and restricted lagoonal deposits.

The Al-Khor Member

This member's type locality is also from the Traina DW 4 well between the drilled depths of 0–30 m plus 20 m of outcrops in the same area totaling 50 m thick. This member encompasses the upper part of the Rus Formation and comprises white to light-grey dolomitic chalky limestone with streaks of gypsum, marls, and clays.

The upper contact is placed at the change from the dolomitic limestone of the Al-Khor Member to the yellow shale of the overlying Dammam Formation. The skeletal grains include shark teeth, foraminifera, mollusks, and echinoid debris. The lithology of this member is relatively consistent all over the country. Its sediments were laid down under normal marine subtidal conditions.

The different lithologic components of the Rus Formation are shown in Fig. 9.18.

Facies Analysis

As detailed above, the Rus Formation comprises a wide range of rock lithologies on the temporal and spatial scales. These lithologies are formed of the following facies:

Skeletal wackestone and packstone

This facies consists of grey to yellow limestone. It forms a major part of the Al-Khor Member but is rarely found in the Traina Member. The main skeletal grains include Alveolinidae, Nummulitidae, Mollusks, echinoderm, shark teeth, molds, and casts of bivalves and gastropods. Its thickness ranges between 20 cm in the Southern Province and 2.5 m in the Northern Province. Some of the rocks contain dolomite rhombs. The wackestone changes gradually to packstone with dolomite and gypsum crystals. There are some rare horizons of grainstone with dolomite at the upper part of the Al-Khor Member.

Chalky limestone

This is a white to light-grey chalky limestone with some local iron staining. Fauna is scarce and represented by ostracods. In some areas, it contains fine quartz grains and gypsum crystals. The thickness ranges from 2 to 4 m and is present mainly in the Al-Khor Member.

Dolomitic limestone

This facies comprises 3.5–4 m of grey to light-yellow, aphanitic dolomite crystals in wackestone and packstone textures and is associated with some gypsum streaks. It represents the basal part of the Sulfate Facies of the Traina Member.

Cherty limestone

This is a hard and massive limestone, light-brown to yellow, with relatively abundant chert nodules ranging in thickness from 1 to 2 m. It is restricted to the Al-Khor Member. Some horizons are formed of cherty dolomitic limestone.

Marl

This is a thin facies 45 cm^{-1} m thick and consists of grey to yellow, friable, and fossiliferous marl. The main fossils are

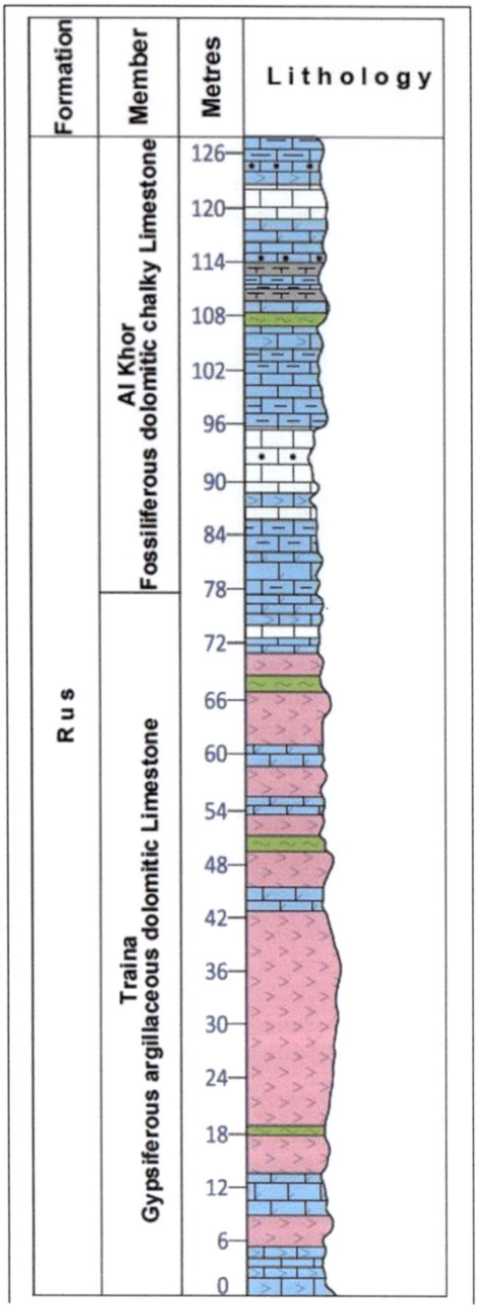

Fig. 9.18 Stratigraphic characteristics of the Rus formation in Qatar (Al-Saad 2003; LeBlanc 2021)

dwarfed pelecypods and gastropods. It reaches its maximum thickness at the Traina Member.

Claystone

This facies is formed of thin bands 20–35 cm thick of grey to green, calcareous, and gypsiferous claystone. It is common throughout the Khor Member and also represents the basal part of the Sulfate Facies of the Traina Member.

Gypsum

Gypsum is the dominant mineral in the Traina Member and is widely distributed in the Southern Province. It is formed of white to silver, relatively hard gypsum intercalated with dolomitic limestone in the Southern Province and associated with the chalky limestone in the Northern Province. It attains a maximum thickness of 25 m in the Southern Province and thinning out to around 20 cm in the Al-Khor Member in the Northern Province.

Dill et al. (2003) studied the early Tertiary strata of the northern sector of the Dukhan field, west Qatar, mainly the Rus and Dammam Formation. They dealt with the two formations as a single depositional regime, dividing it into four units (Fig. 9.19). The lower two units form the Rus Formation:

Unit I: Lower Chalk Member

They divided this unit into two subunits. The lowermost is formed of soft dolomitic marl (unit Ia). The marl accumulates at the base of the mesa structures, forming gentle slopes and filling the spaces between them. The soft sediments of Unit I are exposed in the central part of the area under investigation. A horizon of nodular chalky limestone 1 m thick has been used as a marker horizon. This marker has subdivided the unit into two subunits: subunit Ia1 is beneath the marker, and subunit Ia2 is above it. The dolomitic rocks were originally skeletal and pelletal packstone and wackestone, and although dolomitization is extensive, it did not fully obliterate the original sedimentary textures. The marls of subunit Ia1 alternate with green argillaceous sediments.

Unit Ib is characterized by coarse-grained carbonates. Its limestone forms ledges that protect the underlying softer rocks. The upper half of this unit contains chalcedony concretions and quartz crystal clusters generated from the recrystallization of chalcedony in the upper part of the unit.

Unit II (Upper Chalk Member)

This Unit is also subdivided into two subunits. Subunit IIa consists of chalky limestone and soft marl overlying unit Ib but has no green argillaceous interbeds. Unit IIb comprises thick, hard limestone. These limestone beds are recognizable over the study area and can be used as a datum. Hard limestone forms the mesa slopes and escarpments because of its resistance to weathering and erosion. Sometimes, the unit IIb is formed of thin laminations less than 10 mm. Petrographic investigation shows that these rocks are transitional from packstone to boundstone, where stromatolitic limestone alternates with shell debris composed of abundant foraminiferal skeletons, mollusks, and ostracods.

While the work of Dill et al. (2003) is based on outcrop samples, Rivers et al. (2019) studied the near-surface Tertiary sediments (including the Rus Formation)

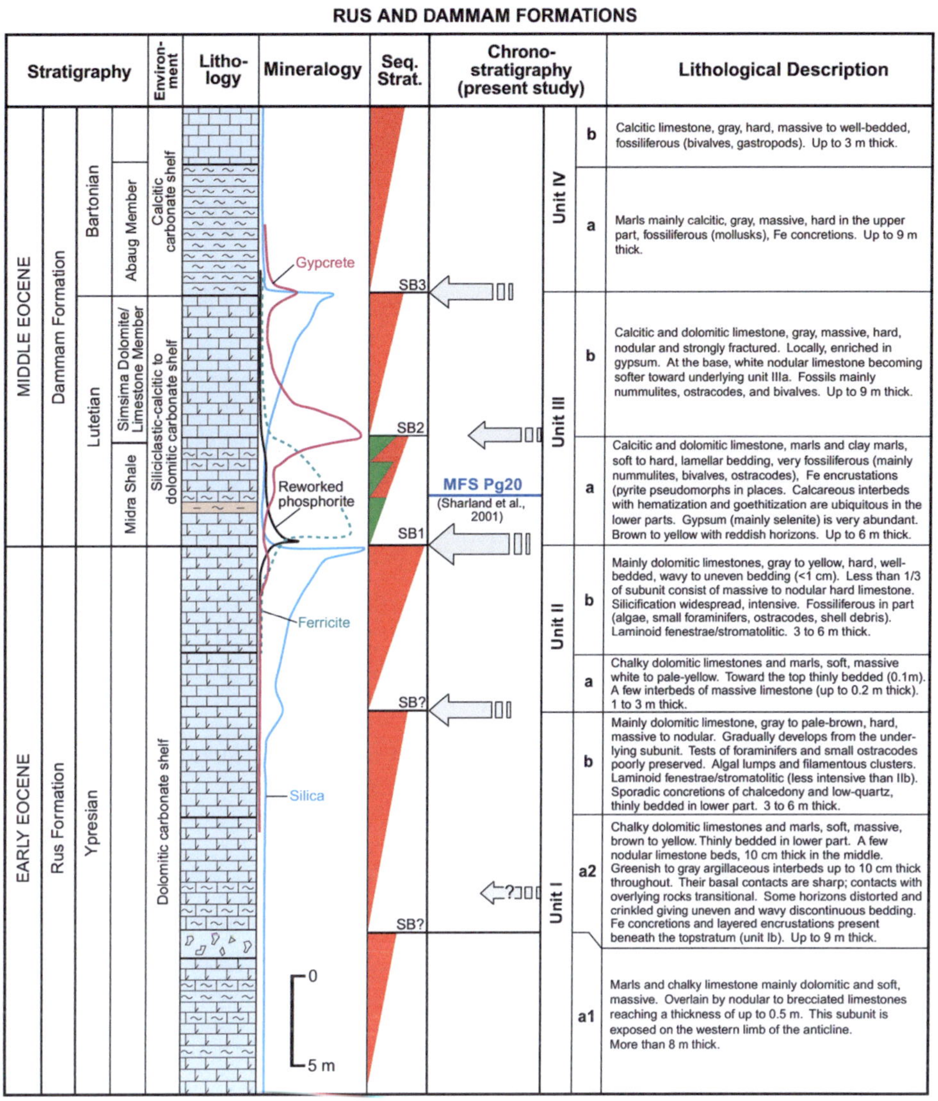

Fig. 9.19 Stratigraphy of the Rus and Dammam formations from the northern parts of the Dukhan anticline, west Qatar (Dill et al. 2003)

from three boreholes in different parts of Qatar. They also treated the whole succession as one stratigraphic unit, including the Umm er Radhuma, Rus, and part of the Dammam formations. Their study involved extensive petrographic, geochemical (including isotopes), and petrophysical investigation.

Borehole 1 (Fig. 9.20)

In Borehole 1, the Rus sediments were divided into two parts. The first is between 61.3 and 29.4 m, and the second is between 29.4 and 14.6 m.

Depth interval 61.3–29.4 m below sea level

This interval consists of meter-scale fine dolomite cycles, followed by gypsum, and capped by green palygorskite beds. Skeletal grains are still recognizable despite dolomitization. These include small benthic foraminifera such as *Rotalia* sp., *Clavulina* sp., and miliolids. There are also some ostracods and echinoderm debris. There are cases of high abundance of a single type, such as dolomitic miliolid grainstone. In some parts, there are alternations of gypsum beds with stingers of dolomitic mudstone. Gypsum is found in beds several meters thick, showing radiating crystals with rare anhydritic layers. The clay beds contain root traces (mangrove?), and the capping surfaces are erosional or bored. The top bed comprises 3 m thick alternations of green, palygorskite, and brown illite clays.

Depth interval 29.4–14.6 m below sea level

This interval is composed of decimeter-to-meter-scale coarse crystalline limestone associated with clay deposits of palygorskite and illite with rhizoliths. Above 22.4 m, the lithology becomes dolomitic mudstone, wackestone, and packstone, containing peloids, small benthic foraminifera, and small mollusks. The top part is a coarsely crystalline limestone with a thin cap of rhizolith-bearing clay bed.

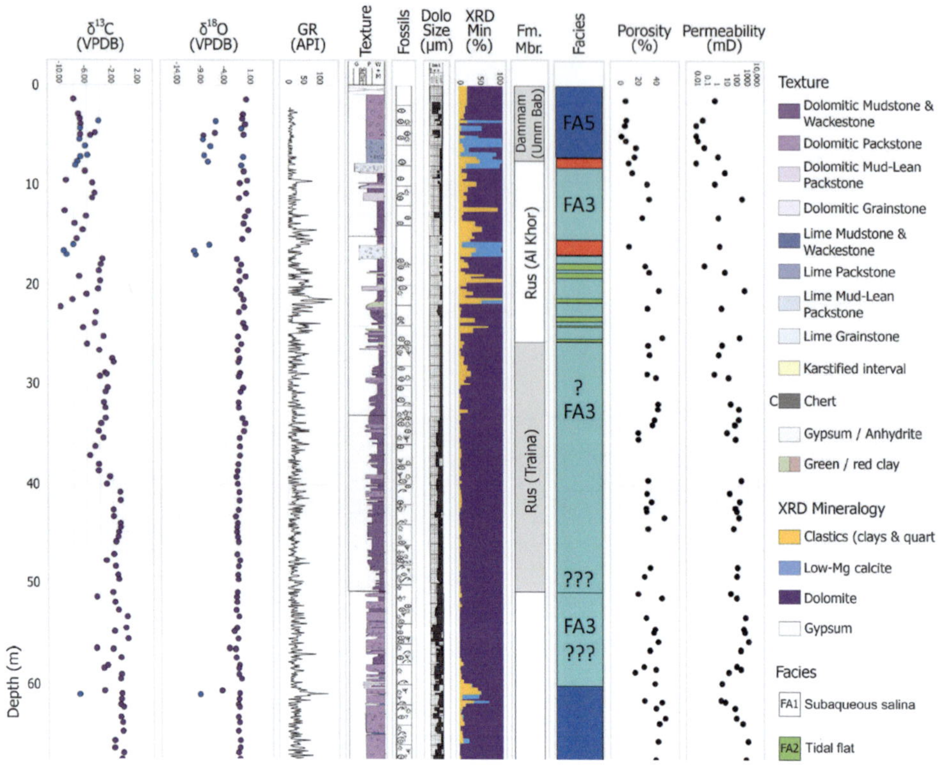

Fig. 9.20 Lithological characteristics of the Rus formation from Borehole 1 in Qatar (Rivers et al. 2019)

Borehole 2 (Fig. 9.21)

The Rus Formation in Borehole 2 is divided into two depth intervals:

Depth interval 62.4–39.0 m below sea level

This interval consists of dolomite with small amounts of chert and calcite crystals. Between 62.4 and 46.0 m, the section is composed of friable sand-sized dolomite crystals interbedded with dolomite rock rubble. This section is overlain by mudstone intercalated with cm-thick wackestone and packstone. Strata contain burrows filled with coarse dolomite. The skeletal grains are obliterated by dolomitization and include mainly dasycladacean green algae with some miliolids.

Depth interval 39.0–14.6 m below sea level

Mixed lithologies of dolomitic mudstone, wackestone, and packstone intercalated with beds of palygorskite and sporadic crystalline limestone form the bulk of this interval. The dolomitic rocks are bioturbated or laminated with some reworking and brecciation associated with the clay beds. Dolomite obliterated most of the skeletal grains. But ghosts of miliolid foraminifera can be recognized in thin sections.

Borehole 3 (Fig. 9.22)

The Rus Formation in this borehole is also divided into two depth intervals:

Depth interval 51.0–26.5 m below sea level

This interval is formed of lighter color rocks, mainly dolomitic packstone and wackestone. It contains burrows, laminations, and possible ripple marks. The clay mineral palygorskite is increasing upward, becoming the second mineral after dolomite. Dolomite is destructive, obliterating most skeletal grains, but it is possible to recognize the nummulites foraminifera in thin sections. Other biological components include small benthic foraminifera miliolids and *Rotalia* Sp., small bivalves, and gastropod molds.

Depth interval 26.5–7.7 m below sea level

This interval comprises mixed lithologies, including bioturbated dolomitic wackestone, mudstone, and less frequent packstone alternating with cm-scale palygorskite and illite clays with two crystalline limestone beds. The beds exhibit frequent erosive surfaces, mud cracks, and mud clasts. Skeletal grains include small benthic foraminifera miliolids and rotalids with mollusk molds.

Depositional Environments

As seen from the nature of the present sediments, the Rus Formation sediments were deposited under two main sedimentary regimes. The first is a supratidal to tidal environment, where a relatively thick succession of gypsiferous limestone was deposited, mainly in the Southern Province.

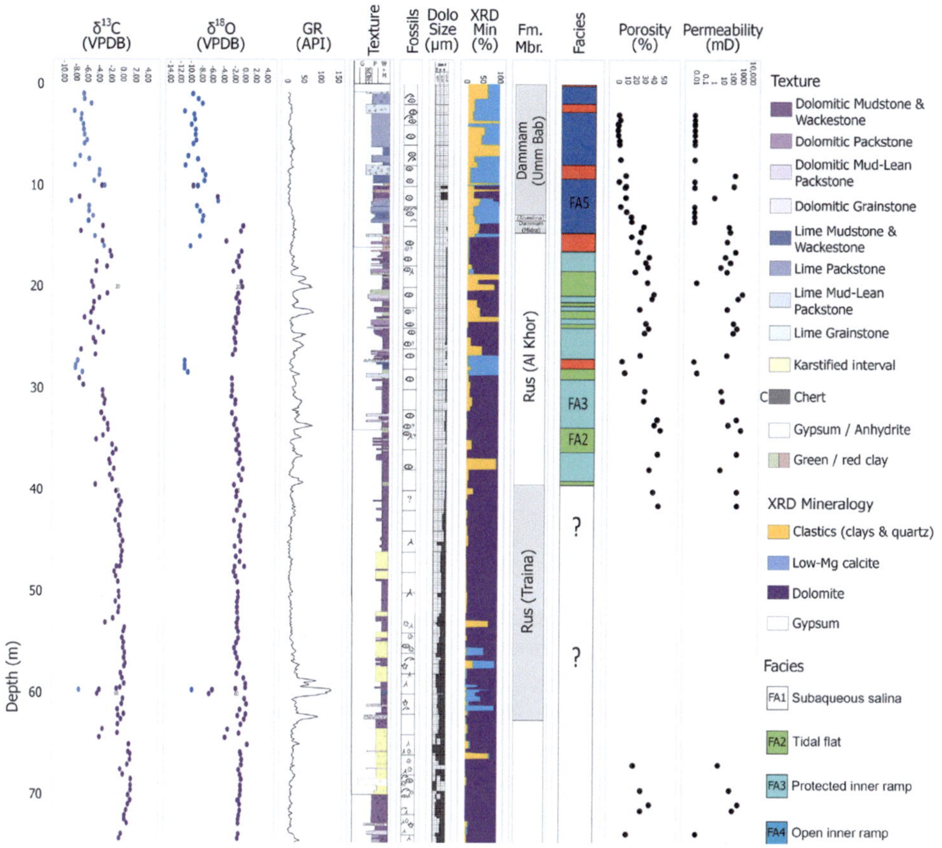

Fig. 9.21 Lithological characteristics of the Rus formation from Borehole 2 in Qatar (Rivers et al. 2019)

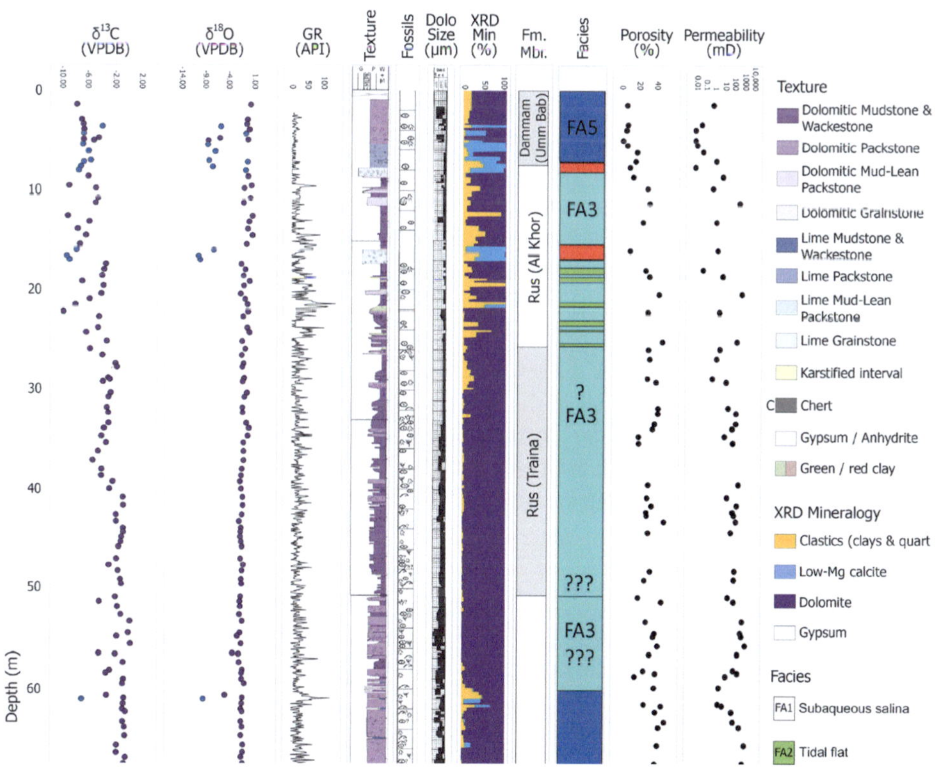

Fig. 9.22 Lithological characteristics of the Rus formation from Borehole 3 in Qatar (Rivers et al. 2019)

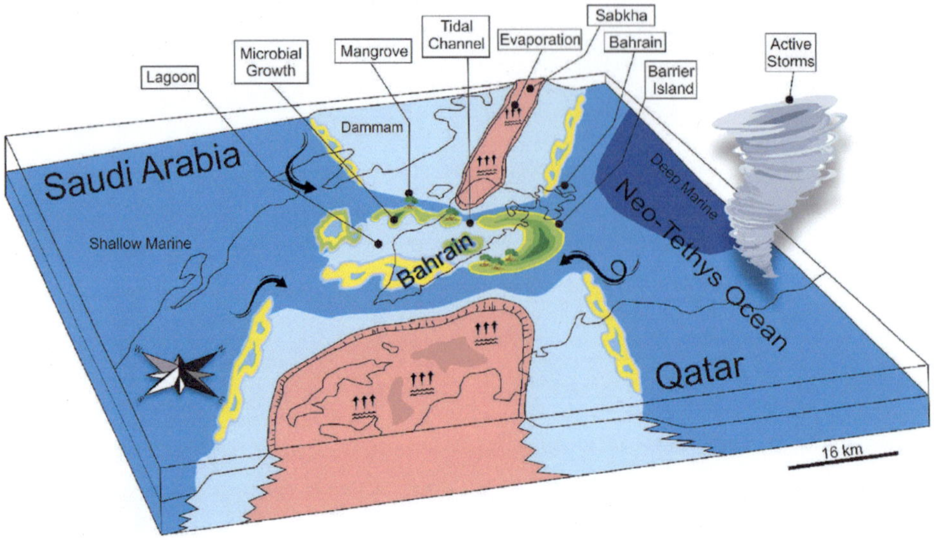

Fig. 9.23 The different depositional environments of the Rus formation in the Arabian Gulf region (Amer et al. 2023)

The second regime is a tidal to subtidal marine setting, which was dominant in the Northern Province. This system deposited chalky and skeletal carbonates with less evaporites. The variations between these two regimes are attributed by some workers (Al-Saad 2003) to the activation of the paleohighs, such as Qatar Arch, which was higher in the Northern Province.

The cyclic nature of sedimentation, where evaporitic sediments alternated with carbonates, can be seen better in the Traina Member of the Southern Province. Such configuration is caused by multiple sea-level fluctuations within a regressive cycle. The sediments of the Al-Khor Member is dominated by carbonates mainly and indicate more stable conditions. The chalky nature of the sediments may suggest direct chemical and biochemical precipitation of calcium carbonates from very shallow, restricted coastal lagoons or an intertidal zone. Amer et al. (2023) suggested that the Rus sediments were laid down in a lagoon setting restricted by barrier islands associated with tidal channels and intertidal and supratidal zones within a sabkha setting. This system was affected by a combination of sea-level fluctuations and weather conditions. These resulted in transgressive lag and storm deposits. Their construction of the paleogeography of the region during the deposition of the Rus Formations points to the existence of two landmasses north and south of Bahrain that restricted the flow of water in the Neo-Tethys sea (Fig. 9.23).

9.3 Middle to Late Eocene: Lutetian to Priabonian (49–33.7 Ma)

This period represents the second part of the "second-order" depositional sequence within the AP10 of (Sharland et al. 2001). They were deposited under the same structural setting of the underlying sediments. This period witnessed the deposition of the Dammam Formation across the eastern parts of the Arabian Plate, the Jaddala Formation in northern Iraq and Syria, and the Pabdeh Formation in Iran (Fig. 9.1). According to Ziegler (2001), the eustatic sea-level curve of Haq et al. (1988) suggests a regional lowering of sea level. Sedimentation of the Pabdeh Formation continued in the Zagros foredeep and Oman; however, the conditions that led to the deposition of the Sachun evaporites in Iran have changed, and the flysch deposition has declined due to reduced sediment supply. This eventually meant margin stabilization and progressive filling of the foredeep. Increasing compression of the northeastern margin of the Arabian Plate led to a basinward shift of the facies belt toward the craton.

Most parts of the eastern shelf of the Arabian Plate were covered by the carbonate deposits of the Dammam Formation, encompassing limestone, dolomite, marl, and shale of around 30 m in Saudi Arabia. Sediments at the base of the formation were laid down in an open marine environment, whereas the upper part suggests shallower marine conditions with siliciclastic input from the west. The Dammam sediments contain nummulites and are dated as upper Ypresian to Priabonian (Ziegler 2001). The contact with the overlying Oligocene sediments seems unconformable (Fig. 9.24).

9.3.1 The Dammam Formation

The Lower and Middle Eocene Damma Formation was named after the Dammam Dome, eastern Saudi Arabia, where Richard Allan Bramkamp (Fig. 9.3) measured the entire succession of the formation from the designated type locality along the old Dhahran—Al Alah road from the

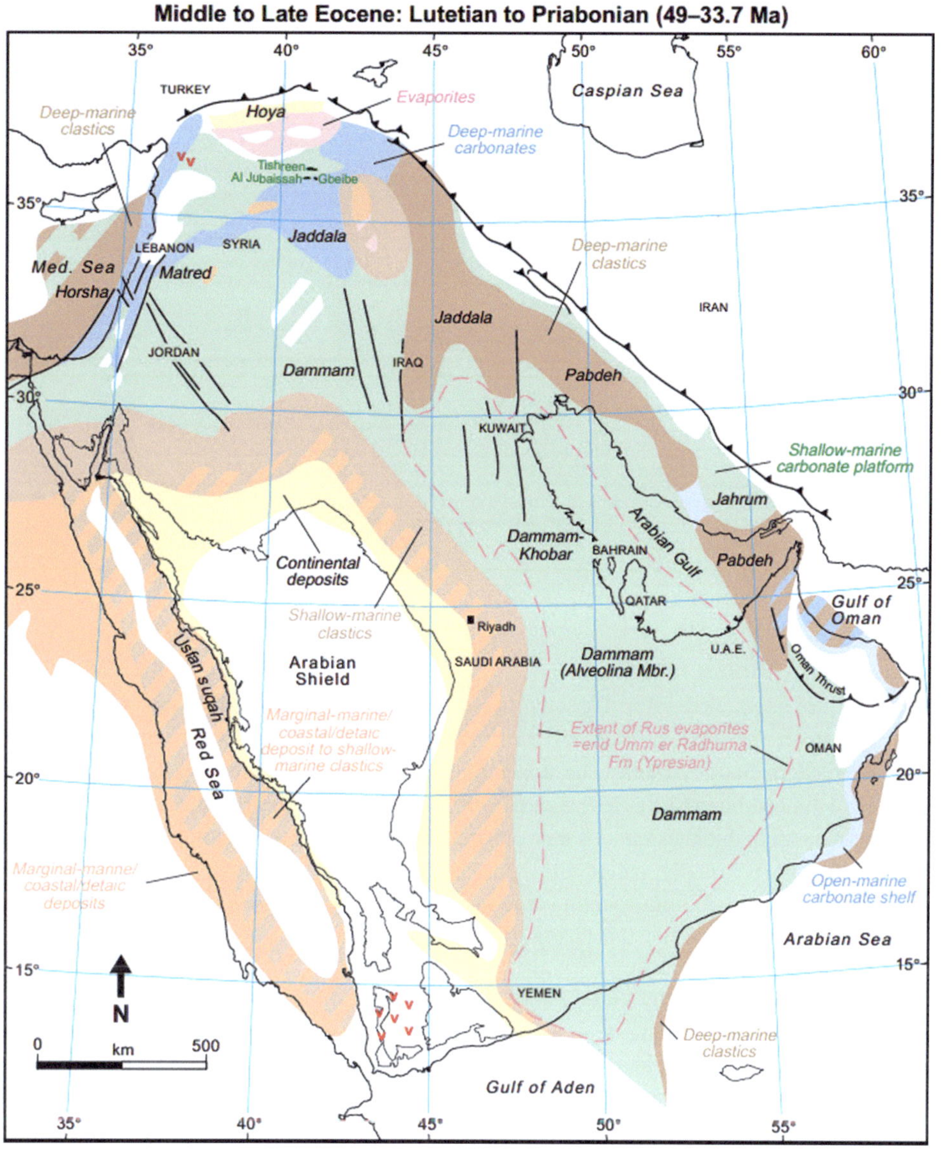

Fig. 9.24 Paleogeographic and facies map of the Middle to Late Eocene: Lutetian to Priabonian of the Arabian Plate (Ziegler 2001)

point where this road intersects the rimrock (Powers et al. 1966). The base of the formation was selected at the change from the limestone of the underlying Rus Formation to the brown shale and marl of the Dammam Formation. The upper contact is unconformable and placed at the change from the clean carbonates of the Dammam Formation with the overlying sandy limestone of the Hadrukh Formation (Fig. 9.1).

The Dammam Formation has five members; two are distinctive and have laterally persistent marl units at the base. These members and their subdivisions are discussed further in descending order in the following sections (Fig. 9.25).

9.3.1.1 The Alat Member

The name Alat is derived from the well Al Alat-1. In this well, the upper strata of this member are exposed in a nearby low hill, and the lower part is encountered between the depths of 0–67 m. This area: the outcrops and the drilled section are considered together as the type locality of this member by Powers et al. (1966). The upper section comprises 9.0 m thick limestone, and the lower succession is 6.0 m marl. The lower marl used to be called the "Orange Marl," but this term is obsolete now. The top of the Alat Member is an unconformable surface separating the chalky dolomitic limestone from the sandy limestone of the Hadrukh Formation. The base is selected at the change from the dolomitic marl above to the dolomite or dolomitic limestone containing Nummulites of the underlying Khobar Member.

The type locality of the pre-Alat members of the Dammam Formation is located southwest of Umm ar Ru'us hill. The other members are the Khobar, Alveolina, Saila, and Midra.

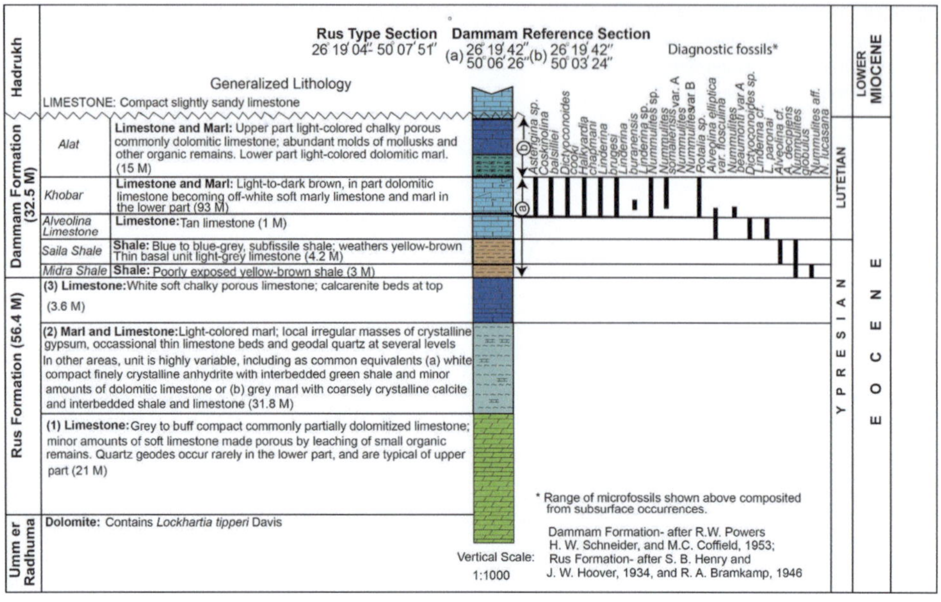

Fig. 9.25 Stratigraphy of the Rus and Dammam formations in their type localities in Saudi Arabia (Powers et al. 1966)

9.3.1.2 The Khobar Member

This member is named after the city of Al Khubar (or Al Khobar) on the eastern flank of the Dammam Dome rimrock, Eastern Saudi Arabia. The upper part of this unit consists of limestone, and the lower part is of marl. At the type section, their thickness is 7.8 and 1.5 m, respectively. The upper part is usually called the "Khobar dolomite," although it may contain limestone in some sections. The top of this member is placed at the change from *Nummulites*-bearing dolomitic limestone and dolomite to the Marls of the overlying Alat Member. The lower contact is at the change from the Nummulitic marl above to the *Alveolina* limestone below.

9.3.1.3 The Alveolina Member

This member is only 1.0 m thick and consists of limestone containing the foraminifera *Alveolina elliptica* (Sowerby) var. *flosculina Silvestri*. The top is selected at the change from marl to limestone and the appearance of the foraminifera. The lower contact is placed at the change from the *Alveolina*-bearing limestone to the shale or marl of the underlying Saila Shale Member.

9.3.1.4 The Saila Shale Member

This member comprises around 0.6 m of limestone overlain by a 3.6 m thick clay shale bed. The upper contact is placed at the change from the *Alveolina*-bearing limestone to the bluish shale below. A thin limestone bed forms the base of the clay shale. The contact between this limestone and the brown shale below marks the boundary between this member and the underlying Midra Member.

9.3.1.5 The Midra Shale Member

This member was designated as the "Shark Tooth Shale" and forms the base of the Dammam Formation. Its top is selected at the first occurrence of yellow–brown clay shale below the limestone unit of the Saila Member. The lower contact is placed by the change from the yellow–brown shale or grey marl above to the chalky limestone or grain-supported limestone of the underlying Rus Formation.

The Dammam Formation has been described in Iraq (Aqrawi et al. 2010); Kuwait (Al-Awadi et al. 1997); Bahrain (Willis 1967); United Arab Emirates; and Oman (Alsharhan and Nairn 1995; Boukhary et al. 2005). The formation is divided into different units in these countries, but most are similar in lithology and fossil contents. This points to the stable tectonic and depositional conditions during the deposition of the Dammam Formation all over the Arabian Plate (Fig. 9.26).

9.3.2 The Dammam Formation in Qatar

The Dammam Formation outcrops in Qatar cover around 70% of the country's surface. These exposures comprise white to grey and yellow shaly and marly dolomitic limestone. They range in thickness between 35 and 52 m. The structural elements of the country, such as the Qatar Arch, and the smaller folds, such as the Dukhan anticline and the Simsima Dome, influence the outcrop distribution (Al-Saad 2005; Sugden 1962) (Fig. 9.27).

The reference section, a full section of the formation in Qatar, was selected between Dukhan and the head of Zekrit

Fig. 9.26 Vertical stratigraphic subdivision of the Dammam formation across the Arabian Gulf region (Alsharhan and Nairn 1995)

FORMATION / AGE	KUWAIT		SAUDI ARABIA	BAHRAIN	QATAR		UAE (ABU DHABI)
D A M M A M (MIDDLE EOCENE)	Unit 3	Chalky Limestone	Alat Member	Alat White Limestone Member	Upper Dammam	Abarug Dolomitic Limestone	D1 (Argillaceous Shaly Limestone)
				Alat Orange Marl Member		Abarug Marl	
	Unit 2	Chalk	Khobar Member	Brown Crystalline Limestone Member		Umm Bab dolomitic Limestone	D2 (Nummulitic Limestone)
		Sandy Limestone					
	Unit 1	Dolomitic Limestone			Lower Dammam		D3 (Marly Limestone)
		Nummulitic Limestone	*Alveolina* Limestone Member	*Alveolina* Limestone Member		Dukhan *Alveolina* Limestone	
		Shale	Saila Shale Member	Shark Tooth Shale Member		Midra and Saila	
			Midra Shale Member			Fhaihil Velaces Limestone	

Fig. 9.27 Outcrops of the Dammam formation in Al Nakhish Hills near Dukhan

Fig. 9.28 The different vertical stratigraphic subdivision schemes proposed for the Dammam formation in Qatar (Al-Saad 2005)

Saudi Arabia Type Section Powers et al. (1966)	Cavelier (1970)		Abu-Zeid and Boukhary (1984)
Alat Member	Abaruq Dolomitic and Marl Member	Upper Dammam Subformation	Abaruq Member
Khobar Member	Simsima Dolomitic and Limestone Member		Umm Bab Member
Alveolina Limestone Member	Dukhan Alveolina Limestone Member	Lower Dammam Subformation	Dukhan Member
Saila Shale Member	Midra and Saila Shale		Midra Shale Member
Midra Shale Member			
	Fhaihil Velates Limestone		

Bay by Sugden and Standring (1975). The total thickness of the formation is 52 m and consists of four lithologic units. These are, from the bottom, 5.5 m of light grey shale with foliate cleavage, marl with some orange to yellow iron staining. This is followed by 33.5 m of light grey limestone alternating with white dolomite and chalk. It contains frequent chert nodules. The third unit is formed of 10.6 m of dolomite, chalk with patches of silica. The top unit is 1.8 m of light grey, rubbly limestone. The underlying formation is the Rus Formation. In some southern parts of Qatar, the formation is overlain by the Dam or the Lower Fars Formation (Sugden and Standring 1975).

Originally, the Dammam Formation was designated as Early/ Middle Eocene by earlier workers such as Powers et al. (1966), Cavelier (1970), and Sugden and Standring (1975). Based on the occurrence of the latest Paleocene/Early Eocene planktonic foraminifera in the underlying Umm er Radhuma formation, the Rus Formation was assigned to the Middle Eocene Lutetian stage. Accordingly, the age of the Dammam Formation in Qatar was revised to be Middle/Late Lutetian. Later, Hewaidy and Al-Saad (1993) assigned the Midra and Dukhan members to the Middle Lutetian and the upper Umm Bab and Abaruq members to the Late Lutetian age.

Cavelier (1970) divided the Dammam Formation in Qatar into two parts. These are the lower Dammam and upper Dammam subformations. Subformation is not a geologic term and was soon removed from usage. He also divided the formation into five members, and so did Abu-Zeid and Boukhary (1984), who revised Cavelier's classification and divided the formation into four formal members. These are the Midra Shale, the Dukhan, the Umm Bab, and the Abaruq (written as Abarug or Abaruk by some workers) members.

The Dammam Formation contains a rich nummulite population that was documented by Casier (1971), Blondeau and Cavelier (1972), Boukhary (1985), Hewaidy and Al-Saad (1993) and Al-Saad (2005). The different vertical stratigraphic subdivisions of the formation are shown in Fig. 9.28.

Kok and Leblanc (2012) introduced a new member within the Dammam Formation: the Bir Zekreet Member. The type locality of the new member is in the Zekreet area, at the back of the Islamic School and opposite the Cuban Hospital. The outcrop is at the bottom of the second mesa. The coordinates of the type section are N25 deg 26′ 32.2″ longitude: E 050 deg 51′ 52.7″.

The Bir Zekreet Member comprises around 2–5 m of thinly laminated and very fissile, ferruginous shale. The shale ranges in color from brown-grey to beige, reddish, or greenish in some places. In some crops, the shale shows the usual "paper-shale" weathering. Unlike the Midra Shale, this shale lacks fossils and has no limestone layers. The dolomitic and fossiliferous limestone of the Umm Bab Member underlies the Bir Zekreet Member. It is overlain by the marly and chalky dolomitic marls of the Abarug Marl Member of Cavelier (1970) (Fig. 9.29).

Differential weathering has removed considerable parts of the Dammam Formation in different parts of Qatar. Al-Saad (2005) has reported that although the outcrops of this formation are nearly complete in the hills around the Dukhan area, where the reference section has been selected, a large part of the succession has been removed in central Qatar and, to

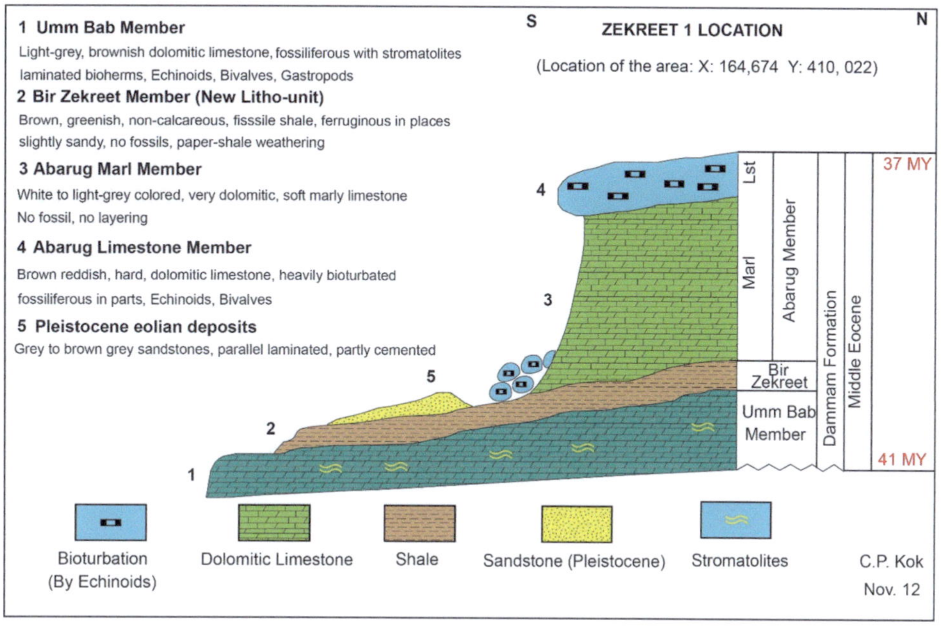

Fig. 9.29 Type locality and lithologic characterization of the newly introduced Bir Zekreet member (Kok and Leblanc 2012)

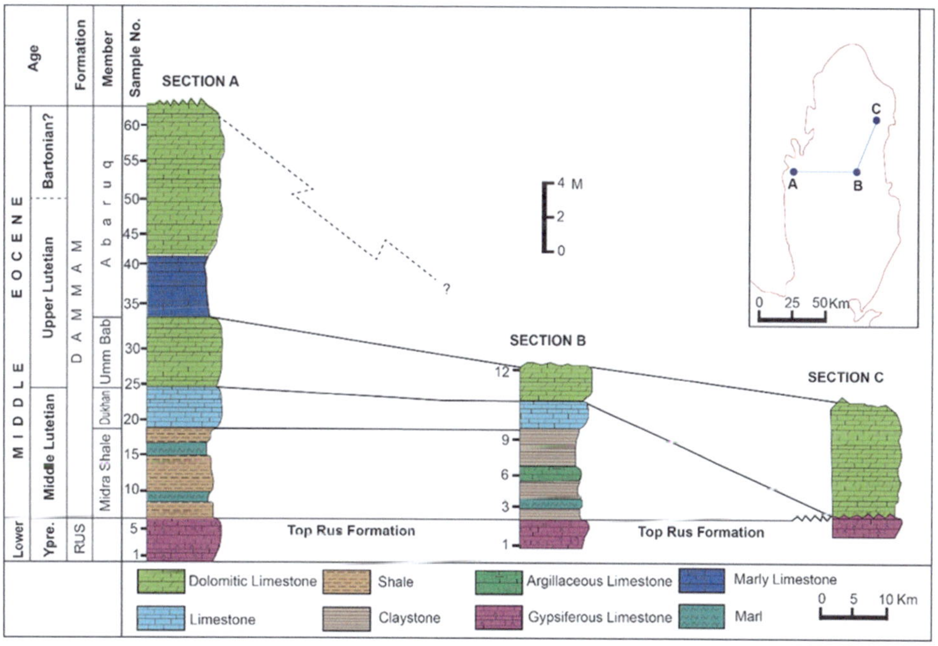

Fig. 9.30 The different outcrop sections of the Dammam formation, as studied by Al-Saad (2005)

a large extent, in the northeastern parts (Fig. 9.30). Differential weathering led to the removal of the soft shale and marl, creating the usual mesa structures characterizing the Dammam outcrops in Qatar (Fig. 9.31).

With such differential removal of large sections of the Dammam Formation in the outcrops, it became difficult to identify the different members of the formation, particularly if they are formed of the same lithologies. LeBlanc (2015, 2019, 2021) and Kok and Leblanc (2012) conducted detailed geologic work on the Tertiary outcrops of Qatar. Kok and Leblanc (2012) combined a composite reference section of the Dammam Formation from five locations (Fig. 9.32). This is very useful because experience has shown that it is difficult to recognize the different formation members in the field. They also included the characteristic fossils of each member. This will make it easier for future researchers to identify these easily. The localities are marked on the figure and include:

Fig. 9.31 A typical mesa structure of the Dammam formation outcrops resulted from differential weathering

A *The Abarug Dolomitic Marl and Limestone Member* with the exposure is at Lati. 25° 26′ 31.56″ N, Long 50° 50′ 57.56″ E.

B *Bir Zekreet Member* with the outcrops, can be seen at Lat 25° 26′ 32.20″ N, Long. 50° 51′ 52.70″.

C The top part of the *Umm Bab Dolomite and Limestone Member* with the exposure can be seen at Naslat Umm Hadidah at Lat. 25° 24′ 5.52″ N, Long 50° 53′ 7.59″ E.

D Thick part of the *Umm Bab Dolomite and Limestone Member* from its base at the QNCC quarry at Lat. 25° 11′ 47.87″ N, Long. 50° 50′ 15.28″ E.

E Outcrop of the basal portion of the *Umm Bab Dolomite and Limestone Member* and the full section of the *Dukhan Alveolina Limestone and Midra (and Saila) Shale members* can be seen at Lat. 25° 17′ 15.79″ N, Long. 50° 48′ 12.60″ E (One kilometer from the QP's Fahahil Plant).

They also followed Cavelier (1970) in classifying the Abarug Member into the Abarug Dolomitic Marl and the Abarug Limestone Member. This is a distinctive feature in the field.

Al-Saad (2005) provided a detailed description of each of the Dammam formation members. The following synthesis is based mainly on his work:

9.3.2.1 The Midra Shale Member

The outcrops of this member can be found in southwestern and central Qatar (A & B locations, Fig. 9.30). It comprises soft, grey to light green, fossiliferous, fissile, calcareous, and gypsiferous shale intercalated with two 0.3–0.5 m thick marl beds. The alternation of shale and marl is related to the sea-level fluctuations during the Eocene caused by the activation of local structures, such as the Qatar Arch in the member's depositional area. The shale sediments were

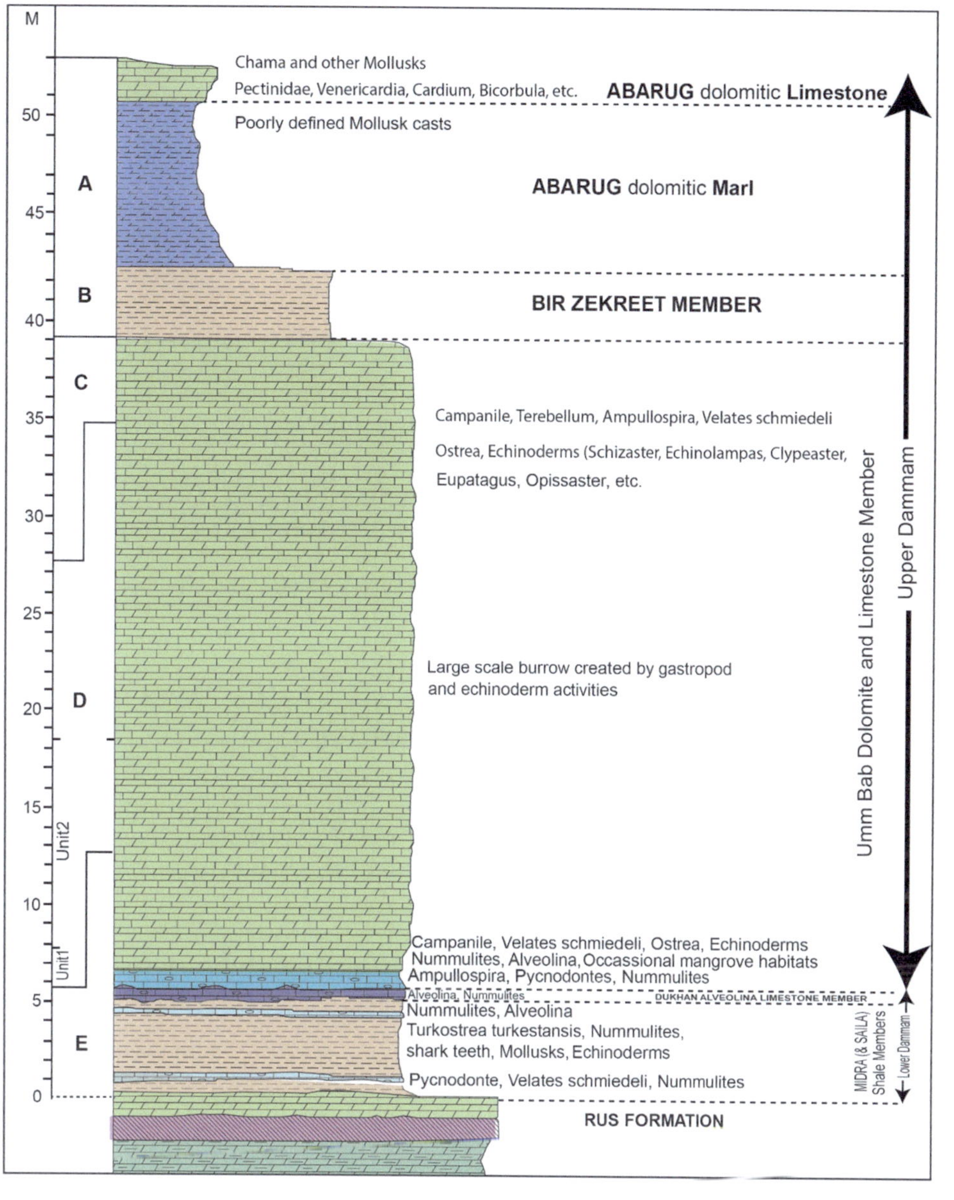

Fig. 9.32 A full, composite section of the different members of the Dammam formation, including the newly introduced Bir Zekreet member with their characteristic fossils (LeBlanc 2021)

probably derived from the nearby land. The thickness of the Madra Shale ranges between 2 and 6 m.

In the central areas of the country, this member consists of 4–9 m of massive, fossiliferous, gypsiferous, and calcareous claystone interbedded with two argillaceous limestone at the base and the middle parts. These units are intercalated in some areas with 2–4 cm of marl bands. Fine sandstone and phosphatic concentrations were also reported.

The main sedimentary structures of the Midra Shale include thin parallel laminations and ironstone nodules. The Midra Shale Member has a rich fossil population. These include shark teeth, large benthic foraminifera (*Nummulites*, *Alveolina*, and *Lockartia*), bryozoa, mollusks, and rare echinoids. The shark teeth fossil is used to distinguish the base of the Dammam Formation from the underlying Rus Formation. The presence of rich mollusks characterizes the top of this member.

Drilled wells through the Midra Shale in the areas of Khor Al Adaid, Al Nakhash near Dukhan, Salwa, and Salwa resort have shown that the carbonate content in the form of limestone and dolomite of the Midra Shale increases westward toward Salwa. These carbonates are also rich in fossils and sedimentary structures such as bioturbation (Fig. 9.33).

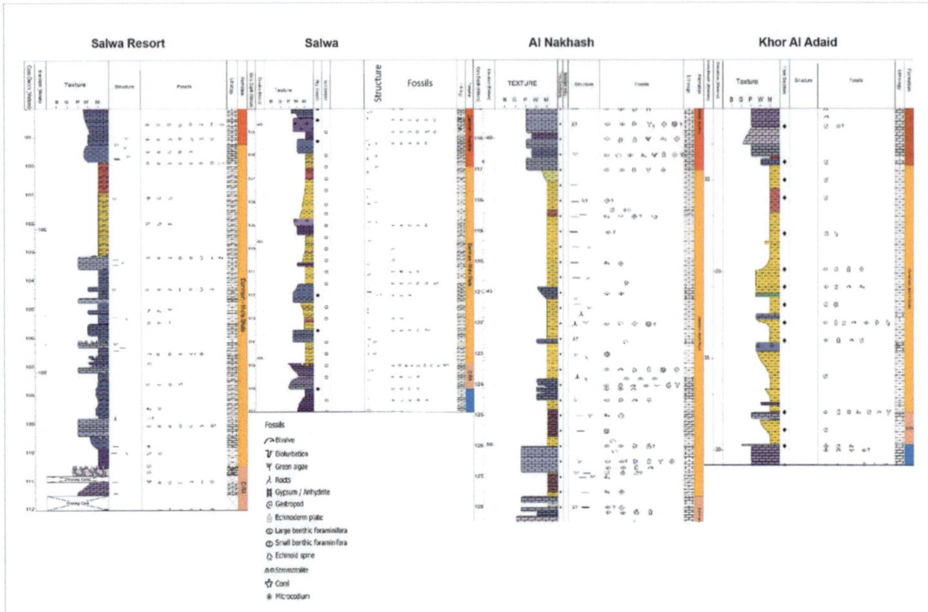

Fig. 9.33 Lithologic variations of the Midra Shale from the Khor Al Adaid in the southeastern corner to Salwa Resort in the central west of Qatar (Courtesy of John Rivers)

9.3.2.2 The Dukhan Member

The outcrops of the Dukhan Member are restricted to the southwestern and central parts of the country (Locations A & B, Fig. 9.30). It comprises massive, nodular limestone rich in *Nummulites* and *Alveolina* fossils. The sediments of this member represent a transgressive tract that starts with mudstone at the base and changes into wackestone and packstone at the top. The main fossils are echinoid spines, gastropods, and large oystrea. This is the thinnest member of the Dammam Formation, and it ranges in thickness between 1 and 2 m.

9.3.2.3 The Umm Bab Member

Area-wise, this is the most exposed member of the Dammam Formation. It consists of 15–30 m massive dolomitic limestone with a thin marl bed at the base and a thin cherty lens at the top. The thin marl bed (0.3 m thick) separates this member from the underlying Dukhan Member. The main fauna in these sediments include *Nummulites*, *Alveolina*, pelecypod, and gastropod casts with echinoderms and mollusks.

9.3.2.4 The Abaruq Member

The Abaruq Member is reported from southwestern Qatar (Location A, Fig. 9.30). It comprises around 12 m of marly limestone at the base, changing gradually to massive, thin-bedded, partially nodular dolomitic limestone. This dolomitic limestone is frequently intercalated with chert zones. The main fauna is rare large benthic foraminifera with mollusk casts and molds.

The different members of the Dammam Formation in Qatar are shown in Fig. 9.32.

9.3.3 Facies Analysis

The Dammam Formation is the last unit of extensive distribution across the Arabian Plate. It is found in the areas extending from southern Iraq to Oman. It represents a carbonate ramp rich in large foraminifera, mainly nummulites. There are no indications of a reef or buildups, and the shallow shelf changes gradually into deeper water where the sediments of the Pabdah formation were deposited in Iran. The main facies are including the following:

Lime mudstone and dolomitic mudstone (Fig. 9.34a–d)

This facies consists mainly of micritic limestone with rare or no grains. Some variants contain evaporite or silica crystals. Others have clotted texture, and there are indications of microbial burrowing and bioturbation. The micritic groundmass has been dolomitized in many areas into aphanitic dolomite, where traces of the original micritic materials are preserved. In some cases, well-developed euhedral dolomite replaces the original texture completely. There are patches of iron staining and organic matter. This facies was deposited in a sheltered coastal lagoon where water became shallow to allow precipitation of evaporites. The abundance of microbial activity suggests a low-energy, protected depositional setting.

Large foraminifera packstone and wackestone (Fig. 9.34e–h)

This is the main building facies of the Dammam Formation and can be traced all over the Arabian Plate. The skeletal grains are mainly the large benthic foraminifera *Nummulites*, *Alveolina* sp., and other similar fossils. Most of them are well-preserved, but others are abraded or even

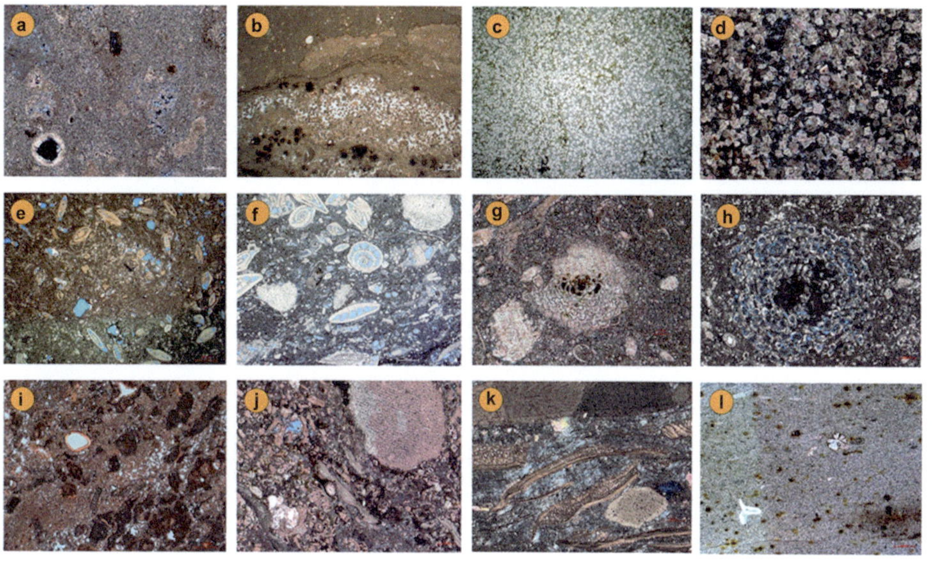

Fig. 9.34 Microfacies of the Dammam formation in Qatar **a–d** Lime mudstone and dolomitic lime mudstone **e–h** Large foraminifera packstone and wackestone and dolomitic packstone **i** Peloidal packstone **j**, **k** Skeletal packstone and wackestone **l** Shale (Courtesy of John Rivers)

deformed. In some cases, their skeletons were dissolved, creating biomoldic porosity. It was also noticed that some of these were replaced by euhedral dolomite rhombs, which were partially dissolved. The dolomite preserves the original shape of the fossil (Fig. 9.34h). This facies is similar to many facies in the carbonates of the Arabian Plate, such as the Cretaceous *Orbitolina*-bearing limestone of the Mauddud Formation. The facies was deposited in a relatively high-energy, open, normal marine shelf. The concentration of the foraminiferal skeleton is related to an extensive sieving process by currents and waves that accumulate these fossils against a baffling barrier, such as topographic irregularities in the old depositional surface.

Peloidal packstone and wackestone (Fig. 9.34i)

The sediment of this facies contains peloids of different shapes and sizes. They are either aggregates of organic matter or micritized skeletal grains. They are also stained with iron oxides and have some local patches of micritic clots that may be produced by microbial activity. In some cases, the matrix is replaced by aphanitic dolomite. Partial dissolution of the matrix or the rare skeletal grains generated a considerable porosity. These sediments are deposited in protected shallow coastal lagoons with extensive microbial activity.

Skeletal packstone and wackestone (Fig. 9.34j, k)

This facies is the second most common after the large foraminiferal-bearing facies. It consists of varied skeletal grains, including mollusks (pelecypods mainly), echinoderms, algae, and rare corals. There are indications of compaction and abrasion with some scattered, well-developed dolomite rhombs. In some cases (Fig. 9.34k), the whole groundmass of the facies is replaced by silica. The sediments of this facies were laid down in a high-energy, open marine carbonate shelf. The presence of coral fragments may suggest the presence of solitary corals distributed across the shelf rather than a real buildup.

Shark teeth-bearing shale (Fig. 9.34l)

This facies makes the main sediments of the Midra Shale Member of the Dammam Formation. It consists mainly of the clay mineral palygorskite and lesser amounts of illite. In the outcrops, it is iron-stained and greenish in color, forming thin bands between the limestone beds. The presence of shark teeth suggests relatively deep marine water. The clay materials were derived from the nearby land and then pushed away from the shoreline by a high-energy storm system.

Mangrove-bearing limestone

Mangrove remains were reported from the ridges encircling the Al-Khor Estuary northeast of Qatar (Sadooni and Al-Saad 2012). These hills range in height from two to three meters and are composed primarily of the Dammam formation sediments. The uppermost two meters comprise nodular limestone with plant remains that belong to the Umm Bab Member. The plant remains are fossilized mangrove trunks and molds. These are composed of numerous cortical laminations encircling a central opening likely occupied by the plant stem (Fig. 9.35). Some of these structures have a diameter ranging from 10 to 20 cm and are filled by friable, bluish-marly sediments. The entire section represents an exposure surface that was covered by mangrove trees. Certain strata are composed of limestone that contain macrofossils, including barnacles and gastropods. This combination of mangroves and fauna is quite prevalent in contemporary mangrove stands, which are present in the adjacent estuary.

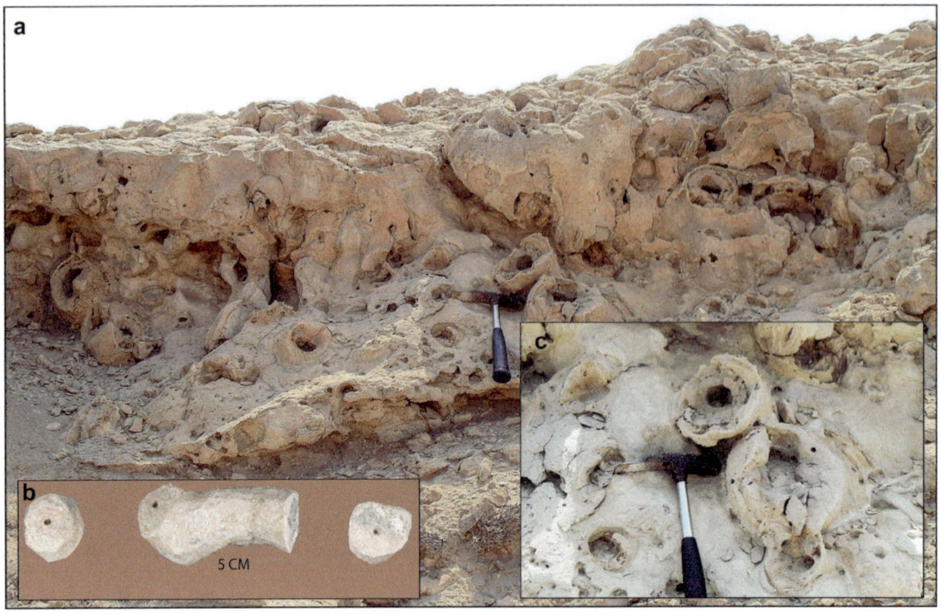

Fig. 9.35 **a** Outcrops of the fossil mangrove stand showing casts and molds, **b** individual petrified stems of the mangrove, **c** close-up of the mangrove molds (Sadooni and Al-Saad 2012)

9.3.4 Depositional Environment

The Dammam Formation sediments were deposited in a sea transgression covering the Rus Formation desiccated, evaporitic basin. The earliest sediments were lime mudstone that filled the depressions in the old topographic surface of the Rus. These depositional settings were semi-isolated coastal lagoons. Microbial activity and evaporite deposition were two characteristic features of such deposits. The transgression gained momentum, covering the whole ramp and leading to the deposition of dark and grey shale with shark teeth. These shale strata may contain the maximum flooding surface MFS Pg20. A relative sea retreat led to the differentiation of the ramp into a set of distinctive depositional environments, including lagoons and inner and middle-to-outer ramps. In these environments, the nummulitic packstone and wackestone were deposited. The last stage of the Dammam Formation was a regional regression that exposed parts of the old ramp and resulted in the karstification of the old carbonates, the precipitation of chemical limestone (the so-called chalky marl or limestone), while the nummulitic limestone continued to be deposited in the deeper inner and outer ramps (Fig. 9.36).

9.4 The Miocene: Aquitanian to Messinian (23.8–5.3 Ma)

In the Arabian Plate, this period is characterized by extensive salt deposits (the saline beds of the Lower Fars Formation) in the northern parts of the plate, the opening of the Gulf of Eden, and the start of the Red Sea rifting leading to the separation of Arabia from the African Plate. On the border with Eurasia, the thrusting of the Sanandaj-Sirjan zone led to the release of large clastic sediments to the Zagros Foredeep Basin. In Qatar, this period witnessed the deposition of the Dam and Hofuf formations (Fig. 9.37). This is a mixed continental (Hofuf Formation) to transitional-marine (Dam Formation) due to the tectonic activities and the shrinkage of the Tethys Basin. The uplift of the western parts of the Arabian Shield and consequent rapid erosion released large quantities of sand and gravel that became part of the Hofuf Formation. These pebbles are formed of igneous and metamorphic rocks and can be seen scattered in the desert of Qatar.

9.4.1 The Dam Formation

The Dam Formation was introduced by Steineke et al. (1958) from the eastern flank of Jebel Dam (Lidam) (26° 21.6′ N., 49° 27.7′ E) and Jebel Lemaigher (26° 17.7′ N., 49° 30.8′ E.), 60 km west of the Dhahran, Saudi Arabia. It consists of 89.8 m of marl and sandy clay of different colors and shades as the main two dominant lithologies, with minor sandstone in the upper part. The lower part is formed of marly and chalky limestone. Fossils are present throughout but poorly preserved. The upper contact is placed at the change from the fossiliferous marl and limestone of the Dam Formation to the overlying sandstone and gravel of the Hofuf Formation. The lower contact is marked at the limestone and marl containing the fossil E*chinocyamus* sp. The main fossils are *Ostrea latimarginata* Vredenburg, *Echinocyamus* sp., and "*Archaias*" sp. Through correlation with the Lower Fars Formation in Iraq, the formation age is considered the middle Miocene. Powers et al. (1966)

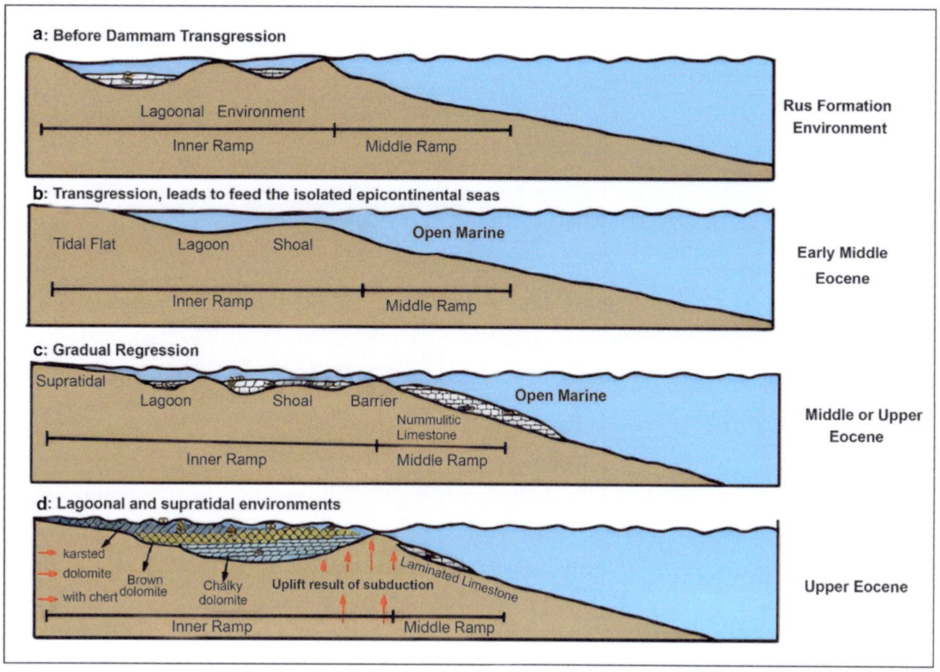

Fig. 9.36 The depositional history of the Dammam formation shows the different stages that created its suits of sediments (Tanoli and Al-Bloushi 2017)

provided a detailed lithologic description of the Dam Formation in its type locality as recorded by Steineke and Koch in 1935. This section appears in Fig. 9.38.

9.4.1.1 The Dam Formation in Qatar

The Dam Formation in Qatar was described for the first time by Cavelier (1970), who divided the formation into lower and upper parts. Abu-Zeid and Khalifa (1983) divided the formation into two members, A (lower) and B (upper). Hewaidy (1991), based on the foraminifera assemblages, suggested a Burdigalian-Helvetian age (Early to Middle Miocene).

The Dam Formation in Qatar outcrops in Jebel Al-Nakhash, near the southern end of the Dukhan anticline. It forms benched hills ranging in height between 15 and 80 m. The total thickness of the formation is around 90 m. A good section of the formation is exposed in the Al-Aeraij quarry (Fig. 9.39).

Al-Saad and Ibrahim (2002) subdivided into two formal members, the Kharrara and Al-Nakhash members. Table 9.3 shows the different classifications used to divide the Dam formation.

The following is a description of the two recognized members. This synthesis is based on Al-Saad and Ibrahim (2002).

Al-Karrara Member

This is the lower part of the Dam Formation. The selected type section is at the Jebel Al-Nakhash, 85 km southwest of Doha, Qatar. The foraminifera *Borelis melo melo* at the base suggests an Early Miocene age. This member is partially exposed, and its upper boundary was placed at the change from its dolomitic limestone to the stromatolitic limestone of the overlying Al-Nakhash Member.

The Al-Kharrara Member comprises around 28 m of claystone alternating with marls and grain-supported limestone capped by dolomitic limestone. In the type locality, the claystone and the sandy facies thin out toward the south and are replaced by dolomitic limestone and limestone. The main fossils include bivalves, molds and casts, fish teeth, and ostracods. The sediments have rich foraminifera fossil assemblages, including miliolids and large arenaceous foraminifera.

Al-Nakkhash Member

This member represents the upper part of the Dam Formation. The type section is selected in the Jebel Al-Nakhash, 85 km southwest of Doha city. The age of this member is Early Miocene due to the presence of the upper part of the foraminifera fossil *Borelis melo melo* Local Range Zone. The upper boundary of this member is selected at the erosive boundary between the gypsiferous limestone of this member and the conglomerates of the Hofuf formation.

This member comprises around 46 m of gypsiferous, chalky, and stromatolitic limestone. The stromatolitic limestone is restricted to the type locality and is replaced by marly limestone in the Al-Kharrara area. Gypsum and celestite are common in the upper part of this member.

Figure 9.40 shows the stratigraphic subdivisions and main lithologies of the Dam Formation in the Al-Nakhash area of western Qatar.

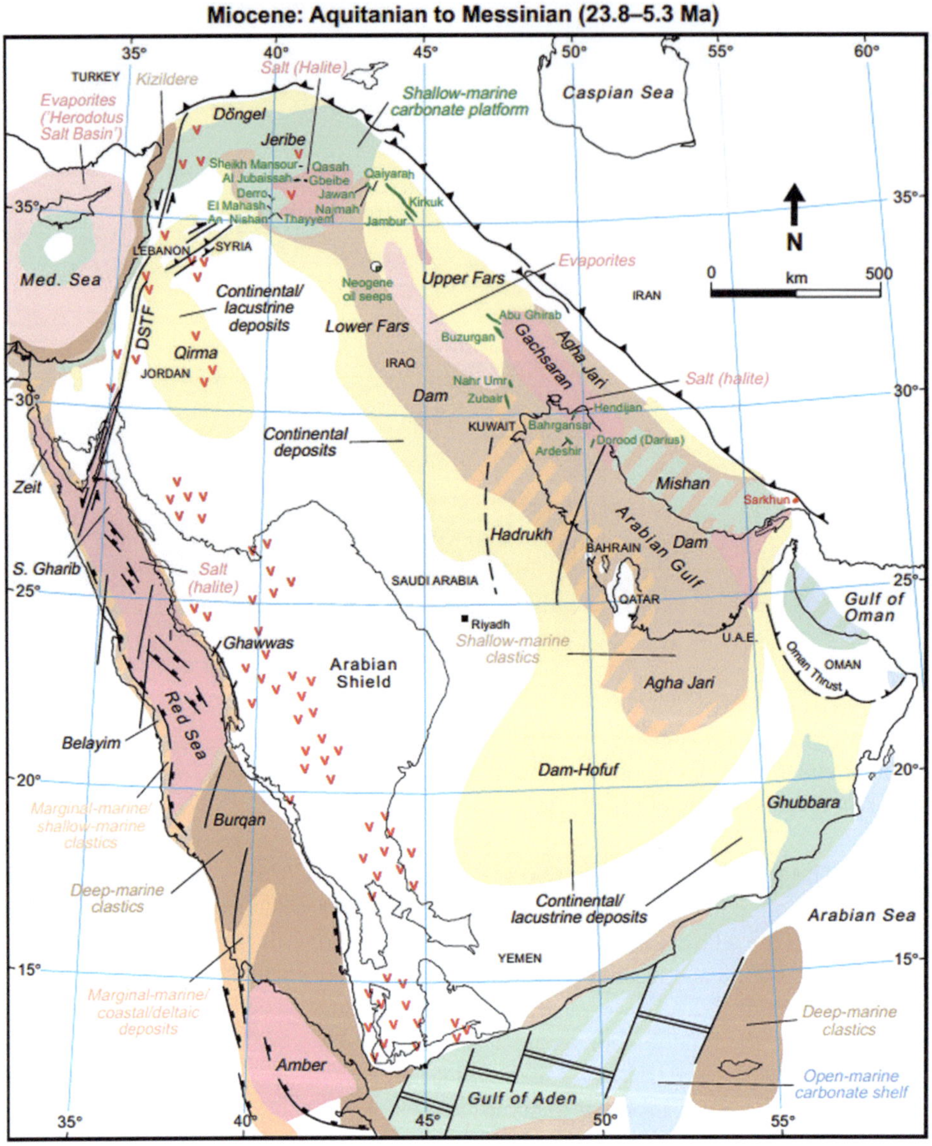

Fig. 9.37 Paleogeographic and facies map of the Arabian Plate during the Aquitanian-Messinian (Ziegler 2001)

9.4.1.2 Facies Analysis

The Dam Formation shows a wide range of lithofacies of a cyclic nature. These were divided into four associations by Al-Saad and Ibrahim (2002). These are carbonates, claystone, marl, and evaporites. This synthesis is based on their work.

Carbonates

They subdivided the carbonates into the following rock types:

Argillaceous limestone: these are thin laminations or beds of yellow argillaceous limestone that range in thickness from a few centimeters to around 4 m and are distributed throughout the succession. They contain some pelecypod shells and foraminifera.

Sandy limestone: This rock type is found mainly in the Al-Kharrara Member and ranges in thickness between 0.35 and 9.0 m. In some cases, the limestone becomes gypsiferous or ferruginous. Fauna includes rare gastropod and pelecypod shell fragments. Thin section analysis shows that these rocks comprise around 40% of fine to medium-grained detrital quartz grains, 40% of micritic matrix, and 20% of grains.

Chalky limestone: This rock-type content increases toward the top of the formation. It consists of 2–4 m of mudstone and wackestone with few pelecypod, gastropod shells, and foraminifera. The micritic matrix also has some clay and quartz.

Oolitic limestone: The oolitic limestone consists of 0.75–1.2 m of medium to coarse grainstone within the Al-Nakhash Member. The ooliths are rounded and well-sorted, with some pellets and peloids.

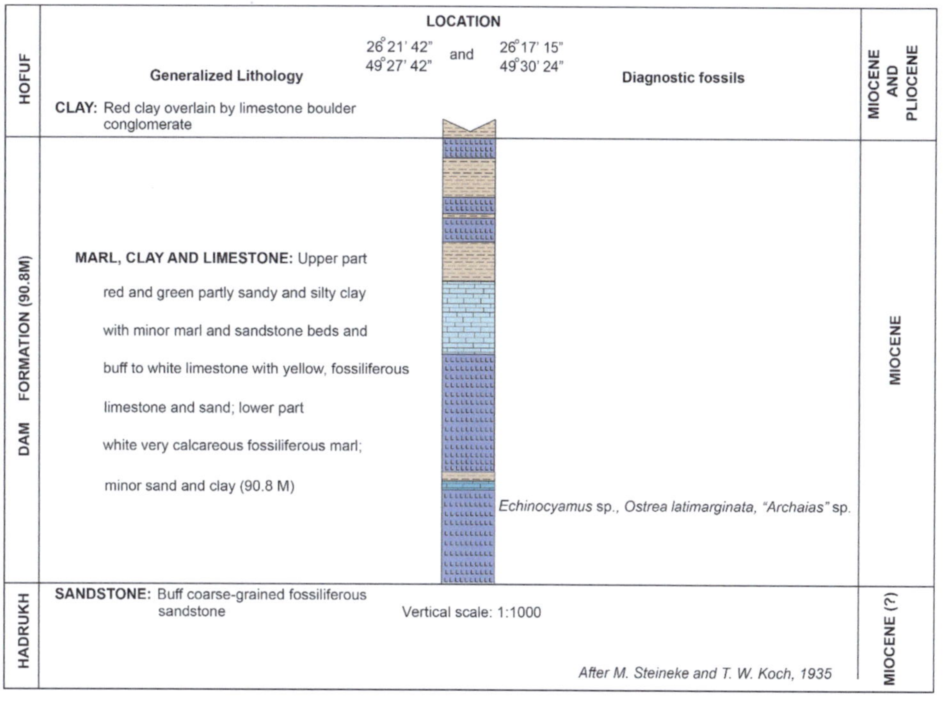

Fig. 9.38 Detailed lithology of the Dam formation in its type locality (Powers et al. 1966)

Fig. 9.39 Outcrops of the Dam formation in the Al-Aeraij quarry. *Photo* by Jacque LeBlanc

Table 9.3 Different subdivisions of the Dam formation (Al-Saad and Ibrahim 2002)

Rock unit	Cavelier (1970)	Abu-Zeid and Khalifa (1983)	Hewaidy (1991)	Al-Saad and Ibrahim (2002)
Dam formation	Upper Dam subformation	Member B	Upper Dam member	Al-Nakhash member
	Lower Dam subformation	Member A	Lower Dam member	Al-Kharrara member

Stromatolitic limestone: This rock type is found in the Al-Nakhash Member and consists of 1 m thick beds of mudstone and skeletal wackestone alternating with thin bands of stromatolites 0.20–0.60 m thick. Some molds of gastropods and algae were reported in some places. These rocks form the base of the Al-Nakhash Member (Fig. 9.41).

Dolomitic limestone: this rock type is formed of 2 m thick, grey, dolomitic limestone intercalated with thin gypsum laminations. In some locations, it caps the Al-Kharrara Member. The original texture is skeletal wackestone/packstone, replaced by well-developed dolomite rhombs.

Claystone (Fig. 9.42)

The claystone lithofacies is the most spectral rock type of the Dam Formation due to its soft texture and red-brown and greenish color. It consists of differentially weathered

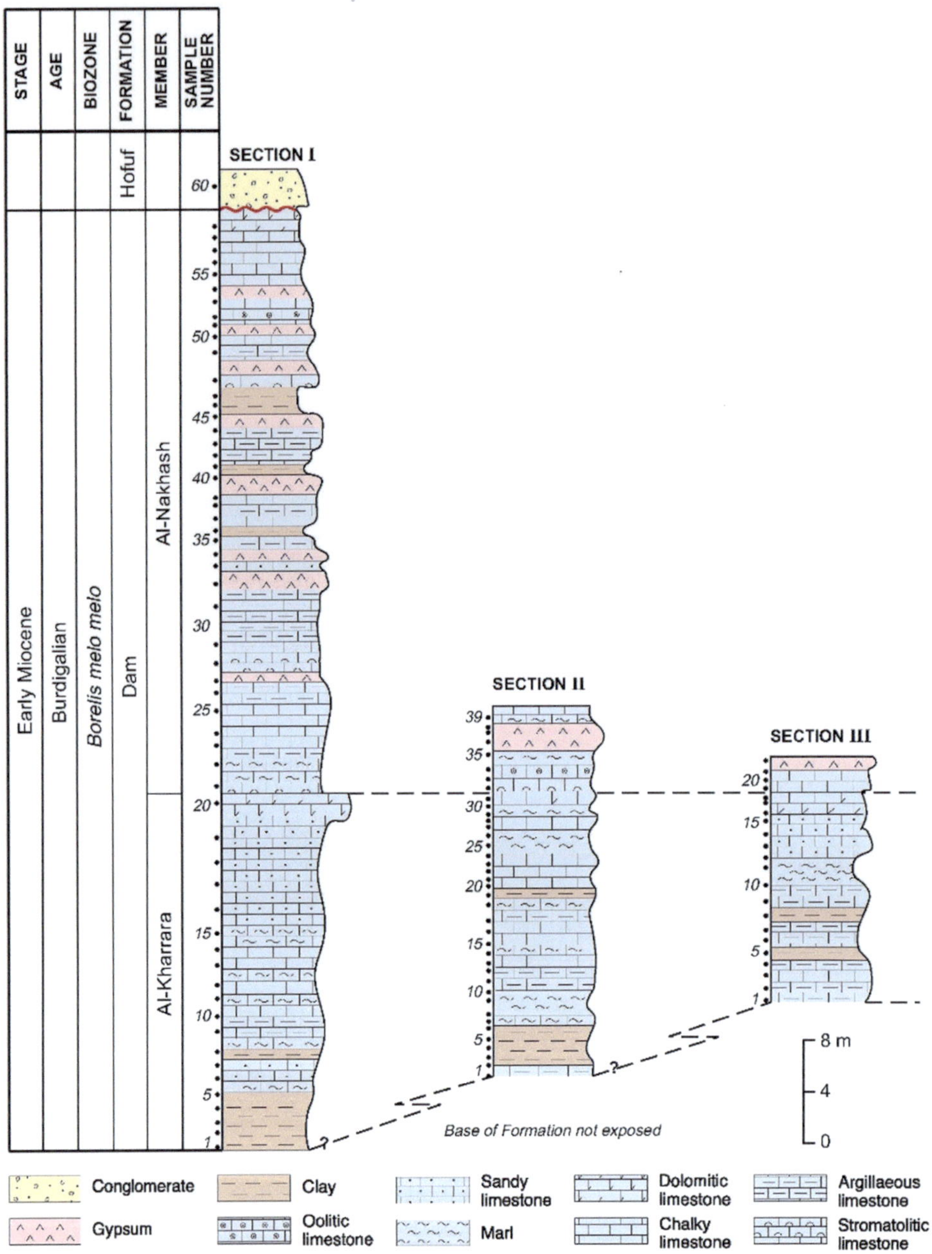

Fig. 9.40 Stratigraphic subdivision and lithologic sections of the Dam formation in the Al-Nakhash area, western Qatar (Al-Saad and Ibrahim 2002)

claystone and silt 1.5–4.2 m thick. The main fauna are small pelecypods and gastropod shell fragments with rare foraminifera. Usually, the claystone is interbedded with gypsum bands ranging in thickness between 2 and 5 mm. This lithofacies is found mainly in the Al-Kharrara Member.

Marl (Fig. 9.43)

The Dam Formation marl is creamy to yellowish in color and ranges in thickness from 0.35 to 2.0 m. It is found at different intervals of formation and is mixed with clay and sand in some cases. Rare small pelecypod shells are also found.

Evaporites

This lithofacies consists of weathered gypsum and celestite, which range in thickness between 0.80 and 1.20 m. It can only be found in the Al-Nakhash Member.

Sandstone

Al-Saad and Ibrahim (2002) did not report this facies, but Dill et al. (2003) described it from the Al-Nakhash Member in the Dukhan anticline and found it also in the Al-Araig Quarry (Fig. 9.44). It consists of fine to medium-grained sandstone

Fig. 9.41 Stromatolitic limestone with bird's-eye porosity in the Dam formation, Al-Araig Quarry

Fig. 9.42 The red claystone lithofacies of the Dam formation from the Al-Araig Quarry

with planar cross-beddings. It is off-white to brown in color and has a honeycomb structure resulting from differential erosion. It forms high and resistant ledges in the outcrops.

9.4.1.3 Depositional Environments

The Dam Formation represents a complex and mixed carbonate and siliclastic depositional system. The carbonates were deposited in a shallow-water carbonate ramp with oolitic barriers. The shorelines were probably intersected by channels that brought clastic materials during humid climatic periods with relatively heavy precipitation to be redistributed by marine waves and tides. If they become abandoned, such channels will be sites for accumulating claystone and other mud-supported sediments as levee deposits. Oolitic and other grain-supported carbonates filled tidal channels in a pattern similar to the present-day sabkha

Fig. 9.43 Yellow and creamy marl of the Dam formation in the Al-Araig Quarry

Fig. 9.44 Off-white sandstone ridge with a honeycomb structure within the Dam formation, Al-Araig Quarry

and oolitic shoals of Abu Dhabi. In such a setting, the stromatolitic limestone was formed in the intertidal zone. The Dam Formation has coral debris in some parts of the Gulf region (e.g., in Saudi Arabia). These are probably solitary corals that did not develop into a real buildup (Bashri et al. 2015) (Fig. 9.45).

9.4.2 The Hofuf Formation

The Hofuf formation was described for the first time by Steineke and Koch in 1935 in an unpublished report (cited by Powers et al. 1966) from a section north of Al Hufuf City, eastern Saudi Arabia. These are continental deposits

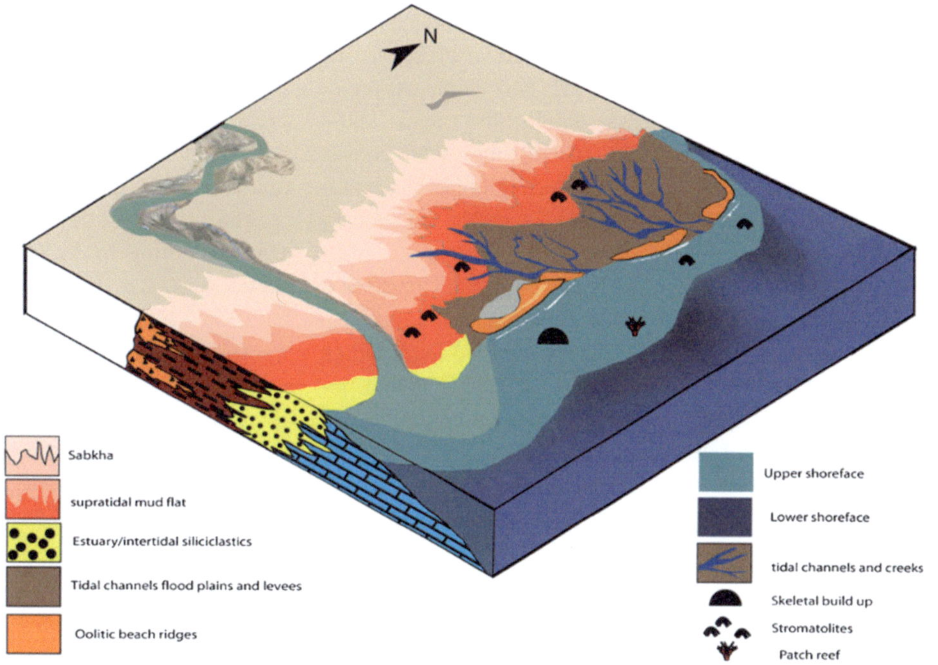

Fig. 9.45 Depositional environments of the Dam formation in the Gulf region (Bashri 2015)

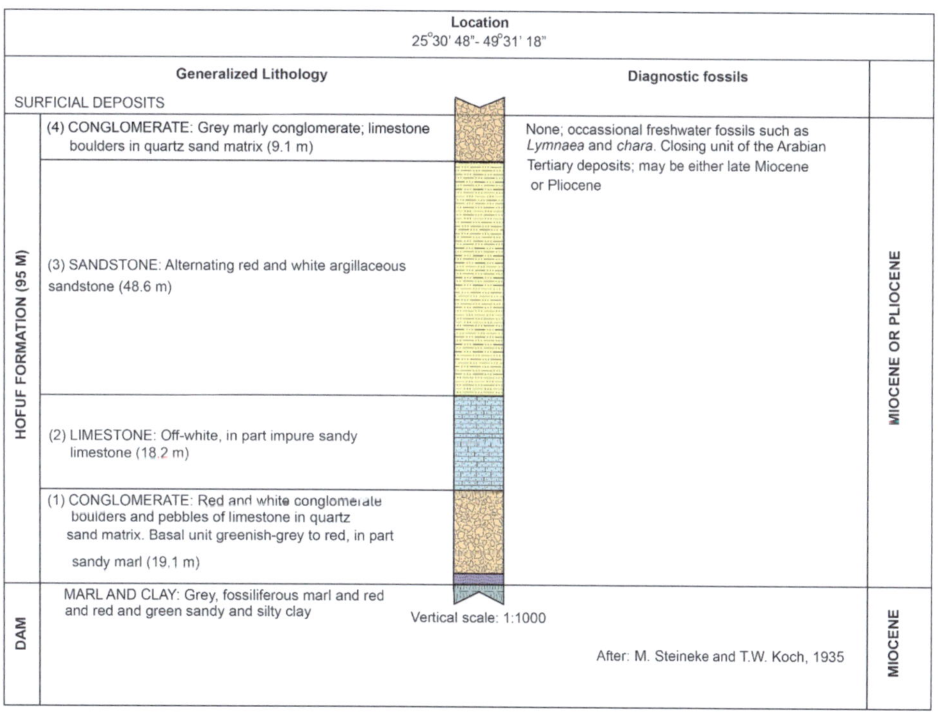

Fig. 9.46 Lithologic components of the Hofuf formation in its type locality, near Al-Hufuf City, Eastern Saudi Arabia ((Powers et al. 1966)

formed of a mixture of conglomerate, sandstone, sandy limestone, sandy marl, and sandy shale. The total thickness of the formation in its type locality is around 90 m (Powers et al. 1966). The wide areal distribution of these deposits all over the eastern parts of the Arabian Plate suggests a strong deformation and rapid erosion of the interior regions after the end of the Dam formation deposition. That tectonic activity provided the conglomerate and other clastic deposits of the Hofuf formation. The gravels are mostly formed of quartz of igneous or metamorphic origin. In some areas, Powers et al. (1966) reported pebbles and boulders derived from the Upper Jurassic Jabal Tuwayq formation. Still, they reinstated that most of these materials were derived from the basement complex. The lithologic units of the Hofuf formation in its type locality are shown in Fig. 9.46.

9.4.2.1 The Hofuf Formation in Qatar

In the south-central and southwestern areas of Qatar, the erosional surface at the top of the Dam Formation comprises erosional products such as mesas, knolls, and ridges. These features are surmounted by thin clastic sediments that consist of gravel-bearing sandstone with fine conglomerate intercalated with silt and clays. These materials form the last Tertiary rock unit.

Cavelier (1970) described the Hofuf Formation for the first time from Qatar. The outcrops of the formation underwent extensive erosion and had also been excavated for building purposes, which makes it difficult to delineate their original distribution. Noweir and Abd-Hameed (an unpublished manuscript) reported that the formation is exposed in seven sites. These are Al Bu'udiyat, Al Tiwar, Al Kharrarah, Al Huwaylah, Um Tuwaym, Ti's Karanah, and El Kakhash. They reported the presence of a washing plant at the Kharrarah to prepare the quarried clastic materials for construction and building purposes. The Hofuf formation may generally cover up to 5% of Qatar's surface area (Fig. 9.47).

The outcrops of the formation in the Al-Subaiha area represent, according to Al-Saad et al. (2002), the best exposed section in Qatar (Fig. 9.48). The sediment thickness ranges between 2 and 18 m. It may represent only the lower part of the formation due to the removal of its upper part compared to the sections described from Saudi Arabia (Thralls and Hasson 1956). The outcrops comprise thin beds of coarse-grained sandstone, gravel, and conglomerates. The pebbles

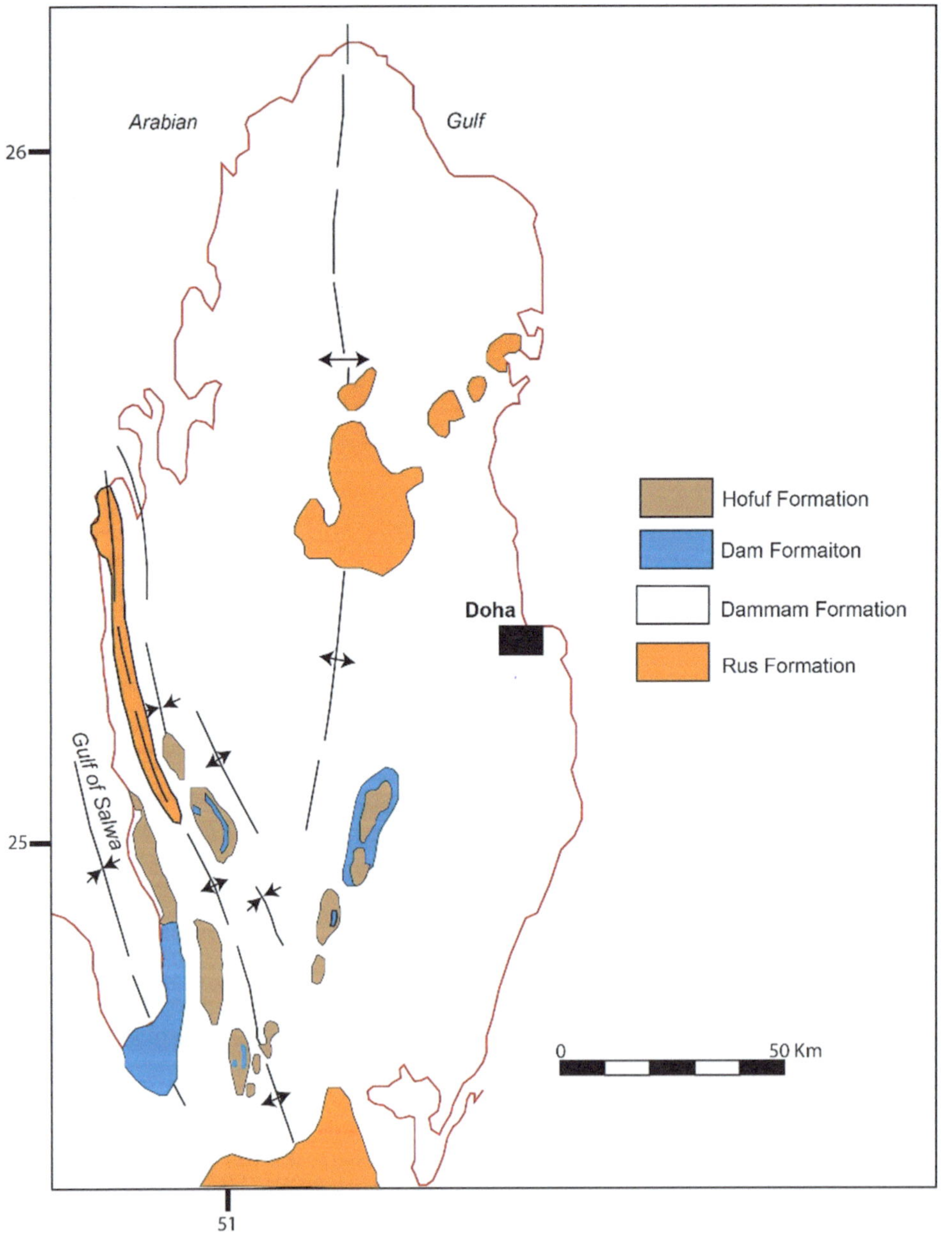

Fig. 9.47 Distribution of the Hofuf formation outcrops in Qatar (Noweir and Abd-Hameed, unpublished)

Fig. 9.48 Alternation of sandstone and conglomerate beds of the Hofuf formation in the Al-Subaiha area, Qatar

are of igneous and metamorphic origin, such as granite, basalt, gneiss, and schist. There are also some quartzite, jasper, limestone, and marl. These clastics overlain unconformably the argillaceous limestone of the early Middle Miocene of the Dam formation, marking a regional unconformity all over eastern Arabia.

Mineralogy and Petrography

The sand part of the Hofuf formation is classified as immature arkosic to sub-arkosic sandstone (Al-Saad and Ibrahim 2002; Pettijohn et al. 1972). The main mineralogical association includes quartz (64–87%), feldspar (5–30%), calcite, mica, mafic igneous rock fragments, heavy minerals, chert, gypsum, and carbonates (Fig. 9.4). Sorting is poor, with the majority of the grains being subangular to subrounded. Quartz is monocrystalline, mainly with a few polycrystalline crystals. Feldspar minerals are albite and orthoclase with minor amounts of microcline. Limestone is present in the form of micrite clots. There is a wide range of rock fragments, including metamorphic, volcanic, granitic, and sedimentary, in the forms of chert and limestone. The major types of rock fragments are granitic and rhyolitic with subordinate basalt.

The heavy minerals range between 0.1 and 0.9 wt%. Most are iron opaque grains such as magnetite and ilmenite with less hematite and limonite. The transparent heavy minerals are mainly zircon (10–33%), tourmaline (10–25%), hornblende (13–38%), pyroxene (18–40%), and epidote (0–5%). Trace minerals include garnet, kyanite, andalusite, sphene, and staurolite. The non-stable heavy minerals include amphiboles and pyroxene.

The conglomerate and gravel parts of the formation are formed of igneous, metamorphic, sedimentary, and quartz pebbles. All parts are in equal proportions. Igneous rock pebbles are formed mainly of rhyolite and granite with subordinate basalt. These pebbles are similar to those of the Arabian Shield (Al-Saad et al. 2002). Metamorphic rock pebbles consist of biotite gneiss and metamorphosed greywacke.

Table 9.4 The main light minerals of the sandstone part of the Hofuf formation (Al-Saad et al. 2002)

Sample	Quartz	Feldspar	Calcite	Mica	Rock fragments
H1	80	15	1	1	3
H2	81	12	1	1	5
H3	86	11	1	1	2
H6	85	10	1	1	3
H7	82	13	1	1	3
H8	85	10	0	1	4
H9	83	10	1	0	6
H11	82	12	0	1	5
H12	71	25	0	1	3
H13	64	30	2	1	3
H14	72	23	1	1	3
H16	65	25	3	3	4
H17	76	15	4	2	3
H20	84	10	2	1	3
H21	87	5	1	2	5

Pebbles of sedimentary rocks are composed of lime mudstone and skeletal wackestone and packstone, similar to the Mesozoic and Cenozoic carbonates of the Arabian Plate.

According to Zingg (1935) classification of pebbles, they are disk, spherical, blade, and rod shapes. The spherical and disk-shaped are the most abundant, followed by the rod and blade shapes.

Textural Characterization

Noweir and Abd-Hameed investigated the compositional and textural characteristics of the Hofuf formation (unpublished). These samples were collected from outcrops of the formation in Al Bu'udiyat, Al Kharrah and Umm Tuwaym. The sample size was representative, weighing between 2 and 5 kg. It comprises gravel with friable materials such as small pebbles, sand, mud, and diagenetic evaporites. The textural properties of the studied seven samples are shown in Table 9.4.

They plotted the size fraction percentages on Folk's ternary diagram (1974). Most of the samples except one were considered gravelly sand inclined toward the sand fraction (Fig. 9.49).

The main textural features of the collected samples were summarized by Noweir and Abd-Hameed (unpublished). This synthesis is based on their work:

(a) The mean size (Mz) of the studied samples ranges between fine pebble (– 2.93 ɸ) to medium sand (1.23 ɸ) with an average mean size of coarse sand (0.10 ɸ). According to the authors, this suggests that most sediment is formed of a relatively narrow range of sand sizes.

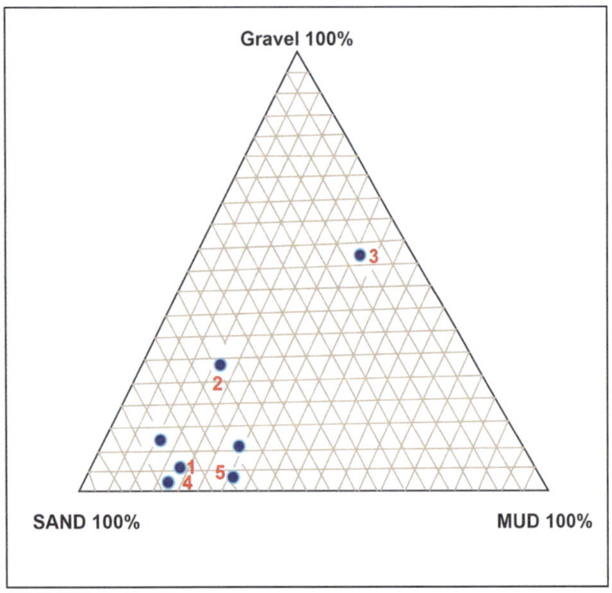

Fig. 9.49 Ternary diagram of Folk (1974) showing the textures of the Hofuf formation (Noweir and Abd-Hameed) (unpublished)

(b) The standard deviation for the studied sediments ranges between moderately well-sorted (0.57 ɸ) to very poorly sorted (2.72 ɸ), with an average standard deviation of poorly sorted (1.52 ɸ).

(c) The skewness values range between strongly coarse skewed (0.56) to strongly fine skewed (0.35), which indicates a wide- range of depositional energies.

(d) The kurtosis of the sediments also shows wider variations ranging from very leptokurtic (2.65) to mesokurtic (1.07) to platykurtic (0.74) (Table 9.5).

Mineralogy and Petrography

Gravel-size fraction

Gravels were either separated from the semi-consolidated beds or collected from the loose or lag gravels that cap the mesas and knolls formed of the underlying Dam formation. Noweir and Abd-Hameed (unpublished) sorted these sediments manually and reported that they were either cobble size (64–256 mm) or pebble size (4–64 mm) (Table 9.6). The cobble-sized pebbles form around 52.6% and consist mainly of igneous and metamorphic rock types (48.73%) and carbonate rocks (40.25%). The pebble-sized fractions form around 47.4% and are made up of quartz (52.5%), carbonates (32.4%), and igneous and metamorphic components (14.1%). The sandstone and chert gravels are rare in both fractions, forming around 8.4% and 0.94%, respectively.

Gravel Mineralogy

Igneous and Metamorphic gravels

Rhyolites are the most dominant variety, forming around 75% of the igneous gravels. Other types include granite, basalt, and porphyritic andesite. Rhyolites are porphyritic with a spherulitic texture. Quartz is in the form of rounded grains or euhedral pyramidal crystals. The plagioclase is mainly an invariably altered oligoclase. Granite is hypidiomorphic granular and consists of quartz, microcline perthite, albite, and altered biotite. Basalt is fine-grained and made up of altered lath-shaped plagioclase and augite with secondary amounts of chlorite and iron oxides. Porphyritic andesite comprises altered andesite in a matrix of plagioclase and actinolite with minor amounts of quartz, chlorite, and iron oxides.

Metamorphic gravels are mainly metagraywacke formed of quartz, feldspar, and rock fragments in a fine groundmass comprising quartz, chlorite, epidote, and iron oxides.

Sedimentary gravels

Sedimentary gravels are formed of carbonates, sandstone, and chert. Carbonate gravels are mainly lime mud with some wackestone and packstone. Sandstone pebbles are quartzite and graywacke. Chert pebbles are composed of microcrystalline quartz and chalcedony.

Table 9.5 The weight percentage of size fraction and size parameters of the Hofuf formation (Noweir and Abd-Hameed unpublished)

Sample number	Percentage of size distribution					Size parameters				
	Gravel			Sand 2–1/16 mm	Silt and clay < 1/16 mm	Md 50 Φ	Mz Φ	σ1	SKI	KG
	Coarse pebbles and cobbles < 16 mm	Fine pebbles 16–4 mm	Granule 4–2 mm							
1	–	4.2	4.0	73.6	18.2	1.1	0.87	1.11	– 0.42	1.81
2	3.8	17.2	8.7	54.0	16.3	0.5	– 0.18	2.14	– 0.44	0.74
3	19.8	32.2	0.6	10.7	36.7	– 3.9	– 2.93	2.72	0.35	2.65
4	–	0.1	0.2	80.2	19.5	1.2	1.23	0.57	0.06	1.07
5	–	7.5	4.9	59.6	28.0	0.7	0.22	1.45	– 0.45	1.33
6	–	3.9	2.2	63.0	30.9	1.2	1.10	1.38	– 0.21	1.67
7	–	7.0	7.4	74.7	10.9	0.9	0.42	1.28	– 0.56	1.24
Average	3.4	10.3	4.0	59.4	22.9		0.10	1.52	– 0.24	1.50

Table 9.6 Percentage by volume of the gravel size fractions of the Hofuf formation (Noweir and Abd-Hameed unpublished)

Type		Cobble size (64–256 mm)	Pebble size (4–64 mm)	Total gravel
Igneous and metamorphic		48.73	14.09	32.3
Quartz gravels		2.54	52.58	26.3
Sedimentary gravels	Sandstone and chert	8.48	0.94	4.9
	Carbonates	40.25	32.39	36.5

Sand-Sized Fraction

The sand-sized samples were collected from the semi-consolidated rock outcrops. They consist of quartz (more than 75%) with fewer amounts of feldspar, rock fragments, opaques, mafics, gypsum, carbonates, and other accessory minerals. Quartz is in the form of subangular to subrounded grains with varied surface textures. Feldspar is the second most abundant constituent, forming around 10% of the sand fraction, and consists of plagioclase, orthoclase, and perthite. Rock fragments of lime mudstone, felsite, chert, and fine-grained quartz originate from acid volcanic rocks. Opaque grains are iron oxides, whereas mafics are formed of amphibole and biotite. Gypsum grains are rare, and they may be of cement origin. Accessory minerals include zircon, apatite sphene, epidote, and garnet.

Facies Analysis

In their study of the Hofuf formation in the Al-Subaiha area, Al-Saad et al. (2002) divided the Hofuf succession, which is around 18 m thick, into eight sedimentary cycles. Each comprises a sandy conglomerate followed by a coarse-grained sandstone and capped by a bioturbated, fine grained sandstone (Fig. 9.50). The thickness of these cycles ranges between 1.2 and 3.3 m. The main sedimentary structures are cross-bedding and bioturbation. They divided the studied sediments into three sedimentary facies.

Clast-supported sandy conglomerate

This facies forms the basal part of each cycle and ranges in thickness from a few centimeters to one meter. The sandy conglomerate is white to brown and dark brown and comprises

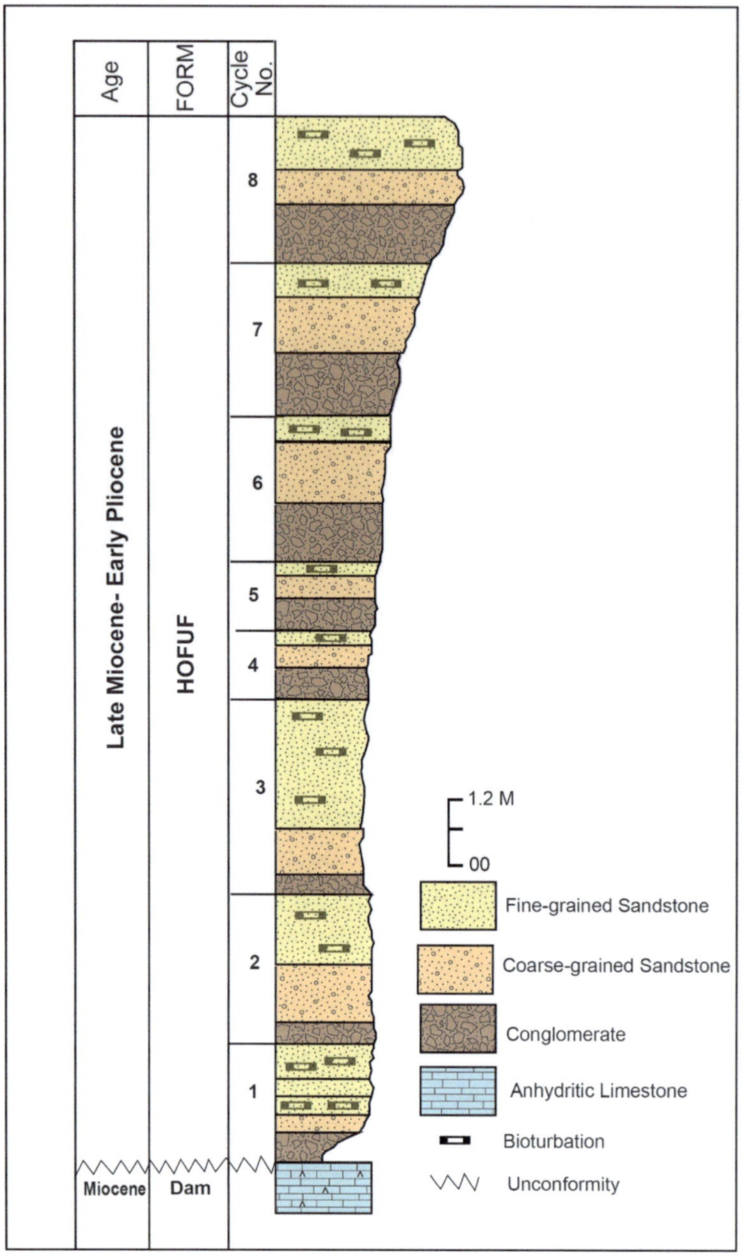

Fig. 9.50 Sedimentary cycles and facies of the Hofuf formation in the Al-Subaiha area, Qatar (Al-Saad et al. 2002)

igneous, metamorphic, carbonate, and quartzite pebbles up to 10 cm in diameter intercalated with sandstone lithoclasts. The majority of the conglomerates are plane-bedded, but some are also grade-bedded.

Coarse-grained sandstone

This sandstone facies is encountered at different levels within the formation. It consists of light brown to light yellow and ranges in thickness from 0.3 to 1.2 m. The main sedimentary structures are planar and tabular cross-bedding.

Fine-grained sandstone

This facies characterizes the upper part of each of the fining-upward cycles. It is composed of light brown to creamy, fine-grained sandstone and ranges in thickness from 0.3 to 2.3 m. The main sedimentary structures are bioturbation (Fig. 9.51) and small-scale planar cross-bedding with the occurrence of plant remains.

Fig. 9.51 Bioturbation in the fine sandstone facies of the Hofuf formation

Depositional Environments

The Cenozoic of Arabia witnessed the Alpine tectonic movement that shaped the Zagros mountains on the borders with Iran and raised the gradient of the Arabian Shield and Platform; hence, extensive areas of eastern Arabia became exposed to large-scale erosion and denudation. According to Al-Saad et al. (2002), Wadi Sahba, at the eastern Arabian Block, was also affected by these tectonic movements and became the sediment source of the Hofuf Formation. During the Pliocene, such tectonic movements also led to the development of the Salwa Syncline and the Dukhan Anticline west of Qatar. Continuous erosion created the mesa structures and deflation surfaces of the upper part of the Hofuf Formation. A major pluvial phase occurred during the late Pleistocene, which coincided with the Last Glacial Maximum LGM). During this period, the present-day Hofuf area was an inland lake with a great river flowing from the Arabian Shield toward the present-day Arabian Gulf. The Greek geographers called this river the Aftan River (Al-Saad et al. 2002; Golden 1984).

To identify the depositional environments of the Hofuf sediments, both Al-Saad et al. (2002) and Noweir and Abd-Hameed (unpublished) constructed a set of cross-plots that are usually used to document depositional environments of clastic sediments. An across-plot of skewness versus soring using the methods of Moiola and Weiser (1968) and Friedman (1967) shows that most of the Hofuf sediments fall in the river domain, with only one sample in the beach environment (Fig. 9.52).

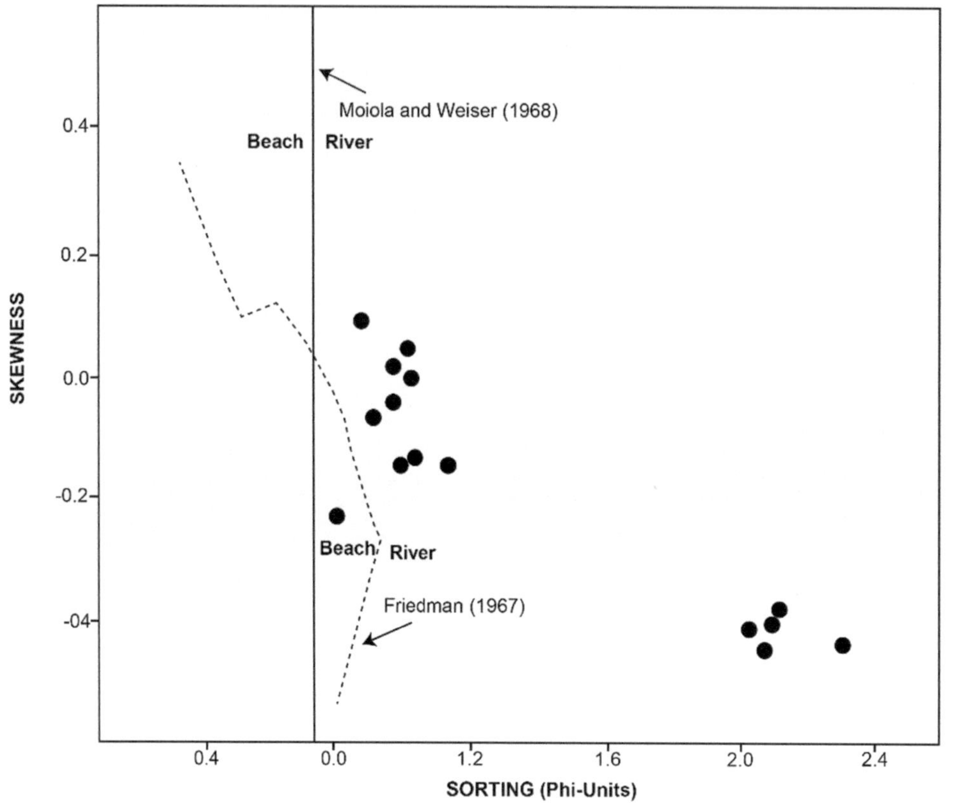

Fig. 9.52 Cross-plot of skewness versus Soring using Moiola and Weiser (1968) and Friedman (1967) methods to identify the depositional environment of the Hofuf formation, Qatar (Al-Saad et al. 2002)

Another cross-plot between the mean diameter and sorting using Moiola and Weiser's (1968) and Fork's (1957) sorting shows that all the studied samples fall in the river domain, with few samples outside the identified depositional domains (Fig. 9.53).

Noweir and Abd-Hameed (unpublished), who studied the Hofuf sediments from other areas, reached the same conclusion. They constructed a cross-plot of standard deviation versus skewness using Friedman's (1961) method and found out that all the samples, except one, fall in the domain of river sand, with one sample in the beach sand. They also constructed a cross-plot of mean size versus standard deviation using the Moiola and Weiser (1968) method, and all the samples except one fall in the river domain and one sample in the beach domain (Fig. 9.54).

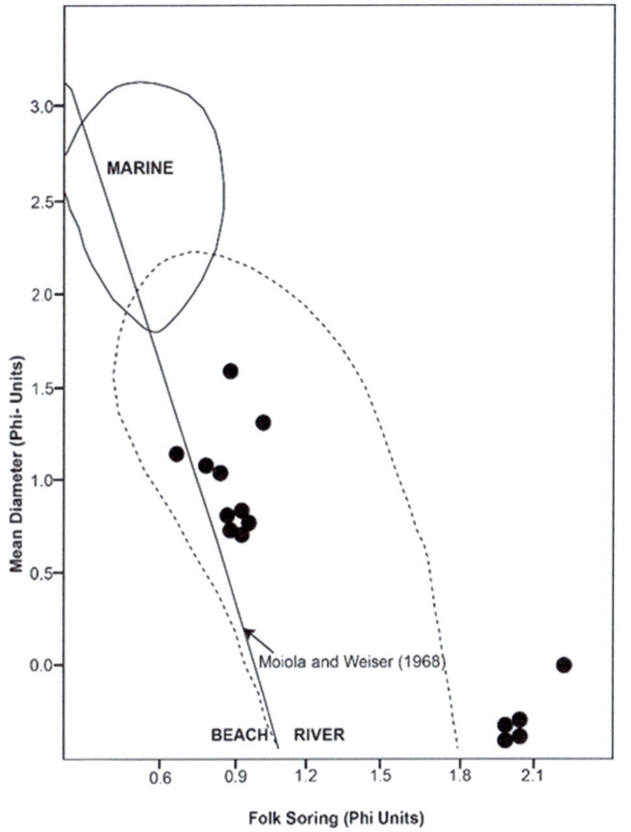

Fig. 9.53 Cross-plot of mean diameter versus Soring using Moiola and Weiser (1968) and Folk (1957) methods to identify the depositional environment of the Hofuf formation, Qatar (Al-Saad et al. 2002)

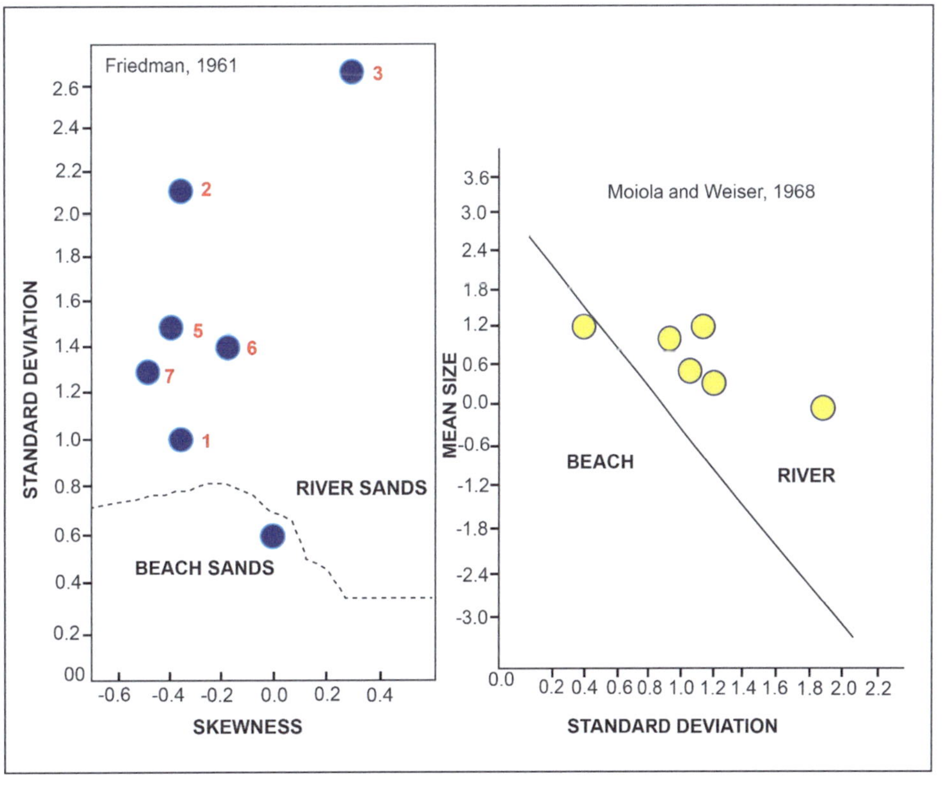

Fig. 9.54 (Left) Cross-plot of standard deviation versus skewness using Friedman (1961) method (Right) cross-plot of mean size versus standard deviation using Moiola and Weiser (1968) method to identify the depositional environment of the Hofuf Formation, Qatar

References

Abu-Zeid MM (1991) Lithostratigraphy and framework of sedimentation of the subsurface Paleogene succession in northern Qatar, Arabian Gulf. New Yearb Geol Paleontol Monthly Magazine Year 1991(4):191–204

Abu-Zeid MM, Boukhary MA (1984) Stratigraphy, facies and environment of sedimentation of the Eocene rocks in the Fhaihil (Gebel Dukhan), Qatar, Arabian Gulf. Rev Paleobiol 3:159173

Abu-Zeid M, Khalifa H (1983) Sedimentological and paleoenvironmental aspects of the Miocene succession in Jebel Nakhash, Qatar, Arabian Gulf. Neues Jb Geol Paläontol Monat 7:334–399

Al-Awadi E, Al-Ruwaih F, Al-Rawdan Z, Ozkaya I (1997) The stratigraphy of the Middle Eocene–Pleistocene sediments in Kuwait. J Arid Environ 37(1):1–22. https://doi.org/10.1006/jare.1997.0268

Al-Hajari SA, Kendall CGSC (1992) The sedimentology of the Lower Eocene Rus Formation of Qatar and neighboring regions. Kuwait J Sci 19:153–172

Al-Saad H (2003) Facies analysis, cyclic sedimentation and paleoenvironment of the Middle Eocene Rus Formation in Qatar and Adjoining areas. Carbonates Evaporites 18(1):41–50. https://doi.org/10.1007/BF03178386

Al-Saad H (2005) Lithostratigraphy of the Middle Eocene Dammam Formation in Qatar, Arabian Gulf: effects of sea-level fluctuations along a tidal environment. J Asian Earth Sci 25(5):781–789. https://doi.org/10.1016/j.jseaes.2004.07.009

Al-Saad H, Ibrahim MI (2002) Stratigraphy, micropaleontology, and paleoecology of the Miocene Dam Formation, Qatar. GeoArabia 7(1):9–28. https://doi.org/10.2113/geoarabia070109

Al-Saad H, Nasir S, Sadooni F, Alsharhan AS (2002) Stratigraphy and sedimentology of the Hofuf Formation in the State of Qatar in relation to the tectonic evolution of the East Arabian Block. Monatshefte 7. https://doi.org/10.1127/njgpm/2002/2002/426

Alsharhan AS, Nairn AEM (1995) Tertiary of the Arabian Gulf: sedimentology and hydrocarbon potential. Palaeogeogr Palaeoclimatol Palaeoecol 114(2):369–384. https://doi.org/10.1016/0031-0182(94)00089-Q

Alsharhan AS, Nairn AEM (1997) Sedimentary Basins and petroleum geology of the Middle East. Elsevier

Amer A, Skyttä P, Ojala A, Heilimo E (2023) New insights into the depositional facies of the Early Eocene Rus Formation, Bahrain. Mar Petrol Geol 154:106338. https://doi.org/10.1016/j.marpetgeo.2023.106338

Aqrawi AA, Goff JC, Horbury AD, Sadooni FN (2010) The petroleum geology of Iraq, vol 424. Scientific Press Beaconsfield

Bashri MAA (2015) High resolution facies analysis and sequence stratigraphy of mixed clastic-carbonate deposits of Miocene Dam Formation Outcrops, Al Lidam Area, Eastern Province of Saudi Arabia. King Fahd University of Petroleum and Minerals, Saudi Arabia

Beydoun ZR (1988) The Middle East: regional geology and petroleum resources. Scientific Press. https://books.google.com.qa/books?id=oU8ZAQAAIAAJ

Beydoun ZR (1991) Arabian plate hydrocarbon geology and potential—a plate tectonic approach. Am Assoc Pet Geol. https://doi.org/10.1306/St33533

Blondeau A, Cavelier C (1972) Les Tertiaires de la presqu ile du Qatar (Golfe Arabique). Donnees de l'Eocene moyen. Bull Geol Soc Fr 14:165–170

Boukhary M (1985) Paleontological studies on the Eocene succession in western Qatar, Arabian Gulf. Rev Paleobiol 4(2):183–202

Boukhary M, Abdelghany O, Bahr S, Hussein-Kamel Y (2005) Upper Eocene larger foraminifera from the Dammam Formation in the border region of United Arab Emirates and Oman. Micropaleontology 51(6):487–504. https://doi.org/10.2113/gsmicropal.51.6.487

Boukhary M, Hewaidy AG, Luterbacher H, Bassiouni ME-A, Al-Hitmi H (2011) Foraminifera and ostracodes of early Eocene Umm er Radhuma formation, Dukhan oil field, Qatar. Micropaleontology 37–60

Buday T (1980) The regional geology of Iraq: stratigraphy and paleogeography, vol 1. State Organization for Minerals, Directorate General for Geological Survey

Casier E (1971) Sur un materiel lchthylogique des Midra (and sailia) Shales du Qatar (Golfe Persique). Bull Inst R Sci Nat Belgium 47:1–9

Cavelier C (1970) Geological description of the Qatar peninsula (Arabian Gulf). Bureau de recherches géologiques et minières

Conchology (2024) Bramkamp, Richard Allan (Phd). https://www.conchology.be/?t=9001&id=13503&srsltid=AfmBOorEdAtx7MFAaFhjBmWBUv8pBcM3i5-5zgUpnBc53kBm4iCjjadj

Dill HG, Nasir S, Al-Saad H (2003) Lithological and structural evolution of the northern sector of Dukhan anticline, Qatar, during the early Tertiary: with special reference to sequence stratigraphic bounding surfaces. GeoArabia 8(2):201–226. https://doi.org/10.2113/geoarabia0802201

Dirks H, Al Ajmi H, Kienast P, Rausch R (2018) Hydrogeology of the umm er radhuma aquifer (Arabian Peninsula). Grundwasser 23(1):5–15

Dunnington H (1958) Generation, migration, accumulation, and dissipation of oil in Northern Iraq. In: Weeks LG (ed) Habitat of Oil. American Association of petroleum geologists symposium, pp 1194–1251

Folk RL (1974) Petrology of sedimentary rocks. Hemphill Publishing Co.

Friedman GM (1961) Distinction between dune, beach, and river sands from their textural characteristics. J Sediment Res 31(4):514–529. https://doi.org/10.1306/74d70bcd-2b21-11d7-8648000102c1865d

Friedman GM (1967) Dynamic processes and statistical parameters compared for size frequency distribution of beach and river sands. J Sediment Res 37(2):327–354. https://doi.org/10.1306/74d716cc-2b21-11d7-8648000102c1865d

Geol C (2024) Nestor J. Sander. http://paleopolis.rediris.es/cg/12/NS/

Golden KW (1984) Artifacts from later pre-Islamic oocupation in Eastern Arabia. Atlal Issue

Haq BU, Hardenbol J, Vail PR (1988) Mesozoic and cenozoic chronostratigraphy and eustatic cycles. In: In sea level changes: an integrated approach, vol 42. Society of Economic Paleontologists and Minerologists, Special Publication

Hasson PF (1985) New observations on the biostratigraphy of the Saudi Arabian Umm er Radhuma Formation (Paleogene) and its correlation with neighboring regions. Micropaleontology 31(4):335–364

Hewaidy A (1991) Contribution to the stratigraphy of Miocene sediments in Qatar. Middle East Res Center Ain Shams Univ Egypt Earth Sci Ser 5:160–170

Hewaidy AG (1994) Biostratigraphy of the Umm Er Radhuma formation in South-East Qatar, Arabian Gulf. New Yearb Geol Paleontol Treatises 193(2):145–164

Hewaidy A, Al-Hitmi H (1999) Paleoecology of the cretaceous—early Tertiary sequence in the Dukhan oil field, west Qatar, Arabian Gulf. In: International conference of the geology The Arab World (GAW 4)

Hewaidy A, Al-Saad H (1993) Surface Eocene stratigraphy of Qatar Peninsula. Al-Azhar Bull Sci 4(1):165–193

James G, Wynd J (1965) Stratigraphic nomenclature of Iranian oil consortium agreement area. AAPG Bull 49(12):2182–2245

Johnson CA (2008) Phanerozoic plate reconstructions of the Middle East: insights into the context of Arabian tectonics and sedimentation. Abu Dhabi International Petroleum Exhibition and Conference

Kok C, Leblanc J (2012) The Bir Zekreet Member: a new lithostratigraphical unit of the Dammam Formation in Qatar

LeBlanc J (2015) A historical account of the stratigraphy of qatar, Middle-East

Leblanc J (2019) The desert roses and other recent gypsum crystals of qatar. Available at: https://sites.google.com/site/leblancjacques/fossilhome

LeBlanc J (2021) Stratigraphic lexicon: a revised guide to the cenozoic surface formations of Qatar, Middle East (excluding the islands). Biosis: Biol Syst

Moiola RJ, Weiser D (1968) Textural parameters; an evaluation. J Sediment Res 38(1):45–53. https://doi.org/10.1306/74d718c5-2b21-11d7-8648000102c1865d

Murris RJ (1980) Middle East: stratigraphic evolution and oil habitat. AAPG Bull 64(5):597–618. https://doi.org/10.1306/2f918a8b-16ce-11d7-8645000102c1865d

Pettijohn FJ, Potter PE, Siever R (1972) Sandstones, sedimentary Basins, and continental evolution. In: Pettijohn FJ, Potter PE, Siever R (eds) Sand and sandstone. Springer US, , pp 544–583. https://doi.org/10.1007/978-1-4615-9974-6_12

Powers RW, Ramirez LF, Redmond CD, Elberg Jr EL (1966) Geology of the Arabian Peninsula: sedimentary geology of Saudi Arabia [Report] (560D). Professional Paper, Issue. U. S. G. Survey. https://pubs.usgs.gov/publication/pp560D

Rivers JM, Skeat SL, Yousif R, Liu C, Stanmore E, Tai P, Al-Marri SM (2019) The depositional history of near-surface Qatar aquifer rocks and its impact on matrix flow and storage properties. Arab J Geosci 12(12):380. https://doi.org/10.1007/s12517-019-4498-6

Sadooni FN, Al-Saad H (2012) Mangrove-bearing limestone from the Eocene Dammam Formation, Arabian Gulf: implications for the mangrove dispersal controversy. Carbonates Evaporites 27(3):243–250. https://doi.org/10.1007/s13146-011-0075-0

Sander NJ (1952) The Eocene stratigraphy along the western shore of the Persian Gulf. II. Biometric test on variations of the genera Lockhartia and Sakesaria. University Paris

Sharland PR, Casey DM, Davies RB, Simmons MD, Sutcliffe OE (2001) Arabian plate sequence stratigraphy, vol 9. Gulf Petrolink

Smout AH (1954) Lower Tertiary foraminifera of the Qatar Peninsula. In: Monograph British, vol 96. Museum Natural History

Steineke M, Bramkamp RA, Sander NJ (1958) Stratigraphic relations of Arabian Jurassic oil. American Association of Petroleum Geologists Symposium, Tulsa, USA

Sugden W (1962) Structural analysis and geometrical prediction for change form with depth of some Arabian plain-type folds. Am Assoc Petrol Geol 46:2213–2228

Sugden W, Standring AJ (1975) Qatar Peninsula. In: Lexique stratigraphique international. Aise Fasc lOa III:120

Tanoli SK, Al-Bloushi AJ (2017) Depositional history of the eocene Dammam formation in Kuwait. In: SPE Kuwait oil and gas show and conference

Thralls HW, Hasson RC (1956) Geology and oil resources of eastern Saudi Arabia. In: International geological congress 20th symposium Sobre Yacimentos de petroleoy gas, Mexico

Tleel JW (1973) Surface geology of Dammam dome, eastern province, Saudi Arabia. AAPG Bull 57(3):558–576

Whittle GL, Alsharhan AS (1994) Dolomitization and chertification of the early eocene Rus formation in Abu Dhabi, United Arab Emirates. Sed Geol 92(3):273–285. https://doi.org/10.1016/0037-0738(94)90110-4

Wikipedia (2024) Max Steineke. https://en.wikipedia.org/wiki/Max_Steineke

Willis RP (1967) Geology of the Arabian Peninsula-Bahrain (2330–7102)

Ziegler MA (2001) Late Permian to Holocene paleofacies evolution of the Arabian Plate and its hydrocarbon occurrences

Zingg T (1935) Beitrag zur schotteranalyse. ETH Zurich

MIX
Papier aus verantwortungsvollen Quellen
Paper from responsible sources
FSC® C105338
FSC
www.fsc.org

If you have any concerns about our products,
you can contact us on
ProductSafety@springernature.com

In case Publisher is established outside the EU,
the EU authorized representative is:
Springer Nature Customer Service Center GmbH
Europaplatz 3, 69115 Heidelberg, Germany

Printed by Libri Plureos GmbH
in Hamburg, Germany